Benchmark Papers in Geology

Series Editor: Rhodes W. Fairbridge
Columbia University

A selection from the published volumes in this series

Volume

17 METAMORPHISM AND PLATE TECTONIC REGIMES / *W. G. Ernst*
18 GEOCHEMISTRY OF IRON / *Henry Lepp*
19 SUBDUCTION ZONE METAMORPHISM / *W. G. Ernst*
29 METALLOGENY AND GLOBAL TECTONICS / *Wilfred Walker*
34 CRYSTAL FORM AND STRUCTURE / *Cecil J. Schneer*
37 STATISTICAL ANALYSIS IN GEOLOGY / *John M. Cubitt and Stephen Henley*
44 MINERAL DEPOSITS, CONTINENTAL DRIFT AND PLATE TECTONICS / *J. B. Wright*
57 MEGACYCLES: Long-Term Episodicity in Earth and Planetary History / *G. E. Williams*
62 OROGENY / *John G. Dennis*
67 ECONOMIC EVALUATION OF MINERAL PROPERTY / *Sam L. VanLandingham*
70 MINERAL EXPLORATION / *Willard C. Lacy*

Benchmark Papers in Geology / 69

A BENCHMARK® Books Series

MINING GEOLOGY

Edited by

WILLARD C. LACY
James Cook University
of North Queensland

Hutchinson Ross Publishing Company

Stroudsburg, Pennsylvania

Benchmark Papers in Geology, Volume 69
Library of Congress Catalog Card Number: 82-968
ISBN: 0-87933-426-6

85 84 83 1 2 3 4 5
Manufactured in the United States of America.

LIBRARY OF CONGRESS CATALOGING IN PUBLICATION DATA
Main entry under title:
Mining geology.
(Benchmark papers in geology; 69)
Includes indexes.
1. Mining geology—Addresses, essays, lectures.
I. Lacy, W. C. (Willard C.) II. Series.
TN260.M57 553 82-968
ISBN 0-87933-426-6 AACR2

Distributed worldwide by Van Nostrand Reinhold Publishing Company Inc., 135 W. 50th Street, New York, NY 10020.

CONTENTS

PART IV: DEVELOPMENT OF RESERVES

PART V: MINING APPLICATIONS

SERIES EDITOR'S FOREWORD

The philosophy behind the Benchmark Papers in Geology is one of collection, sifting, and rediffusion. Scientific literature today is so vast, so dispersed, and, in the case of old papers, so inaccessible for readers not in the immediate neighborhood of major libraries that much valuable information has been ignored by default. It has become just so difficult, or so time consuming, to search out the key papers in any basic area of research that one can hardly blame a busy person for skimping on some of his or her "homework."

This series of volumes has been devised, therefore, as a practical solution to this critical problem. The geologist, perhaps even more than any other scientist, often suffers from twin difficulties—isolation from central library resources and immensely diffused sources of material. New colleges and industrial libraries simply cannot afford to purchase complete runs of all the world's earth science literature. Specialists simply cannot locate reprints or copies of all their principal reference materials. So it is that we are now making a concerted effort to gather into single volumes the critical materials needed to reconstruct the background of any and every major topic of our discipline.

We are interpreting "geology" in its broadest sense: the fundamental science of the planet Earth, its materials, its history, and its dynamics. Because of training in "earthy" materials, we also take in astrogeology, the corresponding aspect of the planetary sciences. Besides the classical core disciplines such as mineralogy, petrology, structure, geomorphology, paleontology, and stratigraphy, we embrace the newer fields of geophysics and geochemistry, applied also to oceanography, geochronology, and paleoecology. We recognize the work of the mining geologists, the petroleum geologists, the hydrologists, and the engineering and environmental geologists. Each specialist needs a working library. We are endeavoring to make the task of compiling such a library a little easier.

Each volume in the series contains an introduction prepared by a specialist (the volume editor)—a "state of the art" opening or a summary of the object and content of the volume. The articles, usually some twenty to fifty reproduced either in their entirety or in significant extracts, are selected in an attempt to cover the field, from the key papers of the last century to fairly recent work. Where the original works are in foreign languages, we have endeavored to locate or commission translations. Geologists, because of their global subject, are often acutely aware of the oneness of our world. The selections cannot therefore be restricted to any one country, and whenever possible an attempt is made to scan the world literature.

To each article, or group of kindred articles, some sort of "highlight commentary" is usually supplied by the volume editor. This commentary should serve to bring that article into historical perspective and to emphasize its particular role in the growth of the field. References, or citations, wherever possible, will be reproduced in their entirety—for by this means the observant reader can assess the background material available to that particular author, or, if desired, he or she too can double check the earlier sources.

A "benchmark," in surveyor's terminology, is an established point on the ground that is recorded on our maps. It is usually anything that is a vantage point, from a modest hill to a mountain peak. From the historical viewpoint, these benchmarks are the bricks of our scientific edifice.

RHODES W. FAIRBRIDGE

PREFACE

Literature dealing principally with the activities and responsibilities of the mine geologist has been separated from that concerned with the exploration for new deposits. The latter is covered in the Benchmark volume *Exploration Geology*. The literature for this volume, *Mining Geology*, has been grouped into six parts. Part I presents a historical perspective of the development of the science and illustrates the evolving role of the mine geologist from a primary concern for maintaining ore reserves to an active input into operational efficiency and management decisions. Part II is a technique section covering mapping and recording the geology of various types of mines in different geological environments. Paper 11 deals with the trend toward computer storage and plotting of geological data, which necessitates a degree of standardization.

Factors influencing the valuation of deposits are covered by selections included in Part III. Geostatistics has had an active role in determining the grade of ore reserves and becomes increasingly important as the value of precious metals increases and lower grades are being mined. Part IV gives guidance in techniques and concepts of the search for additional reserves—determination and evaluation of ore controls. Parts V and VI are concerned with geological contributions to mining and mineral beneficiation.

Many of the papers are relatively old; written in the formative years of a developing science, they set out the ground rules. The tendency in more recent literature is to neglect the "how to" type of article and to emphasize theory. Thus, for the young geologist isolated in a mining camp, there is a serious lack of instructional information. It is hoped that this compilation will help fill the need for basic technique and concept information for the discovery, evaluation, development, and production of ore deposits. Emphasis has been placed on the practical rather than the theoretical aspects, and some of the papers have been severely edited in order to permit broader coverage of exploration geology in a single book. This volume is not intended as a substitute for texts covering mining geology but as supplementary reading.

I would like to thank James Cook University of North Queensland and the University of Arizona for the use of their libraries and facilities, and Ann Archer for her assistance in typing and editing.

WILLARD C. LACY

CONTENTS BY AUTHOR

INTRODUCTION

HISTORICAL PERSPECTIVE

The earliest geologists were philosophers who pondered on the nature and origin of stones, little concerned with practical matters of finding and developing mineral resources.

Aristotle (384–348 B.C.) dominated Greek thought on natural phenomena. He considered that the elements of earth, water, air, and fire are transmutable with production of exhalations. This encouraged the alchemists in their attempts to transmute base metals into gold and silver and the astrologers who concluded that metals were generated by emanations from various heavenly bodies. Astronomical signs for the planet assigned to certain metals survive to the present time. As late as the sixteenth century, Spanish conquistadors concentrated their efforts in the search for gold to the tropics where solar rays were more intense.

During the Middle Ages, thought regarding mineral deposits was dominated by the orthodox followers of Genesis. It took considerable courage for anyone to express concepts departing from this view.

Miners of Cornwall, England, and Saxony, Germany, were little influenced by either the philosophers or the clergy. They learned by careful observation and experience how to discover, follow, and extract metal-bearing ores. In Saxony during the sixteenth century, the state of the art of mining was recorded by Georgius Agricola, born Georg Bauer at Glauchau, Saxony, in 1494. He served as physician, bergmeister (who was responsible for issuing mining permits), and court historian. Agricola abandoned his classical training toward philosophical speculation in favor of careful observation and interpretation and learned much of the character of mineral deposits from his close association with the mines and miners. He commented, "Those things we see with our eyes and understand by means of our sense are more clearly to be demonstrated than if learned by means of reasoning" (Agricola, 1912). Agricola rejected earlier concepts of mineral deposits and set forth the basis upon which practical search for deposits and their extraction could be carried out.

Vannoccio Biringuccio, a contemporary of Agricola born in Siena, Italy, in 1480, did for metallurgy what Agricola did for mining and mining geology. He supplied most of the metallurgical data for Agricola's treatises and presented general principles of mineral economics.

Not until late in the eighteenth century was the basic groundwork for geology as an observational science laid by the Scottish geologist James Hutton (1726–1797). An understanding of geological processes and their role in the genesis, character, and distribution of mineral deposits gradually evolved. These concepts were integrated with the work of French geologist Elie de Beaumont (1798–1874) in his paper *Note sur les emanations volcaniques et metalliferes* (1847).

Many of the concepts of Elie de Beaumont were organized, augmented, and popularized by the U.S. geologist Waldemar Lindgren (1860–1937), who succinctly summarized the ore-forming process:

> As we look back over the wide domain of mineral deposits we perceive a strong tendency to concentration of common or rare elements by magmatic differentiation, by solution or by mechanical transportation; we perceive also cycles of transformations, based upon the laws of stability of chemical compounds. Even when deeply buried, the deposits may suffer changes. Near the surface they are constantly subject to transmutations involving both concentration and dispersion. A few stable compounds are formed while the rest of the elements are scattered by mechanical and chemical transportation. Some constituents are carried down into the earth by the underground circulation of water, perhaps to form new deposits in other rocks. Ultimately erosion sweeps away the wreckage of the old deposits into basins of sedimentation where the elements may be reconcentrated, be it by direct precipitation or by the aid of living matter. The sediments may again be lifted and corrugated, again destroyed by erosion, and new eras of concentration begin. (1933, p. 894)

Lindgren's strong emphasis on igneous processes and associations was the result of the character of deposits in the Western Americas with which he was most familiar. His writings greatly influenced the philosophy and techniques of the U.S.-trained geologists.

The importance of the submarine environment in the formation of mineral deposits was emphasized by John Woodward (1665–1722) but did not gain general appreciation and acceptance until Stanton's book *Ore Petrology* (1972).

In the latter portion of the nineteenth century, professional mining engineers from Cornwall and Saxony, and later the North American mining schools, began to add their greater expertise to mining as operators and consultants. They made appraisals of new mining ventures resulting from accidental discoveries or from prospectors' efforts. They were usually capable applied geologists.

Geological theory was of little assistance to the miner unless it aided in the solution of problems of complex faulting, complex folding, or erratic distribution of ore values. Solution of such problems normally called for a consultant.

The professional mining geologist entered the scene early in the twentieth century, at which time he served largely in the capacity of a consultant. In 1900 a start was made to organize the Anaconda Copper Mining Company's geological department at Butte, Montana, by Horace V. Winchell, and in 1901 Reno Sales became his assistant and later the chief geologist. Winchell and Sales, working with D. M. Brunton, established the essential methods and procedures for the surface and underground mapping of Butte's complex geology. Reno Sales perfected and applied a system of precise geological mapping that proved highly successful in the legal defense of his company's rich ore bodies against encroachment of its neighbors on the basis of the Apex Law. This mapping technique has evolved as the standard adopted throughout the world and remains today in various modified forms as the most effective surface and underground mapping technique of epigenetic mineral deposits.

L. C. Graton noted:

> A gift offered by mining geology to the whole science has been the habit of observing and recording more accurately and on an ever closer scale, and of thinking more quantitatively. Some things need grand sweeps of the eye or mind for their perception; but of a great number such methods would give but a most general notion, with a great chance for error through missing or misinterpreting smaller but immeasureably important details. (1941)

The principal responsibility of the resident mine geologist, initially, was to find sufficient ore within, or adjacent to, the mine to insure its continued operation. This charge entailed mapping the geology to assist in ore search and maintaining records of ore reserves. Geological maps and sections were the most important record since most geological relations of direct value in mining could be expressed graphically, and even more theoretical considerations must be so expressed before they can be applied with precision. The techniques of the mining geologist were summarized in texts by Forrester (1946) and McKinstry (1948). Often, as an escape from his routine mine duties, the mine geologist would examine prospects submitted to his company by a prospector or a promoter.

Current application of geological knowledge and techniques to the search for mineral deposits and to mining operations owes much to the geological surveys and mines departments of the various countries of the world. The government surveys have made substantial contributions through regional mapping programs that have delimited metalliferous provinces and districts, established general laws of mineral deposition, outlined the geological environments in which ore mineralization occurs, and trained technical and field staff.

The exploration geologist, as a part of an organized exploration program, came into existence during and following World War II when metal shortages became a threat to national security and to the industrial economy of all participating nations. Exploration programs were initiated by geological surveys and private companies with attractive tax and subsidy incentives. The effort was given considerable impetus by the uranium boom, and then was extended to other mineral commodities. Consideration was given to regional controls of mineral deposits and to refinement of geophysical and geochemical search techniques. The consequence was a rash of new discoveries in which geological concepts and geochemical and geophysical techniques gradually took over from conventional prospecting.

Within the mine, the responsibilities of the geologist expanded with a shift in emphasis from discovery thinking to operational thinking. This shift was largely brought about by a general change in mining technology from the selective mining of small high-grade shoots that required continued exploration to maintain reserves to the bulk mining of large low-grade deposits that were blocked out well ahead of mining. The mining geologist engages in all activities involving the geology in connection with the total mining and beneficiation operation. His major task is the adaptation of geological knowledge of ore deposits and rock behavior to mining and beneficiation needs and the application of the principles of the science to specific problems of discovery, development, extraction, and beneficiation of the ore. This change in scale and scope is reflected in Peters's (1978) text on mining and exploration geology, and in Kužvart and Böhmer (1978).

Continued development in mining and beneficiation technology requires broadening of geological support. In situ leaching, for example requires notation of permeability and reactivity determinations as well as notation of ore mineral relations to permeability channels. Narrower profit margins require more accurate determination of ore grades, pit slopes, and rock behavior. Environmental restrictions demand accurate impact predictions and control.

REFERENCES

Agricola, G., 1912, *De Re Metallica,* H. C. Hoover and L. H. Hoover, trans., Dover Publications, New York.

Élie de Beaumont, L., 1847, Note sur les emanations volcaniques et metalliferes, *Soc. Géol. France Bull.* **4:**1249.

Forrester, J. D., 1946, *Principles of Field and Mining Geology,* Wiley, New York, 647p.

Graton, L. C., 1941, Ore Deposits, in *Geology 1888–1938,* 50th anniv. vol., Geological Society of America, Boulder, Colo.
Kužvart, M., and M. Böhmer, 1978, *Prospecting and Exploration of Mineral Deposits,* Elsevier, New York, 431p.
Lindgren, W., 1933, *Mineral Deposits,* 4th ed., McGraw-Hill, New York, 930p.
McKinstry, H. E., 1948, *Mining Geology,* Prentice-Hall, Englewood Cliffs, N.J., 680p.
Peters, W. C., 1978, *Exploration and Mining Geology,* Wiley, New York, 696p.
Stanton, R. L., *Ore Petrology,* McGraw-Hill, New York, 713p.

Part I

MINING GEOLOGY COMES OF AGE

Editor's Comments on Papers 1 Through 4

1 **McLAUGHLIN and SALES**
Excerpts from *Utilization of Geology by Mining Companies*

2 **HENDERSON**
Excerpts from *The Role of the Geologist in Underground Mines*

3 **PETERS**
The Role of the Geologist in the Life Cycle of a Mine

4 **OLIVER**
The Geologist's Role in Evaluation Economics

Mines throughout the world began to organize geological departments because of the successful utilization of geological concepts and methods developed at Butte, Montana. Requirements, techniques, and emphasis varied with the nature and geological history of the ore body and host rock, mining and beneficiation methods, and special legal or economic requirements.

Paper 1, by McLaughlin and Sales, reflects the emphasis in the United States upon the epigenetic hydrothermal types of deposits. Alteration effects, favorable rock types and structures, and zoning patterns are critical to the prediction of ore distribution and continuity.

In contrast, Paper 2 by Henderson reflects the different problems and emphasis of mine geology in deposits within highly deformed, strata-bound, syngenetic type deposits that have influenced geological thought and methods in Australia.

Peters (Paper 3) illustrates the changing role of the mine geologist within the life cycle of a single mining operation as it moves from exploration, development, production, technology and production-rate modifications, production decline, and finally salvage. It demonstrates the breadth of background training required by the mine geologist and the necessity for a continuing program of training.

Oliver (Paper 4) states that the function of geology is to serve as a bridge between natural resources and industrial technology. All stages of this bridging are limited by economic parameters, and the geologist becomes a partner in the development within these limitations.

Lacy (1975) emphasizes the changing role of the mine geologist as related to the change in character of economic deposits—from the small high-grade bonanza deposit to the bulk low-grade-type deposits. This entails a change from being a finder and tabulator of ore reserves to an active participant in the design and organization of mining and beneficiation systems. Increasing emphasis is being placed upon evaluation of the physical, hydrological, and metallurgical behavior of the ore and host rock. Recording depends more and more upon photographic or remote sensing techniques that increase the productive capacity of the geologist, and the mining geologist is becoming involved with production planning, financial policy, and taxation problems.

Advances in the science of geology present new ideas and new methods. It is the responsibility of the mining geologist to be familiar with current technological achievements and to make adequate and practical use of them. However, without a sound geological philosophy, supported by carefully prepared maps and records, the mine geologist's work could become empirical and mechanical.

REFERENCE

Lacy, W. C., 1975, The Geologist in an Operating Mine, in *People in the Mining Industry,* Proceedings of the Student Conference (May), Australasian Institute of Mining and Metallurgy, Parkville, Victoria, pp. 31-39

1

Reprinted from pages 683 and 685–686 of *Ore Deposits of the Western States,* The Committee on the Lindgren Volume, ed., American Institute of Mining and Metallurgical Engineers, New York, 1933, 797p.

UTILIZATION OF GEOLOGY BY MINING COMPANIES

Donald H. McLaughlin* and Reno H. Sales†

PART I

GENERAL CONSIDERATIONS

ADAPTATION to mining needs of the rich store of geologic knowledge concerning mineral deposits and the application of principles of the science to specific problems in the finding, development and stoping of particular ore bodies are the major tasks of the mining geologist. On him to an increasing degree is placed responsibility for estimation of ore and appraisal of mineralized ground, for explicit guidance of exploration and development, and for reliable advice with regard to the many geologic conditions that influence mining operations. In the long continued and steady efforts to meet these requirements, a special skill in close observation, recording, and quantitative interpretation has been developed, particularly by geologists on staffs of mining companies, which in many districts has proved to be effective, not only for technical ends, but also for advancement of scientific understanding.

[*Editor's Note:* Material has been omitted at this point.]

* Professor of Mining Engineering, Harvard University.
† Chief Geologist, Anaconda Copper Mining Company.

Most geological relations of direct value in mining operations can be expressed graphically on maps and sections, and even the larger concepts of the science must usually be investigated in terms of features that can be so expressed before they can be applied with exactness to immediate practical needs. There is no acceptable substitute for such records and without them no geologist is entitled to design and recommend exploration except in a very general way. The highest degree of geologic skill is, of course, required to make full use of available data but no amount of brilliance or experience can render specific directions trustworthy that are not based on faithful and persistent observations and on accurate recording of all accessible evidence.

Carefully prepared records, though absolutely essential, are not, however, an end in themselves. In the hands of humdrum workers, they mean little. Interpretation of the facts and expression of the conclusions in terms capable of direct use require inventiveness, judgment and appreciation of operating requirements in addition to adequate scientific understanding. Critical and penetrating thought is indispensable in this phase of mining geologic work, and differences in individual skill, capacity and experience become most sharply emphasized. It is, unfortunately, the point where many good routine workers stall after meeting their immediate responsibilities without realizing the possibilities for further service. The science has not yet reached a stage of perfection where interpretation can be cut and dried, or where a simple answer can be expected automatically from a given statement of conditions. Without a sound geological philosophy that gives an insight into conditions controlling ore deposition, the work becomes empirical and mechanical, and interpretations even from elaborate and carefully prepared records are apt to be inconsistent and unreliable. The advance of science on many fronts and by many means is constantly affording new ideas and new methods of attack on problems of ore deposition. It is the duty of the mining geologist to be familiar with current achievements and to contribute his share toward them, as well as to make practical use of the results.

Investigations of wide scope, that might by some be regarded as of little apparent bearing on immediate needs, have been undertaken in numerous instances by geological staffs, when fundamental relationships were obscure and use of geology was handicapped by lack of basic knowledge. Studies of the mechanics of rock fracturing under particular conditions, of the tops of recent lava flows in relation to permeability of ancient volcanic rocks, of details of rock alteration, and of the distribution of products of regional metamorphism, may be mentioned as random examples of broad inquiries undertaken to throw light on some specific relationship. Such studies are important not alone for more accurate understanding of processes of ore formation but also for more effective use of the local geology in ore finding. Ability to recognize vital problems and to see ways to improve the quality of geologic service by their solution is clearly essential if static, routine and mediocre work is to be avoided. Fortunately, advancement of the profession is not hampered by lack of good problems, for the supply of them still left in a number of older districts promises to outlive the mines.

Geologists closely associated with active operations have opportunities to master the intricacies of individual ore bodies and types of ore occurrence with a thoroughness impossible from casual study. Ability to see obscure but critical details comes only with repeated examinations of exposures and from actually living with the problems of a district. Indeed, the contrast between the meager information that can be gained from first inspection of altered vague material in a mine face and the exact evidence that can be distinguished with confidence after long familiarity forces one to be skeptical of opinions or conclusions, even of experienced consultants or scientists, when based only on brief examinations.

Commercial limitations may to some extent restrict freedom for purely scientific investigations in mining geologic work, but the disadvantages are generally more than offset by benefits gained from the discipline which is provided and enforced by economic requirements. The mining geologist must necessarily keep in mind the useful purpose of his work and avoid attractive digressions that do not serve his major objectives. The expectation that predictions will eventually be tested by subsequent exploration imposes a wholesome restraint on vague and speculative ideas, especially after a few experiences that engender respect for the uncertainties and surprises that can be concealed only a few tens of feet in the solid rock beyond the face of a drift or below an outcrop. An outstanding benefit resulting directly from economic factors is the confirmation or correction of ideas and predictions by the facts actually revealed as mining operations progress; this affords opportunity to test the accuracy of conclusions, which the geologist not associated with technical or commercial activities rarely enjoys. In no better way can thoroughly sound methods of observation and interpretation be built up and the confidence attained that allows steady concentration on progressive and productive efforts.

[*Editor's Note:* Material has been omitted at this point.]

2

Reprinted from pages 263-264, 264-265, and 267 of *Exploration and Mining Geology, with Particular Reference to the Commonwealth of Australia,* L. J. Lawrence, ed., Eighth Commonwealth Mining and Metallurgical Congress Publications, vol. 2, Australasian Institute of Mining and Metallurgy, Parkville, Victoria, 1965, 1126p.

THE ROLE OF THE GEOLOGIST IN UNDERGROUND MINES

By

Q. J. HENDERSON[1]

INTRODUCTION

The major role of the mine geologist is to evaluate the possibilities of any extensions or repetitions of ore. Fundamentally, he is a structural geologist and his success depends squarely on detailed structural geology. He must collect the maximum geological data in an environment where it is slowly exposed and more rapidly destroyed and interpret it, so that as complete a three dimensional understanding of the orebody is known, with the greatest possible accuracy. before it is mined. He must provide by means of plans and sections an ever up-to-date picture of the orebody outline together with the distribution of the metal content as a sound basis for the future planning of development policy and requirements. More specifically he must advise on current mining and operating problems and assist the mining staff to apply the geological information to actual mining.

The success of geology in current mining operations depends entirely upon the degree of co-operation that exists between the geological and mining departments. If a geologist is to fulfil his rightful role as a mine geologist, he must have a thorough understanding of the fundamentals of ore genesis and ore localisation, and to be fully effective in the search for ore, not only does he require imagination and courage to formulate exploration programmes but also tact and understanding to have them accepted. He must have a restless, enquiring and striving mind that is never satisfied with existing interpretations but is always seeking something better and it is only by the painstaking and systematic study of the data gathered from daily routine mapping that the truth can be found.

GEOLOGICAL MAPPING TECHNIQUES

In order to provide the necessary geological data for the solution of current mining problems and to guide the search for new ore, detailed mapping of all available underground faces and backs must be undertaken. However, the greatest effort is not required in collecting the data nor in presenting, but in interpreting the facts and drawing from them the correct conclusions.

The objective in underground geological mapping is the recording of all observable detail on

1. ore characteristics and structure,
2. structure and lithology of country rock, and
3. faults and fractures.

Only conventional methods of geological mapping have been used in past routine practice but specialized studies are now using the statistical techniques evolved by Sander (1930) and developed by Turner and Weiss (1963). The specific methods of mapping under conditions of open and timbered stoping together with all forms of development in ore and country rock, as used at North Broken Hill Ltd., have been described by Parkin (1940).

As it is neither practical nor economic to sample each floor of each stope, grade mapping by eye estimation, supplemented by chip channel sample assays where required, still provides a rapid and economic method of grade control. In general it is a matter of personal opinion just where changes of grade occur as well marked grade boundaries are lacking in the North mine. Grade mapping is only attempted within intervals of 5 per cent lead although it is possible to be more precise.

Even after years of practice there are certain conditions that are difficult to estimate. The two main difficulties are in the zinc rich areas and in the masses of low grade rhodonitic ore. As the Zn:Pb ratio is variable, there are some areas that are zinc rich and because light reflections from some of the sphalerite crystal faces may simulate those of galena, there is a tendency to overestimate the lead content. In the case of the low grade masses of rhodonitic ore, in which sporadic aggregates of galena are dispersed, the exposed face is not necessarily representative of the block, so that not only is eye estimation difficult but even channel sampling is unreliable.

In addition to estimating lead grade the type and proportion of the gangue minerals are recorded for use in more accurate density calculations.

1 Chief Geologist, North Broken Hill Ltd.

LEVEL DEVELOPMENT

Advance plans and sections are provided to the planning section for use in the layout of all development. The data shown includes an outline of the orebodies and the location of all faults and any ground conditions that would require more than a minimum of timbering, in order that the best possible layout may be planned. As development of the level proceeds the data are modified where necessary and kept sufficiently far ahead of the working faces to enable them to be used to the fullest extent.

Drives and crosscuts in country rock and lode are mapped structurally, lithologically and for grade. The grade is further controlled by the weekly sampling of all development in ore.

FAULTS AND FRACTURES

A feature of North Mine geology is the presence of numerous faults and fractures, many of which are pre-ore but have suffered post-ore movement. Although these faults have little effect on the orebodies, they do affect level layout and mine safety. It is therefore necessary that, in order to make accurate projections to new levels, all relevant data relating to the attitude of all fault planes, type of fault filling and, if possible, direction and amount of movement are recorded.

However, one such fault has had a profound influence on the mining history of North Mine and it has been only the steady accumulation of geological evidence over the years that has revealed the true significance of the Cuming fault—a low angle, brecciated overthrust fault with deep leaching. It is now known that the deep leaching (over 3300 ft) in the North Mine, is not the result of post-mine leaching as was suggested by Garretty and Blanchard (1942) but is the result of the natural circulation of ground water through a series of open strike faults in association with the Cuming fault.

[*Editor's Note:* Material has been omitted at this point.]

SAMPLING

Mine evaluation is so closely related to the interpretation of geologic conditions that the mining geologist should direct the sampling of the mineral deposit.

The choice of method is usually governed by the character of the deposit. At North Broken Hill Ltd. all stope samples are taken as shallow channels with a hammer and moil. Empirically, North Mine practice requires as many pounds of sample as feet sampled. Sample limits are controlled by visual estimates of grade to include variations not greater than 5 per cent. Check sampling along some channels shows variations up to 2 per cent, but no greater care is needed since experimental work indicates that no higher accuracy is obtained because the two walls of a drive or crosscut will effect variation even though the number of samples taken is large.

In actual mining a clean face is available and the only care required is to ensure that the sample bags are clean. All samples are taken normal to strike and at mapping height, i.e. 4 ft above the floor. Individual samples are weighted according to true width. When old stopes are being sampled the face must be adequately cleaned by scrubbing with a wire brush and hosing down with water.

In sampling diamond drill core, the core is split longitudinally, one half being retained as the permanent record. When splitting friable ore, care is taken to discard the fine material which accumulates in the bottom channel of the core splitter to avoid up-grading the sample.

Core logging is much facilitated by using the fresh face produced by splitting.

DIAMOND DRILLING

Diamond drilling performs a very important function in the sampling of existing ore reserves, but is used also to test certain possibilities of ore repetition and extensions with a view to discovering as yet unknown ore. More specifically it is used for the economic control of mine development and exploratory development in general.

Too often the drilling of diamond drill holes has been directed with personal opinion or judgment as its sole basis for desirability. The amount and spacing of diamond drill holes for delineating the orebody has not been subjected to rigid economic analysis. The mine geologist should determine what particular footage of drill hole per ton of ore or what three dimensional spacing of drill holes are most profitable.

In any exploration of the northern extension of the North Mine the success of any future programme is tied to development of new drilling techniques that will allow the successful drilling of the broken sericitic zones.

It is the mine geologist's responsibility to

1. recommend precise location and attitude of all diamond drill holes and to decide when holes should be stopped.
2. record all details such as ore and mineralization limits, rock types and ore environment,
3. log all core and supervise the splitting and sampling of any mineralization,
4. interpret the geological data obtained, and
5. prepare reports summarizing the drilling results.

In an area such as Broken Hill, where the rock assemblage is far from simple, it is inevitable that the significance to the geologist of the subtle changes in the various rock units must change significantly with increasing knowledge of the district.

In the study and logging of diamond drill core, it is practically impossible for the geologist to record all the facts that may be observed. Successful observation must therefore depend upon objective selection and emphasis of pertinent details. But selection, at any one time, is governed by the purpose and study bias of the individual geologist.

Unfortunately, it is impossible for the geologist to foresee what subtle characteristics may, at some later date, become of critical importance, but he can insist that all core be preserved as a permanent record.

Therefore, as time goes on, it will be seen that the value of the written log becomes more and more limited while the value of the permanent record becomes more and more evident as new interpretations, based on re-examinations of the core by other geologists, are put forward.

The form on which diamond drill records are to be kept must be designed to suit the various requirements and the different uses to which it will be put. Company directors, for example, are not concerned with details, but rather with a brief statement of ore intersections and their relationship to indicated ore reserves; while the mining engineer is concerned chiefly with location and grade of the ore intersected and their meaning in terms of ore represented.

But the mine geologist is concerned with the geological interpretation of the results and requires detailed descriptions of the ore types and their environment, together with descriptions of the rock types and their structural relation to other parts of the mine. Both the drill foreman and the geologist are concerned with drilling performance because this provides data such as drilling rates and cost per foot, which the mine geologist requires as a basis to plan future drilling campaigns and estimates of the probable cost.

ORE RESERVES

The basic factor in ore reserve estimation is the degree of knowledge involved. At North Broken Hill Ltd. because of the size, general uniformity and high average grade of the orebodies and the development procedure adopted, the tonnage and grade that will be mined from a given stope or pillar is known, within narrow limits, in advance of mining. It must be realized however that ore reserve estimates are seldom as accurate as they would appear because of underlying statistical factors such as incomplete determination of precise shapes and widths and the effect of dilution with waste or low grade material. Because effective planning of the required development and stoping sequence is not possible without a sound geological basis, it is essential that the mine geologist maintains close liaison with the planning section.

CONCLUSIONS

The importance of the role of the mining geologist in successful mining operations cannot be denied. To be successful he must not only have a thorough grasp of geological principles but also a thorough understanding of their practical aspects. Upon his three dimensional understanding of the orebody and consequent estimates of potential ore reserves future development and production policies are based. It is his responsibility to see that every ton of ore within the mine area is discovered. To do this, not only does he require sound technical training but also imagination and courage to formulate the necessary exploration programmes. He must have a restless and enquiring mind that is always striving for a better understanding of the fundamental geological factors controlling the ore occurrence and the possibilities of any repetitions of such occurrences.

In an area such as Broken Hill where metamorphic processes have so changed the original sedimentary structures, it is little wonder that the reading of the structural history has been painfully slow. However, the problem of finding ore still challenges the mine geologist and new concepts of the structural environments of the orebodies are emerging. New possibilities in descriptive analyses and interpretation of structures which have been deformed by intense metamorphism are being opened up by the application of methods based on the geometric properties and tectonic fabrics of the rocks.

[*Editor's Note:* Material has been omitted at this point.]

The mechanisms of flow in solid rocks are varied and imperfectly understood, but studies such as the rheomorphic studies, using laminar flow models will assist in a better understanding of the Broken Hill structures and help draw the right conclusions from the wealth of geological evidence that has accumulated.

REFERENCES

Garretty, M. D., and Blanchard, R., 1942 Post mine leaching of galena and marmatite at Broken Hill, *Proc. Aust. Inst. Min. Met.,* 127.

Henderson, Q. J., 1956. Geological exploration at North Broken Hill Limited. *Proc. Aust. Inst. Min. Met.,* 180 pp. 219-255.

Parkin, L. W., 1940. Underground geological mapping at North Broken Hill Ltd., *Proc. Aust. Inst. Min. Met.* 119 pt. 11, p. 547.

Sander, B., 1930. *Gufugekunde der Gestine* (Vienna).

Turner, F. J., and Weiss, L. E., 1963. *Structural analysis of metamorphic tectonites* (McGraw Hill: New York).

3

Reprinted from *Jour. Geol. Education* **14:**191-193 (1966)

THE ROLE OF THE GEOLOGIST IN THE LIFE CYCLE OF A MINE

William C. Peters, University of Arizona

ABSTRACT: A mining geologist must have an appreciation for the objectives of the particular enterprise. A background in engineering and economics for such an appreciation is commonly developed on a long term basis during a geologist's career; a preliminary framework can, however, be provided at an earlier stage in the classroom through several lectures on the life cycle of selected mining operations. The role of the geologist can be related to the initial stage, youth, maturity, old age, and abandonment of representative mineral-based enterprises. The geologist's effectiveness throughout the cycle can be stressed as depending upon his depth of understanding in relating his science to each important management decision.

Courses in mining geology have the capacity to relate a student's scientific background to the economic factors and objectives of mineral enterprises. By including a few lectures on the theme "the life cycle of a mine" the important role of the geologist in mineral economics can be emphasized and a simple framework can be provided for later on-the-job learning.

Within a few months after graduation, the geologist will find himself collecting and evaluating an overwhelming array of data involving every phase of his academic background from geomorphology to petrology and selecting the specific features important to his objective. Within a few years he may be in a critical decision-making position where he will encounter exploration leases, budget allocations, and production plans. He is not likely to develop an adequate background in all of the associated fields of engineering and economics within such a short time and still keep current in the fast-moving geological sciences. He can, however, build upon a few principles gained from his last academic year and become an effective advisor to other members of the management team.

The geologist's stature in an organization and the acceptance of his recommendations by management depend very largely on his appreciation of the requirements of the enterprise. Part of his job rests upon his ability to present structural trends, hydrothermal alteration, and geochemical anomalies as important and practical matters. His acquaintance with the economic factors involved in the life cycle of a mining operation can help him fill his expected role.

The life cycle of a mine or a mining district (both terms require that something must be extracted for profit) is essentially a result of the depletion process. Even though mines, like people, are individuals with unique physical, technologic, and economic surroundings, there is a basic history of youth, maturity, and old age. One is tempted to use the biological analogy except that there is no reproductive stage. An abstract comparison to the geomorphic cycle brings to mind periods of rapid development, equilibrium conditions, and rejuvenation. Figure 1 illustrates rejuvenation periods in the San Francisco (Oatman) gold mining district, Arizona, as the result of technologic improvements in one instance and as a result of a sudden increase in metal value in a later stage. New concepts and new technology are commonly associated with the rejuvenation in mines and mining districts even though a few mineral-based enterprises have managed to survive periods of change by entrenching themselves in old meandering courses.

The life cycle examples and generalities used in classroom discussions involve elements from various documented mining operations and from the district production records in the U.S. Bureau of Mines yearbooks. State and foreign bureaus furnish similar data. District production records are less clear-cut than individual mine records, but the latter are difficult to obtain. District examples can, however, be related to mine production and history by reference to industrial periodicals.

The time scale in the life cycle is indefinite. One can not really say that a cycle has been completed as long as there is the possibility of geologic reexamination and the discovery of unsuspected reserves. Innumer-

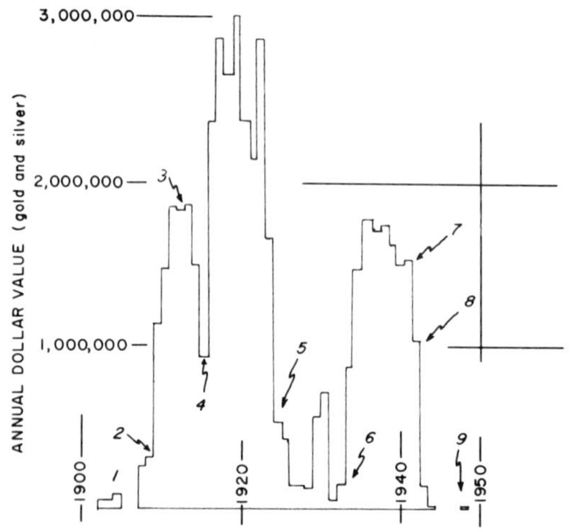

Figure 1. The San Francisco (Oatman) Mining District, Arizona: 1. Early production from exposed ore bodies. 2. Major mine began production from shallow ore bodies (1909). 3. New company formed to develop deeper ore (1913). 4. Major mine began production from deeper ore bodies (1916). 5. Major mine closed. Mills continued operating on custom ore (1924). 6. Price of gold raised from $20.67 to $35 per ounce (January 1934) followed by re-definition of ore reserves and re-opening of mines. 7. Wartime shortages of labor and equipment. 8. Wartime closing of gold mines (1942). 9. Postwar mining attempt (1948).

able mines have had short and colorful life cycles based on exposed mineralization with no apparent roots capable of supporting the cost of deeper development. Many mines have had a discontinuous history lasting for decades or for centuries during which several efforts have failed or have reaped only limited rewards. Important and classic mining operations have supported industrial complexes for 50 years or more in spite of lowering ore grade, increased depth, and tightening economic conditions. In a few remarkable instances, single mines have experienced centuries of continuous production and are likely to continue for some time—the Almaden mercury mine in Spain has maintained such a record since 1499.

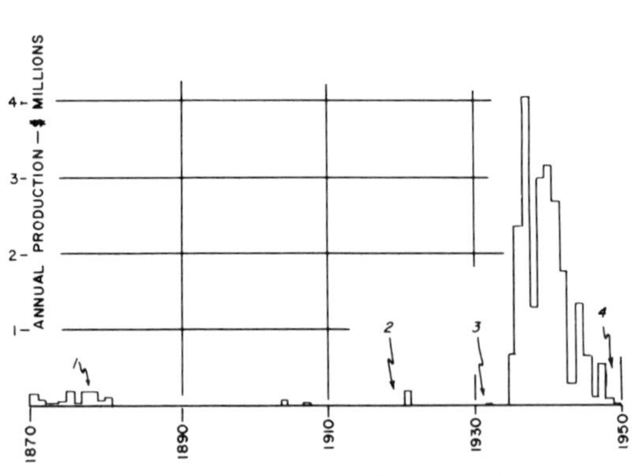

Figure 2. The Mountain City Mines, Nevada: 1. Early production from shallow and oxidized silver ore. 2. Copper characteristics recognized in gossan. 3. Discovery of rich supergene copper ore body. 4. Ore body exhausted.

At the beginning of a representative life cycle, the discovery of mineralization and the recording of some minor high-grade production may antedate the recognition of the main ore bodies by many years. It is quite common (and rather sobering) for a geologist to find old pits and shafts in a target area which he has chosen after painstaking regional study and bold projection of structural trends. The Mountain City, Nevada, area (Figure 2), a noteworthy copper producer from 1935 to 1949, experienced an earlier stage of high-grade silver mining in the 1870's; it was the recognition of copper characteristics in gossan at some distance from the depleted silver mines that led to the discovery of the main deposit. At Mount Isa, Australia, (Figure 3) the gold mines in the 1870's were exhausted within a short time. Copper ore was mined on a moderate scale between 1909 and 1920 and the ore bodies were apparently depleted. The major silver-lead-zinc and copper ore bodies were not recognized until several years later and production did not surpass that of the earlier mines until the 1930's.

At the far end of the life cycle, few mines terminate in such a clear-cut manner as at Mountain City. The decision to abandon an operation is often made after a few years of break-even operations in which the producers are simply "trading dollars for ore" in the hope that their geologists (by this time expected to be philosophers, scientists, and minor prophets) can find some more ore. Even after the main operation has been suspended there may be many years of scavenging operations carried on by low-cost, independent lessees who keep mining until they have blasted "just one more round" and made their reluctant exit.

Between the beginning of the cycle (sometimes with a bang) and the end (often with a whimper), there are the overall characteristics illustrated in Figure 4 and in the following pattern:

A. Discovery in the district—prospects and small mines opened and abandoned without significant production.
B. Repeated examination of the district by geologists, geophysicists, and engineers.
C. Recognition of a potential major ore body.
D. The pre-production interval.
   1. Preliminary estimates of geologic, technologic, and economic conditions.
   2. Preliminary financing on the basis of high risk.
   3. Delineation and testing of the ore body.
   4. Further financing on the basis of reduced risk.
   5. Development of the mine, supporting plants, and townsite.
   6. Employment and training of labor and technicians.
E. Expanding (youthful) production.
   1. Beginning of dividends to investors.
   2. Addition of new tunnels, shafts, pits, and processing plants.

3. Vertical growth toward higher-value finished products through smelting, refining, and fabrication.
4. Horizontal growth toward control of supporting materials and facilities, such as:
   a. Limestone and coal for smelter.
   b. Phosphate rock for use with smelter acid in manufacturing fertilizer.
   c. Power plants, cement plants, railroad system, explosives and machinery plants within the immediate economic area.
   d. Adjacent mines for access as the mine grows larger.

F. Mature production.
1. Innovations in mining and processing to offset lowering ore grade and rising costs.
2. Verification of the limits of the ore body.
3. Innovations to extend the life of the ore body.
4. Increased local exploration for possible extensions and increased "outside" exploration.
5. Cost reduction and extension of machinery life.

G. Declining (old age) production.
1. Sale or lease of assets to a neighboring mine.
2. Cutbacks in local exploration and development with continuing or increased "outside" exploration.
3. Blending of ore from pillars with lower grade ore.
4. Custom processing of ore from other mines.
5. Cost reduction by concentrating on fewer stopes and working benches.
6. Mining of shaft pillars.

H. Mine abandonment, with
1. Salvage of machinery.
2. Departure of most of labor force.
3. Milling or leaching of mine dumps.
4. Custom milling and smelting of ore from other mines.
5. Lessee operations—sporadic as economic conditions change.

The final phase, abandonment, may not be connected with the ultimate depletion of the ore body. A mine may be prematurely abandoned by reason of higher taxes, higher operating costs, or lowering of the mineral price. The cost of meeting such emergencies as large inflows of water may be higher than the value of the remaining ore. Prolonged strikes, expropriation, or war-time closing orders may bring an operation to an untimely end. At any rate, the ore body ceases to be an ore body.

The geologist and his science are concerned in every phase. He enters the scene early and is involved in the recognition of the ore body. His work during pre-production is the focus for financing and development. During the youthful phase, he is constantly gathering data and projecting the ore body while examining nearby ore deposits and non-metallic deposits for acquisition. Mature production brings the most painstaking local exploration and engineering geology problems. With declining production, the geologist's responsibility is spread from an all-out search for the last ore in the mine to a search for a new ore body at some distance away. Abandonment of a particular mine does not release the geologist. He retains a responsibility for evaluating sources of custom ore, for dealing with the hydrologic problems of continued metal recovery from leaching, and for keeping in touch with lessee operations. Geologic advice and direction should be considered as essential throughout the mining life cycle. If there are instances in which such advice and direction are not fully employed, it may be that the geologist and the manager are not "speaking the same language." The communications barrier can be opened in the college classroom. Further mastery in the field of mineral economics must be gained in the years following school and must be placed in context with new developments in the geological sciences.

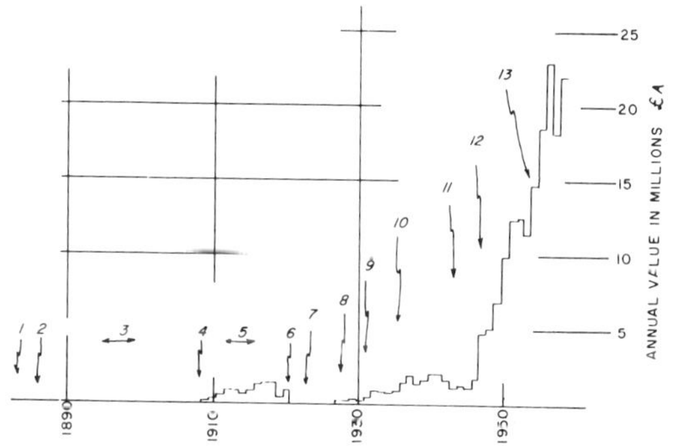

Figure 3. Cloncurry—Mt. Isa Mineral Field, Queensland, Australia (Uranium Production Excepted): 1. Early gold production (1867–1900). 2. Early copper production (1884–1887). 3. Minor gold-copper-silver production. 4. Smelters and railway constructed (1908). 5. Copper and gold production. 6. Rich copper ore depleted, smelters closed (1920). 7. Discovery of the Mt. Isa silver-lead-zinc deposit (1923). 8. Discovery of the Mt. Isa Copper deposit (1928). 9. Silver-lead smelter built (1931). 10. Zinc production begun (1935). 11. Copper production replaced silver-lead-zinc production (1943). 12. Post-war resumption of silver-lead-zinc production (1946). 13. Copper smelter built (1953).

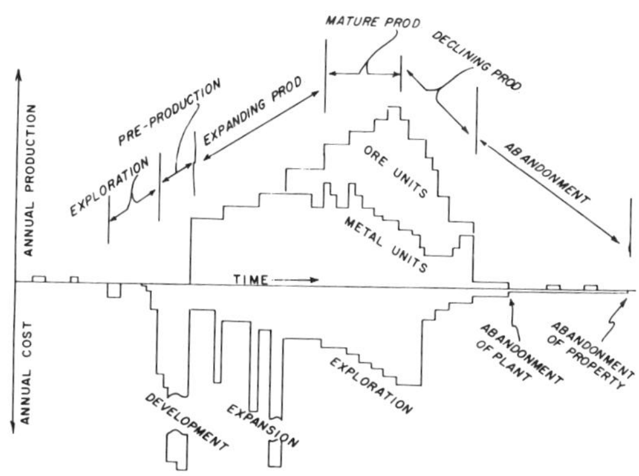

Figure 4. Life cycle of mine.

4

Reprinted from *Am. Assoc. Petroleum Geologists Bull.* **55**:796-800 (1971)

THE GEOLOGIST'S ROLE IN EVALUATION ECONOMICS[1]

W. B. Oliver[2]

Introduction

For several years Carter Oil Company has been actively engaged in acquiring and developing fossil fuels other than oil and gas as a potential source of energy. A substantial part of the new venture effort is in the field of coal. The geologist's role in coal evaluation economics is important and is an integral part of a coordinated program involving diverse professional capabilities.

Good geology is the key to commercial success and poor geology is the source of failure in any exploration-oriented activity. The geologic function is central in any resource-based business venture—or, at least, it so appears to the dedicated professional geologist. This concept is illustrated by Figure 1. Of course, our production and/or mining associates can prove just as readily that all other functions are peripheral to their field. The marketing organization can claim, without challenge, that the absence of a customer would leave all functions idle, the planners can say that without plans there could be no orderly development, and management can rightly assert that without capital and direction, all would flounder. The answer is, then, that all are essential: each is a fundamental part of any natural-resource oriented organization.

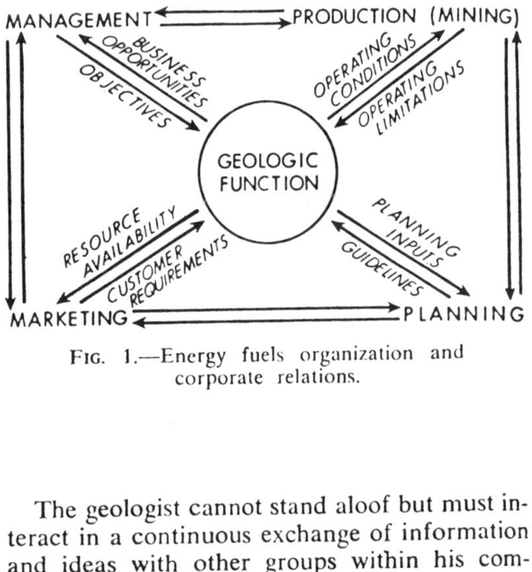

Fig. 1.—Energy fuels organization and corporate relations.

The geologist cannot stand aloof but must interact in a continuous exchange of information and ideas with other groups within his company. Further, he must react in a timely manner to serve their legitimate needs. For instance, his discoveries may point the way to management for new business opportunities. Certainly, his geologic information concerning the characteristics of the discoveries and the potential for new discoveries is a most important input to the planning of future corporate activity. In return, the geologic effort must be guided by the business objectives and planning guidelines conveyed to it.

In similar interaction with the producing or mining function, pertinent geologic information on the extent and characteristics of the mineral body is requisite to the development of efficient mining programs and the selection of suitable

[1] Manuscript received, August 6, 1970; accepted, December 26, 1970. Paper presented before the Association at Calgary, Alberta, June 23, 1970.

[2] Manager, Resource Acquisitions, The Carter Oil Company. The writer is indebted to the management of Carter Oil Company for assistance in the preparation of this paper and for permission to publish.

mining equipment. In turn, the geologist must develop an awareness of the technical and economic limitations of mining that dictate whether a discovery classifies as a commercial or a noncommercial deposit.

In contacts with the marketing group, geologists are apprised of the quantity and the quality of product that can be absorbed by the market, and the marketer is informed of the areas where the company's mineral resources are adequate in quality and quantity to meet present and potential demands.

A common element in most of these interactions is the need for timely, reliable, and quantitative information on the mineral body. The development of this information commences with the reserve's discovery—or, in many instances, even before discovery—and is a continuing process throughout its evaluation, development, and extraction.

The importance of the geologist in the post-discovery phases of evaluation programs has not been emphasized. I believe that the geologic function must continue to be at the center of activity from *discovery* through *development* to *depletion* of the deposit.

Given a situation of the successful culmination of a well-conceived and well-executed exploration program resulting in the discovery of one or more mineral accumulations of guideline quality, what will be required to bring the occurrence to economic usefulness? I believe that many of the same considerations will obtain for any mineral discovery—oil and gas, metals, uranium, coal, tar sand, oil shale, and even sand and gravel. The discovery of a natural concentration of these raw materials is followed by one or more phases of analysis and evaluation to develop the facts required for input into the several functional activities. Not all such prospects merit immediate development; some will never be commercial. The fate of each must rest on detailed analysis in which the geologist plays a key role.

Prospect Analysis

The objective of prospect or deposit analysis is the determination of the relative attractiveness of a mineral deposit (Fig. 2). An orderly procedure for making such an analysis includes the following steps.

1. A preliminary economic evaluation includes the assembly and interpretation of data from the exploration phase.
2. Comparison with other prospects implies multiple successes in the exploration program, or alternate investment opportunities for available capital.

PROSPECT ANALYSIS

OBJECTIVE: To determine the relative economic attractiveness of a prospect

- ☐ PRELIMINARY EVALUATION
- ☐ COMPARISON WITH OTHER PROSPECTS
- ☐ ESTABLISHMENT OF PRIORITIES
- ☐ PREDEVELOPMENT EVALUATION
- ☐ ASSESSMENT OF COMPETITOR POSITION
- ☐ PROVISION OF TECHNICAL INPUT AND SUPPORT FOR OTHER CORPORATE FUNCTIONS

Fig. 2.—Prospect analysis.

3. Establishment of priorities assumes the development of a system for comparing the most important parameters as an aid in selecting prospects for detailed evaluation and/or development.
4. Predevelopment evaluation provides for the generation of additional data on those prospects that appear to be most favorable from the priorities already established.
5. Assessment of competitor position supplies information and an evaluation of industry holdings as an important and continuing part of the analysis.
6. Provision of technical input and support for other corporate functions assure the communication of technical data to other functions on a continuing basis.

How do we actually implement such a prospect analysis? For the purpose of this discussion, I have created a hypothetical coal prospect, "Z," with the idea of illustrating the basic steps involved in evaluation economics in a post-discovery environment (Fig. 3).

Example

The first step is to assemble and interpret all the geologic data on the "Z" prospect. The coal isopach on Figure 3 shows a maximum thickness of more than 7 ft and supports an estimate of recoverable reserve of at least 80 million tons. This interpretation is based on the explor-

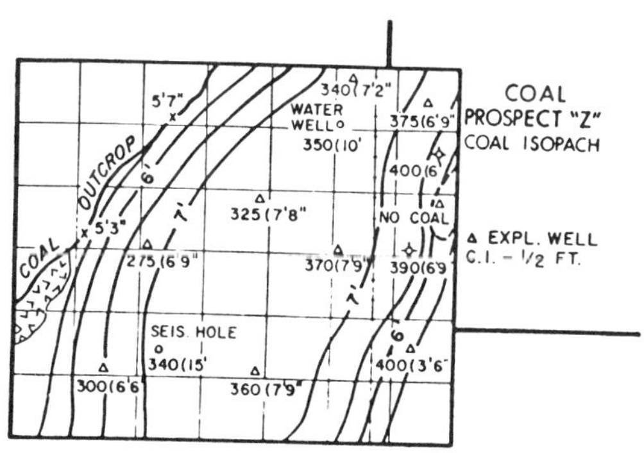

Fig. 3.—Hypothetical coal prospect isopach map

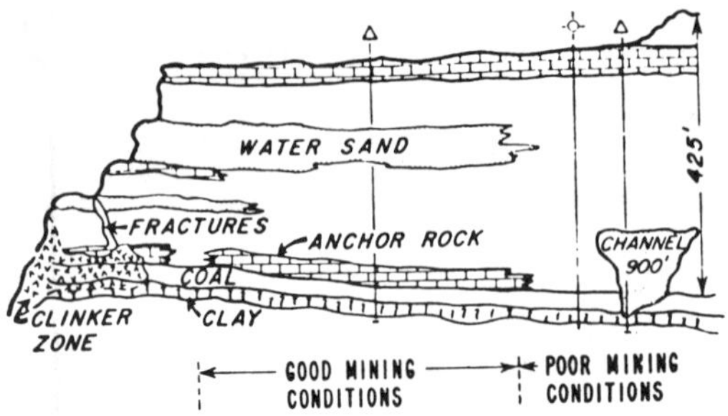

FIG. 4.—Schematic cross section showing geologic conditions at hypothetical coal prospect "Z."

atory core holes, information from wells, surface sections, and regional geologic maps.

Geologic conditions of economic significance are illustrated by the east-west cross section in Figure 4. The thick, competent limestone bed, directly overlying the coal, through the center of the prospect should contribute to good mining conditions. One well, on the eastern side of the prospect, indicates the presence of a channel. Clinker zones (areas of burned coal) are indicated along the outcrop on the west. From sections like this the significant rock units in the section directly above and below the coal are determined. Maps of these units are then prepared and evaluated.

Coal quality data from the exploratory core holes are posted on maps to identify trends and to serve as a basis for estimating average values. The variations in coal quality reflect geologic events, because the environment of deposition and the subsequent metamorphic processes of heat and pressure dictate the present composition of any coal. Hence the geologist is most capable of interpreting local and regional variations.

Each control point is evaluated as to indicated mining conditions and a rating is assigned to those where adequate data are available. These ratings, shown on the map in Figure 5, provide the basis for a preliminary estimate of overall mining conditions. The area through the center of the prospect with the "A" rating, or best mining conditions, appears by inspection to be attractive for further evaluation.

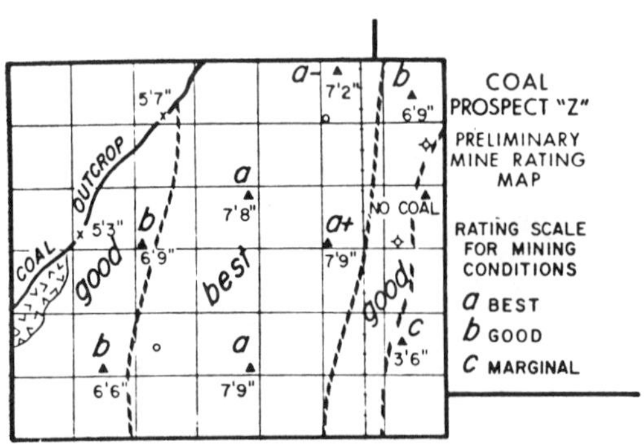

FIG. 5.—Preliminary mine rating map for coal prospect "Z."

All the geologic interpretations are then qualified by assigning a confidence factor. This considers the density and reliability of geologic data available in the prospect area and the indicated geologic complexities of the deposit. The "Z" prospect receives a high confidence rating based on relatively good well control and excellent correlation with regional features. We recognize that the channel problem on the east and the clinker problem on the west could affect conditions within the prospect, but we believe that we have a sound geologic interpretation.

The market potential of the "Z" area is given consideration at this stage. The nearby railroad enhances the value of the prospect for markets tributary to that line.

After consideration of all of the factors, including regional geology, mining conditions, recoverable reserves, acquisition costs, coal quality data, confidence level, and the market potential, the preliminary economic evaluation phase is complete.

Concurrently with the process of evaluating a prospect, a study should be made of acreage held by competitors in the area. The best information available on their blocks, supplemented by sound geologic inferences, commonly permits the development of a reasonably accurate evaluation.

Each prospect resulting from the exploration program goes through this same evaluation process. Priorities are then established for further work. The higher rated prospects are studied to define the requirements for further (predevelopment) evaluation—core holes, improved maps, coal quality data, special geologic studies, *etc.* Cost estimates are made for such an effort and these are used in preparing budget programs for management's consideration. Funds at a level commensurate with the organization's objectives and capital structure are then appropriated to advance one or more of the highest rated blocks to the "ready-for-development" stage.

Figure 6 shows the mine rating map resulting from the predevelopment studies at prospect "Z." The drilling that provided many of the data for the more detailed analysis was planned to define the geologic details of the prospect and was laid out in a regular grid pattern as shown in black circles. The area of best mining conditions, through the center of the prospect, has been delimited. Decreased ratings toward the east are related to poor roof conditions and to the effects of the channel; decreased westerly ratings reflect thinner coal and inferior roof conditions. These are shown on the map as areas of ratings of *b* equals "good" and *c* equals "marginal."

At this stage it is evident that a substantial part of this prospect may be attractive for early development. The actual outline of the area to be developed is arrived at in conference with marketing, mining, and business analysis personnel and is based on the amount of reserves to be dedicated to the mine and a development plan for this particular geologic environment.

Summary

It should be evident that the predevelopment evaluation is not a one-shot effort. Rather, it is a continuing program, where the results of each well or each distinct study phase are incorporated into current work maps. These maps are used as control to identify problem areas and define the ultimate level of work that will be required to attain the necessary confidence level.

Data from the evaluation activity on each prospect are inserted into a feedback loop in the program to ensure a continual reexamination and reevaluation of all prospects. Through this process we minimize the chances of overlooking or bypassing investment opportunities.

The continuing flow of economic and geologic data is fed into the organization's information and control grid. The mining (or production) function is enabled to initiate and update cost estimates on the more attractive prospects, the business function is made aware of long- and short-range business opportunities and the marketing function has an inventory of competitive resources to offer to potential customers.

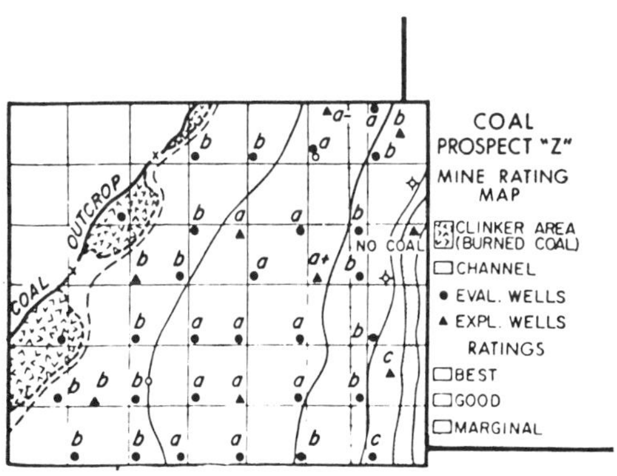

FIG. 6.—Mine rating map for coal prospect "Z."

In following prospect "Z" through each of the steps in prospect analysis we have attained our objective: the determination of the relative economic attractiveness of the prospect. This attainment is, at best, transitory, for a key word in the objective is *relative*. New exploration data or evaluation studies from other areas, changes in market demands or in producing technology, and fluctuating economic and political conditions may each, or in combination, realign the relative evaluation of prospect "Z" and other prospects. Prospect analysis is, therefore, a continuing operation. Not until the decision to develop the property has been made and responsibility for the project turned over to the producing function can the geologist set aside his burden of primary responsibility for the economic evaluation of any prospect.

Conclusions

In conclusion, I want to emphasize the theme which has been woven throughout these comments—the idea that neither geology nor any other scientific discipline is an island unto itself. At one time, about 200 years ago, the accumulated body of scientific knowledge could be assimilated by one man; since then, and at an ever increasing pace, knowledge has expanded until specialties and subspecialties have been created. Experts in these specialties are required to maintain our exponential growth rate in knowledge. The geologist's role in evaluation economics does not ordinarily require the application of detailed knowledge from the frontiers of professional research. Rather, his worth is found in the thoughtful application of basic knowledge from many fields. To "ply his trade" and participate successfully in the functional interactions, the effective economic geologist should have, in addition to his geologic training, a working ability in such diverse fields as economics, data processing, statistical analysis, physics, engineering, political science, and a host of others.

The geologist's role in evaluation economics is definitely important, probably irreplaceable, and a fundamental part of a team effort. The geologic function is the bridge between the riches of nature and the technology of our civilization. Diligent application of the concepts of geology is essential in our energy- and material-oriented world.

We are all thrilled and excited about lunar and outer space exploration and other scientific breakthroughs, but, in my opinion, the basic challenge is the continued development of natural resources to support the economic and physical growth of our society and to provide support for the profound social changes being brought on by a fantastic population growth. Nature has endowed mankind with an adequate storehouse of potentially usable resources. Of importance equal to these physical treasures is the intellect which man has been granted—an intellect which, when properly developed and motivated, permits him to recognize the presence of deposits of minerals and to visualize a way to put them to beneficial use.

Part II

TECHNIQUES OF MINING GEOLOGY

Editor's Comments on Papers 5 Through 11

EVOLUTION OF MAPPING AND RECORDING TECHNIQUES

Both the method of mapping and the nature of parameters mapped change with deposit and technology requirements. Newer developments in mapping and recording do not supercede earlier methods but tend to augment and facilitate. Since no attempt has been made to be complete in presenting techniques in this volume, the reader is referred to Forrester (1946), McKinstry (1948), and Peters (1978) for details.

Detail of scale and the technique of three-dimensional mapping of geological features, of importance to the discovery, development, and extraction of ore deposits, necessitated new procedures for undergound and surface mapping. Geological parameters controlling ore deposition were often subtle, and ore bodies presented small targets. The cost of inaccurate projections was large; accuracy thus became essential.

In Papers 5 and 6 Schmitt outlines both surface and underground mapping methods. Although techniques of ground control of the surface have been modified with the usage of photogrammetry, the geologist often must establish his own ground control for mapping on the scale that may be required. It is perhaps idealistic to expect that "all details capable of being mapped should be recorded." However, most occasions allow only one opportunity so that every effort should be made to record all relevant information. Most mines and mining districts are mapped many times as new concepts evolve and as the geologist becomes able to recognize subtle differences in rock types or alteration. The first mapping, however, has the advantage of more exposures and fresher rock faces.

Methods for mapping complex mineralized fracture systems were developed at Butte, Montana, by Anaconda Company geologists and have become a standard. These are described by Linforth (Paper 7). He includes information on the general organization of geological activities within the mine organization and on the duties of the geological staff.

Where mine workings are on a level or nearly level plane, and where the geological structures are steeply dipping, geological mapping is fairly straightforward. However, when the mine workings are irregular in shape and elevation, and where the features to be recorded are flat lying, new techniques must be used to portray accurately geological relations. In Paper 8, James presents an approach to the mapping of irregular stopes.

The advent of the large open pit mines with rapidly changing benches and continual mining activity in the vicinity of faces that must be recorded has led to special rapid mapping techniques summarized by Metz (Paper 9). The use of computer-oriented recording of geological data and printout of special purpose maps has further expedited open pit mapping and expanded its usefulness in grade control and the determination of mining methods and scheduling of mining activities.

Deposits subjected to intense structural deformation and metamorphism require special pole density techniques to decipher their configuration and tectonic history (Turner and Weiss, 1963). In Paper 10, Davis illustrates the application of these structural analysis techniques to gain an understanding of the deformational history of a syngenetic strata-bound massive sulphide deposit. The simultaneous study of host rock deformational history and that of the contained sulphides aids in genetic interpretation and in the search for extensions and additional ore shoots.

Constraints of geological mapping and of logging diamond drill core have always made it essential for the geologist to resort to some abbreviated method for recording. Various attempts have been made to standardize these abbreviations (Chace, 1956; McKinstry, 1948; Peters, 1978; Berkman, 1976). The need for a complete set adaptable to computer storage, manipulation, and recall is recognized by Blanchet and Godwin (Paper 11), who have proposed a shorthand method for not only geological terms but also geological concepts and characteristics. Such a method has tremendous possibilities for improving quality and for speeding the mapping and logging activities. It has, however, an inherent danger of locking observations into predetermined models that may be false or irrelevant.

REFERENCES

Berkman, D. A., 1976, *Field Geologists' Manual,* Australasian Institute of Mining and Metallurgy, Parkville, Victoria, Aust., 295p.

Chace, F. M., 1956, Abbreviations in Field and Mine Geologic Mapping, *Econ. Geology* **51:**712–723.

Forrester, J. D., 1946, *Principles of Field and Mining Geology,* Wiley, New York, 647p.

McKinstry, H. E., 1948, *Mining Geology,* Prentice-Hall, Englewood Cliffs, N.J., 680p.

Peters, W. C., 1978, *Exploration and Mining Geology,* Wiley, New York, 696p.

Turner, F. J., and L. E. Weiss, 1963, *Structural Analysis of Metamorphic Tectonites,* McGraw-Hill, New York, 545p.

5

Reprinted from pages 717-719, 723-726, 729-732, and 733-736 of *Econ. Geology* **27**:716-736 (1932)

CARTOGRAPHY FOR MINING GEOLOGY

Harrison Schmitt

INTRODUCTION.

THERE are definite signs of revolt in a certain school of geologists against the current methods and philosophy used in mapping the geology of ore deposits. The procedures adequate for regional mapping, non-metallic and petroleum geology are found unsuitable. Technique and methods have been described here and there in the literature, but have not been brought together into a system, neither have certain obvious refinements been generally adopted, such as the use of larger scales, and the "isolation of exposures method." Much of the effect of these is automatic, because by their use many supposed problems of the ground solve themselves. The "isolation of exposures" method [3] (5, pp. 194–5) [4] consists in mapping the solid rock as distinct from the mantle rock, which leaves an accurate record of the outcrops and gives a base for interpretation that can safely be used by geologists not personally familiar with the ground, and also lends meaning to the areas of mantle rock, for hidden dikes, intrusive, gossans or collapsed ground may be defined.

Another obvious but neglected refinement is enlargement of scale. Maps drawn on scales as large as 1 in. = 1000 ft. may be inadequate for directing exploration or recording the data obtained from special studies of gossans, alteration halos, or other features. Indeed, in places maps on scales up to 1 in. = 30 ft. on both the surface and underground are found to be appropriate. This is illustrated by Fig. 1, which shows three maps on different scales, of the same ground at Hanover, New Mexico, all now reduced to the same scale. The first map gave no suggestion of hidden ore, the next enlargement led to the discovery of ore, and the third map added new knowledge and led to the discovery of important additional ore.

The ore at Hanover favors a 50–100 foot limestone horizon and is searched for by diamond drilling. If every foot of ground penetrated is to count the holes must be kept near the intersection

[3] It may be called the "solid method" for convenience.

[4] Numbers in parentheses are references to the bibliography at the end of this article.

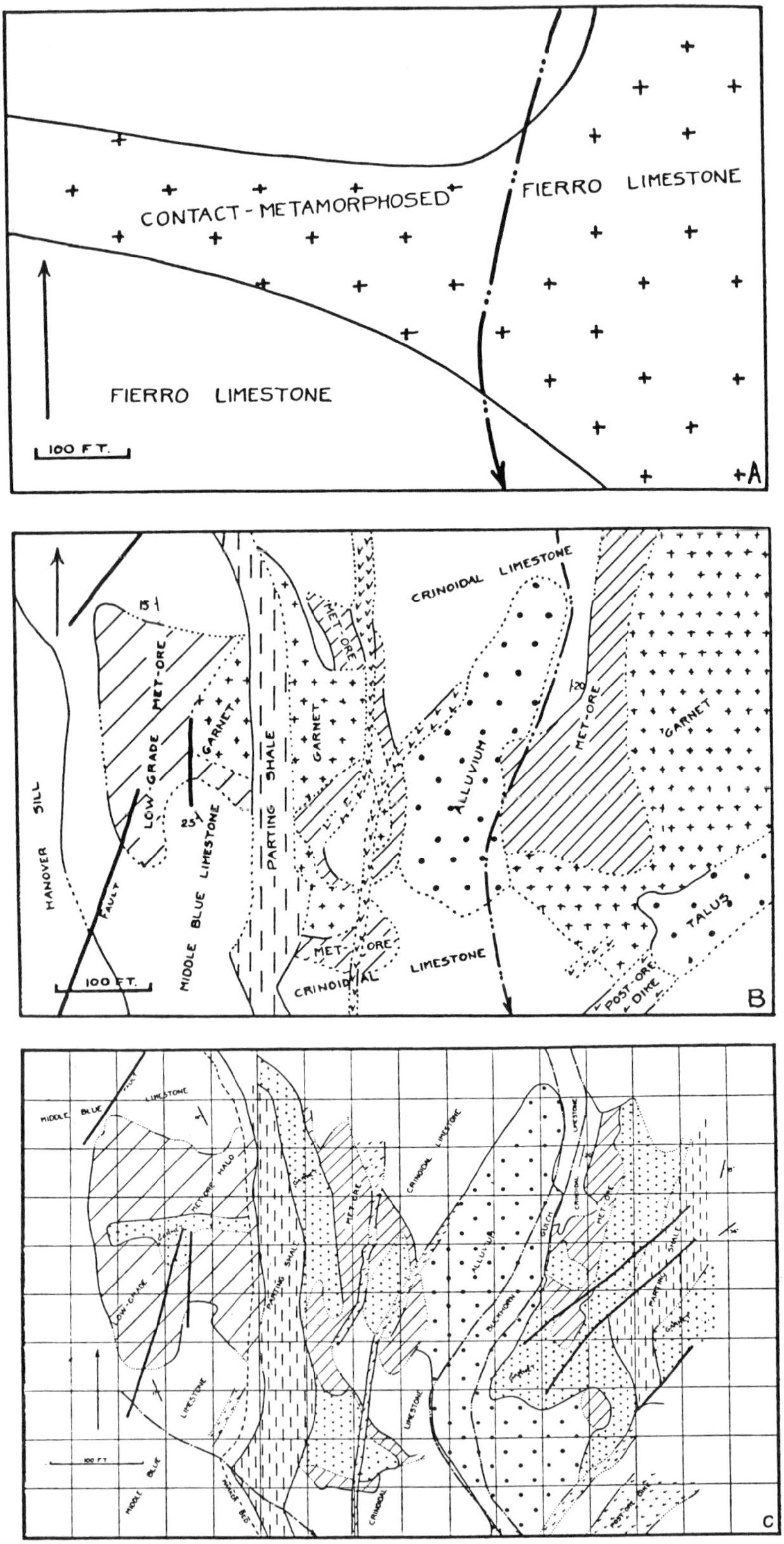

FIG. 1. Three maps of the same area: (*a*) was mapped on a scale of 1 in. = 10416 ft., (*b*) on 1 in. = 100 ft. and (*c*) on 1 in. = 30 ft.

of the favorable horizon with vertical fissures or other small targets. The accuracy needed was obtained by mapping most of the square mile or more of mineralized ground on a scale of 1 in. = 100 ft., but critical areas required a scale of 1 in. = 30 ft.

Some feel that field mapping is and should be largely inductive in method (5, pp. 189–195), for although brilliant and rapid advances have been made by deductive reasoning in geology in general, it is clear that imagination and speculation have had too much play in the mapping of many complex areas. Mining geologists in the West know at least a half-dozen mineralized areas that have been recently remapped and that yielded a picture with little resemblance to the earlier ones. Such contradictions are brought about by inadequate mapping methods and faulty philosophy, not necessarily by careless mapping.

[*Editor's Note:* Material has been omitted at this point.]

Mapping Geology and Topography Together.

When a plane table and stadia method is used for detailed mapping, the topography may be economically mapped with the geology (16), for, with abundant geological detail, enough points will be located for the geology alone to permit the construction of a good topographic map. Only where the geology is simple will extra points be needed for topography. Commonly the points critical for topography are also critical for geology. In rugged topography with a minimum of located points it is good practice to draw the contour lines in the field, but if there is an abundance of points and the detail of the topography is described in the notes, the contour lines may be conveniently and no less accurately drawn in the office. On large scales with abundant points the drawing of contour lines becomes a mechanical process.

Modifications of the Hanover Method.

Grid Method.—When precise detail is wanted on large scales, such as 1 in. to 50 ft. or larger, it is often advantageous to map by what may be called the grid method. The area is laid out in 50 or 100-ft. squares; a numbered stake is located at each corner. The corners are plotted on a tracing, a white print is taken to the field, and the geology is plotted square by square. Additional measurements are made with a Brunton compass and a tape. Fig. 2-C shows an area mapped in this manner. Although more satisfactory for certain kinds of detail mapping than the plane table method, this has the disadvantage of the topographic base map method, *i.e.*, no observations for elevations are available along critical contacts such as key beds. Elevation is accurately known at the stakes only.

Infiltration Method.—When first beginning to map unknown ground, if the geologist does not wish to commit himself to mapping all parts of the area in detail, he may use what may be called the infiltration method. This consists in establishing precise control, covering the ground in reconnaissance, and locating critical points with the stadia rod. The first observations, although scattered, are taken with the degree of precision used for complete detail. This preliminary work will function as a reconnaissance and may indicate areas that need not be mapped further. But if the map requires completion in critical areas, the remaining detail can be taken and tied to the earlier observations (Fig. 3).

Comparison of Methods.—The claim corner base and open-sight alidade methods should be used in detailed study only where the more expensive methods cannot be considered. Of course, they may be adequate on simple ground where the area to be mapped is small or where the scale can be small. On average ground, however, where an adequate appropriation for mapping is available, either the topographic map method or the Hanover method is to be recommended. On unknown ground some expense may be saved by the use of the infiltration method, a modification of the Hanover method. On difficult ground the grid method is a useful auxiliary.

The choice between the topographic map method and the Hanover method depends on local conditions. Where expense is not a critical factor, the former with the modifications suggested will be better, particularly in difficult topography. Simple topographic forms and average geology, however, may be more economically mapped by the second method. Indeed, the cost

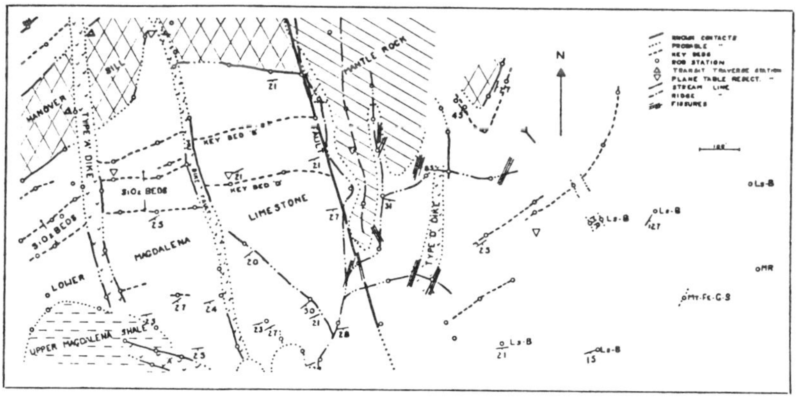

FIG. 3. A map which shows the stages in mapping by "infiltration." The area on the left is completed, that on the right is being invaded.

may be less than half, because with the first the ground must be gone over by both the engineering and the geological parties. The Hanover method is particularly advantageous where many key beds are to be followed.

Certain modifications of the plane table and stadia method and topographic map method have been considered but not tested by the writer. The objection to the plane table and the stadia method, that the geologist does not have a map to sketch upon, might be obviated by having the instrument man signal the coordinates of the stadia points at which the geologist has held the rod. A black-board and crayon could be used for signaling and the geologist could plot the points on a duplicate map and sketch in the detail.

The objection to the topographical map method, that accurate locations are difficult to obtain when the geology is being sketched on the original topographical map, might be overcome by the use of a hand-sized range finder recommended by James Stuart [8] for measuring distances. Presumably he obtains the direction of the sighted point by compass, its elevation by interpolation on the topographical map.

Range finder sights on several located points might obviate the inaccurate compass observations needed with this method. In this case, a triangular or three legged rule with adjustable legs or a single-leg compass could be devised to simplify the plotting of the desired vertex of the resulting triangle. For any given location the legs would be set at the three known distances which would give directly the position of the desired point in relation to the base. If an accurate range finder is unavailable a small stadia telescope mounted on a light tripod (16, p. 780) might be made to serve. In this case permanent circular stadia rods would be set up on some of the prominent known points.

When a vertical angle to a known point exceeds five degrees it becomes necessary to reduce the inclined distance to the horizontal. An inclinometer could be attached to either the range finder or stadia telescope and each leg of the triangular plotting scale could be constructed as a small slide rule with a cosine or stadia scale, as the case may be.

[*Editor's Note:* Material has been omitted at this point.]

General Suggestions.—When a satisfactory section has been measured and the geologist is intimately familiar with the position of each key bed and sedimentary contact, he may start the geological mapping. The location of partly concealed cross-cutting features such as dikes, faults and veins, except those whose trace is parallel to the bedding, will be revealed by following key beds with the rod. The trace of those parallel to the bedding, in turn, will possibly be revealed by following the contacts of the across-bedding cross-cutting features (Fig. 3).

Where a dike or sill is covered by mantle rock, but occasional ribs outcrop, the dip and strike may, in places, be revealed by the "sympathetic" fractures parallel to the contacts. These fractures are caused by flow lines, shrinkage or post-intrusion shear. When mapping a dike, it is advisable to record at each point of observation the width, change in type, and degree of metamorphism and alteration. At a certain "contact-metamorphic" deposit in the Southwest, epidotic alteration of a dike commonly means nearby ore, whereas sericitic alteration is unrelated to ore.

When tracing faults one should not interpret too freely and should remember that commonly they are not regular planes and that their variation may have significance. For example, in at least two silver districts in Mexico, ore shoots are related to concavities in the footwalls of fault-veins.

When tracing veins, all mine openings, the width of the vein from place to place, dip and strike and all changes in mineralization should be measured and recorded. The type and amount of voids, gossans and wall rock alterations are of particular importance. Certain gangue minerals intimately associated with ore are resistant and remain in the outcrop after all of the other minerals have been leached. They may make valuable criteria. For example, in the Parral district, Chihuahua, Mexico, fluorite and voids in an outcrop usually mean ore below (6, p. 287).

Strict attention must be paid to the topography and mantle rock. In the Parral district, Chihuahua, the oreshoots are surrounded by masses of silica, with the result that most of the important ore shoots outcrop on elevated places. The intershoot areas are commonly loci for arroyos (6, p. 287). In the Little Florida Mountains, New Mexico, strips of mantle rock (talus largely) about ten feet thick cover the outcrops of manganese veins collapsed because of solution. The wall rocks are resistant and always outcrop. The veins were discovered not through outcropping ore, but by determining that the strips of mantle rock marked faults. At Hanover, New Mexico, the zinc ore outcrops are covered by mantle rock, the wall rocks are bare. In general then, dikes, sills, intrusions, gossans and collapsed areas are in places disclosed by the patterns of mantle rock that cover them and, in detailed work, mantle rock must be mapped as carefully as any other lithologic unit.

Caliche [9] is locally a criterion for ore in northern Mexico and the Southwest and must be observed and mapped. A patch of caliche was the only surface expression of the Ahumada-Erupcion manto oreshoot. Indeed, a majority of the mines and prospects in the Southwest have caliche as an intimate associate of the gossan; few districts that have been visited lack it. Long narrow bands of caliche, up to 3 feet thick, cap the veins in the Cordero district, southern Chihuahua.

Dips and strikes of beds and contacts comprise a large proportion of observations made and should be taken with care because the accuracy of structural work depends upon them. To obtain a satisfactory mean in any given small area, it may be necessary to take three or more observations. Where the exposure is good, an acceptable observation may be obtained by placing the eye in the plane of the bed and then sighting on the bed with a Brunton compass. Where the strike and trace of the plane of a fault, vein, or contact, do not coincide, it may be advisable to record both and represent them on the map.

When starting to map, much is gained by deferring the classification of the rock units until all the field work and microscopical and chemical work can be done. In the field the units may be called *A, B, C*, etc. Then, when defined (*A* and *B* may be combined or *C* divided) and a representative sample obtained, final identification and classification is carried out. These key letters may be used for symbols on the final map, or to make patterns if the map is not colored.

Detailed mapping, particularly on large scales and complicated ground, commonly requires some " specialization of the moment,"

[9] $CaCO_3$, minor $CaSO_4$.

for if, at the same time, the geologist should attempt to record data on gossan identification, metamorphic or alteration halos and, oxidation collapse, he would find that his ideas would become confused and his work inefficient. There would be too many things to be kept in mind at once to do justice to any one of them. The first mapping ought to be confined to structure, leaving the specialized studies to be taken up one by one on the same ground. The first survey leaves a control of stakes or other marks on the ground to which the later specialized observations can be tied without the need of an engineer at a plane table because a copy of the original map can be used and the specialized detail sketched directly upon it.

[*Editor's Note:* Material has been omitted at this point.]

Synthesis.

Cross-sections.—The most useful cross-section is made through the best known country. Placing sections on a geometrical pattern is advisable only where the geology is simple. When compiling a cross-section, the components of the dips of intersecting planes can be obtained by formula or by the usual charts. If the intersection of one cross-section with another is plotted on both, the interpretations can be readily compared and tied together. Sections are improved if nearby mine workings are projected to them.

Structural Contour Maps.—Most geologists are familiar with structure contour maps. These are especially valuable in mining geology (10, 19), where ore follows definite horizons. To be effective, greater accuracy is needed than for oil-field structure, for example, and the following suggestions are given with that in mind.

When the surface geological map is made and key beds are followed and located at regular intervals with the stadia rod, two purposes are served if as many points as possible are spotted for the determination of dip and strike by the three point method. First, with these accurate dip determinations, a given stratigraphical interval can be obtained with a small limit of error by compiling a number of cross-sections, up to ten possibly, along the strike, all of which are concerned with this one interval only. Large variations from the mean are thrown out and the final mean is accepted as the standard interval. Second, because it is directly dependent on the local dip, the local vertical distance to the contoured horizon can be determined accurately.

At this stage a tracing is made on which are plotted all observed dips and strikes, key horizons with the points located on them, and the traces of all known faults and igneous contacts. The intersections of cross-cutting contacts or faults with the contoured horizon—disclosed by underground development—are shown by distinctive lines projected to the horizontal plane. Then at each point the vertical distance to the contoured horizon is computed by dividing the stratigraphical interval by the cosine of the dip. This distance subtracted from the elevation of the point gives a corresponding elevation on the horizon to be contoured. Elevations obtained in this way and the observed strikes form the basis for drawing the contours.

The final map is usually a print from the original tracing. The contours, key beds, contacts and faults are differentiated by color or by different types of lines. The reading of a contour map, where the fault structure is complicated, is made easier by coloring or patterning a selected bed.

Special Maps, Charts and Curves.—Where the basic data of measurement and observation are quantitative, many useful syntheses can be made which include equal-ore-width maps, curves showing variation in rocks caused by metamorphism, and charts showing variation in gossans or significant gangue minerals such as fluorite and quartz. The intensity of fissuring may be represented by counting the number of fissures per unit of length and plotting a curve.

Equal-ore-width Maps.—Equal-ore-width maps have been previously described (4, p. 321). The known parts of an ore body are represented on any convenient plane of projection by lines connecting "points" of equal width. Such a projection may be useful in predicting ore-shoot extensions or giving information on the "shape-habit" of ore bodies.

Equal-ore-value Maps.—In equal-ore-value maps (4, p. 320), the known parts of an ore body are represented on a convenient plane of projection by outlining areas of equal value. Indirectly, equal-value lines are thus derived. Such a map often gives important information on ore-shoot extensions and the effect of oxidation, leaching and enrichment.

Curves Showing Variation of Metal Content.—Curves showing variation in metal content in various directions in mines could be advantageously used to a greater extent in problems of leaching, oxidation, enrichment, primary zoning and extension of ore shoots (14, p. 410; 6).

ACKNOWLEDGMENTS.

The writer is particularly indebted to Augustus Locke for his continued encouragement of the gathering of the material for this paper and for the several critical readings he gave the manuscript. W. H. Emmons, W. S. Bayley, and Charles Behre, Jr., also read the manuscript and made pertinent suggestions.

El Paso, Texas.

BIBLIOGRAPHY.

1. Breed, C. B. & Hosmer, G. L.: Elementary Surveying. John Wiley & Sons, Inc., New York City, 2 vols., 1918.
2. Hayes, C. W.: Handbook for Field Geologists. John Wiley & Sons, New York, 1921.
3. Lahee, Frederic H.: Field Geology. McGraw-Hill Book Co., New York, 1923, New ed., 1931.
4. Schmitt, Harrison: Extension of Ore Shoots, etc. Trans. Am. Inst. Min. Eng., 1929.

5. GREENLY, EDWARD, AND WILLIAMS, HOWARD: Methods in Geological Surveying. Thomas Murby & Co., London, 1930.
6. SCHMITT, HARRISON: Geology of the Parral Area of the Parral District, Chihuahua, Mexico. Trans. Am. Inst. Min. Eng., pp. 268–290, 1931.
7. ——, Geologic Notes on the Santa Barbara Area in the Parral District of Chihuahua, Mexico. Eng. and Min. Jour., vol. 126, p. 410, 1928.
8. BANCROFT, HOWLAND: Am. Inst. Min. Eng., Bull. 91, 1914.
9. BIRDSEYE, C. H.: Topographic Instructions of the United States Geological Survey. U. S. Geol. Surv. Bull. 788, 1928.
10. FOWLER, G. M. AND LYDEN, J. P.: The Ore Deposits of the Tri-State District. Am. Inst. Min. and Met. Eng., Tech. Pub. 446, 1931.
11. GILBERT, D. C.: Mine Maps. Eng. and Min. Jour., vol. 132, pp. 18–20, 1931.
12. BROCK, R. W.: Scientific Ore Finding. ECON. GEOL., vol. 19, pp. 674–676, 1924.
13. MCKINSTRY, H. E.: A Protractor for Plotting Dips on Inclined Sections, ECON. GEOL., vol. 18, pp. 393–397, 1923.
14. RANSOME, F. L.: The Copper Deposits of Ray and Miami, Arizona. U. S. Geol. Survey, Prof. Paper 115, pp. 144–145, 1919.
15. ——, The Plane-table in Detailed Geological Mapping. ECON. GEOL., vol. 7, pp. 113–119, 1912.
16. PELTON, E. F. AND IRWIN, D. D.: The Plane-table in Geologic Mapping. ECON. GEOL., vol. 7, pp. 778–783, 1912.
17. WEGEMANN, C. H.: Plane-table Methods adapted to Geologic Mapping. ECON. GEOL., vol. 7, pp. 621–637, 1912.
18. HIGGINS, D. F.: The Plane-table in Geological Mapping, with especial reference to graphic horizontal control by intersection methods. ECON. GEOL., vol. 8, pp. 729–751, 1913.
19. HEWETT, D. F.: Geology and Ore Deposits of the Goodsprings Quadrangle, Nevada. U. S. Geol. Survey, Prof. Paper 162, 1931.

6

Copyright © 1936 by McGraw-Hill Publications Company
Reprinted from *Eng. and Mining Jour.* **137:**557-561 (1936)

ON MAPPING UNDERGROUND GEOLOGY

Harrison Schmitt

The author describes a method he believes to be modern practice that records helpful information for the miner

THE unsatisfactory status of the profession of mining geology compared with established ones such as law and medicine is due partly to its newness, partly to the difficulty of valuing its successes and failures, but perhaps largely to its inadequate methods of mapping. I went through the stage of discouragement which many mining geologists have experienced, when it seemed as though geology could be of little help to the miner. My present conviction of the value of geology is the result of my having seen many orebodies discovered by pure geological method. The cynical statements so often heard, "Who ever heard of an orebody being discovered by geologists?" or "Moving dirt is the best way to hunt for ore," can be refuted by the records.

Most recent discoveries of ore by geological methods resulted from detailed studies of structural conditions, and these had detailed large-scale mapping as their base. Few recent college graduates seem to have had training in the mapping methods known to be effective. Many schools whose teaching of theory is excellent leave training in mapping to luck. The student is expected to acquire the needed skill somehow after graduation; yet it may mean his bread and butter for several years. Indeed, he may never learn to map well and therefore be ineffective in exploration.

There is marked contrast between underground and surface geological mapping. (See "Cartography for Mining Geology," *Ec. Geol.* Vol. 27, pp. 716-35, 1932, wherein I have described surface mapping.) The former is usually done under favorable conditions. Every detail of rock structure along the mine openings can be exposed, and, if necessary, the walls can be washed. The chief disadvantage compared with surface mapping is the relatively small area of rock exposed, but this is usually balanced by the greater vertical exposure of the major features of structure.

Before underground detail mapping is begun the geologist should see that the underground transit surveys are in good order and, where possible, closed. I have found instances where transit mine surveying has been done, and is being done, under lax discipline and supervision. Only one of four large mines examined belonging to different companies in a certain district had accurate underground surveys. The others had to be remapped before the geological work could begin. This situation is due partly to the custom of placing newly graduated engineers on surveying work without supervision by experienced men, partly to lax discipline, and partly to inadequate training in the mining schools. Accurate geological sections require accurate elevations, yet in many mines the elevations of survey stations are not determined at all. When the transit surveys are made the elevation of the rail below the station should be measured and recorded along with the elevation of the station itself, for the profile of the rail is the base for constructing the vertical sections.

Most mining geologists with a number of years' experience in mapping believe that all details capable of being mapped should be recorded, including those which at first appear of remote significance. They may become significant when plotted and integrated on the office maps. Details of definite value include the attitude, age, width, kind of filling, and direction of slip of faults; the attitude of shear, fractures, contacts, bedding, and schistosity; the classification of the rocks and their degree and kind of alteration. There is no place for guessing when recording such data. The map must be objective and factual. Correlation and inference, except when obvious, are subordinated until the maps are built up in their final form in the office. The best base for the field maps themselves seems to be strong, clear tracing paper cut to letter size (8½x11 in.) to conform to standard files and notebooks. (Keuffel & Esser No. 197 medium Ionic or similar paper is good.) The outline of the mine workings and the coordinate lines should be traced directly on this paper with black india ink from the master "hardy-copy" (single-mounted heavy-detail paper) composite map of the mine levels such as is standard at most mines. A maximum of detail can be plotted with a 4H pencil (4H "Koh-I-Noor" appears to be the best) on the tracing paper—and the resulting map is sufficiently waterproof under average conditions. The letter-size sheets can be carried in a spring-back letter-size aluminum notebook and this, in turn, in a large-pocketed waterproof game coat or a canvas bag, or, if the mine is extremely warm, in a pocketed skeleton vest. Fine detail should not be sketched underground with colored pencils, for the points are too soft and blunt and unless an excessively large-scale base map is used they make too wide a line. Most shades of meaning can be expressed without undue loss of time by symbols and notes written with a hard, sharp pencil. Immediately after coming out of the mine the penciled field maps should be copied with ink on the trackings of the mine levels.

When mapping a mine level the most convenient horizontal plane of reference is waist high; that is, 3 ft. above the rails. For accurate work the control map of the level should show the irregularities in the mine openings measured at the horizontal plane of reference. Points of observation should be located by measuring from the nearest survey station with a cloth tape.

In areas without local magnetic anomalies the strike and dip of structural planes are usually measured with a Brunton compass. In areas with local anomalies it may be necessary to locate two points of equal elevation on opposite sides of the drift on the plane of mapping and to connect these with a line which is the strike. Or the angle of divergence of the strike with the center line of the drift may be determined with a compass by noting the angle of swing of the compass card around the compass needle. The method is accurate even when the needle diverges 180 deg. from the true magnetic meridian. The compass must be kept at the same point in space when turning the angle. This may be facilitated by mounting the compass on a light camera tripod and "Y" head. If possible the compass card should be graduated to read from 0 to 360 deg. Important critical dips of the vein or other structural planes may be measured accurately by holding a string or wooden straight-edge on the line of dip at the top and bottom of the drift and measuring the inclination with a Brunton. The inclination of the string, however, is best determined with a suspended inclinometer made from a protractor such as is described for use with the "string" or German compass (Fig. 4).

The geological mapping of underground openings which do not lie approximately on a horizontal plane—that is, stopes, raises, and winzes—is a difficult problem. Useful projections of such openings, longitudinal vertical projections along veins, and composite plans of the rooms in *blanket* deposits are often compiled by the engineers for ore-reserve estimations and other purposes and are usually available. Such projections, however, cannot be used for any but the broadest of geological interpretations, because in underground mapping, as in all geological mapping, the primary concern is with the location and attitude of planes in space and the interrelations of these planes.

Location of planes in raises, winzes, and stopes which lie between the principal levels of mapping is best done with the string compass and inclinometer. These instruments (Fig. 4) are designed for quick, accurate determination of the bearing and inclination of a cord tautly stretched between nails which represent stations of traverse. The compass frame may be made of brass sheet or other non-magnetic alloy on which is mounted a Brunton compass with the mirror and sighting bars removed. German compass equipment which may be purchased from the instrument supply houses has no provision for setting off the magnetic declination. The Brunton compass card should be graduated from 0 to 360 deg. to eliminate the possible mistake of reading the wrong quadrant.

The string compass traverse is usually started from a known transit station. Nails are driven into the walls of the opening at each turning point and a mason's cord is stretched tautly between them. The bearing is determined with the suspended compass, the inclination by two readings of the inclinometer each equidistant from the ends of the cord. The average of the two readings is the true inclination with the error from sag compensated. A clearly marked cloth tape is then stretched between the stations, the inclined distance recorded, and offsets to the walls and/or to geologic planes are measured with another tape held normal to the string as distances vertically up and down or horizontally right and left. As a point or points on the structural planes are located, the dip and strike are recorded. The traverse notes

Fig. 1. . . . GEOLOGY in relation to the shaft and mine levels is plotted on this cross-section. Faults found on the surface are projected to those found underground and their characteristics are sometimes found to be similar. The determination of the approximate outline of hypogene alterations has been made on this cross-section, which is indicated in the figure.

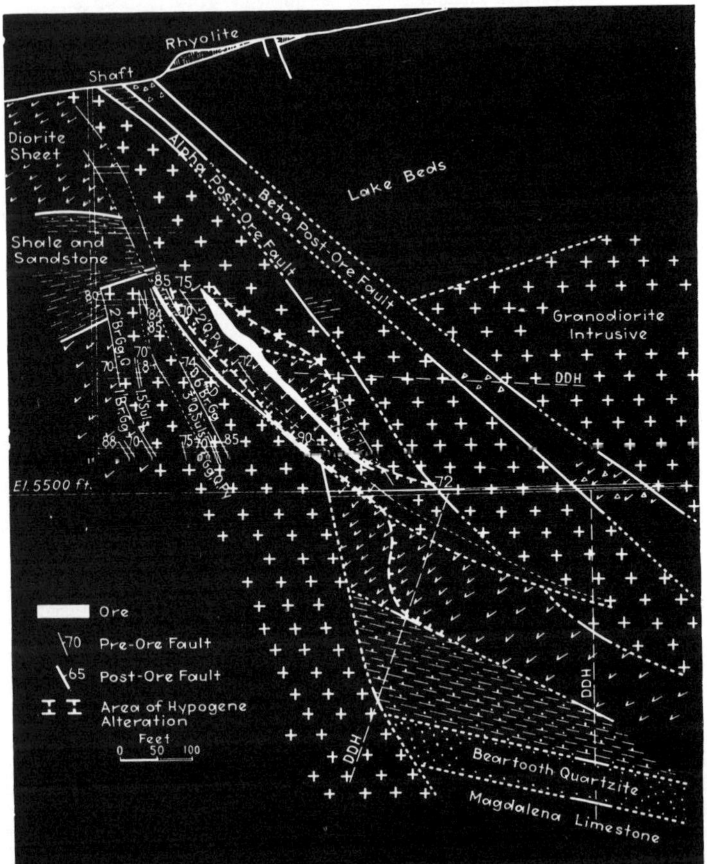

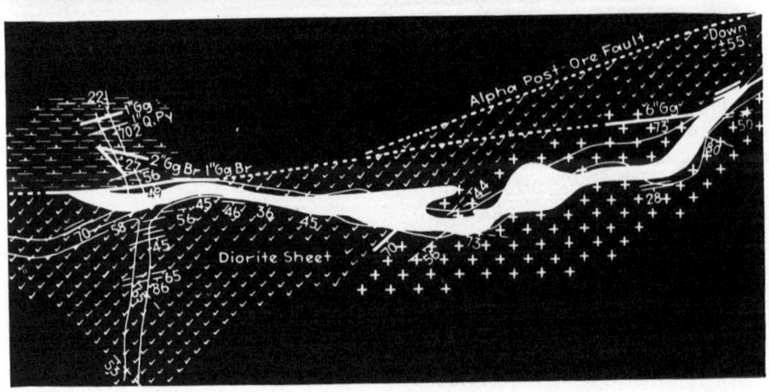

are calculated with a slide rule which has sine and cosine scales and the traverse is plotted as a composite plan with elevations written in at each point. When horizontal or vertical sections are passed through the area of traverse, the structural planes are projected to them and other near-by data appended. I know of no better method of recording and representing data from workings off the normal levels. In accuracy it compares favorably with mapping on the mine levels.

Where the stoping is done by square-setting, and minute detail is wanted, many geologists make geologic maps of each floor and, for filing, bind them together in their proper relative position. At the Ground Hog mine, Vanadium, N. M., it has been found that for accurate location of square-set stope detail measurement of the position of each floor with the string compass is necessary. Beautifully detailed sections can be made, of course, if a geologic map is made of each floor. Such plans and sections are useful in directing the details of stoping and in avoiding the loss of ore in the walls.

The horizontal level plan is essentially a picture of conditions actually observed and, at least in its first stages, contains a minimum of interpretation. Proportionally more interpretation is contained on vertical and horizontal sections excepting those near extensive workings. The vertical sections, when completed, are the basis for further interpretation and extrapolation to the level plans until finally the integrated set of maps gives an approach as nearly as possible to a three-dimension picture. The limit to which the interpretation and extrapolation can be advanced safely is largely a matter of general experience; partly a matter of experience with a specific area. Details of construction of sections will be described later.

Detailed level mapping of the first grade, which is detail on scales of not less than 40 ft. to the inch, is expensive and laborious work in the average mine (Fig. 2). Under many conditions such exact work is not needed, as, for example, in mines which exploit simple, unfaulted veins or where reconnaissance mapping is in order for a mine examination, or for most grades of governmental reports. In most mines, however, where formal geologic studies are being carried on to assist exploration, the highest grade of detail mapping is the only kind worth considering or which will do any good. At the very best the facts recorded will fall short of the number desired.

Usually, the primary character of the rocks and their age relationships, which include the stratigraphic succession, should be studied before formal detail mapping begins. Much of this preliminary work can be done on the surface. Additional knowledge of the rocks and their relationship will be obtained when the detailed underground mapping is carried out. The new information may upset some of the earlier concepts and make necessary a review of some of the ground already mapped. Few geologists who have mapped detail have not felt the need some time or other of remapping ground they have previously covered.

Little need be added to what has already been said about the location of geologic planes. As they are traced out I mark them by smudging with an acetylene mine lamp. They are mapped before other observations are recorded largely because many are lithologic boundaries. Further, where there is much detail, speed is gained and fewer details are missed if the work is subdivided, attention first being fixed on structure (largely planes), then lithology, then alteration and other related data.

The detailed study of faults should receive emphasis. Age classification is especially important, although often complicated by reopening from pre-mineral to post-mineral time. To fol-

Fig. 2. . . . THE MAPPED GEOLOGY of a mine level is a picture showing rock contacts, faults, the vagaries of the ore deposit, and other important geological conditions. Study of these maps by the miner affords important data for planning development and exploitation. It also helps to construct a three-dimensional view of a deposit. Pre-ore and post-ore faults are shown and the figures indicate the dip of the faults and rock contacts. The letters indicate the minerals present and the composition of the various rocks

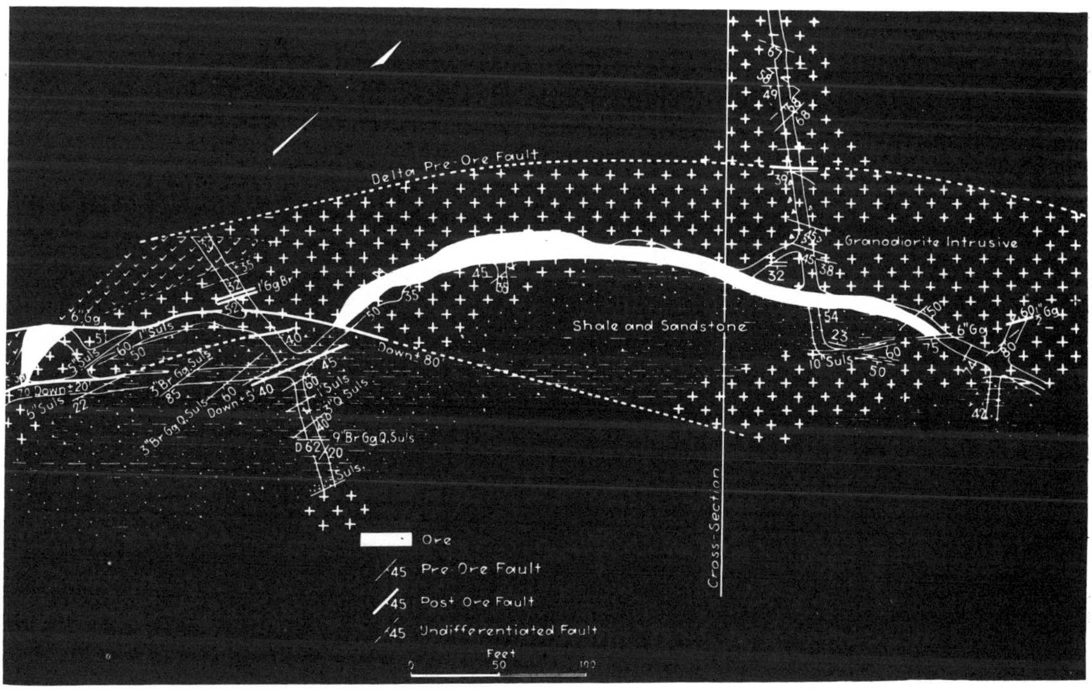

low ore in strongly faulted country one should know particularly two classes of structural information; first, the approximate original position and attitude of the structures which localized the ore, and, second, the position, attitude, direction of relative movement, and amount of offset of the post-mineral faults. The pre-mineral faults or fractures are usually recognized by their effect on the localization of ore, ore minerals, or gangue minerals associated with the epoch of ore deposition. Such mineralization is usually banded in or parallel to the break. If the pre-ore faults and fractures fall in a definite pattern or system, the pre-ore age of a given mineralized fault which is connected with or parallels the pattern or system may at times be inferred.

Post-ore faults within a given mineralized area may be recognized by the character of the breccia fragments, which, if the faulting is important, with the formation of thick gouge, may appear as well-rounded fragments of the harder parts of the ore or of the altered walls associated with the ore. Often such fragments are quartz. Patient digging in the gouge may be necessary to find them. The well-known criteria for post-ore faults such as the drag on the ore where cut off and small offsets of ore by minor shear planes sympathetic to the main fault should be the first evidence sought. The age of a fault may be fixed if it or its sympathetic shear planes cut veinlets of mineralization contemporaneous with the ore. The various strands, branch faults, and sympathetic shears related to a fault are, of course, generally contemporaneous with it.

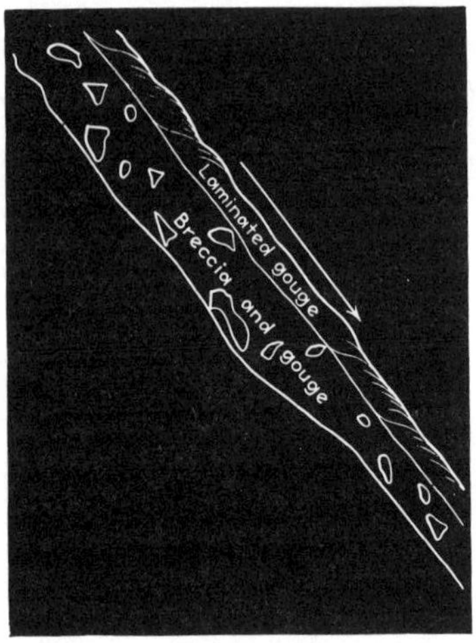

Fig. 3. . . . SECTION OF A FAULT showing attitude of laminations in the gouge

The thickness and character of the gouge or other filling in faults and veins should always be recorded. Within a given area these data are invaluable for dating, correlating, and judging the relative importance of faults and veins. The direction of the striations on many fault planes may be an unreliable criterion for the average direction of slip, because of the heterogeneity of movements which have occurred. On many faults, however, the striations are indicative of the true direction of slip and should be recorded. The direction of relative offset usually can be determined by a careful study of the attitude of the gouge laminations (Fig. 3). I have had little success with the study of chatter marks, tailing of soft material behind a hard projection, and similar criteria.

Fig. 4. . . . A DEVICE for holding a Brunton compass used in taking astromomic bearings in stope surveys can be made with the aid of these details. Small hangers on the protractor permit quick attachment to the strings run through the stopes, and the inclination is readily determined

When mapping dikes, inclusions should receive particular attention. They may determine the age of the dike; for example, they may include fragments of ore, and may give information on the rocks to be expected in depth.

The importance of general studies of the wall-rock alteration cannot be overemphasized. All modern mapping should include classification as to the character and estimates as to the intensity of the alteration. Observations are usually taken at intervals of not greater than 15 to 25 ft. Hypogene and supergene alteration should be distinguished if possible. I estimate the intensity of the alteration as slight, medium, and strong—equivalent to the limits 0-33, 33-66, and 66-100 per cent.

For example, at a certain mine the symbol "Kbd-MD-ALT; Q Ch, Py" means Cretaceous "birdseye" diorite porphry moderately altered with quartz, chlorite, and pyrite as replace-

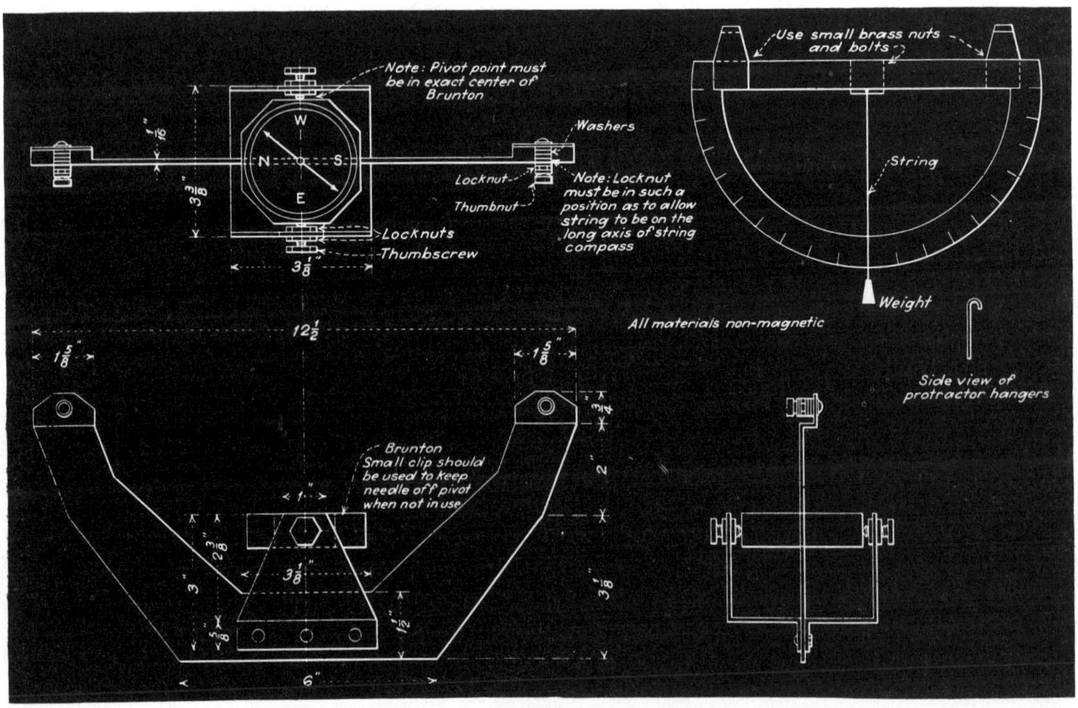

ments whose volume importance is estimated to be in the order given.

Such data may be useful in several ways. For example, since the outline of the boundaries of the alteration along a vein is usually a magnification of the outline of the vein itself, it is possible to judge the position of exploratory drifts and crosscuts in relation to the vein even when these have only slightly penetrated the alteration zone. Further, when a drift has run out of typical wall alteration along the strike it may be useless to crosscut or continue farther for ore unless there is reason to believe one is in an intershoot block. Again, in strongly post-ore faulted country where few observations on the dip of the ore can be obtained, the outside contacts of the alteration zone, when outlined, may aid in plotting the probable attitude of the vein.

At the Pewabic mine, in the Central mining district, in New Mexico, it has been found that the pyrometasomatic silicate alteration and the sphalerite orebodies are mutually associated; that it is useless to explore for ore in the unsilicated limestone. The age classification of dikes is often facilitated by the study and classification of the alteration that has taken place.

The degree of probability of the various observations and interpretations should be estimated and expressed by quantitative symbols. I use solid lines for known contacts and dotted lines for inferred or approximately located contacts, the spacing of the dots being inversely proportional to the degree of probability. An essential of good mapping is clear differentiation between facts and inferences and an adequate representation of this differentiation on the maps.

Assembly of Field Data—The level, stope, and raise data on the letter-size field maps can be plotted on large, master office maps usually made of tracing cloth. These maps as a rule include large-scale sections and level plans on the same scale as the field maps, so the field data can be traced directly. Stope maps, stope sections, and field maps of individual raises, winzes, and small sublevels may be plotted on letter-sized tracings, which are then punched and filed in cloth-bound notebooks. Small-scale—say 100 ft. to the inch—reductions of the more important of the plans and sections will aid in the determination of relationships.

Today the emphasis is more and more turning to oreshoots, their inlets and outlets (roots and gossans), and associated alterations. Obviously, the first step in studying an oreshoot is to make an accurate representation of the known part of it. This is done with equal-ore-width or thickness lines, equal-ore-value and equal-metal-ratio lines usually projected to a longitudinal vertical projection, but if the vein is flat, to a horizontal projection or projection parallel to the plane of the vein. These projections should include the projections of the intersections of pre-mineral and post-mineral faults and veins and of the junctions of branch veins with the main vein. If the vein has been faulted, no effort should be spared to work out the amount and direction of the slip. The inferred position of the faulted segment should be plotted.

All available data from the plans and string compass notes should be plotted on the sections. The trace of the intersections of the various planes should be calculated or derived from the published charts and plotted with their proper component of dip. Correlations of geologic planes from level to level is aided by plotting the angles made by the strike of the planes with the sections and the thickness and character of the filling of the fault and veins (Fig. 2). Wall-rock alteration data should, of course, be plotted on the sections as well as the plans.

The tracing cloth used for the large maps, since it shrinks rapidly for a few days after unrolling, should be cut the proper size and then stretched out and dried on a wall for several weeks before the coordinate lines are drawn on it. Colored india inks should be used for plotting the various data, and they print better if about 1 c.c. of orange ink is added to each 1 oz. bottle, especially to the green, blue, purple, brown, and carmine inks. Lines of various weights, particularly curved lines, can be drawn best in many cases with Barch-Payzant pens. Orange and yellow inks should be avoided when possible except for permanent lines such as coordinate lines, because they dissolve the starch in the cloth and are difficult to erase cleanly. Erased spots should be filled and rubbed smooth with gritless sticks of talc.

Gritless crayons are used for coloring the tracings and prints. (Dixon's and Faber's crayons are good. One set of Faber's has 60 colors.) The color should be laid on evenly and then dissolved and spread smoothly with paper stumps dipped in gasoline, or, better, xylol, which is non-inflammable, and wiped moderately dry with cloth. Surplus paint can be quickly distributed more evenly or wiped off with small wads of cloth. Before prints are made, most of the paint should be removed from the tracings with art gum.

7

Reprinted from pages 695–696, 698–699, and 699–701 of *Ore Deposits of the Western States,* The Committee on the Lindgren Volume, ed., American Institute of Mining and Metallurgical Engineers, New York, 1933, 797p.

APPLICATION OF GEOLOGY TO MINING IN THE ORE DEPOSITS AT BUTTE MONTANA

F. A. Linforth*

INTRODUCTION

The extremely complex structural environment of the ore deposits at Butte has long been recognized. The intricate network of veins and faults, early encountered in the mines, led the operators to realize the necessity of accurate geologic mapping as an aid to the development of the ore bodies. Apex litigation early emphasized the need of accurate geologic knowledge and later became an important influence in the development of a systematic study of Butte geology.

The geological department of the Anaconda Copper Mining Co. was organized in 1900 by H. V. Winchell, who with D. W. Brunton worked out the essential methods of procedure applicable to the Butte problems. In the main, these methods have been followed up to the present time. Since 1906 the geologic work has been under the direction of Reno H. Sales, and its scope and practical usefulness have been greatly enlarged.

EARLIER PRACTICE

The development of a technique and procedure in the application of geology to mining varies considerably with the particular geologic environment of the ore deposits. At Butte the structural features have provided the background upon which the methods were developed. In the early days the study of the structural pattern was the principal part of the work. It brought out the fact that the ores occur in veins as fissure filling essentially related to two sets of fractures which strike east-west and northwest-southeast, respectively. Upon this structure is superimposed a complex system of post-mineral faulting. The earlier problems consisted of deciphering this pattern and advising the operators which way to turn when the ore was unexpectedly interrupted by faults. The method of taking the notes underground and platting them on standard office maps and sections has been abundantly described in the literature.[1] By far the most important feature of this work is the

* Assistant Chief Geologist, Anaconda Copper Mining Company.

[1] Brunton, D. W.: Geological mine-maps and sections. *Trans.* Amer. Inst. Min. Engrs., vol. 86, p. 508, 1906.

Linforth, F. A.: Applied geology in the Butte mines. *Idem,* vol. 46, p. 110, 1914.

Sales, R. H.: Ore discovery and development. *Eng. and Min. Jour.,* vol. 128, 1929.

necessity for careful and accurate observation coupled with the correct estimate of geologic values. It is of the utmost importance that the geologist recognize the relative structural strength of each exposure and record it accurately by a system of standards in the notebook. It is

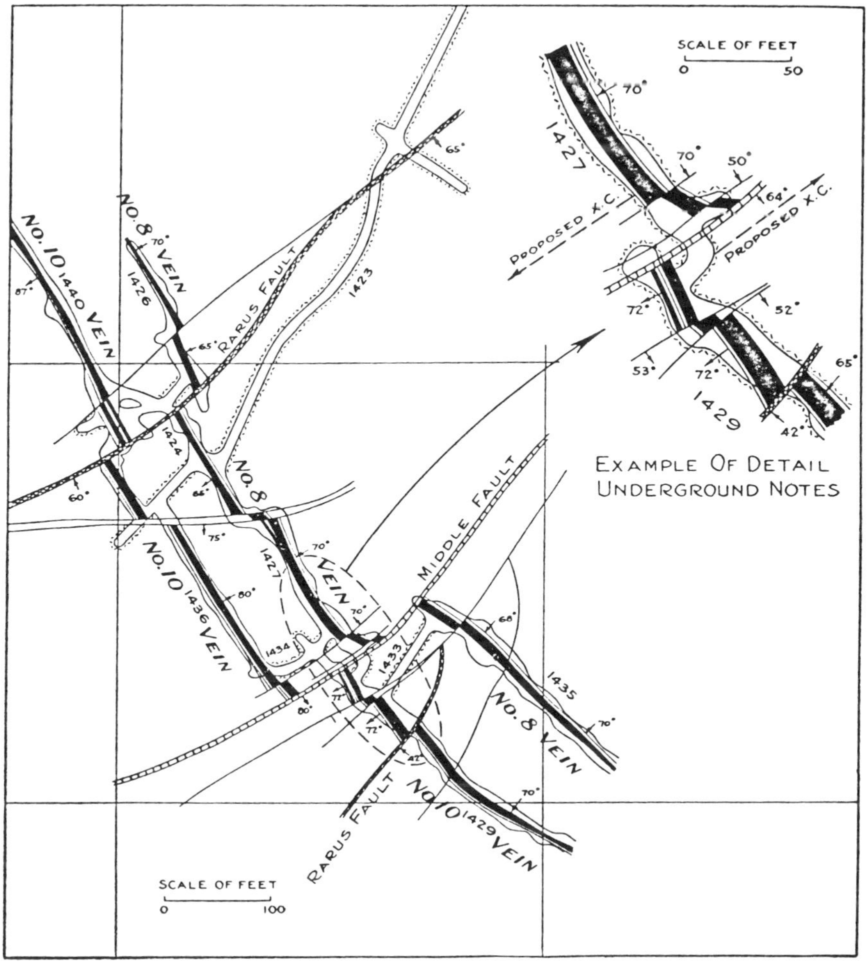

FIG. 1.—Portion of 1,400-ft. level at Mount View mine, showing method of recording data. The original drifting was 1426, 1424, 1427 and 1429, believed to be all on the same vein, The correct interpretation of the structure from the detailed geologic notes at the point shown resulted in the discovery of 1440, 1436 and 1435.

entirely practical to prepare sketches of underground exposures with engineering accuracy and there is no doubt as to the ultimate value of such painstaking endeavor. Failure in this respect spells the chief difference between applied geology that constitutes a reliable aid to the miner and the doubtful, indefinite variety that often does more harm than good.

[*Editor's Note:* Material has been omitted at this point.]

In the study of all these complex structural relations, plan maps and vertical cross-sections are the principal equipment. For the practical man at the mine, however, it has been found desirable to present an easily understandable picture of the combination in three dimensions. For this purpose a simple type of skeleton model has been devised for study of local areas. These models show the drifts, crosscuts and raises with geology painted on them in bright colors, together with an ingenious attachment of celluloid strips which greatly clarifies the correlations. Where veins are interrupted by faults between levels and the trace of the intersection is a sloping line, an accurately made model of this type is of material assistance to the foreman in laying out his raises and intermediates so as to be sure of recovering all the ore with the least possible expense. Although this type of model has also been found useful to the geologist, he usually prefers the separate level glass plate type, especially when raise sections are cut out and inserted between the plates. A well equipped model shop is operated in connection with the geological department.

[*Editor's Note:* Material has been omitted at this point.]

SAMPLING AND ORE ESTIMATION

In 1919 the direction of the mine sampling was placed in the hands of the Geological Department. The complicated internal vein structure and the erratic distribution of the minerals occurring within the ore body made geologic guidance necessary in reliable sampling. The requirements of the operating department became more elaborate. The miner wished to know not only the average width and grade of the material broken but the corresponding figures for that part of it that would pay to ship. As selective mining and underground sorting were further developed in the mines, there was greater need for accurate sampling control of these operations. A carefully standardized system of sampling and averaging based on the geologic study of the internal structure of the veins was devised. It includes a specially designed card for each stope, which shows a plan sketch of each floor on a scale of 20 ft. to the inch with the number and position of each sample. A side column gives the corresponding assay results. A similar attached card shows the assay-foot calculations and the averages for both the width broken and the width of contained ore at 10-ft. intervals. From these cards the metal content of the various bands and streaks of ore minerals and granite composing the stope face may be readily ascertained as well as the average of the entire excavation. In addition to the sampler's stope card, the geologist prepares a corresponding sketch upon which is carefully mapped the structural detail. This combined sample and geologic record of each stope floor has proved of great assistance to the foremen and shift-bosses in controlling ore-breaking operations. Monthly reports compiled from the daily stope-sampling record are prepared for the general superintendent and his assistants.

In addition to sampling and geologic mapping in the stopes, the development work is sampled every round. A copy of the daily report of development sampling is sent to the geologist, who compares the assay results with his geologic notes and aids in deciding on any changes to be made in the direction of advance. Assay results from the sampling of drifts and raises are recorded graphically on a set of curves that show the width and metal content of the ore. Each working place has a separate card. The assay widths and values are platted on one set of coordinates and the advance on the other. The advance is platted at 50 ft. to the inch to correspond with the standard geologic maps. This plan provides complete information with respect to both geologic structure and assay value by superimposing any geologic tracing over the corresponding assay card. It also provides an excellent means of delimiting ore shoots and is of material aid in making ore reserve estimates.

ORGANIZATION

In 1924 the work of the Mining Engineering Department was combined with that of the Geological Department, under the direction of the Chief Geologist and Engineer. This combination has proved very efficient, particularly with respect to coordination of the work. Service maps flow through the organization at regular intervals without delay and all duplication of effort has been eliminated. The routine work of the mining engineers involves measuring the miner's contracts each week and the mine samplers are obliged to assist in this work, temporarily placing themselves under the direction of the engineers. This association has provided a closely woven organization of geologists, engineers and samplers under the supervision of the Chief Geologist and Engineer and his assistants. The mines have been divided into two groups for geologic and engineering work and into three groups for the control of sampling. Each of the engineering groups is directed by a chief assistant engineer; two experienced geologists are in charge of geological work and a chief sampler and three assistants are in charge of the sampling. These men are in daily conference with the chief assistant geologist and engineer, who in turn reports to the head of the combined organization.

In recent years the equipment and organization have become sufficiently complete to establish a practical system of development called the ton for ton program. Combined geologic, engineering and sampling information has made it possible to determine the current rates of finding new ore per foot of development advance in different areas. By this means the geologist recommends the necessary work to be done to maintain any desired tonnage. The result has been the development of a new ton of ore for each ton mined and checks at frequent intervals keep the management advised as to temporary deficits or surpluses, so that the development program can be adjusted. This system has met with marked success and represents a very practical result of the application of geology to mining.

It will, therefore, be apparent that the work of the combined geological, engineering, and sampling departments at Butte is largely a work of service, and great stress is laid upon the part of the geologist in aiding actual mining operations. At the same time the fundamental purpose of the geologic study is to gain the broadest possible knowledge of the geology of the Butte district. Work is continually in progress on problems concerning ore genesis and its relation to igneous activity, on mineralogic distribution with respect to zonal arrangement; on the criteria for predicting extensions of ore shoots; on the mechanics of processes that produced the fracture systems and other related problems of pure geology.

8

Reprinted from *Eng. and Mining Jour.* **147**:74–75 (1946)

Profile Technique Useful In Mapping Stope Geology

ALLAN HARRIS JAMES, Formerly Geologist, U. S. Smelting, Refining & Mining Co.
Present address 77 Massachusetts Ave., Cambridge, Mass.

A STOPE on an inclined bed or on a vein raises the question as to how best to portray the geologic features. If the ore is regular in dip and strike, the simplest picture may be a projection of the stope outline and geology in the plane of the inclined layer, but if the layer is folded or warped, or is to be used in conjunction with plan maps of levels or surface maps, this scheme is unsatisfactory. The alternative, a projection to a horizontal or vertical plane, is more practical but demands a considerable use of descriptive geometry or trigonometry because most of the measurements have to be made along inclined lines.

For mapping room-and-pillar stopes in flattish limestone replacement deposits in New Mexico, I have devised a "profile technique" which can be used, with appropriate modifications, in a wide variety of inclined stopes. Plans and sections can be plotted underground without delaying completion of maps pending office computations. Furthermore, the graphic solution of the three-dimensional problem remains on record in systematic form.

The Procedure

The essential steps of the technique are as follows: (1) Measurements in the stope are made with tapes which are stretched between wall points and temporarily fixed in horizontal or inclined positions, depending on circumstances, and (2) the vertical and horizontal distances between the points are ascertained by plotting a profile at the side of the map as the work proceeds. On the profile, which is not to be confused with a vertical section, the measured lines are plotted in their true length and inclination so that each point on the line appears at its proper elevation. See accompanying illustration, which describes the procedure of mapping a stope on a field sheet.

One small disadvantage inherent in any horizontal projection of a set of inclined workings is that strikes of features, such as the fault shown at points 3, 5, 8, and 9 in the accompanying illustration, do not line up, because they are observed at different levels. Despite this disadvantage, however, it is probably best in the original record to plot such features at the points at which they are observed and with their true strikes. In addition, it may be feasible to measure the bearing of the trace of the fault on some geologic surface and plot this on the projection in a dashed line.

I want to thank J. V. Neuman, chief engineer and geologist, Bayard Unit, United States Smelting, Refining & Mining Co., for suggestions and discussion which have aided in the development of mapping techniques suitable to local conditions, and the management of the company for permission to publish this article.

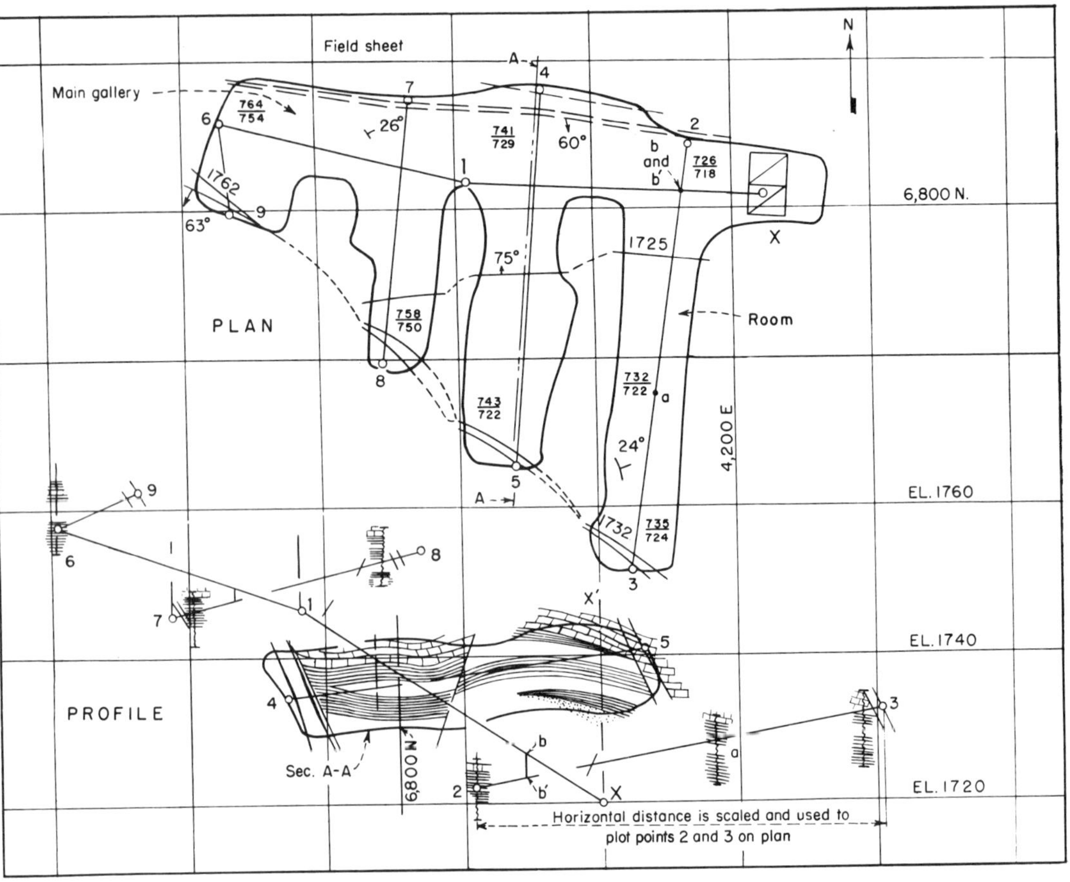

How the Profile Technique Was Used to Map a Stope

1. The distance from X to 1 along the main gallery was measured by stretching a tape between these two points. For survey stations, lead-headed roofing nails were driven into cracks in the rock. The lead head reduces the tendency of the nail to bend, because it cushions the hammer blow.

2. The length and inclination of the line X-1 were plotted in the profile at the bottom of the sheet, and the horizontal distance between X and 1 was scaled from the profile.

3. Using the Brunton bearing of line X-1, the point 1 was plotted in its true position on the plan.

4. Without removing the tape, a second tape was stretched from point 2 to point 3 in the first room. The second tape passed below the first tape at point *b* (shown on the second tape as *b'*).

5. The vertical distance between the two tapes was measured, and the line 2-3 was plotted on the profile after point *b'* had been located.

6. After scaling the horizontal distance 2-3 on the profile and using the measured bearing of line 2-3, the points 2 and 3 were plotted on the plan.

7. It should be noted that line 2-3 could have been plotted from right to left, or from left to right; which direction is followed is not significant except with respect to the particular lines on which they appear. However, consistency increases the clarity of the notes. Here, for example, measurements in the gallery X-1 and 1-6 are plotted from right to left, whereas measurements in the room 2-3, 4-5, and 7-8 are plotted from left to right.

8. Intermediate points of geologic interest along the lines 2-3 were plotted on the profile and then transferred to the plan. Thus at *a* on the profile, there appears a geologic column from roof to floor and the elevations at the roof and floor are shown in their correct vertical positions. These elevations, scaled from the profile, have been recorded numerically on the plan as 732 and 722, respectively, at point *a*.

9. After all data for the first room 2-3 had been recorded, the tape was removed to the second room and the points 4 and 5 were established in a similar manner. At the same time vertical geological sections such as the one through room 4-5 were prepared.

9

Reprinted from *Mines Mag.* **62:**10–12 (1972)

Rapid Geologic Mapping In Large Tonnage Open Pit Mines

*By Robert A. Metz**

DURING the past decade we have witnessed spectacular improvements in productivity of open pit mining equipment. Thousand-horsepower engines, 200-ton trucks and 15-yard electric shovels, which such a short time ago were just dreams on a drawing board, are now realities. More productive equipment and metal shortages have encouraged increased production at existing mines and the development of new larger tonnage operations. Operators and engineers have been able to employ computers and various other types of automation to "stay on top of things." The mine geologist, however, is merely faced with a bigger and more rapidly changing pit to pick in.

Keeping geologic information current in these mines presents entirely new problems to mine geological staffs. To obtain adequate geologic information on the developing orebody requires enlargement of the staff to share the work load or revision of methods in keeping with the times. The former has the disadvantage of loss of the comprehensive knowledge of the pit that a chief geologist likes to have, not to mention management's usual lack of enthusiasm for force increases.

*Mr. Metz is district geologist, Duval Corp. (5357 East Pima, Tucson, Ariz. 85712), and was formerly a pit geologist for the Ray Mines of Kennecott Corp. He earned his Geological Engineering degree in 1955 from the Colorado School of Mines.

CONVENTIONAL GEOLOGIC MAPPING TECHNIQUES

Traditional pit mapping techniques all have certain drawbacks in this new era. There are probably as many variations of these as there are pit geologists, but it might be well to briefly review some of the more common ones. These include compass-tape traverse, compass resection, use of sample stations (or other well established closely spaced control points), specially surveyed geologic control points, and plane table.

Compass-Tape Traverse

This method involves traversing from a survey station or other control point, usually with a cloth tape, along the toe of a bench face. If the entire face is mapped, the location of the crest and toe of the exposure must be estimated and plotted. Geologic features are then plotted with respect to the distance of the tape along that particular leg of the traverse. This method is slow, with a great proportion of the time consumed by the mechanics of surveying, and cannot be used in an active area. It requires established control points near the face sufficient to begin and close a traverse. Magnetic attraction from high voltage power lines, rail, and heavy equipment (and sometimes magnetite) can seriously interfere with this method.

Compass Resection

Compass resection to locate points of geologic importance along the face is somewhat faster, but requires more survey stations for accurate resections; survey stations are not apt to be abundant in a very active open pit. The magnetic deviation mentioned above can be very detrimental to this method. Also, in geologically complex areas, repeated resections become tedious.

Sample Stations

The use of sample stations along or near a face for geologic mapping control can work out quite well in many cases. However, in some operations these are not surveyed or, if surveyed, they may not be plotted in time to be used by the geologist. Also, with the present trend toward the use of blast hole cuttings in ore control, face samples and thus sample stations are disappearing from many operations.

Specially Surveyed Control

A variation of the above method is setting geologic control points at points of interest (faults, veins, contacts, etc.) along the face, having them surveyed and plotted, and then mapping the geology. This is time consuming in that it involves visiting the exposure twice. In a rapidly changing pit, these stakes are not likely to remain long enough to be useful and survey crews may not always have time to shoot them in.

Plane Table

The conventional plane table method involves a two man crew. The rod man (geologist) gives stadia shots at selected points which are numbered identically by him and the instrument man. The rod man records geologic features in a notebook at each point and these are later transferred to the map. This method is considered too time consuming because of the note-taking and transcribing process. It also presents many chances for error or omission.

In summary, it can be seen that none of these methods is well suited to an operation where the face may be inactive for only a few hours or less at any particular time.

THE NEW PROBLEM

Consider the situation presented by an open pit mine covering the better part of a square mile, with rapidly changing topography as the result of the removal of more than 200,000 tons of material a day, where long bench faces will advance 100 feet in a week. With any of the traditional methods, geologic mapping would be limited to abandoned faces or general features of the active ones.

A Solution

A method has been developed to cope with this situation which accomplishes detailed mapping of active faces and involves usually less than one day a week in the field. It permits correlation of rock types, structure, alteration and mineralization, and ore zones as the bench advances. This method combines conventional plane table mapping with voice communication. Equipment used consists of the following:

plane table
high-standard alidade

FIGURE 1: Plane table set up at mine triangulation station. Rod man (circled) transmits details of geology from face; usual working distances are from 300 to 900 feet but can be greater.

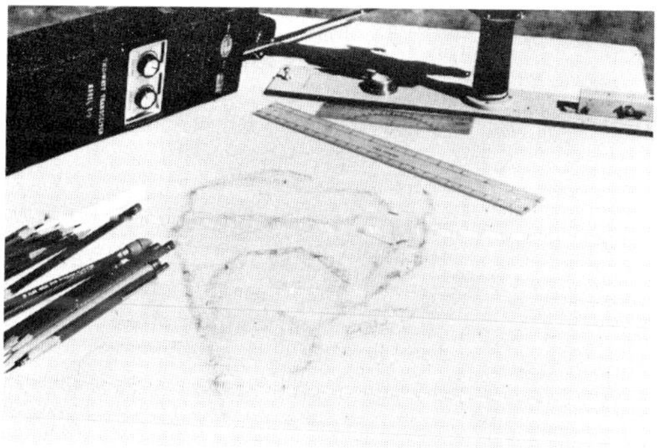

FIGURE 2: Close up of plane table map and equipment. This view faces north, the general direction in which the bench is advancing. Repeated use of the same sheet eliminated duplication of work and facilitates structural correlations in the field.

light weight stadia board with "flexible" rod face
2 two-watt citizens' band transceivers with rechargeable Ni-Cd batteries
one "shoulder talk" unit
Mylar plane table sheet

The mapping procedure is as follows:

The plane table is set up over a pit triangulation station or any convenient point near the face to be mapped (in which case a three-point resection is performed to locate and orient the table). The rod man walks along the crest of the bench giving stadia shots at pertinent intervals to adequately define the topography of the face. He then descends to the toe and proceeds along the face describing the configuration of the exposure, rock type, alteration, etc., over the radio. Faults and contacts are described by points where they intersect the base and crest of the exposure (apparent strike) and their true attitudes (dip and strike), thickness of gouge and breccia, etc. This information is put on the map immediately by the instrument man using a standard set of colors and symbols. As stadia readings usually require only ten to fifteen seconds, delays experienced by either member of the crew are minimal. Vertical angle corrections are seldom necessary in this type of surveying. After completing mapping a face the rod man checks the map for accuracy of representation before the plane table is moved. The operations are illustrated in the accompanying figures.

Advantages

The plane table-radio method offers a number of advantages over the traditional methods. Foremost among these is speed. Data can be gathered and recorded along a face at a rate up to, and at times exceeding, 1000 linear feet per hour. This has been found to be three to five times as productive as two geologists working separately and using one of the other methods.

Because of this speed, mapping can be done on active faces with minimum interference from mining equipment. The shovel area itself can be mapped by setting up the equipment and plotting crest topography before a scheduled shutdown (shift change, lunch period, etc.,) and mapping geology during the idle period.

Accuracy is another important feature. Topography is current and errors in transcription of data are minimized. Because the same field sheet is used repeatedly there is no unnecessary overlap or omission of mapping along a pit face. This also facilitates continuity of geologic projections.

FIGURE 3: After mapping a pit face, the rod man checks the field sheet for accuracy of representation. Note shoulder mounted speaker, antenna, and microphone which permit main transceiver unit to be worn conveniently on the belt. Hi-visibility vest is worn by the rod man for safety and to enable the instrument man to follow his movements more closely.

Although desirable, it is not essential that the instrument man be a geologist. Given a solid briefing on geologic nomenclature and representation, and with the benefits of radio communication, a sub-professional person can be trained to do the work with relative ease.

Not to be neglected is the safety value of a radio-equipped two-man crew. The instrument man is in a position to instantly warn the rod man of hazards unseen by him while working under a high bank. With the large equipment now in use and the attendant noise level in open pit mines, two-way radio communication is especially valuable.

These advantages are felt to greatly outweigh any drawbacks the plane table-radio method might have. Where it has been used it has been felt to successfully accomplish the goal of rapid and accurate gathering of necessary geologic data. Although the method was developed especially for use in large output open pit operations, it should be applicable to smaller operations as well with a proportionate saving in time for field work.

The author wishes to express appreciation to the Duval Sierrita Corporation and Duval Corporation for permission to present this paper and in particular to Mr. F. H. Howell for encouragement in preparation and critical review of the manuscript.

10

Reprinted from pages 634, 635, 637, 641, 642, 643, 644, 645, 646, 650–651, and 653–655 of *Econ. Geology* **67**:634–655 (1972)

DEFORMATIONAL HISTORY OF THE CARIBOU STRATA-BOUND SULFIDE DEPOSIT, BATHURST, NEW BRUNSWICK, CANADA

George H. Davis

Abstract

The Caribou sulfide deposit near Bathurst, New Brunswick consists of fine-grained sulfide lenses which lie conformably within a sequence of low-grade metamorphic rocks. Pyrite is the chief mineral in these lenses and occurs with interstitial quartz, chlorite, and minor calcite. In addition, sphalerite, chalcopyrite, and galena are found in appreciable quantities. The sulfide lenses are nearly vertical and range in thickness from several inches to 50 feet. They are spatially associated with the tight, steeply plunging Caribou fold structure. Strata-bound concordance of the sulfide lenses within highly deformed metamorphic rocks makes it difficult to determine the age of mineralization. An attempt is made here to establish the time of mineralization relative to deposition, deformation, and metamorphism of the associated Ordovician rocks by detailed comparison of the structural histories of the ore and host rock.

Rocks in the Caribou mine area have been subjected to three periods of deformation, one during the Taconic orogeny and two during the Acadian orogeny. The deformation (F_1) which accompanied low-grade metamorphism during the Taconic orogeny produced isoclinal folds with an axial plane schistosity. Bedding was transposed during development of the foliation. The regionally metamorphosed Ordovician rocks were further deformed by kink and parasitic folding (F_2) accompanying the formation of the Tetagouche anticlinorium in Acadian time. A late Acadian pulse (F_3) deformed the northwest limb of the Tetagouche anticlinorium, producing the Caribou fold and modifying the orientation of F_2 folds. Minor folds and axial plane fractures in the hinge zone of the Caribou fold were formed during F_3 deformation.

Megascopic structures reveal that the Caribou deposit was subjected to at least two periods of deformation. Folding and faulting of sulfide lenses in the hinge zone of the Caribou fold indicate that sulfide mineralization preceded development of this host structure (F_3). A reverse parasitic fold, identical in style and orientation to F_2 folds, contains deformed sulfide lenses and demonstrates that sulfide deposition was prior to early Acadian deformation.

Microscopic examination of specimens from the Caribou deposit reveals sulfide foliations which are interpreted to be products of crystallization, slip, and plastic flow during metamorphism. The sulfide foliations are defined by: (1) elongated sutured pyrite aggregates, (2) oriented fragments of broken pyrite, (3) oblate pyrite framboids, (4) laminae of sphalerite, chalcopyrite, and galena, (5) lens-shaped sphalerite grains, (6) microlaminae of aligned platy pyrite crystals, and (7) sulfide "pressure shadows." These sulfide fabrics do not appear to be inherited by replacement of deformed non-sulfides.

The central conclusion is that the Caribou deposit was subjected to at least two periods of Acadian deformation. Not as firmly established is the interpretation that the deposit was regionally metamorphosed during the Tactonic orogeny. Primary depositional textures have been modified or destroyed during the complex deformational history of the deposit, and hence the history of its initial emplacement by syngenetic or epigenetic processes is obscure.

[*Editor's Note:* Material has been omitted at this point.]

Structural Geology

Methods of study

A major objective of field work was to define the orientation of fractures, foliations, and small-scale folds within the Caribou mine area. Accordingly, some 7,000 measurements of planar and linear features were made during geologic mapping of both the Caribou mine area (Fig. 2) and the adit level of the Caribou mine (Fig. 3). Three computer programs were devised for plotting the orientation measurements sterographically on lower-hemisphere equal-area net projections. A "pole-plotting program" was used to plot the orientation of structural elements represented by less than fifty measurements. Each structural element, represented by a single point, was plotted directly using a Calcomp Plotter. A "pole-density program" was used to determine pole-density distributions for particular structural elements. Densities in percent were plotted directly by the Calcomp Plotter on 10-inch diameter circles. The "pole-density program" represents a combination of the "pole-plotting program" and the counting method described by Noble and Eberly (1964). Prefold orientations for particular structural elements were determined through a "rotation program." Through the use of this program, pole density distributions could be replotted according to any designated rotation matrix. The sterographic plots used to represent orientation data in this paper were derived from these programs. All contouring of pole densities was done by hand.

As an aid in distinguishing systematic sets of structural elements, the Caribou mine area and the adit level of the Caribou mine were each subdivided into four domains defined on the basis of uniform foliation trend. The area occupied by each domain in the Caribou mine area is shown in the inset of Figure 7. Specific structural elements (e.g., axes of sinistral kink folds or axial planes of dextral parasitic folds) for each domain were plotted independently by computer.

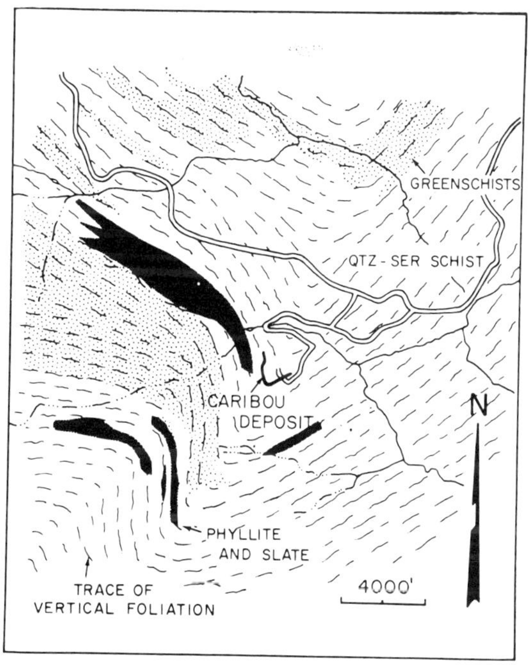

FIG. 2. Generalized geologic map of the Caribou mine area showing the distribution of metamorphic rock units.

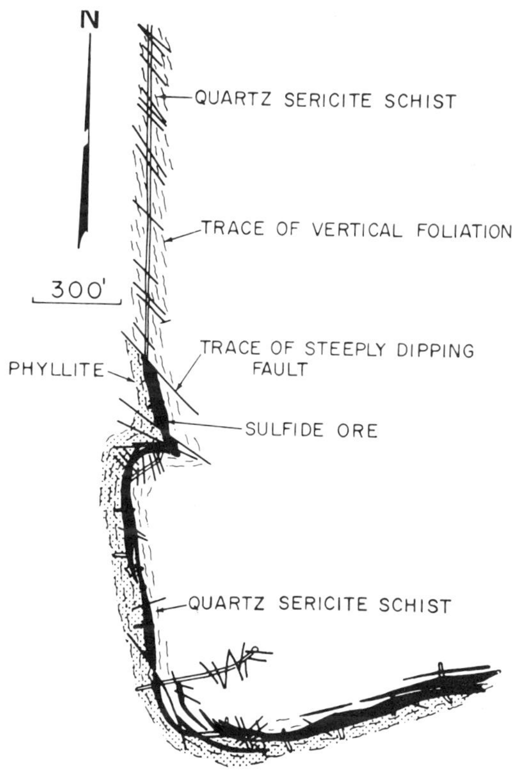

FIG. 3. Generalized geologic map of the adit level of the Caribou mine showing the distribution of metamorphic rock units and sulfide layers.

[Editor's Note: Material has been omitted at this point.]

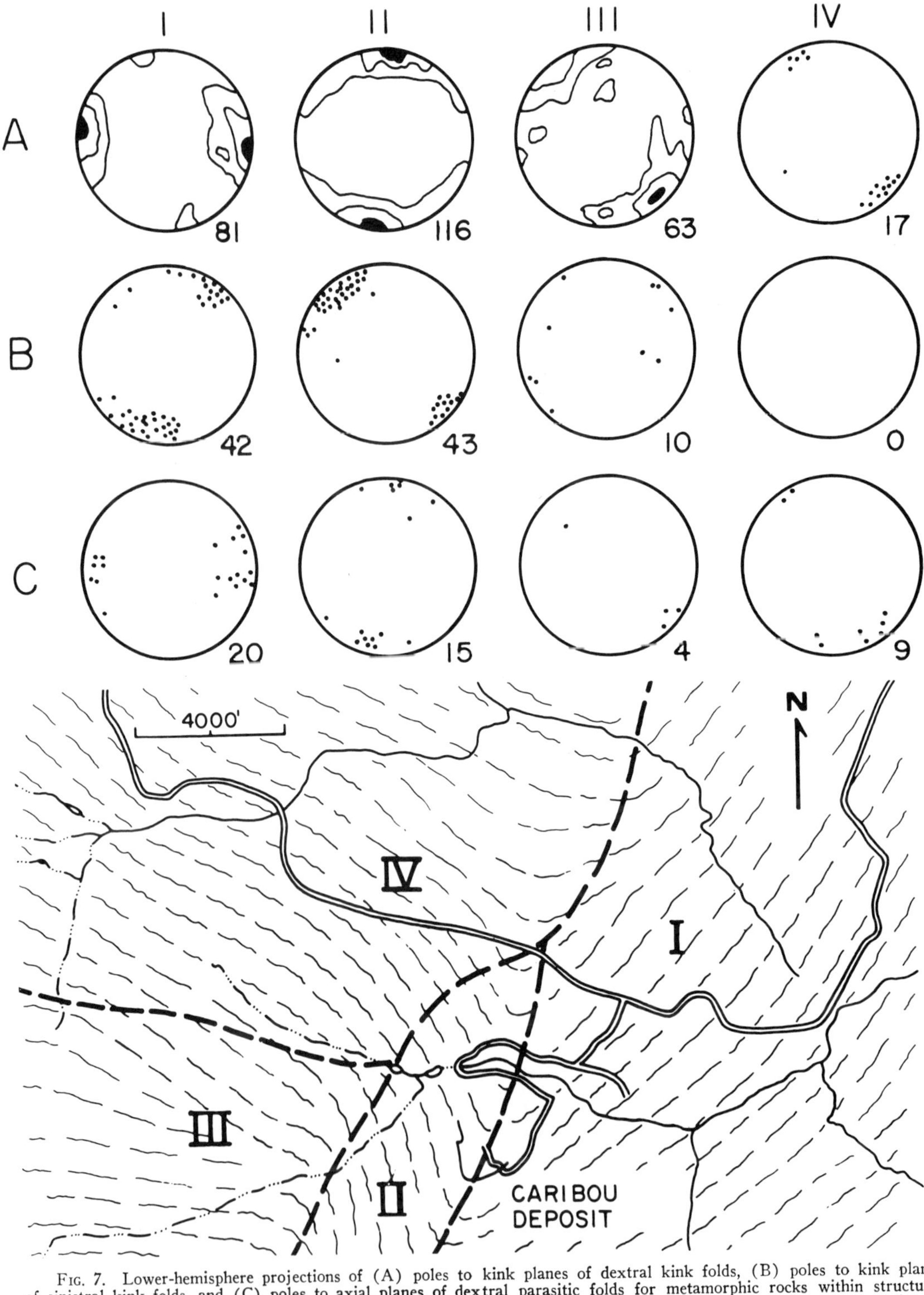

FIG. 7. Lower-hemisphere projections of (A) poles to kink planes of dextral kink folds, (B) poles to kink planes of sinistral kink folds, and (C) poles to axial planes of dextral parasitic folds for metamorphic rocks within structural domains I, II, III, and IV in the Caribou mine area. Contour values are 1%, 5%, and 10%. The number of data points is noted by each diagram.

Sulfide Fabric

Sulfide foliations are common in specimens from the Caribou deposit. The chief varieties of sulfide foliations are defined by: (1) lens-shaped sphalerite grains, (2) laminae and streaks of sphalerite, chalcopyrite, or galena, (3) tabular aggregates of intergrown pyrite grains, (4) tablet-shaped pyrite grains, (5) "pressure shadows"[1] of chalcopyrite or sphalerite around pyrite grains, (6) microlaminae of tiny, elongated pyrite grains, (7) oblate framboids, and (8) aligned pyrite fragments. These sulfide foliations parallel the foliation (S_1) of the metamorphic host rocks.

The sphalerite in sufide-bearing phyllite is generally lens shaped to tabular and aligned strictly parallel to S_1 (Fig. 11A). Such tiny lenses are commonly associated with chlorite, minor quartz, and fine-grained pyrite in narrow zones that parallel S_1. The sphalerite lenses lack internal grain boundaries and therefore appear to be individual grains.

[1] The term "pressure shadow" is used as a descriptive term in this paper. No implication of mechanism of formation is intended.

Laminae and streaks of sphalerite and chalcopyrite were commonly observed. Although irregular in detail, the laminae parallel S_1. Some of the sphalerite laminae contain twinned, elongated grains which are parallel to S_1 (Fig. 11B). In some cases, the twin boundaries themselves are aligned parallel to S_1. A rare textural variety of chalcopyrite is shown in Figure 11C in which chalcopyrite is interlaminated with chlorite laths and exhibits a well-defined sulfide foliation.

Some clusters of equigranular pyrite grains form fine-grained tabular aggregates which are parallel to S_1. Grains within these aggregates display intergrown or sutured boundaries. The lack of apparent orientation of the individual grains contrasts sharply with the composite tabular form of the aggregates. The length/thickness ratio of these aggregates ranges up to 8 to 1.

Pressure shadows of sphalerite or chalcopyrite occur at opposite ends of some pyrite grains (Fig. 11D). These sulfide fringes lie in the plane of the foliation and tend to decrease in thickness away from the pyrite. Pyrite grains with their sulfide pressure shadows form crude sulfide lenses which are parallel to S_1.

[*Editor's Note:* Material has been omitted at this point.]

Etched massive sulfide specimens composed of very fine-grained pyrite may exhibit faint brown-gray streaks.[2] These are subparallel to S_1 and locally curve around coarse-grained pyrite aggregates. At high magnification, the streaks are seen to consist of series of very fine-grained lens-shaped to tabular pyrite grains (Fig. 11E). These grains occur within alternating, very fine-grained and exceedingly fine-grained microlaminae which accentuate the sulfide foliation. The network of pyrite grains is tight with individual grains separated by sharp, planar to curviplanar contacts. No interlocking grain boundaries are evident. The subparallel microlaminae shown in Figure 11F are defined by alternating microlaminae of uniform grain size. Narrow chalcopyrite laminae border a zone containing pyrite grains which are oriented perpendicular to the trend of the microlaminae (Fig. 11F).

Etching revealed the presence of pyrite framboids in many polished sections, and most of these framboids were seen to be subround in trace (Fig. 11G). In several rare cases, oblate pyrite framboids, apparently deformed, were found within zones of microlaminae (Fig. 11H). Where deformed, the framboids were found to parallel S_1.

Various degrees of cataclasis are displayed by pyrite in specimens of sulfide-bearing phyllite and massive sulfide ore. Cataclasis in sulfide specimens varies from simple fracturing and granulation to thorough brecciation. Internal fracturing in pyrite is common (Fig. 12A). Fractures within broken pyrite fragments do not extend beyond the margins of the pyrite into the quartz-rich matrix. This is particularly evident in specimens which display fracture-filling by quartz, sphalerite, or chalcopyrite (Fig. 12B). The veins are restricted to the interior of the pyrite fragments. In many cases, mosaic textures have resulted from displacement of fragments of internally fractured pyrite grains (Fig. 12C). The original "fit" for the pyrite grains can be envisioned through matching fragments and/or internal growth zones.

A common deformational texture consists of isolated, angular pyrite fragments that generally are parallel to the foliation (Fig. 12D). This texture presumably reflects internal fracturing followed by displacement and rotation of the fragments. The oriented fragments themselves define a kind of foliation. Gradations exist from intensely fractured pyrite, to pyrite which is broken along fractures

[2] Sulfide specimens from the Caribou deposit must be etched with nitric acid (concentrated) for approximately 20 seconds, to disclose the growth zones, framboids, and other internal structures of pyrite which are described in this paper.

[*Editor's Note:* Material has been omitted at this point.]

FIG. 11. (A) Sulfide foliation defined by lens-shaped sphalerite grains. The sphalerite (light gray) is elongated parallel to the trace of the foliation and the disseminated pyrite band (white). Matrix (gray) consists primarily of fine-grained lepidoblastic sericite. Reflected light, in air, Spec. C54E. (B) Sphalerite exhibiting twinned, tabular to lens-shaped grains. The direction of twinning and elongation of the sphalerite grains (light gray) generally parallels the trace of the foliation. Sphalerite foliation is accentuated by interlaminated lenses of quartz (black). Reflected light, in oil, Spec. C54E. (C) Sulfide foliation defined by thin chalcopyrite streaks. The chalcopyrite grains (cp) are associated with pyrite (py) in a matrix composed of quartz and chlorite (black). Reflected light, in oil, Spec. C64E. (D) Round to irregular pyrite masses (white) with pressure shadows of sphalerite (gray) and galena (light gray). The trace of the elongated sphalerite and sphalerite/galena intergrowths parallels the trace of the foliation. Matrix (dark gray) consists of chlorite and quartz. Reflected light, in air, Spec. C24F. (E) Microlaminae of very fine-grained pyrite (white) in massive sulfide ore. Individual pyrite grains tend to be elongated parallel to the trace of the microbanding. The microbands are parallel to the foliation direction. Reflected light, in oil, Spec. C77F. (F) Microlaminae of very fine-grained pyrite with chalcopyrite in massive sulfide ore. The thin pyrite bands are composed of very fine to fine-grained pyrite (white). Pyrite grains in the relatively thick band (center) are oriented perpendicular to the trend of the microlaminae. The borders of this relatively thick band are marked by the presence of thin chalcopyrite layers (cp). Reflected light, in oil, C58A-5. (G) Pyrite framboids in chlorite phyllite. The framboids of pyrite (white) are composed of subhedral pyrite grains. Matrix is quartz and chlorite (black). Reflected light, in oil, Spec. C56F. (H) Deformed pyrite framboids. The oblate pyrite framboids are each composed of many subhedral grains of pyrite. These framboids occur within microlaminae in massive sulfide ore which parallel the direction of the foliation. The microlaminae are in contact with medium-grained, zoned pyrite. Reflected light, in oil, Spec. C54I.

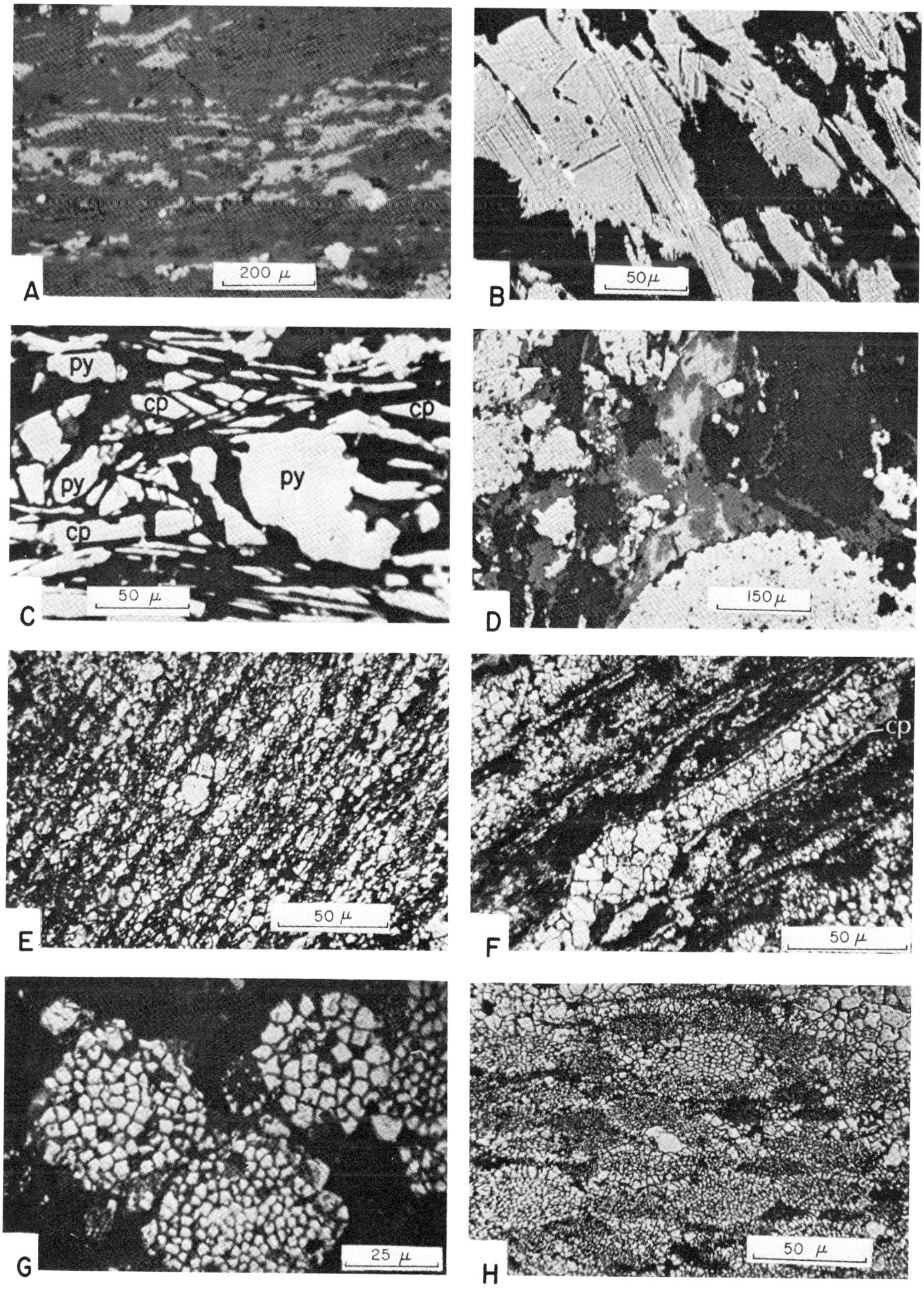
A
200 μ
B
50μ
py
cp
cp
py
py
cp
C
50 μ
D
150μ
E
50 μ
cp
F
50 μ
G
25 μ
H
50 μ

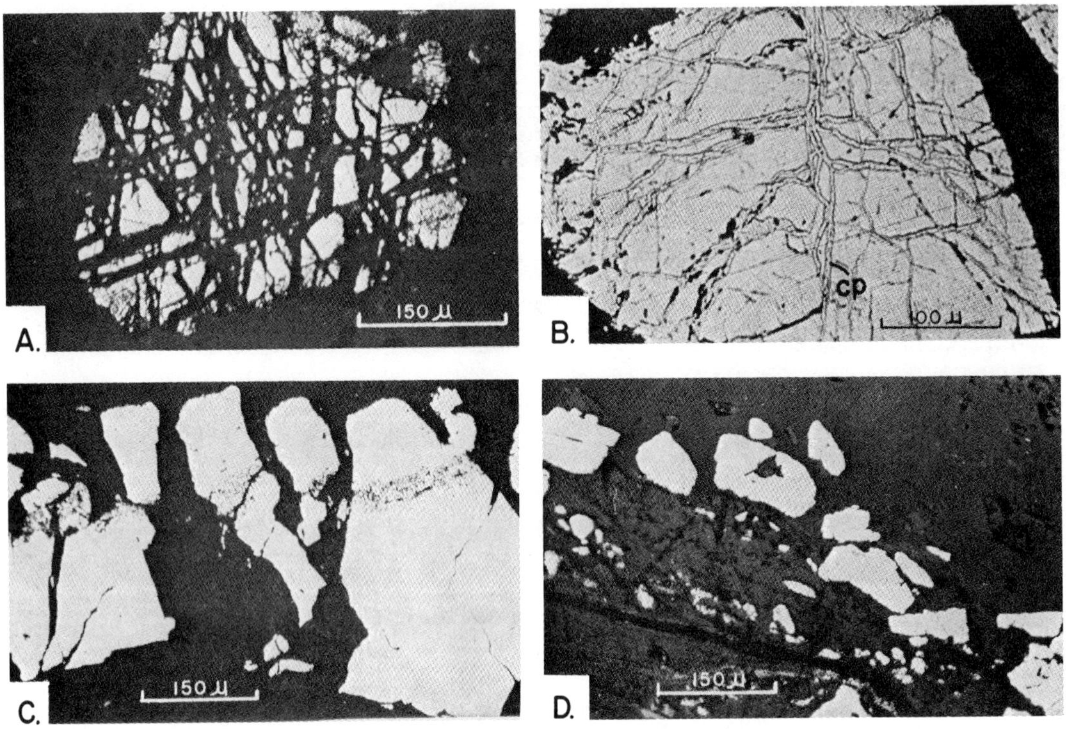

FIG. 12. (A) Internally fractured pyrite. Pyrite (white) is thoroughly cut by hairline fractures which do not extend into matrix of quartz (dark gray). Reflected light, Spec. C70D. (B) Internally fractured pyrite cut by thin chalcopyrite veinlets. The fractures and chalcopyrite veinlets (cp) do not extend beyond the margins of the pyrite into the quartz/sericite gangue (dark gray). Reflected light, Spec. C48D. (C) Mosaic textures defined by broken pyrite grains. Cataclastic pyrite fragments (white) are mutually displaced as a result of movement parallel to the foliation. Pre-deformation inclusion band (gray) aids in determining the original "fit." Reflected light, in air, Spec. C66A-7. (D) Sulfide foliation defined by angular pyrite fragments. The mechanically derived pyrite fragments (white) are elongated in a direction which parallels the trace of the foliation. The pyrite is associated with a gangue of quartz and sericite. Reflected light, in air, Spec. C17.

[*Editor's Note:* Material has been omitted at this point.]

Summary

Structures in rocks within the Caribou mine area reveal a complex history involving three periods of deformation. The diagnostic structural elements used to label these three periods of deformation are listed in Figure 17.

In late Ordovician time, volcanic and sedimentary rocks of Ordovician age were subjected to isoclinal folding (F_1) during low-grade regional metamorphism. The structural features produced include the following:

(1) foliation (S_1) defined by alignment of platy minerals, mineral banding, and subparallel feldspar metaphenocrysts and

(2) transposition structures characterized by rootless intrafolial folds of original bedding (S_0).

Subsequent to Taconic deformation, the resultant

TECTONIC MODEL FOR CARIBOU MINE AREA

STAGE	EVENT	STRUCTURAL ELEMENTS	STYLE
F_3	LATE ACADIAN Development of Caribou Fold on Northwest Limb of Tetagouche Anticlinorium	1) Modification of orientation and distribution of F_2 kink folds and parasitic folds 2) Conjugate faulting in hinge zone of Caribou fold 3) Minor folding and axial plane fracturing in hinge zone	
F_2	ACADIAN Development of the Tetagouche Anticlinorium	1) Dextral parasitic folds 2) Dextral and sinistral kink folds	
F_1	TACONIC Isoclinal Folding and Low-Grade Regional Metamorphism	1) Foliation defined by parallelism of platy minerals, mineral banding, and alignment of feldspar metaphenocrysts 2) Intrafolial folds	

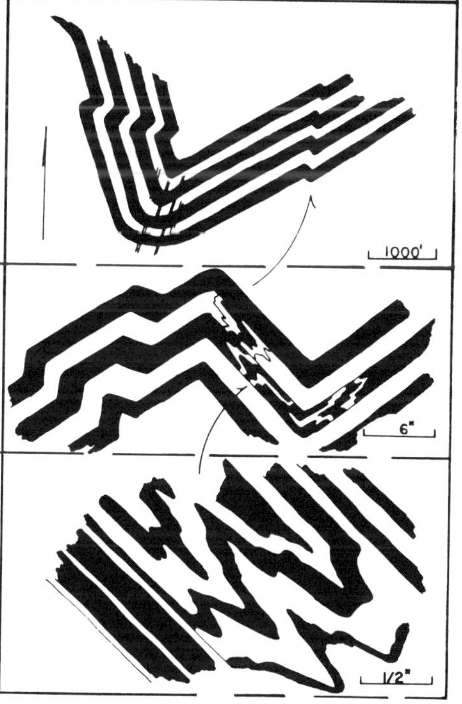

FIG. 17. Tectonic model for Caribou mine area.

metamorphic rocks were subjected to two additional structural events. The Tetagouche anticlinorium developed during Acadian deformation (F_2). Structures present in the Caribou mine area which can be related to Acadian deformation include steeply plunging kink folds and dextral parasitic folds as well as the axial plane fractures associated with these folds.

As a late stage of Acadian deformation, Caribou folding (F_3) resulted in modification of the orientation of F_1 and F_2 structural elements. Contemporaneous with F_3 folding was the development of minor folds and axial plane fractures in the hinge zone of the Caribou fold. Rocks in the hinge zone of the Caribou fold were broken and displaced along conjugate faults.

[*Editor's Note:* Material has been omitted at this point.]

Conclusions

Structural elements provide a basis for establishing a three-stage deformational model for the rocks that host the Caribou sulfide deposit. During the Taconic orogeny, volcanic and sedimentary rocks of Ordovician age were isoclinally folded and subjected to low-grade regional metamorphism (Fig. 19A). The schistosity displayed by rocks of the Tetagouche Group developed as an axial plane foliation. As a result of isoclinal folding and partial transposition of the original bedding, foliation and lithologic layering appear to be parallel on a regional scale. Remnant isoclinally folded beds cut by the foliation indicate that concordant structural discontinuities actually

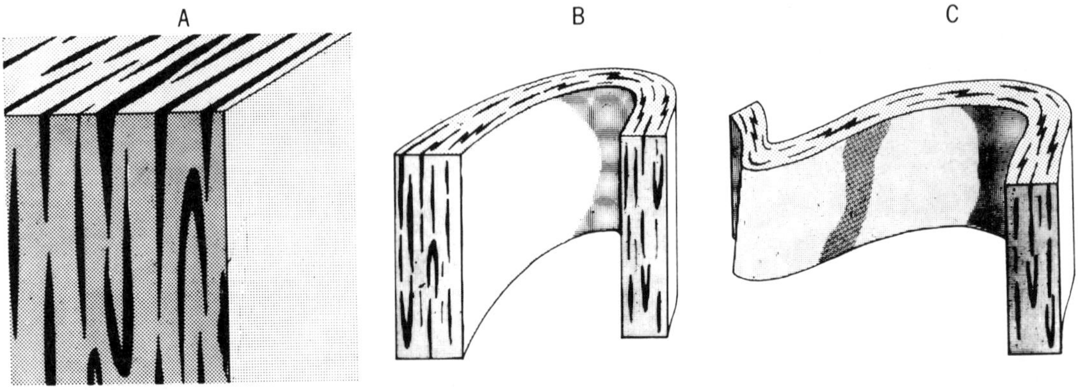

FIG. 19. Schematic diagram portraying three-stage deformational history of Ordovician rocks in a part of the Bathurst district. (A) Transposition of bedding during regional metamorphism, (B) formation of the Tetagouche anticlinorium, and (C) refolding of the northwest limb of the Tetagouche anticlinorium during the Caribou folding event.

interrupt the seemingly continuous Ordovician stratigraphic sequence.

Metavolcanic and metasedimentary rocks in the Caribou mine area were folded and faulted during Acadian deformation. Kink folding and parasitic folding accompanied the development of the Tetagouche anticlinorium (Fig. 19B). In a late stage of Acadian deformation a portion of the northwest limb of the Tetagouche anticlinorium was refolded (Fig. 19C), producing the Caribou fold which now hosts the Caribou sulfide deposit.

At least two periods of deformation have affected the Caribou sulfide deposit. Faulting, folding, and fracturing of sulfide lenses in the hinge zone of the Caribou fold indicate that sulfide mineralization preceded the development of the host structure. Conspicuous sulfide deformation within a breached reverse parasitic fold on the west limb of the deposit documents sulfide deposition prior to Acadian deformation.

The presence of minor discordant chalcopyrite and pyrite veins demonstrates that at least some mineralization post-dates the time of development of the foliation. However, the evidence is strong—but not conclusive—that the textures displayed by most of the sulfide minerals evolved during metamorphism. The sulfide fabric displayed by the sulfide lenses appears to be the result of deformation, plastic flow, and crystallization of sulfides controlled by slip parallel to the developing foliation.

Formation of the Caribou strata-bound sulfide deposit accompanied or preceded regional metamorphism. Therefore, neither the Caribou fold nor the Tetagouche anticlinorium was responsible for controlling the locus of sulfide deposition.

Acknowledgments

This paper stems from a dissertation research project supervised by William C. Kelly of the Department of Geology and Mineralogy at the University of Michigan. I am indebted to William Kelly for demanding careful definition of research goals and for responsible guidance throughout the investigation. Bruce R. Clark provided invaluable assistance in devising computer programs for plotting the structural data. Sincere appreciation is also extended to Edwin H. Goddard and E. William Heinrich who offered many helpful suggestions for improvement.

The Eastern Exploration Division of Anaconda American Brass, Ltd. generously supported this study of their Caribou sulfide deposit. I am grateful to William T. Swenson, Vice President, and Robert E. Schaaf, Chief Geologist, for their advice and encouragement during the project. The National Aeronautics and Space Administration and National Science Foundation kindly provided financial support for my graduate program.

Special thanks are extended to the staff of the Bureau of Audio-Visual Services at the University of Arizona who helped in the preparation of illustrations for the manuscript.

DEPARTMENT OF GEOSCIENCES,
THE UNIVERSITY OF ARIZONA,
TUCSON, ARIZONA 85721,
October 1, 1971; February 9, 1972

REFERENCES

Alcock, F. J., 1935, Geology of the Chaleur Bay region: Canada Geol. Survey Mem. 183, 146 p.

Boyle, R. W., and Davies, J. L., 1964, Geology of the Austin Brook and Brunswick No. 6 sulfide deposits, Gloucester

County, New Brunswick: Canada Geol. Survey Paper 63-24, 23 p.

Davies, J. L., 1966, Geology of Bathurst-Newcastle area, New Brunswick, *in* Guidebook, Geology of parts of Atlantic Provinces: Annual Meetings, Geological Assoc. of Canada and Mineralogical Assoc. of Canada, p. 33–43.

de Sitter, L. U., 1964, Structural geology: New York, McGraw-Hill Book Co., ; 551 p.

Dewey, J. F., 1965, Nature and origin of kink bands: Tectonophysics, v. 1, p. 459–494.

Gill, J. E., 1969, Experimental deformation and annealing of sulfides and interpretation of ore: Econ. Geol., v. 64, p. 500–508.

Graf, J. L., and Skinner, B. J., 1970, Strength and deformation of pyrite and pyrrhotite: Econ. Geol., v. 65, p. 206–215.

Leech, G. B., Lowden, J. A., Stockwell, C. H., and Wanless, R. K., 1963, Age determinations and geological studies (Report 4): Canada Geol. Survey Paper, 63-17, 140 p.

Lowden, J. A., Stockwell, C. H., and Hart, S. R., 1963, Age determinations and geological studies (Report 3): Canada Geol. Survey Paper 62-17, 140 p.

Noble, D. C., and Eberly, S. W., 1964, A digital computer procedure for preparing Beta diagrams: Am. Jour. Sci., v. 262, p. 1124–1129.

Potter, R. R., Jackson, E. V., and Davies, J. L., 1968, Geological map New Brunswick: Department of Natural Resources, New Brunswick map number N.R.-1.

Smith, C. H., 1957, Bathurst-Newcastle map-area: Canada Geol. Survey Map 1-57.

Stanton, R. L., and Gorman, H., 1969, A phenomenological study of grain boundary migration in some sulfides: Econ. Geol., v. 63, p. 907–923.

Stanton, R. L., and Willey, H. C., 1970, Natural work-hardening in galena, and its experimental reduction: Econ. Geol., v. 65, p. 182–194.

Tupper, W. M., and Hart, S. R., 1961, Minimum age of the Middle Silurian in New Brunswick based on K-Ar method: Geol. Soc. America Bull., v. 72, 1285–1288.

Turner, F. J., and Weiss, L. E., 1963, Structural analysis of metamorphic tectonites: New York, McGraw-Hill, 545 p.

Whitten, E. H. T., 1966, Structural geology of folded rocks: Chicago, Rand McNally and Co., 661 p.

11

Reprinted from *Econ. Geology* **67**:796–813 (1972)

"Geolog System" for Computer and Manual Analysis of Geologic Data from Porphyry and Other Deposits

P. H. Blanchet and C. I. Godwin

Abstract

The Geolog System is a computer-oriented method of recording geological field observations, primarily from drill holes, in porphyry-type and other mineral deposits. The System also facilitates manual analysis and interpretation of the data.

Data are compiled on GEOLOG, an 80-column format describing principal zone (cap, supergene, hypogene, fresh rock, fault), interval, rock type, fractures, alteration mineral assemblages, limonites, and ore mineral assemblages. Prescribed headings on GEOLOG are keyed to logical word abbreviations and simple codes to give concise, consistent, and readable descriptions. Because column names are easily modified and various codes and scales are available, GEOLOG is flexible; it therefore facilitates description of many types of deposit, accommodating a variety of logging techniques that may emphasize theoretical examination, routine survey, or rapid reconnaissance. Collecting data with Geolog System appears better than existing or standard logging techniques because information is recorded to be: keypunched directly; retrieved readily; consistent with respect to routine description of presence or absence of particular features; specific by use of prescribed headings, codes and scales; related to an existing porphyry deposit model. Time and effort involved in collecting drill data on GEOLOG are comparable to common logging procedures.

Interpretation of the large amount of complex data on GEOLOG is aided by using a computer to provide manipulation of data and appropriate output. Because GEOLOG is keyed to recently published models of porphyry deposits that show ore location with respect to zoning of mode and degree of dispersion of mineralization, alteration mineral assemblages, and ore mineral assemblages, application of the System has important future use in predicting ore locations from even preliminary data.

Introduction

Typical drill hole data from most types of deposits consist of large quantities of complex observations that probably can only be fully utilized if recorded for, and processed by, a computer. A number of recent papers describing porphyry deposits present a descriptive geological vocabulary and emphasize characteristic, district-scale patterns that define favorable targets for ore deposits (James, 1971; Lowell and Guilbert, 1970; Rose, 1970). These and others in the bibliographic references provided the basis for models, codes, scales, and formats incorporated in Geolog System.

This system, as presented here, attempts to (1) provide an 80-column format called GEOLOG (Fig. 1) for comprehensively recording geologic information in a manner suitable for key punching and subsequent computer analysis, (2) systematize and abbreviate the geological vocabulary relating to mineral deposits, (3) incorporate recently developed porphyry deposit models (Figs. 5, 6, 7) to provide a sound theoretical basis for prediction of ore location, (4) illustrate types of computer output obtainable (Figs. 8 through 12).

Assay data and drill hole location specifications are written on GEOLOG but not within the 80 columns. These data are stored on another format (ASSAYLOG) that is also suitable for key punching. Consequently, it is possible to merge and correlate ASSAYLOG and GEOLOG data in different ways. Various methods of interpretation or quantitative analysis can be applied to GEOLOG and ASSAYLOG data. BENCOM and CUBE, two computer programs for processing exploration data, were available to the authors.

Deposit Description with Geolog System

Geolog System is specifically for field description of diamond, rotary, or percussion drill holes by a geologist with no tools more sophisticated than a binocular microscope. However, specimens or exposures from underground workings, surface outcrops, pits or trenches can also be described. Features that generally must be noted in evaluation of a mineral deposit are shown in Table 1. This table also indicates where these features are listed in GEOLOG, where codes for the features are to be found, and which porphyry deposit models are

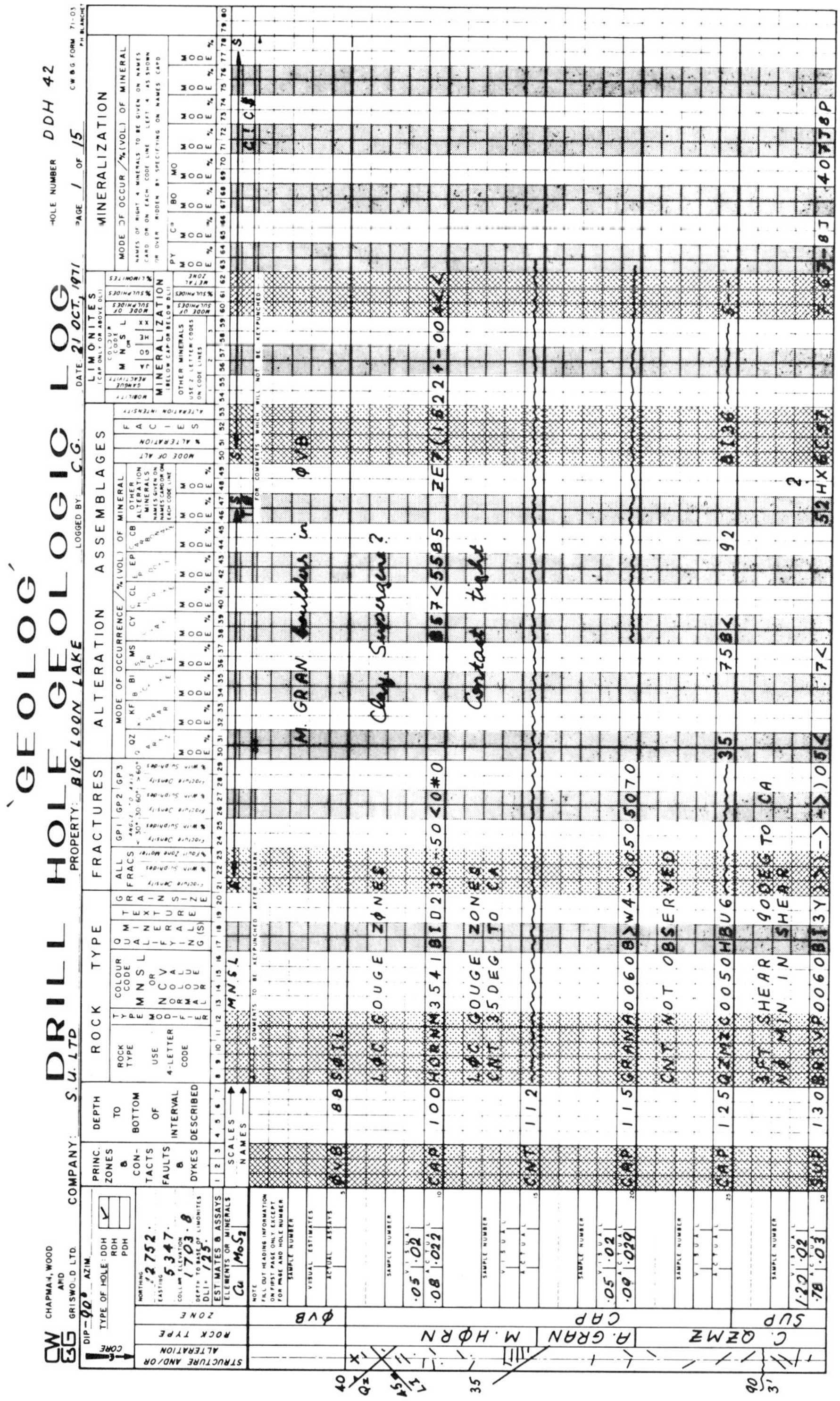

FIG. 1. GEOLOG, a drill hole geologic logging format. Summary columns are cross-hatched and relate to models of zoning in porphyry deposits.

TABLE 1. Location of Main Features Describing a Deposit

Feature	Location in GEOLOG (Fig. 1)	Location of Codes	Summary Refers to Following Porphyry Model
1. Interval described	Columns 4 to 7		
2. Principal zone (cap, supergene, hypogene, fault, fresh rock)	Columns 1 to 3	Table 4	
3. Rock type	Columns 8 to 20 (summary columns 8 to 12)	Figures 2 to 4 and Tables 2 and 3	
4. Structure	Visual log and columns 21 to 29 (summary columns 21 to 23)	Table 5	
5. Alteration	Columns 30 to 53 (summary columns 50 to 53)	Tables 5 through 7	Figures 5 and 6
6. Limonites	Columns 54 to 62 in capping only (summary columns 60 to 62)	Figure 2	
7. Mineralization	Columns 63 to 78 in capping; below capping also columns 54 to 62 (summary columns 60 to 62)	Tables 5, 6 and 8	Figures 5 and 7
8. Remarks for keypunching	Columns 8 to 30, written between rows: 5, 10, 15, 20, 25, 30	Tables 2, 3, 4	
9. Comments not for keypunching	Columns 31 to 78, written between rows: 5, 10, 15, 20, 25, 30	Tables 2, 3, 4	

significant with respect to Summary columns in GEOLOG. Each feature is described in detail under GEOLOG column headings by use of codes described in Figures 2, 3, 4 and Tables 2 through 8. Codes, besides being short, help to make observations quantitative, specific, and consistent. Enhanced consistency is maintained by the geologist or among geologists for a property or a number of properties.

Data in the 80 columns of GEOLOG are ready for key punching and subsequent computer processing. GEOLOG, however, can be interpreted visually after short acquaintance with its format and the shorthand codes.

Visual interpretation is facilitated by

1. Use of four-letter rock codes (Table 2), two-letter mineral codes (Table 3), and one-, three-, or five-letter abbreviations (Table 4).
2. Simplicity of codes and scales (Figs. 2, 3, 4, and Tables 5 through 8).
3. Use of one row (a single computer card) for all geological information relating to a given drill hole interval.
4. Assignment of a particular variable to specific columns so that simple scanning provides a rapid means of investigating changes with depth.

Geolog System also has built-in flexibility which permits adaptation to different types of deposits by minor changes to some column heading in GEOLOG and the selection of more appropriate scales. This can be done as required by writing the changes in the heading lines labeled "NAMES" and "SCALES." The interval described by a code line does not have to be uniform although it can approximate a constant vertical scale. To facilitate correlation with ASSAYLOG, however, a code line corresponding to the bottom of every assay interval should be filled in. Approximate sections can be constructed by joining consecutive pages together so that they can be unfolded into a continuous strip. Copying and filing are no problem because GEOLOG is printed on legal size paper. Intuitive comments or information which the geologist prefers not to code are retained as notes, as computerized remarks, or as symbols in the visual log. Theoretical studies can be pursued as desired by selecting highly detailed instead of general codes; for example, see descriptions of breccia, conglomerate, and porphyry (Table 2).

Normally each feature included in GEOLOG is described in a detailed and disciplined way. In addition, a condensation of the detail is made in the cross-hatched summary columns of Figure 1 (also see Table 1). For example, alteration minerals and their modes of occurrence are recorded in columns 30 to 49. Part of the summary is "facies" (column 52) that can be defined by reference to Table 7. These summaries require careful judgments to be made by the geologist describing the hole. The data collected in the summary columns can be analyzed in the same way as detailed data. In addition, summaries have three other purposes:

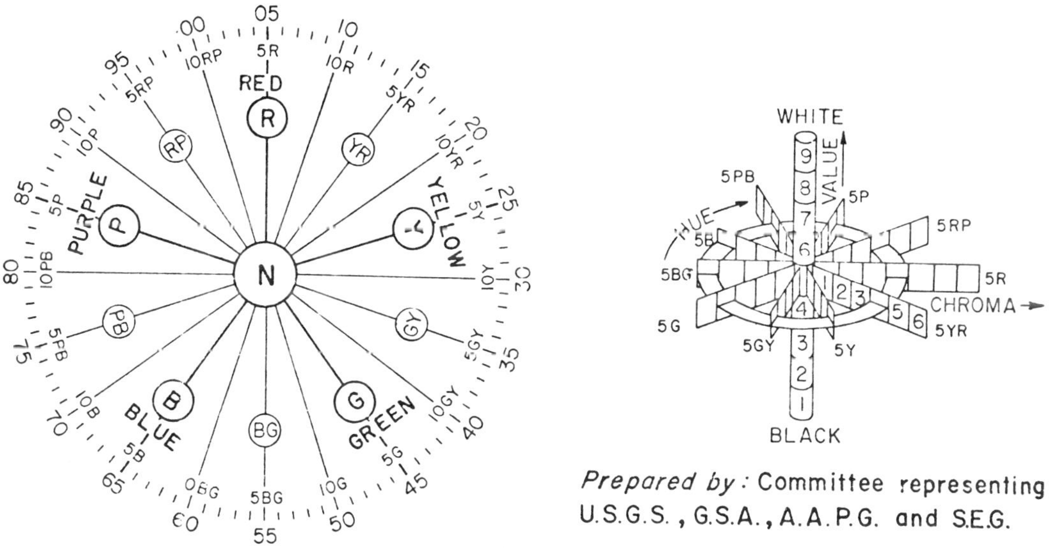

FIG. 2. Rock Color Chart and Munsell System Codes.

1. They require an "on the spot" interpretation by the geologist logging the hole.
2. They aid rapid logging by providing a check list of key features that might be particularly useful in themselves, for example, during a brief property examination.
3. They allow, for porphyry deposits, immediate correlation to the models of Figures 5, 6, 7 so that exploration or development can be guided.

Geolog System provides a readable and compact format, and a precise, consistent and flexible technique for describing a mineral deposit. It is constructed to suit a variety of interests ranging from theoretical studies, to routine examination and to rapid property evaluation.

Predictions Aided by Models and Geolog System

An ore body might be in a specific location within a zonal sequence characterized by variations in factors such as rock type, alteration, structure, and mineralization type. To be useful in decision-making these zones must be neither too large nor too small in comparison to ore body size. For example, McKinstry (1948, p. 210) states "It is evident that if an orebody is 100 feet in diameter, an irregular target ring two miles across would not be of great help. On the other hand, a ring only ten feet larger than the orebody would be at least as hard to find as the orebody itself."

Thus if an appropriately larger zone indicative of significant mineralization can be recognized, the most probable targets for ore can be predicted. The accuracy of this prediction also depends on the quality of mapping and drill hole logging, and on the validity of the models upon which predictions are made.

TEXTURE CODES FOR

I. IGNEOUS

(a) Inequigranular (Porphyritic, Conglomeratic)

Structure or Framework	Percentage of Large Fraction				
	<10	10-25	25-50	50-75	>75
Closed, intact		1	2	3	4
Open, disrupted	6	7	8	9	

(b) Equigranular (Non-porphyritic)

A	amygdaloidal	K	poikilitic (poikiloblastic)
C	cataclastic	P	porphyritic (porphyroblastic)
W	foliated	Q	spherulitic
!	graphic	T	tuffaceous
[	intersertal ophitic, diabasic	U	uniform
/	lineated	V	vesicular
¤	miarolitic	X	other

II. SEDIMENTARY

B	bedded	Ø	oolitic, pisolitic
Z	cross bedded	.	pelletoidal
S	fissile	R	sparry
F	fossiliferous, bioclastic	T	tuffaceous
:	graded bedding	U	uniform
L	laminated, varved	X	other

III. METAMORPHIC

#	augen	K	poikiloblastic, sieve (poikilitic)
D	banded	P	porphyroblastic (porphyritic)
W	foliated	H	schlieren, nebulitic
G	granulose	Y	slaty
/	lineated	U	uniform
>	lit-par-lit	X	other
M	migmatic		
N	mylonitic		

FIG. 3. Texture Codes.

GRAIN SIZE CODES FOR

EQUIGRANULAR GRAIN SIZE — CHART I

IGNEOUS, METAMORPHIC & CHEMICAL										
NOMENCLATURE	GLASSY	APHANITIC	FINE GRAINED		MEDIUM GRAINED		COARSE GRAINED		PEGMATITIC	MEGA-PEG
SEDIMENTARY										
NOMENCLATURE	CLAY	SILT	SAND: v. fine | fine	SAND: med | crs	SAND: very coarse	GRANULE	PEBBLE: small	PEBBLE: large	COBBLE	BOULDER
SYMBOLIC CODE	0	1	2	3	4	5	6	7	8	9

Grain size boundaries:

GRAIN SIZE IN MM	2^{-8} (1/256)	2^{-4} (1/16)	2^{-2} (1/4)	2^{0} (1)	2^{1} (2)	2^{2} (4)	2^{4} (16)	2^{6} (64)	2^{8} (256) MM
PHI SCALE	8	4	2	0	−1	−2	−4	−6	−8

INEQUIGRANULAR GRAIN SIZE — CHART II

FINER FRACTION GRAIN SIZE

	0	1	2	3	4	5	6
IGNEOUS, METAMORPHIC & CHEMICAL	GLASSY	APHANITIC	FINE GRAINED		MEDIUM GRAINED		COARSE & LARGER
SEDIMENTARY	CLAY	SILT	SAND: fine	SAND: med	SAND: coarse	GRANULE	PEBBLE & LARGER
MM (upper boundary)	1/256	1/16	1/4	1	2	4	MM
PHI (upper boundary)	8	4	2	0	−1	−2	PHI

SINGLE-CHARACTER AND TWO-CHARACTER SYMBOLIC CODES

COARSER FRACTION GRAIN SIZE: Igneous, metamorphic & chemical	Sedimentary	MM / PHI	Code	0	1	2	3	4	5	6
MEGA-PEG	BOULDER	256 MM / −8 PHI	9	< 90	[91	(92	# 93	) 94	] 95	> 96
PEGMATITIC	COBBLE	64 / −6	8	V 80	W 81	X 82	Y 83	Z 84	& 85	+ 86
COARSE GRAINED	PEBBLE large	16 / −4	7	P 70	Q 71	R 72	S 73	T 74	U 75	- 76
COARSE GRAINED	PEBBLE small	4 / −2	6	J 60	K 61	L 62	M 63	N 64	Ø 65	
MEDIUM GRAINED	GRANULE	2 / −1	5	E 50	F 51	G 52	H 53	I 54		
MEDIUM GRAINED	V. COARSE & SMALLER		4	A 40	B 41	C 42	D 43			

GRAIN SIZE CODES → 0 1 2 3 4 5 6

FIG. 4. Grain Size Codes.

TABLE 2. Four-Letter Rock Type Code

The first four letters of a rock name form its preferred code, with the fourth letter, if a vowel, being replaced with the next consonant. One letter of a double consonant group, occurring with the first four letters, is usually excluded. Agglomerate becomes AGLM.

Some rock names in the following list are followed by a two-character contraction or short form. These are in brackets and preceded by an equal sign. They are used, as will be seen, to form four-character codes for compound rock names or to facilitate print-out.

ALAS	alaskite
APLT	Aplite
ADAM	adamellite = quartz monzonite
ANØR	anorthosite
ANDS	andesite (= AN)
ARGL	argillite
ARKS	arkose
AGLM	agglomerate
BASL	basalt (= BS)
BRXX	breccia, general Third character X: X alphabetic = genetic origin X numeric = provenience* (see lower left tables, below) Fourth character X: Sphericity-roundedness** (see chart at lower right, below)
CGXX	conglomerate, general Third character X: Fourth Character X: (as above, see tables and chart below)
CØQN	coquina
CASI	calc-silicate rock
CATC	cataclasite
CYSN	claystone
	clay (= CY)
	carbonaceous (= C:)
C:SS	carbonaceous sandstone
C:SH	carbonaceous shale
C:RK	carbonaceous rock
CARB	carbonatite
CHER	chert
DACT	dacite
DIAB	diabase (= DB)
DIØR	diorite (= DR)
DUNT	dunite
DØLM	dolomite (= DØ = D.)
D.LS	dolomitic limestone
DYKE	dyke or (= D/)
DIKE	dike, as preferred, or
D/XX	dyke, where XX is 2-character contraction or rock type, as
D/BS	basaltic dyke
D/G.	granitic dyke
D/DR	diorite dyke
D/DB	diabase dyke
D/LM	lamprophyre dyke, etc.
EVAP	evaporite
	feldspathic (= F:)
	feldspathoidal (= F.)
F.XX	feldspathoidal rock, where XX = 2 character contraction of rock type
F.AN	feldspathoidal andesite
F.BS	feldspathoidal basalt
F.DR	feldspathoidal diorite
F.GB	feldspathoidal gabbro
F.LT	feldspathoidal latite
F.MZ	feldspathoidal monzonite
F.SY	feldspathoidal syenite etc.
FELS	felsite
F:SS	felspathic sandstone
	grano = granu (= GR)
	green (= G#)
GABR	gabbro (= GB)
GRAN	granite (= G:)
GRFL	granofels
GRLT	granulite
GNES	gneiss (= GN = G*)
G#S#	greenschist
G#SN	greenstone
GREY	greywacke
GRIT	grit
GRDR	granodiorite (= GD)
GRNT	granite igneous rock (= G.)
	hornblende (= HB)
HBIT	hornblendite
HØRN	hornfels
	igneous (IG)
IGRK	igneous rock
	iron (= IR)
IRSN	ironstone
IGNM	ignimbrite
LAMP	lamprophyre (= LM)
D/LM	lamprophyre dyke
LATT	latite (= LT)
	lit-par-lit (= L>)
L>GN	lit-par-lit gneiss
L>LS	lit-par-lit limestone
	limey (= L:)
LISN	limestone (= LS)
L:DØ	limey dolomite
LAVA	lava (= LV)
MØN	monzonite (= MZ)
MIGM	migmatite
	mixed (= MX)
MXGN	mixed gneiss
MARB	marble
MYLN	mylonite
	mudstone
METM	metamorphic rock
NØRT	norite
	ortho (= Ø:)
	oolitic (ØØ)
Ø:GN	orthogneiss
ØØLS	oolitic limestone
PLUT	plutonic rock (= P.)
PEGM	pegmatite (= PG = P:)
PERD	peridotite
PYRX	pyroxinite
PHØN	phonolite
PARG	paragneiss
PHYL	phyllite
PYRC	pyroclasite
PHØS	phosphorite
PPZZ	porphyry, general Z may be replaced by the 2-letter code of the dominant mineral or mineral combination: e.g. PPKF, PPK, PPQZ. Qualifying mineral (Cols. 17 & 18) is less dominant than above, unless Z not specified.
	quartz-quartzose (= QZ
QZDR	quartz diorite (= QD) = tonolite
QZMZ	quartz monzonite (= QM) = adamellite
QZGB	quartz gabbro (= QG)
QZBS	quartz basalt (QB)
QZLT	quartz latite (= QL)
QUAR	quartzite
QZSS	quartzone sandstone
RHYL	rhyolite (= RH)
RHYD	rhyodacite
ROCK	rock (= RK)
	stone (= SN)
SASN	sandstone (= SS)
SLSN	siltstone (= SL)
	silicate (= SI)
F:SS	feldspathic sandstone
SERP	serpentinite
SCHS	schist (= S#)
SHAL	shale (= SH)
SYEN	syenite
SEDM	sedimentary rock
TONL	tonalite = quartz diorite
TACT	tactite
TUFF	tuff
TRAC	tractyte
UNCN	unconsolidated material (soil, clay, silt, sand, gravel)
VEIN	vein (= V/)
V/XX	vein, general where XX may be replaced by the 2-letter code of the dominant mineral. Qualifying mineral (cols. 17, 18) is less dominant than XX above
VOLC	volcanic rock

TABLE 2.—(*Continued*)

THIRD CHARACTER TABLES OF

GENETIC ORIGINS (Alphabetic Values of Third Character)		PROVENANCE* (Numeric Values of Third Character)	
C	collapse-solution	1	> 50% plutonic
D	dyke		
F	fault	3	> 50% metamorphic
M	metamorphic	5	> 50% volcanic
P	plutonic		
S	sedimentary	7	> 50% pyroclastic
T	clastic		
V	volcanic	9	> 50% sedimentary
Y	pyroclastic		
X	unspecified	0	> 50% mixed provenience

FOURTH CHARACTER
CHART OF
SPHERICITY AND ROUNDEDNESS
with single character and
two-character symbolic codes

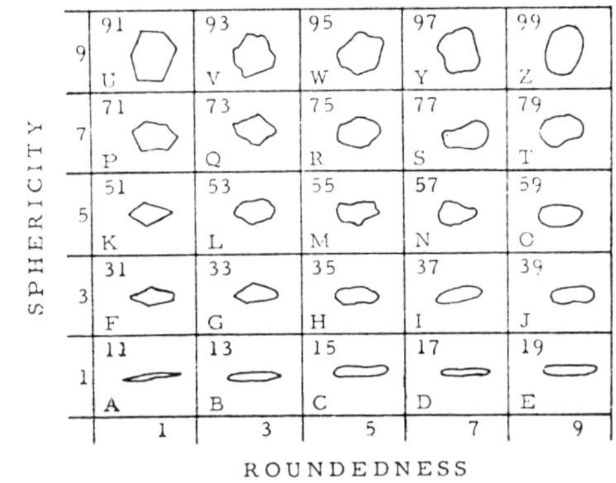

* Provenance = origin or source, any type of rock.
** Authors prefer roundedness to "roundness."

Recent studies of a large number of porphyry deposits have resulted in recognition of unifying patterns based on (1) rock type (James, 1971; Lowell and Guilbert, 1970); (2) mode and degree of dispersion of ore minerals and possibly alteration minerals (Lowell and Guilbert, 1970; (3) alteration zoning (Lowell and Guilbert, 1970; Rose, 1970; (4) sulfide, oxide and native mineral assemblages (Lowell and Guilbert, 1970; Meyer and Hemley, 1967; Rose, 1970). Most of these features are described in Figures 5, 6 and 7 which are adapted from the generalized porphyry model developed by Lowell and Guilbert (1970, Fig. 3, p. 379). GEOLOG summary columns (see Table 1) relate directly to Figures 5, 6 , and 7.

Zoning patterns from mapping exposures and hole logging of a deposit might correlate with the models of Figures 5, 6, and 7 which roughly coincide and can be superimposed. Since the location of an "ore shell" (Fig. 7) is most favorable in specific zones, the prediction of possible ore location can be made even with limited data, such as surface geology and one or two drill holes. For example, this approach was important to the discovery of the Kalamazoo orebody, Arizona (Lowell, 1968). Characteristics peculiar to deposits in the area being investigated might modify the models.

Computer Output from GEOLOG

Data on GEOLOG is in a convenient form for keypunching and processing by a computer for effective display. Constant scale printout of GEOLOG data is shown in Figure 8. Assay data from the separate file called ASSAYLOG has been added to this output. Room remains for notations to be added. This format, then, provides a detailed, graphic strip-log of a drill hole.

A program called BENCOM averages drill hole data by bench interval. Geological data are treated in the same way as assay data. BENCOM provides listings of drill hole average over bench intervals (Fig. 9) and bench plans (Fig. 10). The listing is a convenient condensed drill hole strip-log for the variables in the headings. The bench plan shows the location of the drill hole at the mid-point of the bench interval with an asterisk; the hole is identified to the right of the asterisk. Bench averages for four variables are plotted in the four quadrants surrounding the asterisk. These four variables can be hand contoured as required.

Figure 11 illustrates the printout of a bench plan from a more elaborate computer program called CUBE (Coode and Chen, 1971). In this program the assay or geological values are plotted in three dimensions as averages weighted with respect to the distance from plotting point to data points in drill holes. The symbols printed on the plan effectively contour the data. Data summaries for these plans list tons of each category within each bench and the reliability of that estimate. For example, Figure 12 is a listing of the tons of rocks involved within each category of fracture intensity. Analysis of these data

Table 3. Two-Letter Mineral Code
The first two letters of the mineral name form its preferred code.
Note particularly the two-mineral combinations.

AC actinolite
AD adularia
AL alunite
AB albite
AA andalusite
AG angelsite (68% Pb)
AH anhydrite
AK ankerite
AN anorthite
AP apatite
AR aragonite
AS arsenopyrite (45% As)
AØ asbestos
AU augite
AZ azurite (58% Cu)
See also MX = malachite/azurite combination

BA barite
BE beryl
BI biotite
BX biotite/hornblende mineral combination

BX	proportions unspecified
BI	biotite alone
B>	biotite > hornblende
B#	" = "
B<	" < "
HØ	hornblende alone

BS bismuthanite (70% Bi)
BØ bornite (63% Cu)
BR brochantite (56% Cu)

CA calcite
See also DX = dolomite/calcite combination
CB carbonate(s)
CE cerussite (77% Pb)
CH chalcanthite (25% Cu)
CC chalcocite (80% Cu)
C! chalcocite on gangne
C$ chalcocite alone or on economic mineral(s) e.g., CC on CP

CP chalcopyrite
CL chlorite
CD chloritoid
CR chromite (46% Cr)
CS chrysotile
ØL chrysolite - See olivine
CK chrysocolla (K as in Coke) (36% Cu)
CN cinnabar
CY clay - See also CX = clay/muscovite combination
CX clay/muscovite

	mineral combination
CX	proportions unspecified
C>	clay > muscovite
C#	clay = "
C<	clay < "
MU	muscovite alone

CZ clinozoisite
CU copper
CØ cordiorite
CV covellite (66% Cu)
CI cuprite (89% Cu)

DC dickite
DG digenite
DI diopside
DØ dolomite
DX dolomite/calcite

	mineral combination
DX	proportion unspecified
DØ	dolomite alone
D>	dolomite > calcite
D#	" = "
D<	" < "
CA	calcite alone

EN enargite (47% Cu)
EP epidote
ES enstatite
ER erythrite (30% Co)

FA famatinite
FL feldspars
FR ferberite (61% W)
FM ferrimolybdite (40% Mo)
FU fluorite
FD feldspathoids

GL galena (86% Pb)
GX galena/sphalerite

	mineral combination
GX	proportion unspecified
GL	galena alone
G>	galena > sphalerite
G#	galena = "
G<	galena < "
SL	sphalerite alone

GA garnet
GS glass(es)
GC glaucophane
GN glauconite
GØ goethite
GR graphite
GK greenockite (78% Cd)
GY gypsum
GD gold

HE hematite (70% Fe)
HX hematite/magnetite

	mineral combination
HX	proportion unspecified
HE	hematite alone
H>	" > magnetite
H#	" = "
H<	" < "
MG	magnetite alone

HS hematite, specularite
HØ hornblende - See also BX
HU huebnerite (61% W)
HM hydromica
HY hypersthene

IL illite
IM ilmenite (32% Ti)

JA jarosite

KA kaolin
KY kyanite
KF K-spar, orthoclase

Table 3.
(Continued)

KX K-spar/plagioclase mineral combination
KX proportion unspecified
KF K-spar alone
K> K-spar > plagioclase
K# K-spar = "
K< K-spar < "
PF plagioclase alone

LU leucite
LI limonite(s)

MA magnesite
MG magnetite - See also MX = HE/MG combination
MC malachite (58% Cu)
MX malachite/azurite mineral combination
MX proportion unspecified
MC malachite alone
M> " > azurite
M# " = "
M< " < "
AZ azurite alone
MN manganite
MZ monazite
MT marcasite
MR mariposite
MI micas, general
MM montomorillonite
MU muscovite
MØ molybdenite (60% Mo)
M[illegible] molybdenite rosettes
M: molybdenite disseminate
M/ molybdenite vein(lets)

NP nepheline
NI nicolite (44% Ni)

ØL olivine, chrysolite
ØP opal
KF orthoclase, K-spar
See also KX = KF/PF combination
ØX oxide, general
ØQ opaques, general

PF plagioclase
See also KX = KF/PF combination
PH phlogopite
PY pyrite
PL pyrolusite
PX pyroxene
PR pyrrhotite
PP pyrophillite
PØ powellite (58% Mo and W)
PT platinum

QZ quartz
QA quartz, agate
QC " , chert
QV " , vein, massive
QX " , crystal(s)
QR " , rutilated

RN rhodonite
RC rhodochrosite
RU rutile

SA sanadine
SC scapolite
SH scheelite (64% W)
MS sericite, muscovite
SE serpentine
SD siderite

SI sillimanite
SV silver
SØ sodalite
HS specularite, hematite
SL sphalerite
SP sphene
ST staurolite
SB stibnite (72% Sb)
SX sulfides, general

TA talc
TL tellurides, general
TN tennantite (50% Cu, 26% Sb and As)
TE tenorite (80% Cu)
TT tetrahedrite (50% Cu, 26% Sb and As)
TZ topaz
TØ tourmaline
TR tremolite

UR uraninite (92% U)

VA vanadinite, (73% Pb, 11% V)
VE vesuvianite

WF wolframite (62% W)
WN wulfenite (56% Pb, 26% Mo)
WD wad
WØ wollastonite

ZE zeolite(s)
ZI zircon
ZØ zoisite
ZZ any mineral

Table 4. One-, Three- and Five-Character Abbreviations

Used particularly for remarks and comments on GEOLOG, these abbreviations are now finding far wider use in the System. The first three letters are again the preferred form, with the third letter, if a vowel, being replaced with the next consonant. However, common usage or principal letters that readily suggest the sound of the word often take precedence.

Compound words are usually first letters, as in DDH for diamond drill hole. The preferred method of pluralizing is by adding 's or *s to the 3 letter abbreviation, thereby converting it to 5 characters. On key punching and other computer input/output, the apostrophe, if used, will be converted to the asterisk because the use of the apostrophe in computer language has been pre-empted for other purposes.

Both upper and lower case letters are permissable with lower case being preferred. Computer input/output will of necessity be in upper case.

abb	abbreviation
acs	accessory
acs*s	accessories
alt	alteration = altered, or
alt*d	altered
a	a = an
&	and
ang	angle = angled, or
ang*d	angled
anl	analysis
anl*s	analyses
apx	apex
app	approximate = approximately = approximation
Apr	April
asy	assay
asy*s	assays
Aug	August
bnk	bank
brn	barren
bln	baseline
*:	because
bed	bed = bedding = bedded
bed*s	beds
blk	black
blu	blue
bwx	boxwork
brx	breccia
brx*d	brecciated
bkn	broken
brw	brown
cap	cap = capping = capped
cer	certain(ly)
crg	charge = chargeability
clr	clear
color	color = colour
crs	coarse

Colors

yel	yellow
grn	green
blu	blue
orn	orange
red	red
prp	purple
tan	tan
brw	brown
wte	white
blk	black
gra	gray

Color Modifiers

D	dark
L	light
V	very
X	extremely
M	moderately = medium
F	fairly to faintly
V.D	very dark
V.L	very light
X.D	extremely dark
X.L	extremely light
M.D	moderately dark
M.L	moderately light

(Add sh to color abb to form - ish, such as:

yelsh	yellowish
grnsh	greenish

col	column
col*s	columns
cf.	compare = comparison
cdr	conductor
cnt	contact, or
C/	contact
contu	continue
contd	continued
cnr	corner
crk	creek
xtl	crystal
xtl*s	crystals
D	dark
dat	data
dtm	datum elevation
day	day

Days of Week

Mon	Monday
Tues	Tuesday
Wed	Wednesday
Thr	Thursday
Frd	Friday
Sat	Saturday
Sun	Sunday

/d	per day
deg	degree
dduct	deduct = deduction
dduce	deduce
den	density = dense
ddh	diamond drill hole
ddh*s	diamond drill holes
dik	dike, or
dyk	dyke; also
D/	
dto	ditto = same as above
dwn	down

Table 4.
(Continued)

=	equal, or
#	equal
≠	not equal
el.	elevation
end	end = ended = ending
end-d	end of data
eoh	end of hole
est	estimate(d)
eg.	example
eg.*s	examples
xcept	except = exception
X	extreme = extremely = exceptionally
etc.	etc.
expol	extrapolation = extrapolate
fac	face
fac*s	facies
/	fault, or
flt	fault - faulted, or
flt*d	faulted
flt	faulting, or
flt*g	faulting
Feb	February
fel	felsenmere
flk	flank
flo	float, general
flofb	float, frost boil
flogl	float, glacial
flors	float, residual
flotl	float, talus
fwd	forward
ft.	foot = feet
frc	fracture = fracturing = fractured, or
frc*d	fractured
frc*s	fractures
frg	fragment = fragmenting = fragmented, or
frg*d	fragmented
frg*s	fragments
frx	fresh rock
Fri	Friday
fne	fine

gam	gamma(s)
gen	general = generally
gos	gossan, general
gosex	gossan, exotic =
gosfr	gossan, fringe
gosin	gossan, indigenous
g.	gram(s)
>	greater than
grn	green
gra	gray
grd	ground
gnd	grained
hrd	hard
hvy	heavy
h*t	heat
ht.	height
hi.	high
hor	horizontal
hr.	hour
hrs	hours
hyp	hypogene
ID.	identity = identification
imp	important
in.	inch(es)
ie.	that is = in other words
iface	interface
ipb	improbable(ly)
ips	impossible(ly)
inpol	interpolation = interpolate
interp	interpret = interpretation
int	interval
int*s	intervals
Jul	July
Jun	June
jux	juxtaposition
kg	kilogram(s)
km	kilometre(s)

lrg	large
Lat	latitude
lhs	left-hand side
<	less than
lne	line
lnm	lineament
lnr	linear
lna	lineation
L	light
lng	long = length =
Lng	longitude
loc	location = locate = locality
Mar	March
mater	matter = material
mtx	matrix
May	May
m	metre(s)
M	medium = middle = middling = moderate(ly) or
mod	moderate(ly)
mid	middle = middling = mid
min	minute(s)
msc	miscellaneous
mgl	milligal
mm.	millimetre(s)
ml.	millilitre(s)
Mon	Monday
mth	month
med	medium
Month	of Year
Jan	January
Feb	February
Mar	March
Apr	April
May	May
Jun	June
Jul	July
Aug	August
Sep	September
Oct	October
Nov	November
Dec	December
Msl	Munsell
MCC	Munsell colour code

N	no = none = not any = nil
not	not
no.	number
no*s	numbers, or
nos	numbers
Nov	November
neg	negative(ly)
OCX	Outcrop, general
oce	outcrop, excellent
ocg	outcrop, good
ocf	outcrop, fair
ocp	outcrop, poor
ocs	outcrop, subcrop or inferred
ocr	occur = occurrence
Oct	October
onl	only
op.	open pit
opt	option = optional
orn	orange
ovb	overburden
oxd	oxide = oxidized zone
peg	pegmatite = pegmatitic
//	parallel
per	per, or
/	per
ppd	perpendicular
pdh	percussion drill hole
phx	phenocryst, or
pheno	phenocryst
pic	picture = photo
pic*s	pictures = photos
rdh	rotary drill hole
rem	remark
res	resistivity
rhs	right-hand-side
rt.	right
rox	rock(s)
rxtyp	rock type, or
rtp	rock type
Sat	Saturday
Sep	September
sol	solution
slb	soluble
sos	same old stuff
sec	second (time)
2nd	second (order)
ps.	polished section
ps:	post script
pt.	point
(+)	positive(ly), or
+++	positive(ly)
posit	position
prp	purple
pos	positive(ly)
prb	probable(ly)
psb	possible(ly)
(?)	questionable(ly), or
???	very questionable(ly)
slp	since last point
slk*s	slickensides
sg.	specific gravity = density
stb	stable
stn	station
stw	stockwork, or
stkwk	stockwork
st.	straight
str	structure
sup	supergene, general
supox	supergene, oxide
supsl	supergene, sulphide
sface	surface, or
sfc	surface
s/x	survey, or
S/XXX	survey, general
S/LGL	legal
S/EMA	electromag, airborne
S/EMG	electromag, ground
S/GRV	gravity
S/MGG	mag, ground
S/MGA	mag, airborne
S/GCS	geochem, soil
S/GCT	geochem, silt
S/GCR	geochem, rock
S/GCB	geochem, bio
S/GCG	geochem, gossan
S/GEO	geological
S/FRC	fracture
Sun	Sunday
TS.	thin section
Thr	Thursday
typ	type = typical
T/	transition = transition zone
Tus	Tuesday
:.	therefore
tot	total
uncer	uncertain
vod	void
vod*s	voids
vol	volume
V/	vein
v	very
w	with
w*o	without
w*n	within
wte	white
X	extreme = extremely = exceptionally
xtl	crystal
yel	yellow

TABLE 5. The Single Character, Single Significant Factor S Scale
One of many scales developed for coding various estimated geologic factors

Symbol or Code	Definition & Assigned Value	Another Definition of Symbols in Column I	Symbol or Code	Definition & Assigned Value	Symbol or Code	Definition & Assigned Value	Symbol or Code	Definition & Assigned Value	Another Definition of Symbols to 9 in Column IV	Other Definitions of Symbols in Columns IV or V	Symbol or Code	Definition & Assigned Value	Symbol or Code	Definition & Assigned Value	Symbol or Code	Definition & Assigned Value
I			II		III		IV				V		VI		VII	
!	0.01		I	0.1	1	1.0	—	10		Completely . . . ed	$	100	Z	1,000		
.	0.009	Slightly <0.01	H	0.09	Q	0.9	9	9	White	Extremely well . . . ed, excellent extremely high	+	90	Y	900	Δ	>9,000
			G	0.08	P	0.8	8	8	Very light gray	Very high, greater than high, very well . . . ed	>	80	X	800	8	8,000
:	0.007	Definitely <0.01	F	0.07	Ø	0.7	7	7	Light gray	High, heavy, large, well . . . ed	]	70	W	700	%	7,000
			E	0.06	N	0.6	6	6	Medium light gray gray	Above medium, fairly large, fairly well . . . ed	)	60	V	600	\	6,000
	0.005	Several specks visible	D	0.05	M	0.5	5	5	Medium gray	Medium, moderate, mid, middling, central	#	50	U	500	*	5,000
			C	0.04	L	0.4	4	4	Medium dark gray	Below medium, fairly poorly . . . ed fairly low	(	40	T	400	/	4,000
	0.003	Odd speck visible or trace	B	0.03	K	0.3	3	3	Dark gray	Low, small, poor, poorly . . . ed	[	30	S	300	☼	3,000
	0.002	May be present	A	0.02	J	0.2	2	2	Grayish black	Very low, less than low, very poorly . . . ed	<	20	R	200	@	2,000
?	0.001	Questionally present	!	0.01	I	0.1	1	1	Black	Extremely poorly . . . ed, extremely low	—	10	$	100	Z	1,000
O	0.000	absent, nill, not	O	0.00	O	0.0	0	0		Completely un . . . ed	O	0	0	0	O	0

This S Scale forms a nearly complete one-significant figure series from 0 and 0.001 to >9000, as well as seven sub-sets: 0 to 0.01, 0 to 0.1, 0 to 1.0, 0 to 10, 0 to 100, 0 to 1,000 and 0 to >9,000.

TABLE 6. Mode and Degree of Dispersion of Mineralization. V, E, D and P refer to the modes of occurrence: veins, envelopes, disseminations and pervasive, respectively

Mode of Occurrence Mineralization			Degree of Dispersion	Mode of Occurrence Alteration Assemblages			
Veins and macro - veins including stockwork and gouge	V	V	1	V	V	or V	veins
Veins, veinlets, fracture fillings & minor disseminations	$\frac{1}{8}D + \frac{7}{8}V$	$D \ll V$	2	$E < V$ or $P \ll V$	$\frac{2}{8}E+\frac{6}{8}V$	$\frac{1}{8}P+\frac{7}{8}V$	veins and moderate envelopes or minor pervasive
Veinlets and some disseminations	$\frac{2}{8}D + \frac{6}{8}V$	$D < V$	3	$E = V$ or $P < V$	$\frac{4}{8}E+\frac{4}{8}V$	$\frac{2}{8}P+\frac{6}{8}V$	envelopes and veins equal or veins and moderate pervasive
Veinlets with moderate disseminations	$\frac{3}{8}D + \frac{5}{8}V$	$D \leq V$	4	$E > V$ or $P \leq V$	$\frac{6}{8}E+\frac{2}{8}V$	$\frac{3}{8}P+\frac{5}{8}V$	envelopes with some veins or pervasive with moderate veins
Veinlets and disseminations more or less equal	$\frac{4}{8}D + \frac{4}{8}V$	$D = V$	5	E or $P = V$	$\frac{8}{8}E$	$\frac{4}{8}P+\frac{4}{8}V$	envelopes or pervasive equal to veins
Disseminations and moderate veinlets	$\frac{5}{8}D + \frac{3}{8}V$	$D \geq V$	6	$P < E$ or $P \geq V$	$\frac{2}{8}P+\frac{6}{8}E$	$\frac{5}{8}P+\frac{3}{8}V$	pervasive with some envelopes or moderate veins
Disseminations with some veinlets	$\frac{6}{8}D + \frac{2}{8}V$	$D > V$	7	$P = E$ or $P > V$	$\frac{4}{8}P+\frac{4}{8}E$	$\frac{6}{8}P+\frac{2}{8}V$	pervasive and envelopes or with some veins
Mostly dissemination with minor veinlets or micro-veinlets	$\frac{7}{8}D + \frac{1}{8}V$	$D \gg V$	8	$P > E$ or $P \gg V$	$\frac{6}{8}P+\frac{2}{8}E$	$\frac{7}{8}P+\frac{1}{8}V$	pervasive with some envelopes or minor veins
Disseminations	D	D	9	P	P	P	pervasive

might, for example, be a guide to zoning classification or an important consideration in mining or pit design.

Development of Geolog System

Development of Geolog System was initiated early in 1970. It has evolved through several versions,

TABLE 7. Alteration Facies in Porphyry Deposits

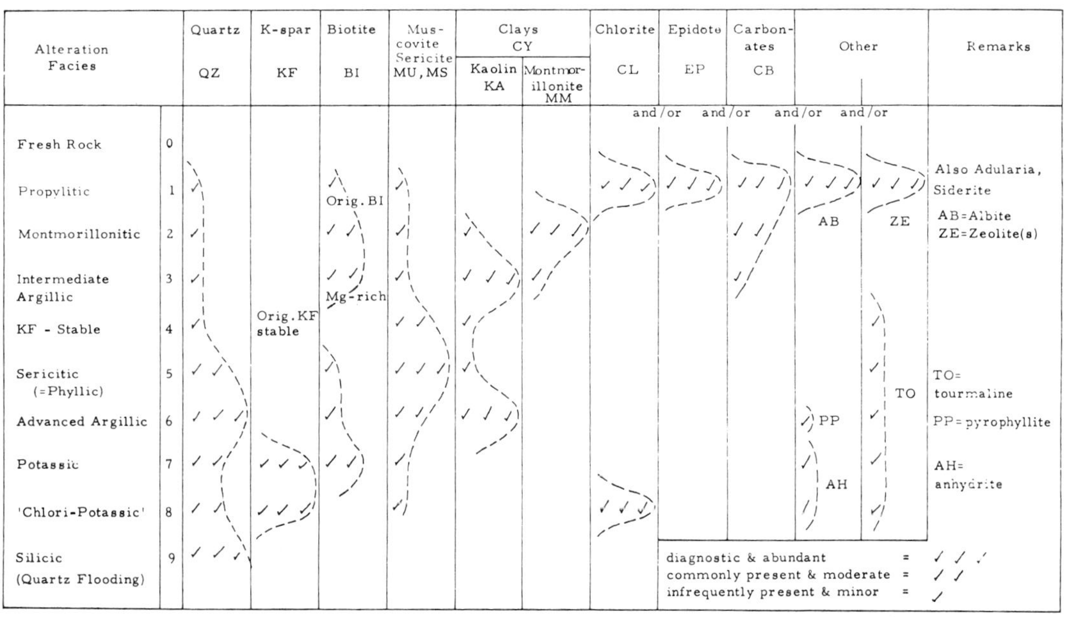

TABLE 8. Sulfide, Oxide, and Native Mineral Assemblages in Porphyry Deposits

Mineralization Zones		Pyrite PY	Chalco-pyrite CP	Molyb-denite MO	Wolframite Covellite Digenite WF, CV, DG	Chalcocite CC	Hematite Magnetite HX	Galena-Sphalerite GX (Gold-Silver occurrence)	Pyrrhotite PR	Native Copper CU	Remarks
Shroud	1	✓				✓✓	✓✓	✓	✓✓	✓	
Peripheral Shell	2	✓				✓	✓	✓✓	✓✓	✓	
Low Pyrite Shell	3	✓✓✓				✓		✓			
Deep Ring	4	✓✓✓					✓✓✓✓✓				
Pyrite Shell	5	✓✓✓✓✓	✓✓	✓	✓✓	✓✓					
Ore Shell	6	✓✓	✓✓✓✓	✓✓✓	✓✓	✓✓					
Low-Grade Core	7	✓	✓✓	✓✓			✓✓				
Deep Core	8	✓	✓				✓✓				

diagnostic & abundant = ✓✓✓✓✓
frequently present & substantial = ✓✓✓✓
usually present & moderate = ✓✓✓
sometimes present & minor = ✓✓
infrequently present & trace to minor = ✓

mainly as a result of field experience with the Casino Creek porphyry copper-molybdenum deposit, Yukon Territory. The present version of the System is being used on a porphyry molybdenum-copper deposit in the southwestern United States.

At Casino an earlier version was used to log 36,922 feet of split diamond drill core from 49 holes, and 17,481 feet of rotary drill chips from 35 holes. Phillips and Godwin (1970, p. 49) noted on logging of diamond drill core that "Under good conditions and after some initial experience a minimum of about 15 minutes was required for the examination of each

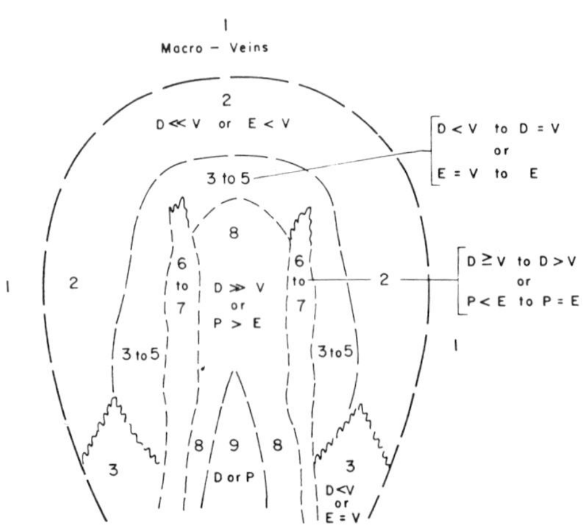

FIG. 5. Model for mode and degree of dispersion of mineralization in porphyry deposits. (After Lowell and Guilbert, 1970.) 1 indicates a very low degree of dispersion of a mineral—that it occurs in veins almost entirely. 9 indicates a very high degree of dispersion—that it occurs pervasively or very well disseminated in the wall rock. See Table 6 for definitions and details.

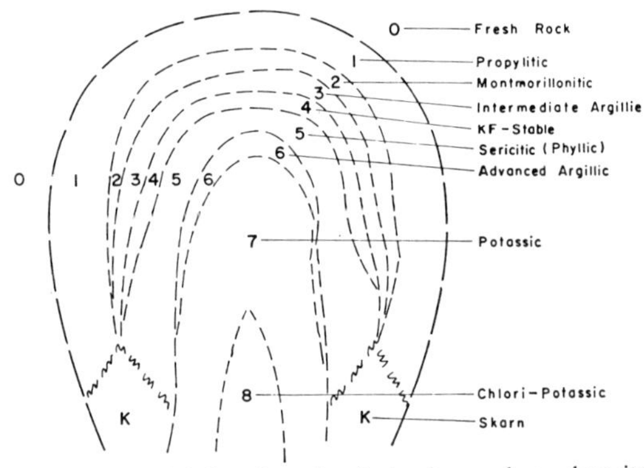

FIG. 6. Model for alteration facies in porphyry deposits. (After Lowell and Guilbert, 1970.) Note that numbers increase inwards toward the deep core, as in Figure 5. Compare also with Figure 7. See Table 7 for details and definitions.

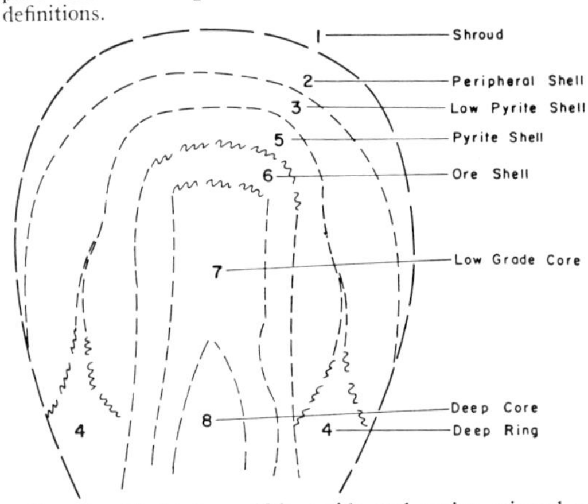

FIG. 7. Model for sulfide, oxide and native mineral assemblages in porphyry deposits. (After Lowell and Guilbert, 1970.) See Table 8 for details and definition.

TABLE 9. Some Comparisons between Geolog and "Standard" Geological Logging Techniques

Standard	GEOLOG
1. *Computability*	
—written geologic information is not formatted or quantified, and therefore requires post-field interpretation and rewriting if it is to be keypunched and manipulated with a computer experience to date indicates that it is usually unsatisfactory to attempt to transcribe from standard logs to GEOLOG; it is better to re-log	—ready for keypunching and subsequent storage, retrieval and computer manipulation and display —"soft" geologic information and word descriptions have been converted on-the-spot to "hard" or reasonably hard geologic data by the geologist making the observations and estimates, in the field, on-the-spot quantitative estimates, however rough, are far better and of far greater usefulness than any later, in-the-office guess from a word description of some geologic feature or condition
2. *Readability*	
—handwriting is often extensive and sometimes illegible, making information retrieval tedious and difficult —assay logs, visual or graphic log, rock type descriptions and alteration interpretations are often on different lines, written at different scales, or on separate sheets; this makes comparisons laborious and subject to error	—printing in boxes or prescribed spaces promotes legibility —all geological factors pertaining to a given interval or piece of core are recorded on the same line so that comparisons for that interval are made along a single line
3. *Accessibility of Specific Information*	
—data has to be retrieved manually —required data is often hidden in unpredictable locations within lengthy written comments	—data pertaining to a specific interval or to a specific geologic factor can be retrieved readily with a computer —such data are also easily retrieved manually by scanning rows (lines) for information about a specific interval, or by scanning columns for data pertaining to a specific feature
4. *Recording the Absence of a Feature*	
—unusual features get preferred attention, while frequently recurring ones tend to pass without comment; no comment gives little assurance that a feature is absent, particularly if its presence is noted in the intervals immediately above and below the interval in question	—labeled columns serve as ready checklists so that absence or presence of a geologic factor is routinely noted
5. *Codes and Scales*	
—scales are not usually prescribed and codes or abbreviations or word descriptions, like: "considerable fracturing," "very large," "heavily mineralized," "less than trace," have no defined values; these are difficult to interpret and quantify —there is often little consistency in meaning of descriptive terms between geologists because of varied experiences or interests	—codes and scales are prescribed and have defined values and ranges; for some factors, several scales are available, and the user is free to choose the one most appropriate for the range of values anticipated for that geologic factor in any particular deposit —for consistency, ease of recall and readability, rock names are always contracted to 4 letters; mineral or mineral combination names to 2 letters; and comment abbreviations to 1, 3 or 5 characters
6. *Summaries*	
—summaries are seldom routine but occur at the discretion of the geologist; therefore, to reconstruct the overall appearance of the hole requires post-logging interpretation	—specific summary columns refer to predictive models and require "on the spot" judgments that might require intuition; the reason for logging is constantly before the geologist
7. *Flexibility and Discipline*	
—too much flexibility makes it very difficult for the work of one geologist to fit together with the work of his successor —many drilling programs develop in stages; the geologic logging done in Phase 3 often does not fit well with that done in Phase 2 and may bear little relation to the work done in Phase 1	—although not at first readily apparent, a high degree of flexibility has been built into Geolog System and the programs developed to process GEOLOG data —however, a corresponding degree of discipline is required but this is not particularly onerous
8. *Required Training*	
—there is normally no manual of instruction or level of competence required of the geologist doing the logging —inconsistency and bias in description is fostered by the varied backgrounds and experiences of geologists	—it is easier to instruct geologists in Geolog System because all the rules, though not rigid, are definite; each has its degree of flexibility but the limits of variability are closely defined and must be observed

```
                              G   E   O   L   O   G

                              F O R M A T   2

                                                    DRILL HOLE B-1
                                                    --------------

DATE STRIKE  DIP LENGTH  DBR SIZE 1 SIZE 2 NORTHING   EASTING  ELEVATION
DEC70  0:0 -90:0 1149.0  15. NC  40        136075.00 87565.00   6150.00

-------------------------------------------------------------------------------
 D   P  T   H   M    T  F  F D  % S ALTERATION         F    MINERALIZATION
 E   R  Y   O   UC   E  L  R E  F U          ASSEMBLAGES AA
 P   I  P   S   NO   XG T  ATN  R L                    C.
 T   N  E   T   SL   TR    CUS  AWP                    II
 H   .              EO   UA Z  -RI  CIH QZ KF MS CY CL EP CB EN  PY CP MO GL SL AS
     Z  M   R   LU   RI O   ET  STI                    ST
     O  O   O   LR   EN N A  Y A HD M  M  M  M  M  M  M  E  M  M  M  M  M  M
     N  D   C           E LG   LG E O  O  O  O  O  O  O  N  O  O  O  O  O  O
     E  .   K    C   S  LPG  LPG  D  D  D  D  D  D  D  S  D  D  D  D  D  D
     S           O   I M 1PG  1PG E  E  E  E  E  E  E  1  E  E  E  E  E  E
                 D   Z A  2P   2P %  %  %  %  %  %  %  T  %  %  %  %  %  %
                 E   E T   3    3                      Y                    %MOS2  %CU
-------------------------------------------------------------------------------
SCALES                E CDDD EEEE NE NE NE NE NE NE NE    NE NE NE NE NE NE
  0 OVB
  5 OVB
 10 OVB
 15 OVB
 20 HYP Q HRFL  2571 T3 3  546  999 41    1E 91              35                          .005  .070
 25 HYP Q HRFL  2571 T3 4  545  999 41    1E 91              35                          .006  .080
 30 HYP Q HRFL  2571 T3 4  546  999 41    1E 91              35                          .006  .080
 35 HYP Q HRFL  3551 T3 7  423  999 41    1E 91              35  31                      .014  .070
 40 HYP Q HRFL  3551 T3 7  423  999 41    1E 91              35  31                      .014  .070
 45 HYP Q HRFL  3551 T3 3  351  999 41    1E 9F       1B     29  31                      .018  .030
 50 HYP Q HRFL  3551 T3 3  351  999 41    1E 9F       1B     29  31                      .018  .030
 55 HYP   GRNT  1562 I5 6  243  999 43    91                 55  3E    1A                .003  .010
 60 HYP   GRNT  1562 I5 6  243  999 43    91                 55  3E    1A                .003  .010
 65 HYP   GRNT  1564 I5 7  264  999 43    91                 60  3E    9A                .001  .010
 70 HYP   GRNT  1564 I5 7  264  999 43    91                 60  3E    9A                .001  .010
 75 HYP   QZLT  3561 E3 0  111  000          5E       9E     15  9E    9A                .006  .020
 80 HYP   QZLT  3561 E3 0  111  000          5E       9E     15  9E    9A                .006  .020
 85 HYP   QZLT  3561 E3 0  111  000          5E       9E     15                          .005  .010
 90 HYP   GRNT  0070 I5 2  452  999 43    91                 60  9E    9A                .005  .010
 95 HYP   GRNT  0070 I5 5  562  999 43    91 5E              55  3E    7A                .020  .020
100 HYP   GRNT  0070 I5 5  562  999 43    91 5E              55  3E    7A                .020  .020
105 HYP   GRNT  0070 I5 5  562  999 43    91 5E              55  3E    7A                .009  .020
110 HYP   GRNT  0070 I5 5  562  999 43    91 5E              55  3E    7A                .009  .020
115 HYP   GRNT  0070 I5 6  362  999 43    91 5E              59  3E    7A                .006  .060
120 HYP   GRNT  0070 I5 6  362  999 43    91 5E              59  3E    7A                .006  .060
125 HYP   GRNT  0070 I5 7  362  999 43    91 5F              59  3E    7A 1A             .011  .010
130 HYP   GRNT  0070 I5 7  362  999 43    91 5F              59  3E    7A 1A             .011  .010
135 HYP   GRNT  0070 I5 8  462  999 43    91 5F              59  3E    7A 1A             .008  .020
140 HYP   GRNT  0070 I5 8  462  999 43    91 5F              59  3E    7A 1A             .008  .020
145 HYP   GRNT  0070 I5 8  573  999 43    91 5F              59  3E    7A 1A             .018  .010
150 HYP   GRNT  0070 I5 8  573  999 43    91 5F              59  3E    7A 1A             .018  .010
155 HYP   GRNT  0070 I5 6  3W2  999 43    91 5F              59  3E    7A                .004  .010
160 HYP   GRNT  0070 I5 6  3W2  999 43    91 5F              59  3E    7A                .004  .010
165 HYP   GRNT  0070 I5 4  472  999 43    91 5F              59  3F    3A                .009  .020
170 HYP   GRNT  0070 I5 4  472  999 43    91 5F              59  3F    3A                .009  .020
```

FIG. 8. Drill hole strip-log produced by computer from ASSAYLOG and GEOLOG data.

15-foot sample." One hundred feet of rotary cuttings were readily logged in one hour. The cuttings had been divided into coarse chips, fine cuttings, and panned heavy concentrates with magnetic fraction removed. These were glued with "Bondfast" onto plywood boards so that they formed a drill "section" suitable for visual and binocular microscope examination while logging. Time and effort involved in collecting drill data on GEOLOG are comparable to standard logging techniques.

Several mining companies and government agencies have provided drill hole logs thought by the authors to be representative of current logging practice. These are hereafter called "standard" logs. Analysis of these standard logging procedures indicated areas needing improvement and served as guides to what data should be collected. Some aspects of these standards logging procedures are compared to Geolog System in Table 9. The examples of logging obtained were submitted as excellent, but the authors believe that more useful data would have been obtained if the same amount of effort had been expended in filling out GEOLOG. Several attempts were also made to transcribe data from standard logs to GEOLOG. This approach was found to be impractical and, consequently, the standard logs are not readily adaptable to computer analysis. In summary, the main problems with standard logs are as follows:

1. Concrete information was retrieveable only by laborious searching and only after doubtful interpretations of some of the recorded observations.

2. The absence of consistent records for the non-occurrence of a specific feature was particularly distressing.

3. Codes and scales were often vague or meaningless to anyone other than the geologist doing the logging.

4. Features described in the models of Figures 5, 6, and 7 were only partly described; consequently, postfield interpretation was required and considered less useful than an "on-the-spot," although rough, quantitative estimate.

Consistency in detailed descriptions by geologists, individually or collectively, allows correlation and comparisons to be made within a property or among properties. Barnes and Parry (1970) demonstrated that once porphyry copper geologic data were classified and coded they could be operated upon by the computer for retrieval and for calculation of

CHAPMAN WOOD AND GRISWOLD, LTD. HOLE ID: DDH-21

WEIGHTED AVERAGES BY BENCH INTERVALS

BENCH HEIGHT 50.0 FT

DATE STARTED	AZIMUTH	DIP	LENGTH	DEPTH BEDROCK	DRILL HOLE SIZES SIZE 1	SIZE 2	GRID COORDINATES OF COLLAR NORTHING	EASTING	ELEVATION
JUN69	-0: 0	-90:-0	2191.0	10.0	NC TO	NX TO	135789.00	89300.00	5884.00

BENCH INTERVAL FROM	TO,FT	MID-POINT OF HOLE BETWEEN BENCHES NORTHING	EASTING	ELEVATION	GEOLOGICAL DATA EP MODE	EP %	CB MODE	CB %	PY MODE	PY %	MO MODE	MO %	REMARKS
5874.0	5850.0	135789.0	89300.0	5862.0	.00	.00	.00	.00	3.00	3.50	.00	.00	
5850.0	5800.0	135789.0	89300.0	5825.0	.00	.00	.00	.00	3.00	3.28	.00	.00	
5800.0	5750.0	135789.0	89300.0	5775.0	.00	.00	3.14	7.57	3.00	2.78	1.00	.10	
5750.0	5700.0	135789.0	89300.0	5725.0	.00	.00	2.96	7.04	3.00	2.30	1.00	.10	
5700.0	5650.0	135789.0	89300.0	5675.0	.00	.00	.00	.00	3.00	3.10	1.00	.10	
5650.0	5600.0	135789.0	89300.0	5625.0	.00	.00	.00	.00	3.00	3.50	1.00	.10	
5600.0	5550.0	135789.0	89300.0	5575.0	.00	.00	.00	.00	3.00	2.30	1.00	.10	
5550.0	5500.0	135789.0	89300.0	5525.0	.00	.00	.00	.00	3.00	2.70	1.00	.10	
5500.0	5450.0	135789.0	89300.0	5475.0	.00	.00	1.00	1.50	3.00	1.50	1.00	.10	
5450.0	5400.0	135789.0	89300.0	5425.0	.00	.00	.00	.00	3.00	2.36	1.00	.10	
5400.0	5350.0	135789.0	89300.0	5375.0	.00	.00	.00	.00	3.00	3.50	1.00	.10	
5350.0	5300.0	135789.0	89300.0	5325.0	.00	.00	.00	.00	3.00	3.50	1.00	.10	
5300.0	5250.0	135789.0	89300.0	5275.0	.00	.00	.00	.00	3.80	3.50	1.00	.10	
5250.0	5200.0	135789.0	89300.0	5225.0	1.00	.10	.00	.00	3.00	3.50	1.00	.10	
5200.0	5150.0	135789.0	89300.0	5175.0	.00	.00	1.00	1.00	3.24	3.50	1.00	.10	
5150.0	5100.0	135789.0	89300.0	5125.0	9.00	3.50	4.20	5.98	3.00	3.50	1.00	.14	
5100.0	5050.0	135789.0	89300.0	5075.0	9.00	3.50	3.64	10.00	3.00	3.50	1.00	.10	
5050.0	5000.0	135789.0	89300.0	5025.0	9.00	3.50	4.00	10.00	3.00	3.50	1.00	.10	
5000.0	4950.0	135789.0	89300.0	4975.0	9.00	3.50	.00	.00	3.00	3.50	1.00	.10	
4950.0	4900.0	135789.0	89300.0	4925.0	.00	.00	.00	.00	3.00	3.50	1.00	.10	
4900.0	4850.0	135789.0	89300.0	4875.0	9.00	3.50	.00	.00	3.00	3.50	1.96	.10	
4850.0	4800.0	135789.0	89300.0	4825.0	.00	.00	.00	.00	3.00	3.50	3.00	.10	
4800.0	4750.0	135789.0	89300.0	4775.0	.00	.00	.00	.00	3.00	3.50	2.84	.10	
4750.0	4700.0	135789.0	89300.0	4725.0	.00	.00	.00	.00	3.00	4.80	2.36	.10	
4700.0	4650.0	135789.0	89300.0	4675.0	.00	.00	1.00	.80	3.00	3.50	3.00	.10	
4650.0	4600.0	135789.0	89300.0	4625.0	.00	.00	1.00	6.55	3.00	3.50	3.00	.10	
4600.0	4550.0	135789.0	89300.0	4575.0	.00	.00	.00	.00	3.00	3.50	3.00	.10	
4550.0	4500.0	135789.0	89300.0	4525.0	.00	.00	1.00	7.32	3.00	3.50	3.00	.10	
4500.0	4450.0	135789.0	89300.0	4475.0	.00	.00	1.00	3.50	5.24	3.50	2.50	.10	
4450.0	4400.0	135789.0	89300.0	4425.0	.00	.00	.00	.00	3.00	3.50	1.56	.10	
4400.0	4350.0	135789.0	89300.0	4375.0	.00	.00	.00	.00	3.00	3.50	3.00	.10	
4350.0	4300.0	135789.0	89300.0	4325.0	.00	.00	.00	.00	2.50	4.80	2.55	.10	
4300.0	4250.0	135789.0	89300.0	4275.0	9.00	3.50	1.00	10.00	3.00	3.50	3.00	.10	
4250.0	4200.0	135789.0	89300.0	4225.0	9.00	1.74	.00	.00	3.00	3.50	3.00	.10	
4200.0	4150.0	135789.0	89300.0	4175.0	8.00	2.00	.00	.00	3.00	3.50	3.00	.12	
4150.0	4100.0	135789.0	89300.0	4125.0	1.00	1.50	.00	.00	3.00	3.50	3.00	.13	
4100.0	4050.0	135789.0	89300.0	4075.0	.00	.00	.00	.00	3.00	3.23	3.00	.13	
4050.0	4000.0	135789.0	89300.0	4025.0	.00	.00	.00	.00	3.00	3.50	3.00	.16	
4000.0	3950.0	135789.0	89300.0	3975.0	.00	.00	1.00	.40	3.00	3.50	3.00	.10	
3950.0	3900.0	135789.0	89300.0	3925.0	.00	.00	.00	.00	3.00	3.50	3.00	.10	
3900.0	3850.0	135789.0	89300.0	3875.0	.00	.00	1.00	.26	3.00	3.50	3.00	.10	
3850.0	3800.0	135789.0	89300.0	3825.0	.00	.00	1.00	.33	3.36	3.30	3.00	.10	
3800.0	3750.0	135789.0	89300.0	3775.0	.00	.00	.00	.00	3.00	2.18	3.00	.10	
3750.0	3700.0	135789.0	89300.0	3725.0	.00	.00	.00	.00	6.33	1.56	3.00	.10	
5874.0	3693.0				7.88	2.32	2.36	5.78	3.13	3.32	2.04	.10	OVERALL AVERAGES
					210.0	210.0	317.0	317.0	2150.	2140.	1976.	1976.	FOOTAGES 'ASSAYED'

FIG. 9. BENCOM output of GEOLOG data as bench interval weighted averages in table form.

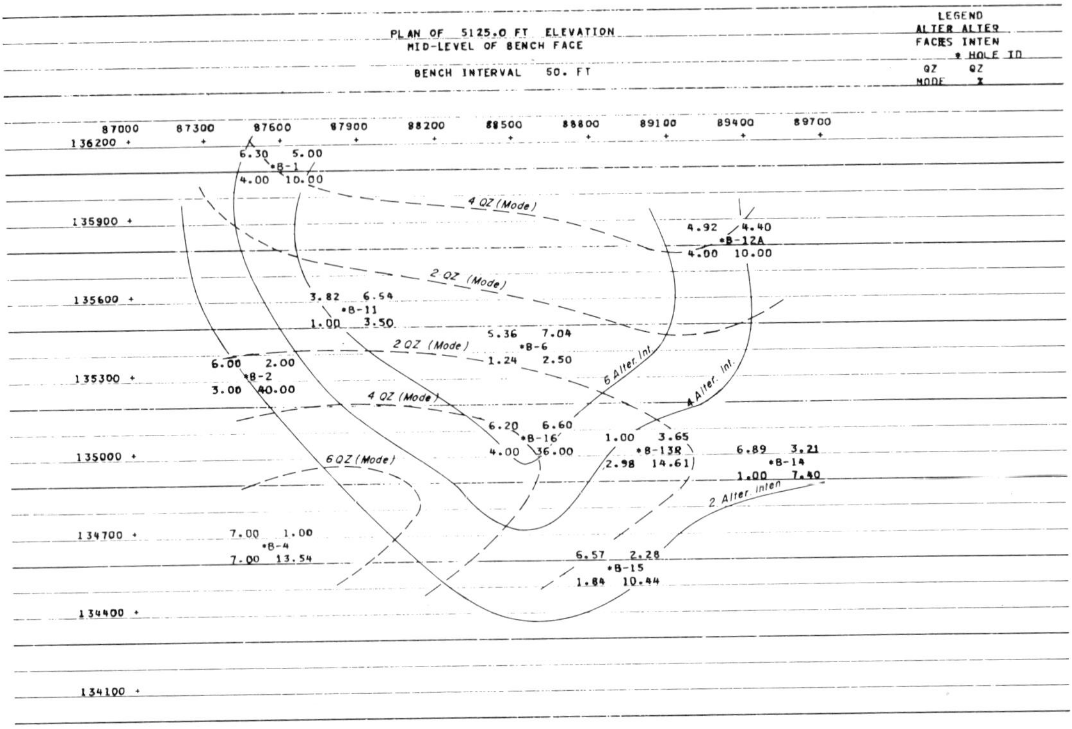

FIG. 10. BENCOM bench plan printout of GEOLOG data showing piercing point of drill holes and four geological variables.

statistics so that probability models could be built. They found, however, that ". . . one of the most severely limiting features of the study is the lack of uniformity and detail in the published information from which the input was derived." Thus, widespread adoption of Geolog System would improve the quality of information collected and its interpretation so that published information on properties could be more useful.

Conclusion

A geologist describing a drill hole, specimen, or rock exposure with the aid of the Geolog System obtains data that are both suited for key punching and subsequent computer processing, and easily readable and interpretable for manual use. Use of the prescribed headings in GEOLOG together with specific shorthand codes results in quantified, systematic, consistent, and comprehensive data collection.

Flexibility of Geolog System allows descriptions to vary, as required, for different studies that may be exhaustive, routine, or brief. Readily modified headings and scales permit application of the System to a wide variety of mineral deposits.

Comparison of information in GEOLOG summary columns to theoretical mineral deposit models, particularly those established for the porphyry type, might be useful in predicting ore locations, even from preliminary drill results. GEOLOG might also provide a basis for accumulating data for the development of better models.

All details of specific codes and output developed for Geolog System have not been presented in this article, but examination of the figures and tables should show substantially what is being attempted in the way of recording and manipulating drill hole information.

Geolog System is thought by the authors to be superior to most descriptive and interpretive schemes in current use. In particular, its predictive potential for locating ore bodies justifies more effort in drill hole description.

Acknowledgments

The authors thank Chapman, Wood and Griswold Limited for permission to publish this paper and for their help in its preparation. J. A. Wood guided the project in its early stages and has reviewed the manuscript. The late Dr. J. A. Gower and Dr. A. J. Sinclair of the University of British Columbia have also kindly reviewed the paper. Others who kindly added advice include M. P. Phillips of Archer, Cathro & Associates Limited, and Dr. J. M. Carr of Brameda Resources Limited.

Appreciation is extended to Brameda Resources Limited, the National Advisory Council on Research in the Geological Sciences (grant GSl-70), and The University of British Columbia for financial assistance to Godwin.

Department of Geology,
The University of British Columbia,
Vancouver 8, B. C., Canada,
November 15, 1971; April 4, 1972
P. H. B.
Formerly, Senior Research Geologist,
Chapman, Wood and Griswold Limited,
145 East 15th Street,
North Vancouver, B. C., Canada,

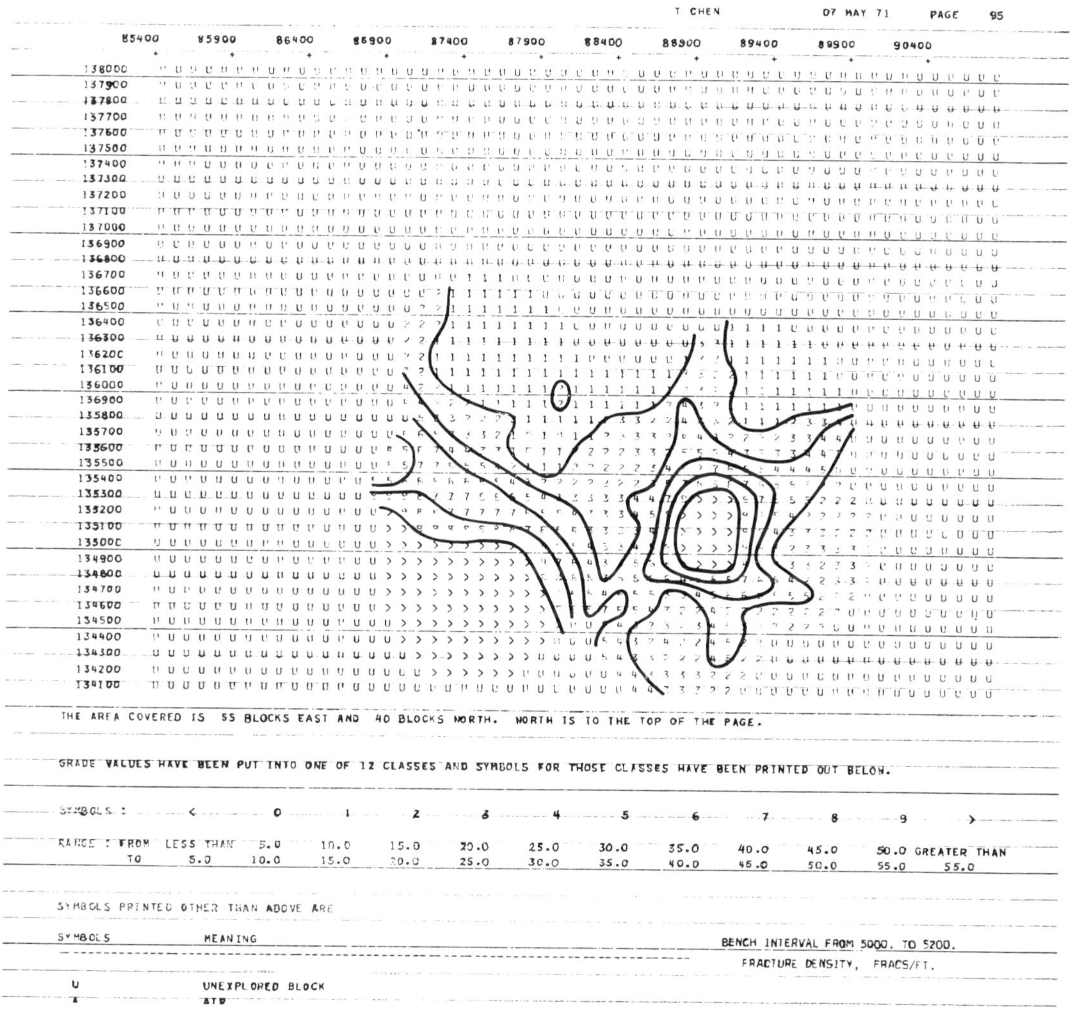

FIG. 11. CUBE bench plan printout of three-dimensionally weighted fracture density data from GEOLOG.

TONNAGE AND GRADE ESTIMATES

FRACTURE DENSITY
BENCH INTERVAL FROM 5000. TO 5200.

IN THE FOLLOWING TABLE THE GRADE VALUES FOR EACH BLOCK HAVE BEEN PLACED INTO DIFFERENT RANGE OF GRADE CLASSES.

RANGE OF GRADE		PROVEN & PROBABLE			PROVEN		PROBABLE		INFERRED	
FROM	TO	NO. OF BLOCKS	NO. OF TONS X1000	AVERAGE GRADE	NO. OF TONS X1000	AVERAGE GRADE	NO. OF TONS X1000	AVERAGE GRADE	NO. OF TONS X1000	AVERAGE GRADE
GRADE BELOW	10.0	3	522.	9.1	348.	8.7	174.	9.9	-	-
10.0	20.0	142	24708.	14.7	7482.	13.9	17226.	15.0	19314.	14.9
20.0	30.0	90	15660.	23.9	5046.	23.0	10614.	24.3	8178.	24.8
GRADE ABOVE	30.0	165	28710.	56.3	7656.	67.1	21054.	62.3	15834.	53.7
TOTAL :		400	69600.	33.9	20532.	35.9	49068.	33.0	43326.	31.0

IN THE FOLLOWING TABLES. THE RATIO OF TONS BELOW CUTOFF TO TONS ABOVE CUTOFF IS SHOWN AS RATIO.

TABLE OF RATIOS FOR PROVEN AND PROBABLE RESERVES

TRIAL CUTOFF GRADE	NO. OF BLOCKS ABOVE CUTOFF	TONS ABOVE CUTOFF X1000	AVERAGE GRADE ABOVE CUTOFF	NO. OF BLOCKS BELOW CUTOFF	TONS BELOW CUTOFF X1000	AVERAGE GRADE BELOW CUTOFF	RATIO
30.0	165	28710.	56.3	235	40890.	18.1	1.42
20.0	255	44370.	44.8	145	25230.	14.6	.57
10.0	397	69078.	34.0	3	522.	9.1	.01

FIG. 12. CUBE summary table output, in assay terminology, of data from CUBE bench plan (Fig. 11).

REFERENCES

Barnes, M. P., and Parry, W. T., 1970, Porphyry copper deposits—a computer analysis of significance of geological parameters: in print.

Blanchard, Roland, 1968, Interpretation of leached outcrops: Nevada Bur. Mines, Bull. 66, 196 p.

Brisbin, W. C., and Ediger, N. M., eds., 1967, A national system for storage and retrieval of geological data in Canada: Geological Survey of Canada. A report by the *ad hoc* committee on storage and retrieval of geological data in Canada, 175 p.

Coode, A. M., and Chen, T., 1971, Weighted averages versus trend surface in grade reserve estimation: Decision-Making IV Conference, Vancouver, February 1971, Centre for Continuing Education, Univ. of British Columbia.

Hutchison, W. W., and Roddick, J. A., 1968, Recording geological data: Western Miner, v. 41, no. 2, p. 39–43.

James, A. H., 1971, Hypothetical diagrams of several porphyry copper deposits: ECON. GEOL., v. 66, p. 43–47.

Krumbein, W. C., and Sloss, L. L., 1951, Stratigraphy and Sedimentation: San Francisco, W. H. Freeman Co., 497 p.

Lowell, J. D., 1968, Geology of the Kalamazoo Orebody, San Manuel District, Arizona: ECON. GEOL., v. 63, p. 645–654.

—— and Guilbert, J. M., 1970, Lateral and vertical alteration-mineralization zoning in porphyry ore deposits: ECON. GEOL., v. 65, p. 373–408.

McKinstry, Hugh Exton, 1948, Mining Geology: New York, Prentice-Hall, 680 p.

Meyer, Charles, and Hemley, J. J., 1967, Wall rock alteration, *in* Barnes, H. L., ed., Geochemistry of Hydrothermal Ore Deposits: New York, Holt, Rinehart and Winston, p. 166–235.

Meyer, Charles, et al., 1968, Ore deposits at Butte, Montana, *in* Ridge, J. D., ed., Ore Deposits of the United States, 1933–1967: New York, Am. Inst. Mining, Metall. Petroleum Engineers, p. 1375–1416.

Peterson, D. W., 1958, AGI data sheet 23: Geotimes, v. 5, no. 6, p. 30–36.

Pettijohn, F. J., 1957, Sedimentary Rocks, 2d ed., New York, Harper & Bros., 718 p.

Phillips, M. P., and Godwin, C. I., 1970, Geology and rotary drilling at the Casino Silver Mines property: Western Miner, v. 43, p. 43–49.

Rose, A. W., 1970, Zonal relations of wallrock alteration and sulfide distribution at porphyry copper deposits: ECON. GEOL., v. 65, p. 920–936.

Titley, S. R., and Hicks, Carol L., eds., 1966, Geology of the Porphyry Copper Deposits of Southwestern North America: Tucson, University of Arizona Press, 287 p.

Travis, R. B., 1955, Classification of rocks: Colorado School Mines Quart., v. 50, no. 1, 98 p.

Whitten, Graham, 1965, Leached outcrops, *in* Lawrence, L. J., ed., Exploration and mining geology: Commonwealth Mining Metall. Cong., 8th, Australia-New Zealand, v. 2, p. 193–204.

Editor's Comments on Papers 12, 13, and 14

12 PARKS
Sampling Practice

13 COX
Excerpts from *The Use of Comparative Sampling Methods at Cleveland Mine, Tasmania*

14 KOCH and LINK
Excerpts from *Procedures and Precision of Ore Estimation from Assays of Vein Samples*

SAMPLING PROCEDURES

An economic evaluation of a mineral deposit and effective production control depend heavily upon (1) total metal value contained in the mineralized rock, (2) percentages of the values that are recoverable, and (3) cost of metal recovery. The object of a sampling program is to supply this information within an acceptable margin of error at a minimum cost. Many mines have faced bankruptcy because of insufficient sampling or testing; but much wastage has also occurred due to oversampling. A 10 percent error in tonnage estimates may alter the life of a mine by a year or so, but a 10 percent error in grade estimate may result either in bankruptcy or the loss of a profitable mining opportunity.

The higher the unit value and the larger the size of the valuable mineral component, the more difficulty is the sampling procedure. For example, evenly disseminated copper values as small grains of chalcopyrite in some intrusive-hosted porphyry copper deposits may be confidently sampled by drill holes on a 150 meter grid, while a limestone-hosted porphyry copper deposit with irregular pockets of chalcopyrite may require a 30 meter grid spacing of assay holes. Sampling of deposits with erratic, coarse gold presents a great challenge, and diamond pipes can be accurately sampled only by a record of production over several years of operation. A statistical evaluation is required for each deposit to determine optimum sample frequency, size, and method. The objective of this section is to introduce ideas as to sampling techniques, control of sample quality, and evaluation of sample values.

Parks (1949), in his book *Examination and Valuation of Mineral Property,* brings together into concise form the sampling techniques that had been perfected for three types or ore deposits: (1) high-grade base metal veins at Butte, Montana; (2) low-grade porphyry copper deposit at Ajo, Arizona; and (3) banded iron formations in the Lake Superior region. However, in the edited portion of Chapter 6 of the book, included as Paper 12 in this volume, only the first type is included since it presents basic principles.

Early sampling procedures tended to conform to inflexible instructions issued by the mining company, often without analysis as to applicability to various sections of the mine or individual segments of an ore body. Cox (Paper 13) compares results of groove, channel, split core, and bulk samples in a deformed stratiform, massive sulphide, copper-tin deposit. This type of analysis and check on sampling efficiency has become common practice, carried out for each mine and major section of a mine in order to establish the best method for grade control during production and in deposit evaluation.

When has enough sampling been done? Koch and Link seek to answer this question in Paper 14 by examining a heavily sampled vein structure at the Frisco Mine, Chihuahua, Mexico. Their paper brings out the necessity to maintain current records of the correlation between sample assay predictions and production results.

Sampling of low-grade gold lodes requires extreme care and control. As the value of gold continues to escalate, the problems become more severe.Techniques developed at the Rand Mine of South Africa for the evaluation of check samples are outlined by Rowland and Sichel (1960). The Witwatersrand region in which the mine is located has an advantage of very fine-grained gold and good lateral continuity of values. The methods outlined might be less applicable where gold distribution conditions were less reliable.

A comprehensive review of sampling methods and analysis of sample data has been compiled by Jones (1972) and is a valuable reference for the mine geologist's office.

REFERENCES

Jones, M. L., ed., 1972, *Mining and Metallurgical Sampling,* Institution of Mining and Metallurgy, London, 268p.

Parks, R. D., 1949, *Examination and Valuation of Mineral Property,* Addison-Wesley, Reading, Mass. 504p.

Rowland, R. St. J., and H. S. Sichel, 1960, Statistical Quality Control of Routine Underground Sampling, *South African Inst. Mining and Metall. Jour.* **60:**251–284.

12

Reprinted from pages 76-85 and 87-90 of *Examination and Valuation of Mineral Property*, 4th ed., Addison-Wesley, Reading, Mass., 1949, 504p.

SAMPLING PRACTICE

R. D. Parks

Following are descriptions of sampling methods and procedures of the Anaconda Copper Mining Company at their Butte mines, of the Phelps Dodge Corporation at their New Cornelia property, and as generally followed by the Lake Superior iron ore industry. These have been selected as well-established practices, developed through years of use to meet conditions found in certain types of ore formation. Porphyry, vein, and bedded deposits are represented as well as single metal and complex ores. The practices outlined, though used primarily by the respective companies for control of mining operations, will serve as guides to the engineer confronted with the task of estimating mineral values and reserves. These selected examples are not intended to cover the entire field of sampling; to do so would be a comprehensive study in itself.

COMPLEX ORES — BUTTE DISTRICT [1] — The principal ores in the Butte mines are copper, zinc, zinc-lead, and manganese, with minor amounts of complex copper-zinc and zinc-lead-manganese ores. Each type of ore contains an appreciable amount of silver with a few cents in gold. All veins are sampled to obtain information necessary for control in mining operations and for estimating ore reserves.

---

[1] Digest of special report "Sampling Methods and Procedures in the Butte Mines of the Anaconda Copper Mining Company" prepared by Edward P. Shea, Mines Geologist, Butte, Montana, July 1947.

Representative samples are desired but the veins are not homogeneous and it is recognized that a single sample represents only a relatively small amount of material. Therefore, accuracy is restricted to practicability.

Veins Definite veins, where the ore minerals are confined within well-defined walls, are sampled from wall to wall and the country rock broken on either wall is measured and sampled for overbreak control. Where veins are composed of bands of different grade ore, the bands are sampled separately, and respective assays are combined by calculations to obtain a total width and average grade for the exposure.

Disseminated Deposits Disseminated deposits or deposits containing small veinlets or blebs of ore-bearing minerals are sampled in a manner similar to that for the definite vein deposits. The entire face is included and where it contains bands of ore of various grades the individual bands are sampled separately and the assays are combined mathematically to obtain the average grade of the exposure. Because of the irregular distribution of the ore-bearing minerals in such deposits, to obtain the most representative sample, cuts are made from top to bottom and across the entire face.

Car samples, or grab samples taken from the cars as the ore is being loaded from this type of deposit, are considered to be more representative than cut samples, as material from the entire excavation is represented rather than that from the face or outside edge. Also, car samples are weight samples and consequently differences in specific gravity are taken into account.

The grade of ore in disseminated deposits is difficult to estimate by observation and sampling is the only medium by which to obtain the value of the ore. As these deposits are not bounded by definite walls but gradually decrease in value from ore in the center to waste on the sides, sample results supply the only dependable mining control.

Low-Grade Deposits Because of the limited range between ore and waste, marginal low-grade deposits require more careful sampling than the good grade ores. To get a representative sample large sacks are used, and samples approximating five pounds per foot in width are cut. The specific gravity of each sample is determined and weighted averages are obtained by combining the width, grade, and specific gravity. The following example shows the effect of specific gravity on an average of three samples:

OMITTING SPECIFIC GRAVITY

Sample No.	Width	% Cu.	Oz. Ag.	W. × % Cu.	W. × Oz. Ag.
1	4.5′	1.2	0.3	5.40	1.35
2	0.5′	14.0	2.1	7.00	1.05
3	5.0′	1.2	0.4	6.00	2.00
Total & Average	10.0′	1.84	0.44	18.40	4.40

INCLUDING SPECIFIC GRAVITY

Sample No.	Width	% Cu.	Oz. Ag.	Specific Gravity	W. × S.G.	W. × S.G. × % Cu.	W. × S.G. × Oz. Ag.
1	4.5	1.2	0.3	2.80	12.60	15.12	3.78
2	0.5	14.0	2.1	4.16	2.08	29.12	4.37
3	5.0	1.2	0.4	2.80	14.00	16.80	5.60
Total & Average	10.0	2.13	0.48	2.87	28.68	61.04	13.75

Spotty ores, or those having wide differences in copper content, vary somewhat in specific gravity; therefore, the latter factor has more effect on ores of this type than on those having more uniform grade, especially if the gangue has a relatively high content of quartz.

Diamond Drilling Holes are diamond drilled to sample veins and an assay record is made of the cores and sludges. Where core recovery is good, core samples are used in preference to sludge samples. Good core sections through veins are considered to be more representative of the vein cut than the sludge because of losses entailed in slimes. Drillers pull the core when the cuttings indicate a change in the material being drilled, record the distance in the hole, and place a marker in the core box. Sampling of the core is done in a manner similar to crosscut sampling in that different types of material are sampled separately. The length of hole is determined from the driller's markers and the recovered core is measured in the core box. When cores are used to obtain a weighted average value of the material cut, recovered core is used and not length of hole. For example:

RECOVERED CORE

Dist. in Hole	Length of hole	Core Feet	% Cu.	Oz. Ag.	Length of Rec. Core × % Cu.	Length of Rec. Core × Oz. Ag.
101.0′-105.2′	4.2′	3.0	5.7	2.0	17.1	6.0
105.2′-110.6′	5.4′	3.8	3.0	1.0	11.4	3.8
101.0′-110.6′ Av.	9.6′	6.8	4.2	1.4	28.5	9.8

Sludge samples are unsatisfactory and are not used except when core recovery is low. In these cases both core and sludge are saved, weighted, and assayed separately, and results are then combined to get an average. Sludge samples are unsatisfactory for the following reasons:

1. Sludge and heavy sulphide cuttings are lost in crevices in the hole.
2. In soft ground, caving from the walls of the hole contaminates the sample.
3. Source of sludge is not definitely known. (Soft material from any uncased location in the hole may contaminate the cuttings.)
4. Sulphides, because of their weight, are difficult to wash from the hole and salt cuttings beyond ore bands.

In addition to the above, the obtaining of sludge samples is impractical, as it slows the drilling and adds to the cost of the hole.

Face Sampling Before cutting samples, the face, sides and back of a working place are examined to determine if and where samples should be taken. Back samples are taken where face samples have been omitted or where the face samples do not represent the advance from the previous face tested. Drift faces are sampled after every blast and crosscuts are sampled on both sides wherever any vein material is exposed. In disseminated type ore deposits, crosscuts are sampled on both sides for the entire length. Raises are sampled on both ends of every floor in vein deposits and on ends and sides in disseminated ore bodies. Sidesets are sampled on every face that shows any mineralization, except those parallel to the structure.

In general, sample cuts are horizontal and horizontal widths are reported to supply proper data for ore reserve calculations as plotted on vertical plane longitudinal sections. In stopes and raises where the dip of a vein is flatter than 45°, samples are cut at right angles to the vein, and are so reported. In this case, slope distance is used in estimating ore reserves. This condition is illustrated in Fig. 11, where, on the first and second floor, with dip less than 45°, the samples are cut normal to the vein. These are reported as "true width" samples.

In sampling veins containing streaks of varying grade and hardness, it is difficult to cut a single sample composed of equal quantities per foot of the dissimilar streaks or layers. Therefore, each streak is sampled separately and the face average is calculated by combining the individual widths and assays of the several parts. This method gives a weighted average by volume only; specific gravity differences are not considered. Experience has shown that weighting by specific gravity is a refinement not necessary, generally, for operations control.

Streaks are sampled by starting at one edge and uniformly cutting or chipping small pieces continuously across to the other edge. This process is repeated until a sufficiently large sample, at least one pound per foot of width, is obtained. Sep-

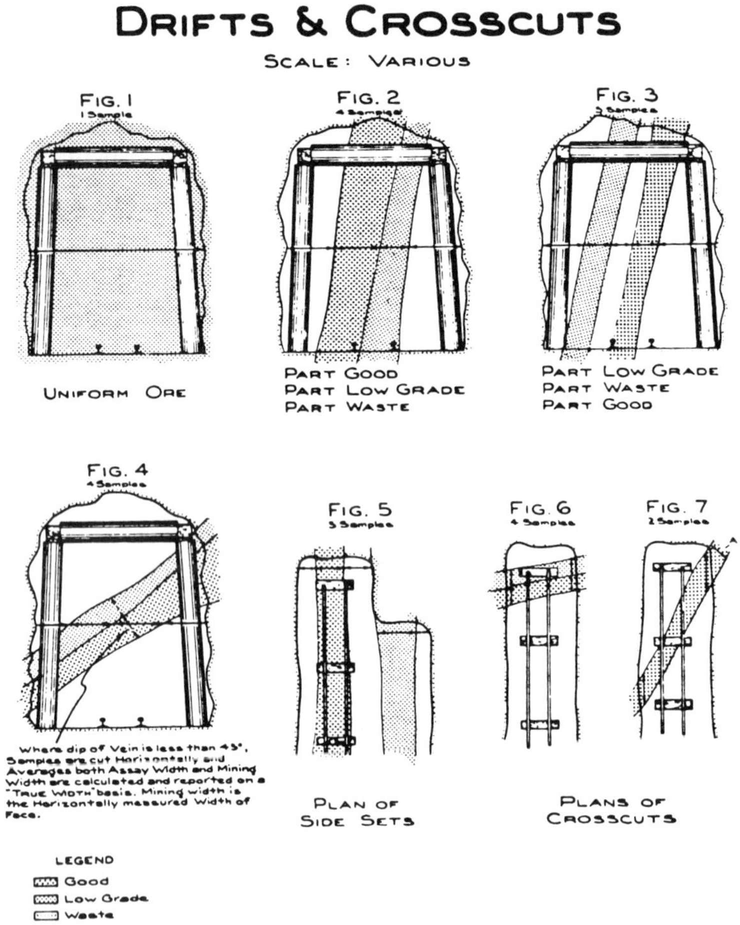

Fig. 13. Sampling of drifts and crosscuts.

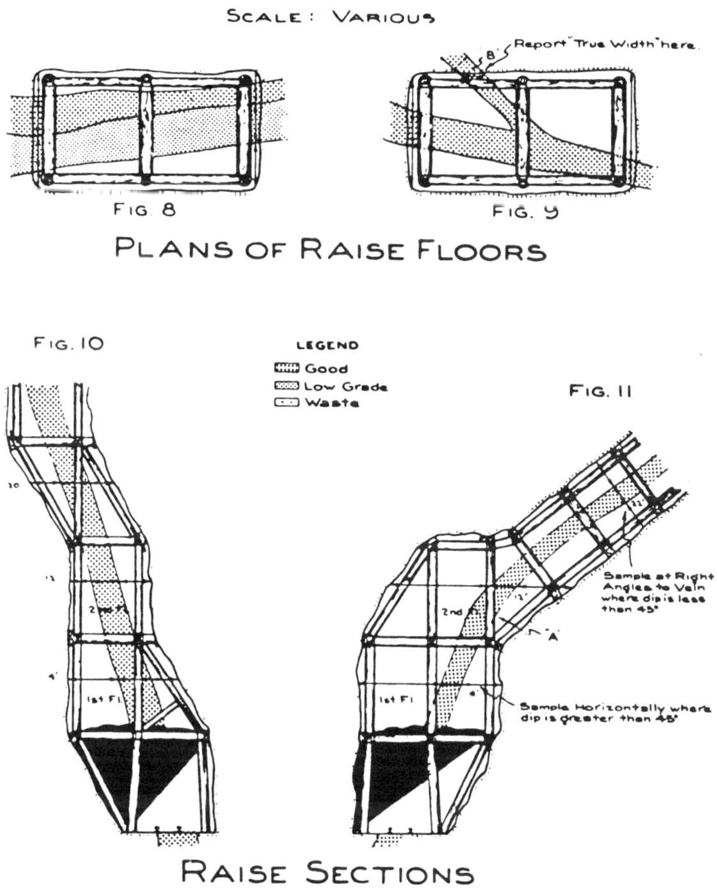

Fig. 14. Sampling of raises.

arate sampling of streaks also gives specific data on location of high- and low-grade layers.

In special cases, where the ore is spotty and low-grade, larger samples of five pounds per foot of width are cut.

Ore samples are limited to five feet maximum width, or length of sample cut; in timbered workings to one set which is equivalent to 5.33 feet. The entire face is sampled in every case, as shown in sketches 1 to 5 and 8 to 11, inclusive, of Figures 13 and 14.

Any section of the face outside the walls of the vein is sampled to provide data for overbreak control even though it may be waste. The sampler checks his assay returns against his estimate of grades and where the difference is too great, the place is resampled.

Judgment, practical experience, and knowledge of principles dictate the sampler's procedure in obtaining representative samples. Where strict adherence to rules would not represent conditions truly, he deviates from prescribed procedure to cut the sample he judges best. For instance, if for some reason a decided change in grade occurs within a short distance, he will cut a sufficient number of samples to show the complete situation. In some cases, this may require separate samples across top and bottom of the face; in disseminated deposits, it may require chipping the entire face at regular intervals, wall to wall and top to bottom, instead of the usual horizontal sample cut four feet above the floor.

Stope Sampling Stopes are sampled at intervals of 10 feet. Ends of stopes are always sampled. In rill stopes the samples are cut across the back.

Silver determination is not made for stope samples. Development, sill, and raise samples are assayed for silver and the silver values in stopes are estimated from the copper-silver ratios in nearby development workings.

When a stope is stopped as unprofitable, the complete back

is sampled at intervals of five to ten feet to check the type of material remaining and to provide a full record for future reference.

Car Sampling Mine car grab samples are taken in special cases. The procedure is to take one pound of "fines" from odd-numbered small (0.7 ton) cars and one pound of coarse material from even-numbered cars. This material is accumulated to form one sample from the total number of cars loaded from each working place during a shift.

When large cars (3.5 tons) are used, five pounds of "fines" and five pounds of coarse material are taken from alternate cars and combined for each working place. The five pounds of material from each of the large cars is composed of five one-pound grabs, one pound being obtained from each corner and one pound from the center of the top of each car.

At the zinc and manganese mines, in addition to the regular cut samples, a sample is also taken from the cars or skips as the ore is hoisted. Where cars are hoisted, approximately one pound per ton is obtained and accumulated until each sample represents 40 to 50 tons. At the mines where skips are used, a five-pound grab is taken from each six-ton skip and the grabs from each eight skips combined into one sample. At the manganese mines where the ore is hauled by truck to the railroad, a sample of about five pounds is taken from each 12-ton truck, and four such samples combined to represent one railroad carload.

Dump Sampling Old mine dumps containing large tonnages are concentrator-tested in a preliminary way by using material obtained by grab sampling from numerous small pits located at regular intervals on the top and sides of the dump. If this sampling indicates profitable possibilities, a large-scale sampling program is inaugurated and power-shovel cuts are excavated at regular intervals. Generally, all material from such cuts to a maximum total of 25,000 tons is shipped, completely sampled and concentrator-tested, and from these results the advisability of shipping a portion or the entire remaining material is determined.

Calculations To obtain the average of ore in the face of a drift or stope, assays of the respective samples are combined on an *assay-width* and *mining-width* basis. Except in special cases they are combined by volume only, with no regard to specific gravity, the results being sufficiently accurate for operating purposes. The assay-width average consists of a combination of the ore streaks and the included waste that cannot be sorted economically from the ore, assuming theoretically perfect mining. The mining-width average is made up from the assays of all the material in the face, that is, ore and waste combined. The assay-width average is made to show the width and grade of the ore in the face, regardless of the width of the mine working. This average is used in estimating ore reserves. The mining-width average shows how the mining is actually being done, thus a comparison of the assay-width average to the mining-width average furnishes a good control on mining operations.

[*Editor's Note:* Material has been omitted at this point.]

Records Complete office records are kept on all sampling. These are for two purposes, (1) operational control, and (2) estimation of ore.

Of particular interest in connection with valuation are those forms and records used for estimation of ore. Form cards are made up for each stope; one for floor plan, one for vertical section, usually longitudinal. Cards are ruled in part with suitable co-ordinates for sketching the desired view to scale. Location of each sample is platted on the sketch of the stope or floor outline as of respective measurement dates. Each stope sketch is tied to mine co-ordinates for location.

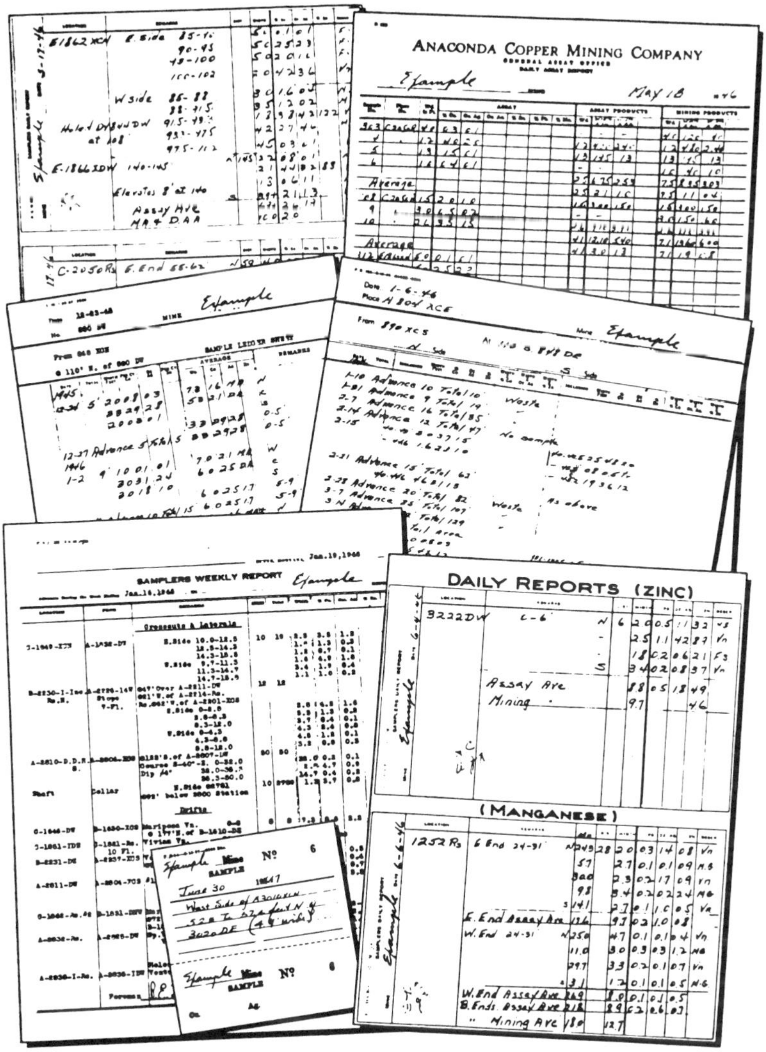

Fig. 15. Sampling records and computation forms.

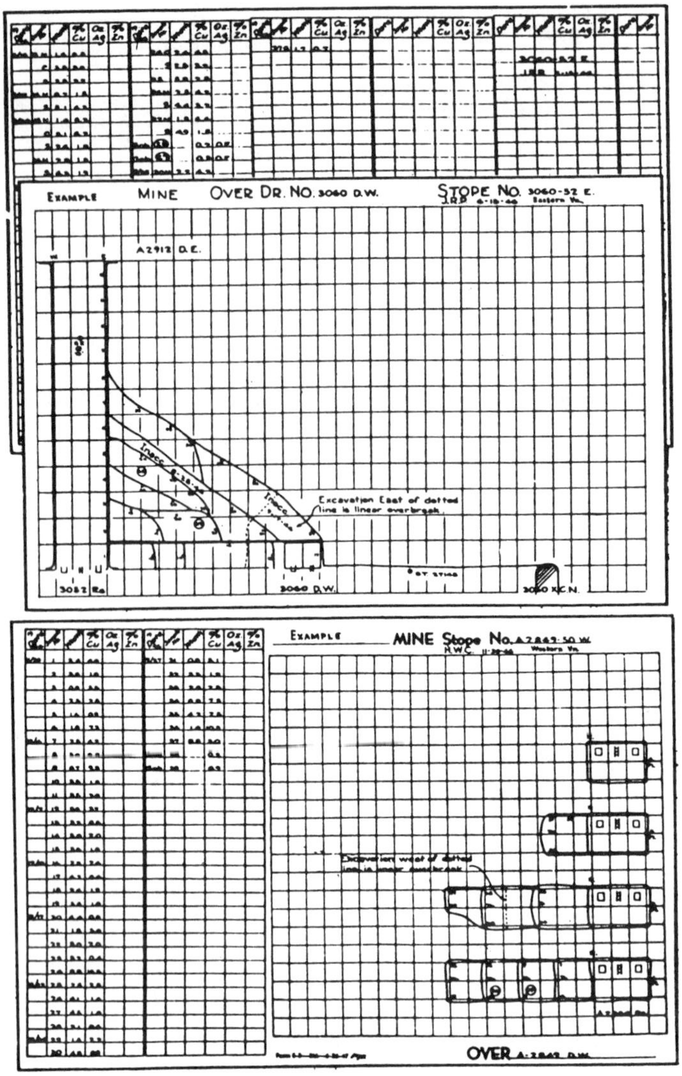

Fig. 16. Stope sampling and records.

Date, sample number, width, and assay are recorded and space is provided also for calculated assay and mining widths (feet × % metal = product) and for amounts of overbreak (waste) or underbreak. Where necessary, as in longitudinal sections of rill stopes, an additional data card may be clipped to the sketch card to provide space for the full record.

Records for operational control consist of sample tags, daily reports from sample and assay departments, ledgers for record, and samplers' weekly reports. Daily reports go to foremen on mine production, to the head of the sampling department, and to the geologist. Weekly reports, showing development, crosscuts and laterals, drifts, raises and winzes, shafts, and diamond drill holes together with total advances and all pertinent computed averages, are forwarded to all interested mine officials and are available to the geologist.

Stope analysis reports are made weekly. These consist of calculated assay-width and mining-width averages of the grade of ore broken in the stopes during the week. This estimate is a weighted average of the total tonnage excavated according to contract measurements in each stope and in all stopes. A length-height factor is determined by dividing cubic feet excavated by mining width. This factor multiplied by average assay-width and mining-width grade gives a product of length × height × width × % copper for both assay-width and mining-width averages. The total L × H × W × % Cu divided by the total L × H × W gives the average % Cu in the ore broken in all stopes during the period. These figures provide good control of operations by showing assay- vs. mining-width averages for each stope and location and quantity of waste broken and shipped with ore.

Each month the Geological Department is furnished a list of stopes shut down, together with mine foreman's reasons for doing so. These are each analyzed against sampling and geological records as to ore possibilities.

13

Reprinted from pages 17-20 and 21-27 of *Australasian Inst. Mining and Metallurgy Proc.* **226:**17-27 (1968)

THE USE OF COMPARATIVE SAMPLING METHODS AT CLEVELAND MINE, TASMANIA

By

R. COX,[1] ASSOCIATE MEMBER

ABSTRACT

The use of comparative sampling methods during the evaluation of the tin-copper deposits at Cleveland mine yielded closely checking average results from groove, channel, AXT split diamond drill core, and bulk sampling for both tin and copper content of the ore.

The weighted mean tin and copper grades and values for the groove, channel, and split core samples showed small percentage differences from bulk sample results, and the *F*-ratio test for single population origin of the four types of samples was accepted at the 99·5 per cent level of confidence.

These results indicate that grade estimates based on the assays from AXT split diamond drill core and/or groove samples would be reliable within practical limits.

INTRODUCTION

The Aberfoyle Tin Development Partnership secured the leases of the Cleveland mine in north-west Tasmania in September, 1961, and immediately commenced a programme of systematic exploration lasting through to 1966. Ultimately, the programme involved 25,207 ft of diamond drilling and 2,980 ft of underground driving and cross-cutting. This resulted in the establishment of indicated and inferred ore reserves totalling 2,850,000 long tons with an average assay of 1·02 per cent Sn and 0·43 per cent Cu. However, it was realized early in the programme that the ore bodies were complex both from the point of view of geological structure and local variation in tin-copper mineralization.

In order to investigate more fully the complex geological structures and to ascertain the degree of reliability that could be placed in the tin and copper grades obtained from sampling diamond drill core, a programme of underground development and comparative sampling was undertaken, on the 1300 ft R.L. Qa adit level, during the period April 1964 to October 1965. This programme involved

1. 2980 ft of driving and cross-cutting over a strike length of 1338 ft,
2. 36 AXT diamond drill holes having a total footage of 3118 ft,
3. taking 756 groove samples from development drives,
4. taking 336 groove samples over 816 ft in 8 cross-cuts,
5. taking 318 channel samples over 756 ft in 8 cross-cuts,
6. taking 346 split diamond drill core samples over 704 ft on 21 cross-sections, and
7. taking 72 bulk samples, having a total weight of 311 tons, over 315 ft in 8 cross-cuts.

In addition to achieving its prime objects, the underground programme provided

1. valuable material for milling and metallurgical research purposes,
2. an opportunity to investigate the nature of the ground with respect to competence, strength, hardness, friability, permeability, and other allied factors important to future mining operations,
3. an opportunity to determine density factors for both broken and unbroken ore for future estimation and operational use, and
4. an opportunity to analyse variations in lode width, tin and copper grade, chert content, and mineralogy of the orebodies over a reasonable strike length.

GEOLOGICAL SUMMARY

The complex geological setting of the Cleveland orebodies was described by Cox and Glasson (1967), and only a brief summary is given here in order to allow an understanding of the sampling problem.

The orebodies are situated in rugged terrain covered with dense rain forest, and outcrop on the steep southern slope of Crescent Spur, in the valley of Deep Creek, between 1100 and 1500 ft R.L. Tin and copper occur with iron sulphides in a lode bed which is stratigraphically comformable with the enclosing rocks. The lode bed varies from less than 1 ft thick to 15 ft thick and occurs within a series of interbedded fine grained, grey shales and grey-chocolate brown cherts, known collectively as Hall's Formation. This formation has a thickness of up to 100 ft in the mine area, and lies between the older Crescent Spur Mica Sandstone and the younger Deep Creek Basic Volcanics. These three formations constitute the Crescent Spur Group of the Cambrian system.

The major ore structure is the overturned SE. limb of a SW. plunging anticline. The lode bed is one stratigraphic horizon which has been displaced along several major NW. dipping axial plane faults into a series of separate lenses. These orebodies occur en echelon, increasing in depth to the south-east. Minor

1 Resident Geologist, Cleveland Tin N.L., Luina, Tasmania.
Original manuscript received at The Institute 27th July 1967. Revised manuscript received 20th November 1967.

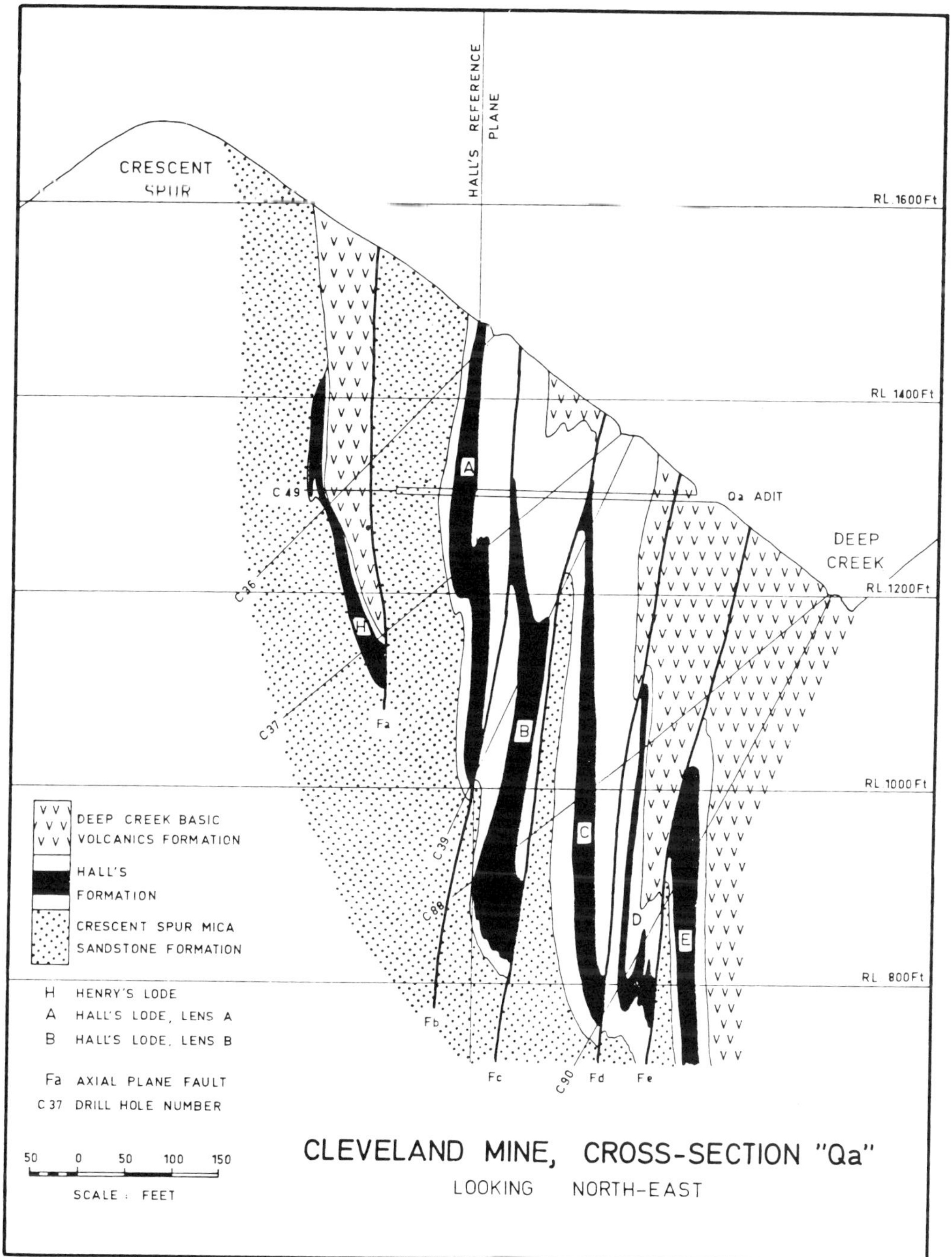

FIG. 1—A typical cross section through the Cleveland mine.

faulting, complementary to the major movement, further complicates the overall structure. A typical cross section of the deposit is shown in Fig. 1.

In any particular intersection an orebody consists of a number of repetitions of the lode bed separated by infolded and/or infaulted bands of chert wall rock. Overall lode width ranges up to 60 ft, and included chert averages approximately 30 per cent by volume. Due to these minor structural complexities the two walls of a cross cut through an orebody may have little resemblance to one another, as may be seen from Fig. 2.

The average lode bed contains approximately 35 per cent sulphides, chiefly pyrrhotite with minor amounts of pyrite, chalcopyrite, marcasite, arsenopyrite, sphalerite, tetrahedrite, and stannite, in a quartz-carbonate groundmass containing minor amounts of chlorite, tourmaline, fluorite, and sericite. Tin occurs chiefly as cassiterite, with only a small proportion in the form of stannite. Copper occurs chiefly as chalcopyrite.

The stratiform ore is uniform in the broad sense only; in detail it is far from uniform. Grain size, concentration, and relative amounts of cassiterite and the various sulphides and other gangue minerals vary rapidly within a few inches. Sulphide mineralization varies from fine disseminations to irregular, massive aggregates and fracture fillings. Cassiterite is very fine grained, averaging approximately 0·2 mm across, and is rarely seen with the naked eye. The ore is usually well banded, hard, and tough.

Sampling, therefore, has to contend with rapidly changing tin-copper content in association with variations in lode bed mineralogy and rock type.

SAMPLING METHODS

In order to determine the relationships between assay results from different types of samples for the Cleveland ore, eight cross cuts (designated K, L, N, P, Qa, R, T, and V), spaced at 120 ft intervals along strike, were driven at right angles to the strike through Hall's Lode, Lens "A", on the 1300 ft R.L. Qa adit level. Close geological mapping and control was an essential part of the programme. Both walls and the back of each cross cut were mapped in detail at scale 1 in. to 5 ft.

Four types of sample were taken on each of the eight cross sections. From each cross cut groove and channel samples were cut along both walls; a horizontal AXT diamond drill hole was drilled as close as practicable to one of the walls at the same elevation as the groove-channel sample location; and a bulk sample was cut as a slot from the back.

The comparative sampling programme was based upon sound sampling principles, of which the following points should be noted.

1. All sampling was carried out with geological control. Standard lengths of cut were never used, but each sample length was laid out on the basis of rock type and mineralization, with the aim of achieving a high degree of uniformity within each length. Sample

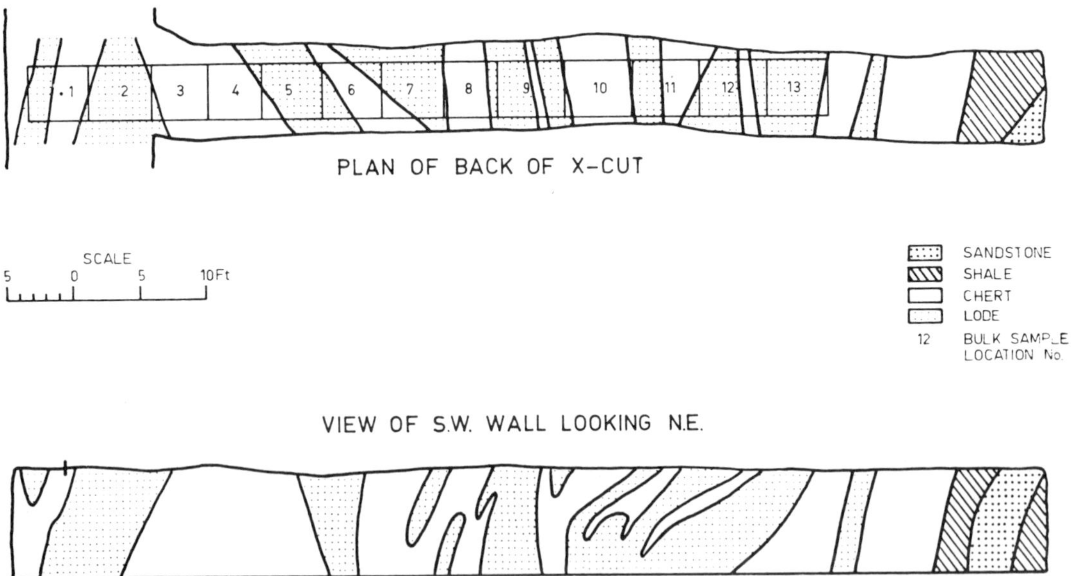

FIG. 2—Plan and section of V cross cut through Hall's Lode, lens "A", on 1300 ft R.L. Qa adit level, Cleveland mine.

lengths therefore varied from a few inches up to a maximum of 48 in. This factor is particularly important in groove and channel sampling, where even an imperfectly cut sample will tend to produce a relatively reliable result if each sample length traverses reasonably uniform material. This ensures that confidence can be placed in the weighted average grade. This condition was not applicable to the taking of bulk samples.

2. All sample locations were carefully marked with paint and are referable to the mine survey grid.

3. All sampling was carried out with thorough supervision. Prior to sampling, all sample locations were thoroughly cleaned. After sampling, all sample cuts were inspected before the samples were accepted for preparation and assaying.

4. All groove, channel, and split core samples were taken by the same reliable sampler and were placed in a new, clean canvas bag with the identification number marked on the outside in ink, and on an aluminium foil marker placed inside the bag. This system insured well against samples becoming unidentifiable.

5. Groove and channel samples were cut from the same sample location in order to facilitate direct correlation of assays and evaluate the reliability of groove sampling for future mining requirements.

These factors ensured that reliable and representative samples were obtained, and that "sampling errors" were kept to the minimum practical limit.

Groove samples

Groove samples, consisting of a 1 in. wide by ½ in. deep groove, were cut by hand using a hammer and moil from both walls of each cross cut, over the full width of the ore body. A total of 518 ft of groove samples were cut from the walls of the eight cross cuts through lens "A". Just under ¾ lb of sample were obtained per foot of cut.

Channel samples

Channel samples, consisting of a 6 in. wide by 2 in. deep channel, were cut by hand from the same sample locations as used for groove sampling. A total of 494 ft of channel samples were cut from the walls of the eight cross cuts. Approximately 17 lb of sample were obtained per foot of cut.

The cuttings from groove and channel samples were caught in a specially designed canvas sheet held firm by the sampler's body against the cross cut wall, thus leaving both hands free for the task of cutting.

Split core samples

Split core samples were taken from AXT diamond drill holes drilled as close as practicable to one wall of each cross cut, and at the same horizon (4 ft above rail level) as the groove/channel sample locations. Every effort was thus made to ensure that split core and groove/channel sample locations penetrated near-identical ground. This was essential if good correlations between rock type, mineralization, and tin and copper grade were to be obtained, in view of the earlier comments regarding the complicated minor structures within each orebody. A total of 273 ft of split core samples were taken from the eight drill holes through lens "A". AXT core has a diameter of 1·28 in., and just under 1 lb of sample was obtained per foot of split core. Core recovery at Cleveland mine is consistently good, and sludge samples are not taken. Core recovery for the eight intersections of lens "A" was between 95 and 100 per cent for all drill runs.

Bulk samples

Bulk samples were bored and fired from the back of each cross cut, and collected on a series of tightly-fitting sollar boards. Each sample was taken separately, and measured 4 ft wide by 2 ft 6 in. deep; the length varied between 3 ft 6 in. and 5 ft.

Sixty-six bulk samples having a total footage of 292 ft were taken from the eight cross cuts through lens "A". Approximately 2000 lb of sample were obtained per foot of sample length. Bulk sample locations were marked up with paint, then bored, fired, and hand-bogged under close control and supervision. Each truck of broken material was carefully identified and then weighed at the adit portal before being transported to the sample reduction plant.

[*Editor's Note:* Material has been omitted at this point.]

ANALYSIS OF COMPARATIVE SAMPLING RESULTS

The prime objective of the comparative sampling programme was to investigate the relationship between the assay results obtained from different types of samples. For evaluation purposes, the relationship between the assay results obtained from the sampling of split diamond drill core and the values obtained from bulk samples had to be investigated. For possible future mining operations procedure, the relationship between the assay results obtained from groove and channel samples and the values obtained from bulk samples had to be investigated. The Cleveland programme bears many similarities to the one carried out at the Kilembe copper-cobalt deposit in Uganda, described by Davis (1962).

As a result of the complex minor structures of the Cleveland orebodies two sample locations, which penetrate an orebody and are separated by only a few feet, have not only different grades but also different overall widths. Fig. 2 illustrates this point. For a further example, the overall horizontal width of lens "A" measured in groove SW. wall, groove NE. wall, diamond drill hole and bulk sample locations is 283 in., 301 in., 180 in., and 318 in. in the Qa cross cut. These locations are all within 12 ft of the centre line of the cross cut. One further point to note here is that the bulk sample slots were cut to include approximately 6 in. of barren wall-rock at each end. It was considered preferable to overbreak and include a little wall-rock, so diluting the lode grade by a small percentage, rather than underbreak. Overbreak varied between 0·8 per cent for the V cross cut and 4·9 per cent for the L cross cut.

In estimating ore reserve grade for this type of deposit, the weighted average grade and overall lode width are equally important. The important value for each sample intersection of an orebody is therefore the metal content or value, where "value" = product (width × grade). In analysing the results of the comparative sampling programme, this point should be borne in mind.

The comparative sampling results for Hall's Lode, lens "A", in each of the eight cross cuts on the 1300 ft R.L. Qa adit level are summarized in Tables 1 and 2.

Comparison of weighted means for four types of sample

The weighted mean tin and copper grades and values for groove samples (both walls individually and combined), channel samples (both walls individually and combined), AXT split diamond drill core samples, and bulk samples obtained from the eight cross cuts are analysed in Table 3. These data were extracted from Tables 1 and 2.

The mean tin and copper grades and values for the various groove, channel, and split core samples are compared with those obtained for the bulk samples, which were the largest samples it was practicable to obtain. The following points are noted.

1. For the overall lode tin grade (per cent Sn), the percentage difference from the bulk sample grade has a range from −7·4 per cent (channel samples, NE. wall) to +12·3 per cent (channel samples, SW. wall).

2. For the overall lode copper grade (per cent Cu), the percentage difference from the bulk sample grade has a range from −3·3 per cent (groove samples, SW. wall) to +10·0 per cent (split core samples).

3. These differences in lode grade, as mentioned earlier, include variations in overall lode width for the different types of sample cut from near identical locations. Thus a more meaningful result is obtained from an analysis of the relationships between tin and copper content, or value, obtained from the four types of sample.

4. For the overall lode tin value, the percentage difference from the bulk sample value has a range from −13·7 per cent (channel samples, NE. wall) to +7·3 per cent (channel samples, SW. wall). The groove (average both walls), channel (average both walls), and split core samples show differences from

TABLE 1

Comparative sampling data—overall lode value Halls lode, lens "A" on 1300 ft R.L. "Qa" adit level

Cross-section	Groove						Channel						Split core		Bulk	
	Lode value (Product W × G)						Lode value (Product W × G)						Lode value		Lode value	
	SW. wall		NE. wall		Av. both walls		SW. wall		NE. wall		Av. both walls					
	Sn	Cu	Sn	Cu	Sn	Cu	Sn	Cu	Sn	Cu	Sn	Cu	Sn	Cu	Sn	Cu
V	190·86	84·82	92·13	93·15	141·50	88·99	292·70	102·76	107·18	100·91	199·94	101·84	382·55	139·80	285·60	128·52
T	383·73	115·79	383·73*	115·79*	383·73	115·79	342·85	112·93	342·85*	112·93*	342·85	112·93	301·84	153·86	363·54	109·56
R	480·46	145·08	390·61	118·08	435·54	131·58	507·50	149·92	473·18	142·64	490·34	146·28	453·91	140·83	461·70	159·60
Qa	248·50	125·60	337·50	163·70	293·00	144·65	285·94*	179·70*	285·94	179·70	285·94	179·70	249·35	129·19	286·20	178·08
P	430·62	189·60	410·74	222·69	420·68	206·15	455·26	188·85	371·44	170·00	413·35	179·43	396·00	180·70	503·04	212·22
N	528·30	158·06	494·76	123·82	511·53	140·94	541·02	148·04	503·68	146·31	522·35	147·18	506·01	158·76	512·46	135·78
L	50·36	46·04	69·52	16·13	59·94	31·09	71·32	46·38	99·72	55·52	85·52	50·95	137·42	47·06	109·44	41·76
K	473·35	107·08	379·49	169·78	426·42	138·43	585·36	153·08	292·72	153·42	439·04	153·25	382·84	129·76	349·32	115·02
Total	2786·18	972·07	2558·48	1023·14	2672·34	997·62	3081·95	1081·66	2476·71	1061·43	2779·33	1071·56	2809·92	1079·96	2871·30	1080·54
Mean	348·27	121·51	319·81	127·89	334·08	124·70	385·24	135·21	309·59	132·68	347·42	133·95	351·24	135·00	358·91	135·07
Total width	3387 in.		3312 in.		3349½ in.		3405 in.		3312 in.		3358½ in.		3276 in.		3501 in.	
Av. width	423½ in.		414 in.		418½ in.		425½ in.		414 in.		420 in.		409½ in.		438 in.	
Wt. mean assay	0·82 %Sn	0·29 %Cu	0·77 %Sn	0·31 %Cu	0·80 %Sn	0·30 %Cu	0·91 %Sn	0·32 %Cu	0·75 %Sn	0·32 %Cu	0·83 %Sn	0·32 %Cu	0·86 %Sn	0·33 %Cu	0·81 %Sn	0·30 %Cu

* Samples not taken: values from other wall are employed.

TABLE 2

Comparative sampling data—overall lode grade Halls lode, lens "A" on 1300 ft R.L. "Qa" adit level

A Groove

Cross-section	SW. wall			NE. wall			Av. both walls		
	length (in.)	%Sn	%Cu	length (in.)	%Sn	%Cu	length (in.)	%Sn	%Cu
V	717	0·27	0·12	595	0·15	0·16	656	0·22	0·14
T	485	0·79	0·24	485	0·79*	0·24*	485	0·79	0·24
R	571	0·84	0·25	544	0·72	0·22	557½	0·78	0·24
Qa	283	0·88	0·44	301	1·12	0·54	292	1·00	0·50
P	409	1·05	0·46	408	1·01	0·55	408½	1·03	0·50
N	433	1·22	0·37	431	1·15	0·29	432	1·18	0·33
L	140	0·36	0·33	145	0·48	0·11	142½	0·42	0·22
K	349	1·36	0·31	403	0·94	0·42	376	1·13	0·37
Total	3387	—	—	3312	—	—	3349½	—	—
Average	423½	—	—	414	—	—	418½	—	—
Wt. av.	—	0·82 %Sn	0·29 %Cu	—	0·77 %Sn	0·31 %Cu	—	0·80 %Sn	0·30 %Cu

B Channel

Cross-section	SW. wall			NE. wall			Av. both walls		
	length (in.)	%Sn	%Cu	length (in.)	%Sn	%Cu	length (in.)	%Sn	%Cu
V	717	0·41	0·14	595	0·18	0·17	656	0·30	0·16
T	485	0·71	0·23	485	0·71*	0·23*	485	0·71	0·23
R	571	0·89	0·26	544	0·87	0·26	557½	0·88	0·26
Qa	301	0·95*	0·60*	301	0·95	0·60	301	0·95	0·60
P	409	1·11	0·46	408	0·91	0·42	408½	1·01	0·44
N	433	1·25	0·34	431	1·17	0·34	432	1·21	0·34
L	140	0·51	0·33	145	0·69	0·38	142½	0·60	0·36
K	349	1·68	0·44	403	0·73	0·38	376	1·17	0·41
Total	3405	—	—	3312	—	—	3358½	—	—
Average	425½	—	—	414	—	—	420	—	—
Wt. av.	—	0·91 %Sn	0·32 %Cu	—	0·75 %Sn	0·32 %Cu	—	0·83 %Sn	0·32 %Cu

Cross-section	Split core			Bulk		
	length (in.)	%Sn	%Cu	length (in.)	%Sn	%Cu
V	702	0·54	0·20	714	0·40	0·18
T	511	0·59	0·30	498	0·73	0·22
R	584	0·78	0·24	570	0·81	0·28
Qa	180	1·39	0·72	318	0·90	0·56
P	406	0·98	0·45	393	1·28	0·54
N	421	1·20	0·38	438	1·17	0·31
L	133	1·03	0·35	144	0·76	0·29
K	339	1·34	0·47	426	0·82	0·27
Totals	3276	—	—	3501	—	—
Averages	409½	—	—	437½	—	—
Wt. av.	—	0·86 %Sn	0·33 %Cu	—	0·81 %Sn	0·30 %Cu

* Samples not taken: values from other wall employed

TABLE 3

Comparative sampling data—overall lode grade and value, Halls lode, lens "A" on 1300 ft R.L. "Qa" adit level: analysis of weighted means for four types of sample from eight cross-sections

Types of samples		Weighted mean grade %Sn	Percent difference from bulk sample grade	Weighted mean grade %Cu	Percent difference from bulk sample grade	Weighted mean value Sn	Percent difference from bulk sample value	Weighted mean value Cu	Percent difference from bulk sample value
Bulk		0·81	—	0·30	—	358·91	—	135·07	—
DDH S/C		0·86	+ 6·1	0·33	+10·0	351·24	− 2·1	135·00	− 0·1
Groove	SW.	0·82	+ 1·2	0·29	− 3·3	348·27	− 3·0	121·51	−10·0
	NE.	0·77	− 4·9	0·31	+ 3·3	319·81	−10·9	127·89	− 5·3
	Av. both	0·80	− 1·2	0·30	Zero	334·08	− 6·9	124·70	− 7·7
Channel	SW.	0·91	+12·3	0·32	+ 6·7	385·24	+ 7·3	135·21	+ 0·1
	NE.	0·75	− 7·4	0·32	+ 6·7	309·59	−13·7	132·68	− 1·8
	Av. both	0·83	+ 2·5	0·32	+ 6·7	347·42	− 3·2	133·95	− 0·8

the bulk sample value of −6·9 per cent, −3·2 per cent and −2·1 per cent respectively.

5. For the overall lode copper value, the percentage difference from the bulk sample value has a range from −10·0 per cent (groove samples, SW. wall) to +0·1 per cent (channel samples, SW. wall). The groove (average both walls), channel (average both walls), and split core samples show differences from the bulk sample value of −7·7 per cent, −0·8 per cent and −0·1 per cent respectively.

The significance of these means and percentage differences can be ascertained by testing for the assumption that the four types of sample (groove and channel, both walls averaged, split core, and bulk) have equal means.

Test for assumption that the four types of sample have equal means

The F-ratio can be used (Hazen, 1958) to test for significant differences among two or more means. When four random samples, each containing several items, are taken from a perfectly homogeneous population the means of the four samples will not be the same due to variations within the parent population. It is expected, however, that the between-sample variation will not be significantly greater than the within-sample variation. In analysing the variance, the sum of the squares of the deviations of the items from the mean and the number of degrees of freedom for each source of variability are computed.

The first estimate of variance is made by the pooled method. For a set of four samples containing several items, the formula for the pooled estimate of the within-sample variance is

$$S_p^2 = \frac{\Sigma(X_1 - M_1)^2 + \Sigma(X_2 - M_2)^2 + \Sigma(X_3 - M_3)^2 + \Sigma(X_4 - M_4)^2}{n_1 + n_2 + n_3 + n_4 - 4}$$

where n_1, n_2, n_3, n_4 are the number of items in each of the four samples and $\Sigma(X_1 - M_1)^2$ equals the summation of the squares of the differences between the individual items and the mean for the first sample. This formula has been simplified to

$$S_p^2 = \Sigma X_{iT}^2 - \left(\frac{T_1^2}{n_1} + \frac{T_2^2}{n_2} + \frac{T_3^2}{n_3} + \frac{T_4^2}{n_4}\right)$$

where ΣX_{iT}^2 represents the sum of the squares of the total number of all items in the samples and T_1^2 equals the square of the total of all the n_1 items in the first sample.

If we assume that the groove, channel, split core, and bulk samples represent four random samples taken from a perfectly homogeneous population, then for the overall lode tin values for groove (average both walls), channel (average both walls), split core, and bulk samples given in Table 4

$$S_p^2 = 1{,}158{,}367 + 1{,}068{,}026 + 1{,}122{,}715 + 1{,}084{,}320 - \left(\frac{(2871)^2}{8} + \frac{(2674)^2}{8} + \frac{(2779)^2}{8} + \frac{(2810)^2}{8}\right)$$
$$= 4{,}433{,}428 - 3{,}876{,}483 = 556{,}945.$$

The second estimate of the variance is made by directly computing the variance of the means. This is the variation between sample means and can be computed by the formula:

$$S_m^2 = \frac{T_1^2}{n_1} + \frac{T_2^2}{n_2} + \frac{T_3^2}{n_3} + \frac{T_4^2}{n_4} - \frac{T_T^2}{N}$$

where T_T^2 is the square of the total of all the N items in all four samples. Again from Table 4, for the overall lode tin values

$$S_m{}^2 = \frac{(2871)^2}{8} + \frac{(2674)^2}{8} + \frac{(2779)^2}{8} + \frac{(2810)^2}{8} - \frac{(11134)^2}{32}$$
$$= 3{,}876{,}483 - 3{,}873{,}936 = 2{,}547$$

Total variation equals the sum of the variation between means and the variation within sample groups.

$$\text{Total variation} = \Sigma X_{iT}{}^2 - \frac{T_T{}^2}{N}$$

For overall lode tin values,
total variation = 4,433, 428 − 3,873,936 = 559,492

The number of degrees of freedom for the between-sample means is 3, or one less than the number of sample means. The number of degrees of freedom for the within-sample groups variation is 28, which is (N-K) or 4 less than the total number of items N, where K equals 4, which is the number of samples. The mean square is found by dividing the sum of squares by the degrees of freedom. These mean squares are the two estimates of the variance of the parent population. The mean square obtained from the between-sample means is always used as the numerator when calculating the F-ratio, because the means are rejected as significant only if they are more spread out than would be expected for samples from the population. The data for the overall lode tin and copper values and grades are summarized in Tables 4, 5, and 6.

If the 99·5 per cent confidence level is selected for rejection, there is evidence for rejection if the F-ratio is larger than that which would be due to chance 99·5 per cent of the time. The computed F-test ratios of 0·04 and 0·10 for the tin and copper values respectively, and 0·41 and 0·37 for the tin and copper grades respectively, are less than the 99·5 per cent value of 5·32, taken from tables (Hazen, 1958, Table 29E). Therefore the assumption of equal sample means, or single population origin of the samples, is accepted for tin and copper grades and values, subject to the limitations of the initial assumptions.

The F-ratio test for single population origin of the samples has been employed, although the original volumes of the four types of samples were not equal, because all sample were reduced to a pulverized sample of 1-2 lb. This pulverized volume is the effective sample volume, and the variance in the assay data is based on the actual pulverized volume and not on the original sample volume taken underground.

CONCLUSIONS

The evaluation of the complex tin-copper deposits at Cleveland Mine presented challenging sampling and geological problems which were investigated with a detailed programme of underground development and comparative sampling. The most important conclusions drawn from this programme may be summarized as follows.

1. The mean tin and copper grades and values for groove, channel, and AXT split core samples showed small percentage differences from bulk sample results which, due to their large size, were employed in the capacity of referee samples.
2. The F-ratio test for single population origin of the four types of samples was accepted at the 99·5 per cent confidence level for the mean tin and copper grades and values.
3. Ore reserve estimates based largely on the re-results of assays from AXT split diamond drill core and/or groove samples would be reliable within close practical limits, and the application of any sort of correction factor to such estimates would not be required for the ground tested.
4. The value of the results of an involved programme of evaluation depends enormously on the quality of the data obtained. The strict control and supervision of all aspects of sampling and sample preparation, and the practical steps taken to ensure that a minimum number of errors were introduced into the samples from these sources, contributed in large measure to the success of the programme.
5. The complex geological structure was more fully appreciated, thus allowing geological interpretations and ore reserve tonnage estimates to be completed with a high degree of confidence.

ACKNOWLEDGEMENTS

The author wishes to express his thanks to Cleveland Tin N.L. for permission to publish this paper, and to Mr A. A. C. Mason (Manager, Aberfoyle Tin Development Partnership, and Project Director, Cleveland Tin N.L.), Mr K. R. Glasson (Consultant Geologist), and Scott W. Hazen, Jnr. (Chief, Mine Systems Engineering, United States Bureau of Mines) for helpful advice.

The author, in the capacity of Resident Geologist for the Aberfoyle Tin Development Partnership (A.T.D.P.), was directly responsible to the manager for all geological control and all sampling, except bulk sampling.

The success of the programme was due to the active cooperation of the small staff of A.T.D.P. which also included Messrs N. A. Gilberthorpe (Project Manager), H. S. Fraser (Project Supervisor), J. Cox (Junior Mining Engineer), and W. Baldock (Electrician/Mechanic).

The cutting and preparation of all groove, channel, and split core samples were carried out by Mr G. Keygan, whose integrity and reliability were much valued.

REFERENCES

Cox, R., and Glasson, K. R., 1967. The geology and mineralization of Cleveland mine, in *The Geology of Western Tasmania* (Symposium, University of Tasmania, Hobart, November, 1967).

Davis, G. R., 1962. Results of comparative sampling methods at Kilembe mine, Uganda, during the exploration phase, *Bull. Instn. Min. Metall.*, No. 673: 145-164.

Davis, G. R., 1963. Observations on sample preparation, *Bull. Instn. Min. Metall.*, No. 674: 255-267.

Hazen, S. W., Jnr., 1958. A comparative study of statistical analysis and other methods of computing ore reserves, utilizing analytical data from Maggie Canyon manganese deposit, Artillery Mountains region, Mohave County, Ariz., United States Bureau of Mines, Report of Investigations No. 5375.

TABLE 4

Comparative sampling data—overall lode value Halls lode, lens "A" on 1300 ft R.L. "Qa" adit level: test for assumption that several samples have equal means

Cross-section	Bulk	Groove	Channel	Split-core
	Sn value	Sn value	Sn value	Sn value
V	286	142	200	383
T	364	384	343	302
R	462	436	490	454
Qa	286	293	286	249
P	503	421	413	396
N	512	512	522	506
L	109	60	86	137
K	349	426	439	383
Totals	2871	2674	2779	2810
	Cu value	Cu value	Cu value	Cu value
V	129	89	102	140
T	110	116	113	154
R	160	132	146	141
Qa	178	145	180	129
P	212	206	179	181
N	136	141	147	159
L	42	31	51	47
K	115	138	153	130
Totals	1082	998	1071	1081

TABLE 5

Comparative sampling data—overall lode grade, Halls lode, lens "A" on 1300 ft R.L. "Qa" adit level: test for assumption that several samples have equal means

Cross-section	Groove	Channel	Split-core	Bulk
	Sn grade	Sn grade	Sn grade	Sn grade
V	0·22	0·30	0·54	0·40
T	0·79	0·71	0·59	0·73
R	0·78	0·88	0·78	0·81
Qa	1·00	0·95	1·39	0·90
P	1·03	1·01	0·98	1·28
N	1·18	1·21	1·20	1·17
L	0·42	0·60	1·03	0·76
K	1·13	1·17	1·34	0·82
Totals	6·55	6·83	7·85	6·87
	Cu grade	Cu grade	Cu grade	Cu grade
V	0·14	0·16	0·20	0·18
T	0·24	0·23	0·30	0·22
R	0·24	0·26	0·24	0·28
Qa	0·50	0·60	0·72	0·56
P	0·50	0·44	0·45	0·54
N	0·33	0·34	0·38	0·31
L	0·22	0·36	0·35	0·29
K	0·37	0·41	0·47	0·27
Totals	2·54	2·80	3·11	2·65

TABLE 6

Comparative sampling data—overall lode Halls lode, lens "A", eight cross-sections on 1300 ft R.L. "Qa" adit level: test for assumption that the four sample sizes have equal means

	Tin values				Copper values			
	Sum of squares	Degrees of freedom	Mean square	F-ratio	Sum of squares	Degrees of freedom	Mean square	F-ratio
S_m^2	2547	3	849	$F = \frac{849}{19890} = 0{\cdot}04$	610	3	203	$F = \frac{203}{2137} = 0{\cdot}10$
S_r^2	556945	28	19890	$F_{0.995}(3–28) = 5{\cdot}32$	59,848	28	2137	$F_{0.995}(3–28) = 5{\cdot}32$
Total	559492	31			60,458	31		
	Tin grades				Copper grades			
S_m^2	0·1210	3	0·0403	$F = \frac{0{\cdot}0403}{0{\cdot}0975} = 0{\cdot}41$	0·0230	3	0·0077	$F = \frac{0{\cdot}0077}{0{\cdot}0208} = 0{\cdot}37$
S_r^2	2·7287	28	0·0975	$F_{0.995}(3–28) = 5{\cdot}32$	0·5825	28	0·0208	$F_{0.995}(3–28) = 5{\cdot}32$
Total	2·8497	31			0·6055	31		

14

Reprinted from pages 1, 3, 4, 5, 7, 8–9, 13–17, and 25 of *Computer Short Course and Symposium on Mathematical Techniques and Computer Applications in Mining and Exploration,* vol. 1, University of Arizona, College of Mines, Tucson, Ariz., 1962, pp. L1-1-29

PROCEDURES AND PRECISION OF ORE ESTIMATION FROM ASSAYS OF VEIN SAMPLES

G. S. Koch, Jr. and R. F. Link

Desirable though it would be to know the exact grade and metal content of an orebody as soon as it is found, for reasons of expense it is only possible to take samples and to make an estimate. With this estimate of grade and metal content in hand, one mines the orebody and only afterwards learns the true values. The calculation of average values from assays of vein samples is considered in this article through the interpretation of assays for gold, silver, lead, copper, and zinc made on samples from the Frisco mine, Chihuahua, Mexico. Conclusions are drawn which can be applied to veins elsewhere, to other types of orebodies, and to geological entities in general.

[*Editor's Note:* Material has been omitted at this point.]

The Frisco mine works some 50 fissure veins that contain the sulfide minerals sphalerite, galena, chalcopyrite, pyrite, and arsenopyrite. At least part of the gold occurs in the native state; the silver-bearing mineral is unknown. The mine occupies a block of ground about 3 km long, 500 m wide, and 800 m deep (Koch, 1956).

Two measures of metal in a vein are used, grade and metal content (width multiplied by grade). Grades of precious metals are measured in grams per ton and grades of base metals in percentage.

[*Editor's Note:* Material has been omitted at this point.]

The assays investigated in detail were taken from the Brown vein exposed in a 756 m long drift on Level 9 and a 930 m long drift on Level 10. This vein was chosen because the writers had assays available from intermediate drifts and stopes as well as from level drifts. Results from another vein (Don Pablo vein, Level 14) chosen at random were consistent.

For each original sample taken at a 2 m interval, an IBM card (Figure 1) was punched to record the following: vein width; assays of gold, silver, lead, copper, and zinc; elevation; level; assay sheet number; serial number; vein number; and mine coordinates. Metal contents were then computed by machine (Koch and Link, 1960).

[*Editor's Note:* Material has been omitted at this point.]

Table 1. Means of Samples Taken at Various Intervals--Frisco Mine, Brown Vein, Level 9 Drift

Sample Interval	No. Samples	Width	Au	Ag	Pb	Cu	Zn	Au x w	Ag x w	Pb x w	Cu x w	Zn x w
2 m	378	1.59	1.70	271	10.7	0.56	7.2	2.73	440	17.4	0.97	11.8
4 m	189	1.61	1.79	272	11.0	0.56	7.1	2.90	441	18.1	0.96	11.8
	189	1.58	1.61	271	10.3	0.56	7.3	2.59	439	16.7	0.97	11.8
8 m	95	1.58	1.55	285	11.4	0.56	6.9	2.42	446	18.4	0.97	11.3
	94	1.65	2.04	258	10.7	0.55	7.3	3.31	436	17.9	0.96	12.3
	95	1.59	1.34	245	9.9	0.59	7.0	2.40	401	16.4	1.00	11.2
	94	1.56	1.88	297	10.7	0.54	7.6	2.79	477	17.1	0.94	12.3
16 m	48	1.56	1.46	295	12.0	0.57	7.5	2.40	440	18.8	0.95	12.2
	47	1.60	1.64	274	10.8	0.56	6.5	2.41	451	17.9	0.98	10.3
	47	1.61	1.70	239	10.9	0.58	7.2	2.64	401	17.9	0.99	11.8
	47	1.69	2.38	278	10.5	0.52	7.4	3.99	471	18.0	0.94	12.8
	48	1.53	1.15	255	10.2	0.61	7.4	1.96	402	16.4	1.00	11.6
	47	1.64	1.53	235	9.6	0.56	6.6	2.84	400	16.3	1.01	10.8
	47	1.55	1.92	339	10.4	0.51	6.9	2.61	561	16.7	0.87	11.1
	47	1.58	1.85	255	10.9	0.58	8.3	2.96	393	17.4	1.02	13.6
32 m	24	1.56	1.88	334	12.8	0.43	7.3	3.35	528	21.5	0.75	12.5
	24	1.55	1.04	257	11.2	0.70	7.7	1.44	352	16.1	1.15	11.8
	24	1.57	2.17	291	11.1	0.65	6.1	3.07	463	17.9	1.11	9.9
	23	1.64	1.09	257	10.5	0.47	6.8	1.78	439	17.9	0.85	10.8
	24	1.62	2.08	241	10.6	0.49	7.3	3.06	420	18.1	0.88	12.1
	23	1.59	1.30	236	11.2	0.67	7.1	2.20	381	17.6	1.10	11.5
	24	1.72	3.13	307	11.6	0.64	7.1	4.96	516	20.8	1.14	12.4
	23	1.65	1.61	248	9.3	0.39	7.8	2.97	424	15.8	0.73	13.3
	24	1.59	0.63	234	10.8	0.73	7.1	1.20	380	17.9	1.17	11.8
	24	1.48	1.67	277	9.6	0.49	7.7	2.73	423	14.9	0.83	11.4
	24	1.62	1.88	239	10.3	0.53	6.6	3.55	388	17.0	0.95	10.3
	23	1.67	1.17	230	8.8	0.60	6.6	2.09	414	15.6	1.07	11.2
	24	1.36	2.50	233	10.0	0.44	6.9	3.15	309	13.8	0.67	9.8
	23	1.74	1.30	450	10.9	0.57	6.9	2.04	824	19.8	1.07	12.5
	24	1.66	2.50	261	9.6	0.56	7.6	3.98	420	16.0	0.99	12.7
	23	1.49	1.17	248	12.3	0.60	9.0	1.90	364	19.0	1.04	14.5

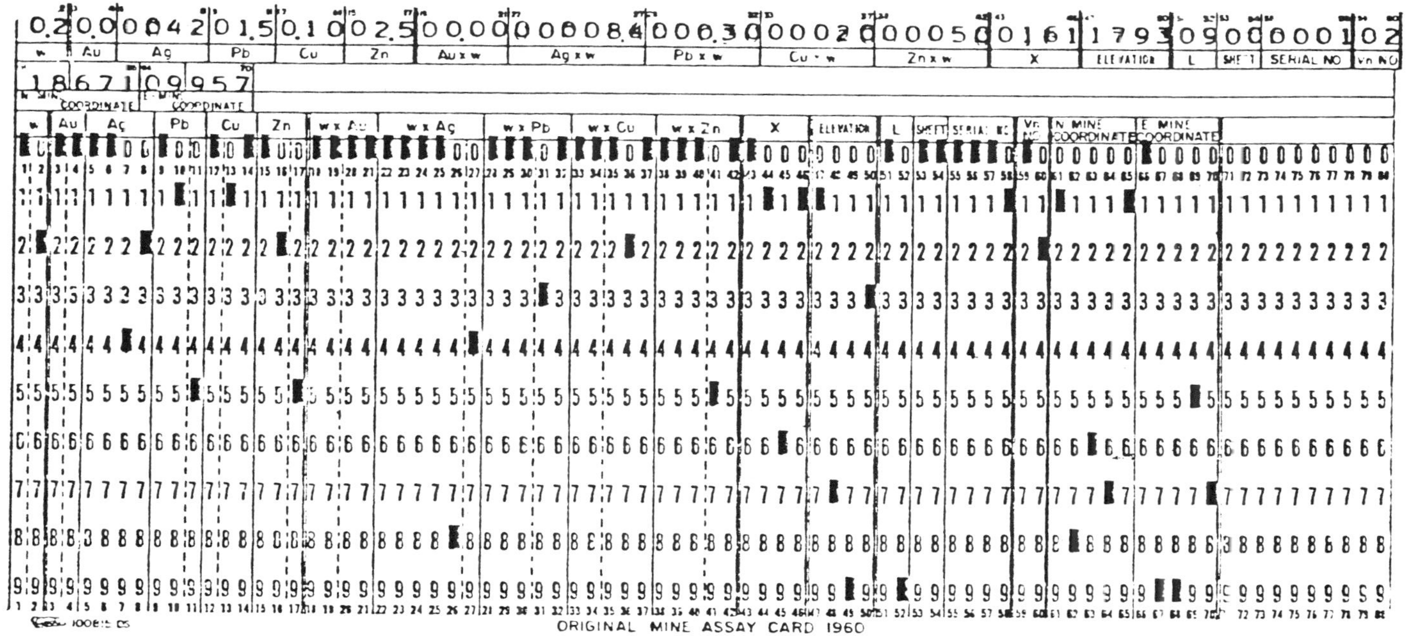

Figure 1. Mine Assay Card

[*Editor's Note:* Material has been omitted at this point.]

For the Brown vein on both Levels 9 and 10, mean vein widths, grades, and metal contents were calculated for the initial assays of samples taken at 2 m intervals. The cards representing individual samples were next split into two equal groups by taking every other card and means for 4 m intervals were calculated. This splitting process was then repeated to obtain means for the four groups of every fourth sample taken at 8 m intervals, for the eight groups of every eighth sample taken at 16 m intervals, and for the sixteen groups of every sixteenth sample taken at 32 m intervals.

[*Editor's Note:* Material has been omitted at this point.]

The precision of a metal value obtained by averaging assays is not perfect, a point that is well-known but often ignored. For example, when a drift is resampled to check the results from an original sampling, the assays and calculated means are never equal to those first obtained, not necessarily because anything was wrong in either the first or second sampling operation but simply because of the variable nature of samples. This is shown quantitatively by Table 8 and Figure 4, which were compiled by taking selected data to focus attention on lead as a representative variable. When groups of assays for 23 or 24 samples spaced 32 m apart were averaged, the estimated lead grades ranged from 9.3% to 12.8% with a range of 3.5%. The 95% confidence interval, a more useful and accurate statistical measure based on the standard deviation, leads to the conclusion that one would expect the true metal content to lie in the range from 8.2 to 13.1.

[*Editor's Note:* Material has been omitted at this point.]

Three main economic considerations affect the setting of acceptable precision. First, the expense of taking individual samples depends of course primarily on the accessibility of the vein, or other geological body, although other factors such as labor costs, and costs of assaying play a part. Where a vein is first exposed underground in a drift, it is completely accessible for sampling, though later it may be covered by lagging or may be stoped out. Where found at the surface a vein may or may not be readily accessible, and where found by diamond drilling, samples from additional drill holes are always expensive.

Secondly, the cut-off point for profitable mining influences the acceptable precision. Referring again to Table 8, if the cut-off were 8%, then the precision obtained from 23 or 24 samples would be acceptable if the mean were to be used only to determine if the drift were mineable. (In the Frisco mine, the cut-off depends of course on combining values for five metals, but the principle is the same.)

Thirdly, the acceptable precision depends on the use to be made of the mean metal values. If a vein is sampled in an exploration program, it may be enough to know that it is above or below cut-off grade without knowing how much. Particularly, if an orebody is below cut-off grade this can be determined with a limited number of samples. Of course, this principle has been used for many years in the common practice in a mine examination of taking a few samples and then resampling if the results of the first samples are encouraging, but data in this paper allow quantitative estimates to be made specifying appropriate numbers of samples.

Table 8. Variation of Lead Estimate with Changing Sample Interval--Frisco Mine, Brown Vein, Level 9 Drift

Sample Interval	No. Samples	Lowest Value	Highest Value	Observed Range	Calculated s	Confidence Limits Lower	Confidence Limits Upper
2 m	378	10.7	10.7	0	-	10.1	11.3
4	189	10.3	11.0	0.7	0.62	9.8	11.5
8	94-95	9.9	11.4	1.5	0.61	9.4	11.9
16	47-48	9.6	12.0	2.4	0.89	8.9	12.4
32	23-24	9.3	12.8	3.5	1.07	8.2	13.1

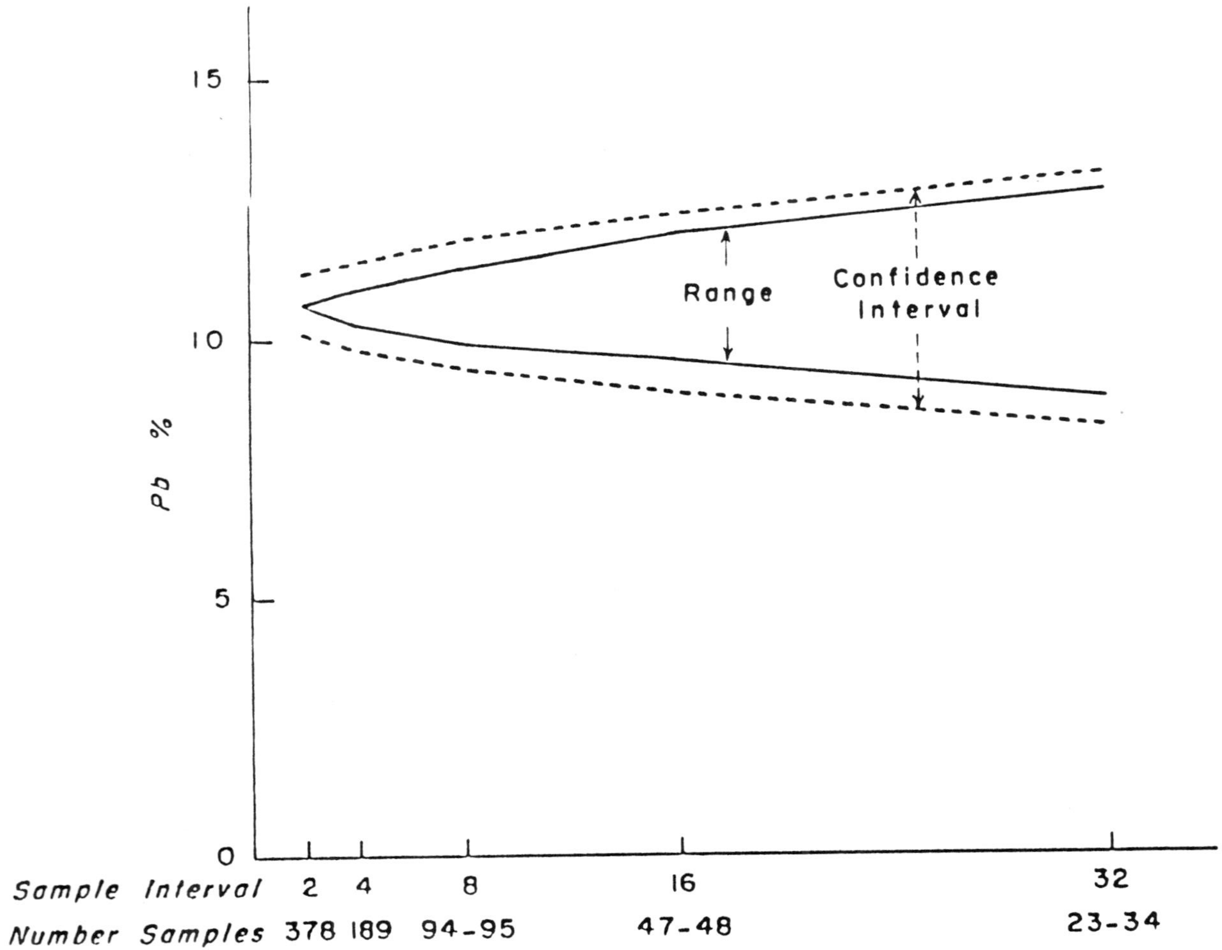

Figure 4. Ranges and confidence intervals of means for Pb computed from assays spaced at various intervals. Level 9 Brown vein drift.

Actually, in exploration, the best procedure would be a sequential sampling program in which initially a few samples would be taken followed by subsequent additional samplings of specified numbers of additional samples if the results of the first sampling were encouraging. This procedure is analogous to an exploration program in which a drilling program is laid out in stages depending on results from the first holes. Hazen (1961) has discussed sequential sampling programs somewhat and the writers are developing such a program suitable for exploration of veins similar to those at the Frisco mine.

On the other hand, if a vein is sampled in an operating mine, the major objective of this sampling is usually to obtain information useful for blocking ore. In this connection it is instructive to compare data from the Frisco mine levels with data from the intermediate levels driven about 6 m above, as it does not seem useful to obtain more precise data in the levels than can be checked in the stopes above the intermediate levels.

[*Editor's Note:* Material has been omitted at this point.]

CONCLUSIONS

This article shows that for comparable orebodies regardless of size, providing there are no strong trends in the mineralization, about the same number of samples need to be taken for equally reliable grade estimation. In general, it would seem to be sufficient to take samples over wider intervals than 2 m. This is largely because the less precise results obtained from taking fewer samples at wider intervals are nonetheless more precise than needed owing to the natural variation of metals, e.g., as calculated in drifts compared to corresponding intermediate drifts.

Many mines do not take regular drift samples because of the expense of sampling at closely spaced intervals. This article shows that results can be obtained from samples spaced at wide intervals from which useful ore estimates can be made.

ACKNOWLEDGEMENTS

The writers thank San Francisco Mines of Mexico, Ltd., for its generous permission in allowing use of its assay data. They thank the National Science Foundation for financial support through Research Grant NSF-G 14189. They also acknowledge the aid of A. Schroeder and J. Jaqua in assisting with the calculations.

REFERENCES CITED

de Wijs, H. J., 1953, Statistics of ore distribution: Reprint from *Geologie en Mijnbouw,* November, 1951, pp. 365-375 and January, 1953, pp. 12-23.

Hazen, Scott W., Jr., 1961, Using techniques of statistical analysis to plan sampling programs: *Internat. Symposium on Mining Research,* University of Missouri School of Mines and Metallurgy, Rolla, 45 p.

Koch, G. S., Jr., 1956, The Frisco mine, Chihuahua, Mexico: *Econ. Geology,* v. 51, pp. 1-40.

Koch, G. S., Jr., and R. F. Link, 1960, Data processing by machine-asset at the mine site: *Mining Eng.,* September, 1960, pp. 1005-1007.

Krige, D. G., 1960, On the departure of ore value distributions from the lognormal model in South African gold mines: *Jour. of the South African Inst. of Mining and Metallurgy,* v. 61, pp. 231-244.

Parks, R. D., 1949, *Examination and Valuation of Mineral Property:* Cambridge, Mass., Addison-Wesley Press, 504 p.

Sichel, H. S., 1952, New methods in the statistical evaluation of mine sampling data: *Trans. Inst. of Mining and Metallurgy,* v. 61, pp. 261-288.

Stanton, R. L., 1958, Abundances of copper, zinc, and lead in some sulfide deposits: *Jour. Geology,* v. 66, pp. 484-502.

Stanton, R. L., 1961, The abundances of lead, zinc, copper, and silver at Broken Hill: *Proc. Australasian Inst. of Mining and Metallurgy,* No. 198, pp. 309-367.

Whitten, E. H. T., 1961, Quantitative areal modal analysis of granitic complexes: *Bull. Geol. Soc. America,* v. 72, pp. 1331-1360.

Part III

VALUATION OF DEPOSITS

Editor's Comments on Papers 15 Through 21

Geologists customarily speak of *estimation* of ore reserves rather than *calculation* of reserves. This difference recognizes the fact that so many indeterminate factors go into the making of ore that reserves can, at best, only be estimated. Since these estimates often become the basis for investment decisions, taxation, trade agreements, and political and military decisions, some agreement must be reached in the usage of terms. In Paper 15, Blondel and Lasky report a consensus in the usage of the terms *reserves* and *resources* on behalf of an international committee of the Society of Economic Geologists. Memories are short and constant reminders of the distinction between these two terms have been essential. To counteract a drift toward loose usage of terms by various political, social, professional, and industrial pressure groups concerned with mineral and energy abundance, the United States Geological Survey and the United States

Bureau of Mines again emphasized the distinctions that had been adopted in 1944. This summary from the *Mining Magazine* is included as Paper 16.

Patterson (Paper 17) outlines the techniques that had developed for the estimation of ore reserves on the basis of cut-off grades and geometry, taking into account geological characteristics to modify reserve classification. The contribution this paper makes, other than a summary of methods in use, is the emphasis it places upon geological factors as a modifier to any ore classification.

The most commonly used methods of computing ore reserves and grade based upon diamond drill core sampling or rotary or percussion drill cuttings are by cross sections, polygons, and triangles. Tonnage and grade may be determined by these methods with acceptable accuracy provided a sufficient number of samples are obtained. However, no simple means exists of determining when an adequate number have been taken or if the deposit has been sampled excessively. Statistical methods can be of assistance in this area.

Irregular and irregularly zoned deposits with a variety of components, having different values, requiring multiple beneficiation treatments, and often requiring selective mining, present a difficult problem to the ore estimator as well as to mining engineer and the metallurgist. Pegmatite deposits, particularly the complexly zoned variety, present this type of challenge. Norton and Page (Paper 18) outline some of the complications involved.

Placer deposits have their own set of difficulties in the sampling for minute quantities of high unit-value materials. Most of the valuation of this type of deposit in the past has been done by specialist consultants who have avoided publication of their procedures. Increasing interest in the reevaluation of large, low-grade, placer gold deposits at the present time makes the article by Daily (Paper 19), giving a how-to and how-not-to approach, a very valuable contribution to the library of the mining geologist. High unit values of gold and platinum make the task of sampling placer deposits particularly difficult.

As has been emphasized in Papers 17, 18, and 19, the commodity and the character of a deposit as well as the contained metal value influence those factors that must be considered in valuation and blocking of ore reserves. The complexity entailed in valuation of iron deposits is not usually recognized outside the industry. Ohle (1972) reviews the variables that must be weighed for reserve estimation; the article has been reprinted in VanLandingham (in press). The suggested evaluation procedure involves reduction of each variable of the ore to a monetary value, so that the pluses and minuses can be totaled, and the relative economic potential of different deposits compared.

Use of statistical analysis for the evaluation of assay data was, as might be expected, first introduced in gold mines where a high unit-value product was irregularly distributed through the ore and where visual estimations could not be relied upon. Resampling would often lead to quite different assay results. South Africans took the lead in the use of statistical analysis in predicting values ahead of production. Forward exploration in the Witwatersrand district is expensive. As early as 1919, Watermeyer advocated the use of previous mine sampling data to determine a percentage value for the frequency of occurrence of an assay value. He proposed using this percentage frequency as an additional weighting factor in determining grade. Sichel (1947) applied statistical methods to check bias introduced in the cutting of samples; however, the work of Krige (1951) stimulated the use of statistical analysis in mine valuation. It could be applied through an essentially mathematical approach as outlined by Krige and by Matheron (1963) or it could be applied rather simply through the contouring of moving averages as outlined in Paper 20 by Krige and Ueckermann.

Matheron introduced the important concept of the variogram that recognizes the variation of value distribution between different types of deposits, between deposits of the same type, and even between different parts of the same deposit. Consequently, the application to ore reserve calculations entailed (1) computation of the variogram of the deposit under study, (2) fitting the proper mathematical mode, (3) computation of all the relevant variances, and (4) kriging the calculations.

The largely theoretical discussion of geostatistics by Matheron is brought into focus in a case history of its utilization at the Eagle Copper Vein, B.C., by Sinclair and Deraisme (Paper 21). However, such an analysis is suspect since it lacks information as to vertical continuity of values. Mining history is replete with examples (rarely reported in the literature) where failure to establish three-dimensional continuity of metal values has led to financial disaster.

REFERENCES

Krige, D. G., 1951, A Statistical Approach to Some Basic Mine Valuation Problems in the Witwatersrand, *Chem. Metall. Mining Soc. South Africa Jour.* **52:**119–139 (reprinted as Paper 16 in VanLandingham, in press).

Matheron, G., 1963, Principles of Geostatistics, *Econ. Geology* **58:**1246–1266 (reprinted as Paper 13 in VanLandingham, in press).

Ohle, E. L., 1972, Evaluation of Iron Ore Deposits, *Econ. Geology* **67:**953-964 (reprinted as Paper 4 in VanLandingham, in press).

Sichel, H. S., 1947, An Experimental and Theoretical Investigation of Bias Error in Mine Sampling with Special Reference to Narrow Gold Reefs, *Inst. Mining and Metallurgy Trans.* **56:**403.

VanLandingham, S. L., ed., in press, *Economic Evaluation of Mineral Property,* Benchmark Papers in Geology, vol. 67, Hutchinson Ross, Stroudsburg, Pa.

Watermayer, G. A., 1919, Application of the Theory of Probability in the Determination of Ore Reserves, *Chem. Metall. Mining Soc. South Africa Jour.* **19** (Jan.):97-108.

15

Reprinted from *Econ. Geology* **51**:686–697 (1956)

MINERAL RESERVES AND MINERAL RESOURCES

F. BLONDEL AND S. G. LASKY

On Behalf of and Approved by
A Committee of the Society of Economic Geologists [1]

ABSTRACT

This report is designed to clarify the meaning of the terms *reserves* and *resources* and to recommend, for common acceptance, rules and definitions to be observed in estimating reserves and resources.

All estimates for mineral reserves and resources rest on geologic appraisal, which contains many indeterminate elements. If the miner thinks the indefiniteness is too great, he reduces it by new exploration or development, but always there remains a residuum of uncertainty. For appraisals of whole countries or industries, fields of inquiry that require economic assessment, the geologic indeterminates remain large and dominant. Moreover, whereas the miner thinks in terms of the economic present or of the economic near future, the economist dealing with resource appraisal thinks of the more distant future and therefore takes into consideration deposits well beyond the miner's present interest. In doing so he must set arbitrary economic limits more far reaching than the geologic uncertainties. Thus, the miner and economist view the matter of reserves and resources within different frames of reference having different boundaries. The terms created by the miner to describe the ore reserves of his mine cannot appropriately be transported into the larger and more indefinite field of inquiry of the economist.

We recommend the following equation as defining the difference between "reserves" and "resources," the term "reserves" being limited to mineral material considered exploitable at present, and "potential ores" consisting of those deposits demanding more favorable conditions:

$$\text{Resources} = \text{reserves} + \text{potential ores}$$

But the conditions needed for the exploitation of "potential ores" vary from conditions only slightly different than at present to those that are widely different, so that a more complete equation would be as follows:

$$\text{Resources} = \text{reserves} + \text{marginal resources} + \text{submarginal resources} + \text{latent resources}$$

The appraiser can be guided by either equation, depending on the information available to him and on the purpose of his appraisal.

The above is a classification of resources. The reserve category itself has to be broken down. For the mine operator we recommend continuation of the traditional breakdown into "proved," "probable" and "possible," but according to the definitions given by C. K. Leith. For the person who would assess the reserve or resource position of an industry or of a region or of a nation, we recommend common acceptance of the classification

[1] Federico Ahlfeld (So. America); F. Blondel (Europe, Chairman); D. M. Davidson (U. S., Member at large); K. C. Dunham (Great Britain, Member at large); J. L. Farrington (Africa); N. H. Fisher (Australia); J. R. Gonzales (No. America); S. G. Lasky (U. S., Member at large); and M. B. Ramachandra Rao (Asia).

(proposed and adopted by the U. S. Bureau of Mines and the U. S. Geological Survey) of "measured," "indicated," and "inferred" reserves, modified by combining "measured" and "indicated" into a category of "demonstrated" reserves.

A word of caution is necessary. Estimates of reserves and resources require constant scrutiny and periodic review. Moreover, such estimates are only a partial answer to those who use them, for they state only the quantity of material that a mine or a region or a country may contain. The user of the information would like to know, in addition, the rate of production the mine or region or country could support under variable economic conditions over some extended period into the future. Such information we as geologists and engineers cannot, of course, furnish.

WHAT exactly is meant by the question: "What are the iron ore reserves of such and such a country?" And along with this, "What rules must be observed in estimating these reserves?" Such questions have been the subject of numerous discussions.[2] The answer to the first has depended in the past on who asked the question and on who furnished the answer; the answer to the second has depended on who made the estimate. Yet clearly it is desirable that there be a single answer to each. We suggest such answers in this report. They are not completely new. In one way or another they have been proposed in the past; we simply rephrase them here in a context designed to be acceptable to all.

SUPPLEMENTARY POINTS OF VIEW

The crux of the confusion lies in the fact that the basic notions, and the terms used, do not have the same meaning for everyone, and that these variations in sense arise largely from the circumstances of their usage.[3] In general, two spheres of usage can be distinguished, first, that of the miner and, second, that of the economist.

Before engaging the considerable sums required to open a mine, the miner first estimates the tonnage of ore or quantity of metal that can be taken out profitably. He may or may not estimate the total amount—commonly he is satisfied with determining only that enough is present to justify the proposed investment. This estimate he calls the "estimate of reserves" of the deposit, and he keeps it up to date throughout the life of the operations—as costs and prices fluctuate, as known ore is mined out and new ore found, and as mining and processing technology improve—in order that he may know constantly, as fully as is practicable, how much ore he can continue to count on.

The point of view of the economist comes into play when the mineral potential of a region or of a country, as contrasted with that of a single mine or group of mines, becomes the item of interest. Everyone knows more or less clearly the importance of minerals in modern economy. The layman knows also, at least vaguely, that individual deposits do not produce indefinitely. Broadly speaking, different minerals come from different deposits differently distributed geographically, and the location of the sources of any

[2] See bibliography.
[3] See particularly Lasky, 1945.

particular material varies with time—slowly, perhaps, but steadily. Inversely to agriculture, which reproduces almost in the same spot, mining consumes little by little the riches that Nature offers to it, and is obliged to turn without cease to new deposits, or to new parts of old deposits. Because of this fact and because there is no end to civilization's demand for a continuing supply of mineral products, efforts are continuously made to visualize the future mineral capabilities of whole countries or region . Manufacturers, financiers, statesmen and their economic advisors are all concerned with this problem, and the conclusions they reach and the decisions they must make are sound only in proportion as they are reliably informed.

The two approaches—that of the miner and that of the economist—are conceptually different, but they are not necessarily in conflict. In reality they are supplemental. The first has limited application, and the second depends on the first. Each is based on its own set of considerations.

THE GEOLOGIC SIDE OF APPRAISAL

Estimating the amount of ore that may be present in the ground involves so many factors that it is a difficult matter at best. Yet because of the nature of the subject the estimates must generally be made in quantitative form. and the very fact of utilizing numbers gives a false impression of precision.

The quantity of ore that will be extracted from the ground is not a definite figure, an objective reality that two equally competent persons will appreciate equally. Of course, once a deposit is exhausted, we usually know what it has given, but an attempt to forecast this figure beforehand must involve many assumptions: First, about the geologic type, quality, and total size of the deposit, and second, about future technology and economics. Accordingly, estimates of the quantity of ore that eventually may be extracted from the ground, as made by different observers, are rarely comparable. One estimate may be ten or even a hundred times another, yet both may be valid, each in its own context.

Before total exploitation, only an uncertain approximation can be made of the tonnage that can be extracted from a deposit—approximate, in the sense that the estimate always involves a margin of error; uncertain, in the sense that even the margin of error is not clear, as will be explained later.

Estimates of mineral potentialities—whether dealing with a single deposit or with a region in general—rest initially on geologic observation of the surface, supported, of course, by whatever mining works or drill-hole data may be present. This appraisal is based both on general geologic knowledge of the region and on a comparison of the particular deposit or area under study with other deposits or areas with which the geologist is in some way familiar. In particular and whether done consciously or unconsciously, he always classifies the deposit as to geologic type as a necessary first step in appraisal.

These intellectual operations—for we must frankly admit that that is what they are, however skillfully they may be performed—give rise to the rough estimates and to the uncertainties about which we have already spoken. The margin of error and the uncertainties are too great for the miner; he therefore executes a series of mining works (trenches, bore-holes, pits, adits) designed to eliminate them. He does not care whether the geologist's assumptions and hypotheses are correct or false; he simply wishes more positive information. But such works tend to be costly, and so he undertakes them only if the geologic estimate has been favorable, and he continues to carry them out only to the point of being assured that he will have enough ore to support operations for some reasonable period; he knows that his investment can be worthless if its recapture is deferred too far into the future.

As to studies bearing on a whole region, these must remain satisfied with the geologic estimate only, as mining works extending over a whole region would be prohibitively costly. Herein appears the first divergence in the way the problem is approached by the miner and by the economist. The miner, interested in single deposit, can reduce the uncertainties of an estimate by further work, or not, as he desires; the economist engaged in long-term assessment of a region or country must accept them.

THE ECONOMIC UNCERTAINTIES

A mineral deposit is rarely a homogeneous mass having neat outlines. Its limits are most of the time undecided, and within it the grade is variable. The entire deposit is not economically exploitable, so that in order to estimate the tonnage that *is* exploitable, the estimator is obliged to choose more or less arbitrary boundaries and to make a number of arbitrary economic assumptions.

For the miner these economic assumptions are relatively precise, for he makes his estimate with major reference to the economic and technical possibilities then prevailing, using cost limits consistent with normal risk-taking and commercial production. The person making a long-term economic appraisal does not do this. He does not reason in the immediate present or even for five or ten years. Moreover, he does not deal with individual deposits, but with countries or industries. He is obliged to think of longer terms and so must take account of deposits that the miner would consider unexploitable. In fact, he takes into account deposits whose possible utilization is so far in the future that the miner often considers him quite impractical.

Moreover, from the time the appraiser ceases to base himself on current economic conditions he no longer has a fixed frontier of reference. Should he consider cost-price relationships superior to the present or inferior, and if so, by how much? Conceptually he is under no compulsion to impose any cost-price limits except as he desires. This factor involves arbitrary decision much more influential than the geologic uncertainties.

The miner and the economist clearly have a different appreciation of this subject. The economic factor exists for the miner, it is true, but it is relatively simple; moreover, for him it can be discussed objectively, with a reasonable chance of agreement. The arbitrary assumptions as to the market situation in a far-away future made by the economist, on the contrary, are rarely likely to be discussed with any measure of agreement.

BORROWED TERMINOLOGY

The first estimates of mineral productivity were made by miners—geologists or engineers; the economist's requirements appeared only recently. Consequently, for the earliest estimates the geologic uncertainties were felt particularly, because the miner had to try to reduce them by costly work, and it was natural that they should have influenced the terminology.

Thus were introduced into practice the terms almost universally used—*proved, probable* or *possible* reserves, or analogous terms. "Possible" reserves are a geologic estimate. If exploration or development verifies the geologic hypotheses or exposes the ore more completely even though not verifying the geologic reasoning, the "possible" reserves are transformed into "probable" reserves; finally, in order to define the conditions for exploitation, the miner executes still more mining work that practically eliminates the uncertainty and transforms the "probable" reserves into "proved" reserves.

The economist, on the contrary and as already indicated, must content himself, save in exceptional cases, largely with a geologic estimate, for he is dealing mainly with resources that are to be exploited in a future that extends beyond the life of the "proved" or even "probable" reserves. In addition, the economic assumptions he introduces make his estimates more tenuous than are even the miner's estimates of "possible" reserves. In making these estimates, however, the terminology created for the appraisal of the mineral *reserves* of a deposit has been converted into a terminology for the mineral *resources* of whole regions and countries.

The significance of this situation is brought out by the quantitative relationships. In the miner's usage, "possible reserves" usually range from about half as much as "proved" and "probable" combined to two or three times as much. But in the economist's usage, the term "possible" is applied to material that may add up to a figure more than a hundred times as large as "proved" and "probable" reserves combined (1, p. 171).

There is still another aspect that both miner and economist, as users of estimates of mineral potentialities, usually fail to appreciate—namely, the margin of error inherent in regional estimates. Such estimates, as explained later, are now commonly expressed in terms of *measured, indicated,* and *inferred* ore. Measured ore, the most certain of these categories, has by definition a margin of error of as much as 20 percent. In practice, this allowable maximum tends to become the rule. The margin of error for indicated ore is, of course, even higher, and for inferred ore higher still. Thus, for estimates that include all three categories, the margin of error is commonly on the order of 50 percent, and may be plus or minus. In the estimates used by the economist—wherein the amount of ore envisioned as being eventually used is so many times the amount in the proved and probable categories, or in the measured and indicated categories—the margin of error is even greater. Nevertheless, this does not nullify the utility of the estimates—it means merely that the user must be conscious of their frontiers and of the terms he employs in describing them.

THE CONDITIONING VARIABLES

From the foregoing, we see that the estimate of the mineral productivity of a deposit or of a region depends on at least two factors:

(1) The limit accepted for the economic possibilities of exploitation. For a single mine, this is usually expressed as a minimum minable, or cut-off, grade; for a region, in which each mine may have its own cut-off grade, the limitation has to be expressed in more general terms.

(2) The degree of certainty desired for the estimate itself—that is, whether the estimator wishes to include only "proved" reserves (as is done for petroleum, for example), or would include also the progressively less certain categories of "probable" and "possible" reserves.

Obviously, for the same deposit or the same region, at a definite period, the estimate increases (a) when a higher price or lower cost is obtainable, or (b) when a lower degree of certainty is accepted. It increases also as technology of recovery is improved, but this factor can be considered as encompassed by the cost-price relationship.

Thus, there is no such absolute entity as "the" reserves or resources of a mine or an area; there are only estimates applicable within particular economic limitations and degrees of certainty. For the sake of clear understanding between the estimator and the user of the estimate, these conditioning factors should in each case be specified.

A CLASSIFICATION OF RESOURCES—SOME NECESSARY DISTINCTIONS

Notice that in the estimate of reserves the miner generally assumes economic conditions roughly equivalent to present conditions. He considers that if more stringent conditions arise they will ultimately resolve themselves, and that he has the right to begin exploitation on the assumption that the present cost-price relationship will apply, on average, for the life of the reserves he then estimates. The economist, on the contrary, assumes that the economic circumstances of his more distant future will *not* be the same as the present.

Keeping this essential difference in mind, and in order to foster clear understanding, we suggest (following the distinction put forward by Lasky in 1949) using the term *reserves* for mineral material considered as being exploitable under existing conditions, including cost, price, technology, and special local circumstances. The term "reserves" would thus be used only in the sense that the miner understands it.

Reserves are thus only part of the total resources. The rest consists of that mineral material which, to be exploited, demands conditions more favorable than those currently existing, as well as further exploration to bring into consideration undiscovered deposits not currently included under the terms "possible" or "inferred." For this remainder we suggest the term "potential ores," although recognizing that realism will usually impose a limit in the evaluation of such material; it would ordinarily serve no purpose, of course, to include those mineral masses whose exploitation, for any reason—

nature of ore, tenor, geographical location—is difficult to visualize even in a faraway future—for example, rock in the United States containing only 15 percent of iron.

Accordingly, we have the equation:

$$\text{Resources} = \text{reserves} + \text{potential ores} \tag{1}$$

It is useful, and for some countries and some commodities possible, to introduce a subdivision of "potential ores." For example, there are many mineral masses not exploitable at present yet exploitable under conditions only slightly better. Some may have been sporadically worked in the past under exceptional conditions. Following the economist's terminology, we might refer to these as "marginal" resources. Beyond the "marginal" deposits are those masses that require conditions even more favorable but still, in the opinion of the miner and the metallurgist, within reach, and to these we assign the term "submarginal" resources, again using a term fixed in both mining and economic terminology. The rest—that part of the "potential ores" that may be exploited in the more distant future of the economist—we propose to call "latent" resources.

Equation (1) then becomes

$$\text{Resources} = \text{reserves} + \text{marginal resources} + \text{submarginal resources} + \text{latent resources} \tag{2}$$

The appraiser can be guided either by equation (1) or by equation (2) depending upon the information available to him and upon the purpose of his appraisal.

Naturally, a deposit or a mineral mass does not have a fixed quantity in one or another of these four categories of resources, or fixed ratios among them. As economic conditions and technology improve, or as more is learned about the deposit, a particular mineral mass or parts of it may very well pass from a lower category to a higher one.

The opposite is also possible, though less likely—that is, material may pass from a higher category to a lower one. For example, the low-grade residue in the native copper deposits of Michigan in the United States is a "marginal resource," but if there occurs the secular increase in general costs that some economists envision, without a corresponding increase in the price of the metal, this material may become "submarginal"; and if the mines should be shut down and the equipment be withdrawn, this material may even cease to be considered a resource at all, as has happened at other mines.

This kind of situation emphasizes the restrictions we would place on the term "reserves." We think that masses or deposits susceptible to profitable exploitation only during periods of high prices should not be considered as constituting "reserves"—not even during the period of exploitation—but only as "marginal resources." In doing this we reason partly as does the miner himself, partly contrary to him. A future exploitation is normally not based on possible exceptional high markets, but on average prices stretched out over a period. If a mineral mass is exploitable at the present time only because of exceptional market prices, it should not be considered a genuine "reserve" until average market values become sufficiently high for exploitation to continue as a matter of course.

Classification of reserves

When the miner estimates reserves he needs a classification that meets three specifications:

(1) It will provide as nearly as possible a positive measure of the quantity of ore or metal he can rely upon, at least the minimum necessary in making his plans as a businessman.

(2) It will provide as nearly as possible a standard procedure for estimating this minimum quantity, and

(3) It will provide a way of expressing the inherent geologic uncertainties.

The miner's usual classification is of *proved, probable,* and *possible* ore, in decreasing order of certainty. It is impossible to get a common definition of these terms. Authorities seem never to have agreed, a fact that reflects the difficulty of defining reserves in a way that will satisfy all or most conditions of mining and all or most kinds of deposits. The following definitions represent our understanding of the way the terms are most often thought of:

Proved ore (positive ore, blocked-out ore, assured ore). Ore exposed on three sides and where there is, therefore, practically no risk of failure of continuity.

Probable ore. Ore exposed on two sides and where there is, therefore, some risk, yet warranting reasonable assumption of continuity.

Possible ore (prospective ore). Ore that cannot be included in the above classes but the presence of which can be reasonably expected, because of exposure by a shaft or drift or bore hole, or for some geologic reason.

These definitions lean strongly toward the rigidly geometric, and various modifications of them have appeared from time to time in order to introduce the geologic factors more adequately. Definitions that meet the specifications outlined above have been given by Leith (9) and we recommend their common acceptance. They are as follows:

Proved or assured ore. Ore blocked out in three dimensions by actual underground mining operations or by drilling, but it includes in addition minor extensions beyond actual openings and drill holes, where the geological factors that limit the ore body are definitely known and where the chance of failure of the ore to reach these limits is so remote as not to be a factor in the practical planning of mine operations.

Probable or semiproven ore covers extensions near at hand, where the conditions are such that ore will probably be found but where the extent and limiting conditions cannot be so precisely defined as for proved ore. Semiproven may also mean ore that has been cut by scattered drill holes, but too widely spaced to assure continuity.

Possible ore. Ore is classed as possible or prospective where the relations of the land to adjacent ore bodies and to geologic structures warrant some presumption that ore will be found but where the lack of exploration and development data precludes anything like certainty of its actual location or extent. Often it is not desirable to assign figures to "possible" tonnages, but they may be designated by terms like "small" or "large."

But this classification applies only to the need of the miner. There is also needed a classification for the man who would assess the present reserve position or the resource position of a whole industry or of a region or a nation. Such assessments are composed of a summation of estimates of individual mines and districts, expanded by geologic and economic judgment. They are made not by the miner, but by civil servants or others who are not privy to the reserve estimates for individual mines. Moreover, in making them it is not necessary that the data on any single deposit be as well measured as for estimating proved or probable ore—indeed, such data are rarely extant for the undeveloped deposits. The need in the national appraisal is not for accurate measurement, but for an appreciation of magnitude.

For this purpose, the staffs of the United States Bureau of Mines and the United States Geological Survey, in assessing the mineral position of the United States in 1944 (13) used the classification of *measured, indicated,* and *inferred* reserves. In fact, they were forced into creating a new classification because the Government exploration rarely yielded the geometric information needed to estimate reserves as proved or probable. The definitions of the terms are as follows:

Measured reserves are those for which tonnage is computed from dimensions revealed in outcrops, trenches, workings and drill holes and for which the grade is computed from the results of detailed sampling. The sites for inspection, sampling, and measurement are spaced so closely and the geologic character is so well defined that size, shape, and mineral content are well established. The computed tonnage and grade are judged to be accurate within limits which are stated, and no such limit is judged to be different from the computed tonnage or grade by more than 20 percent.

Indicated reserves are those for which tonnage and grade are computed partly from specific measurements, samples, or production data and partly from projection for a reasonable distance on geologic evidence. The sites available for inspection, measurement, and sampling are too widely or otherwise inappropriately spaced to permit the mineral bodies to be outlined completely or the grade established throughout.

Inferred reserves are those for which quantitative estimates are based largely on broad knowledge of the geologic character of the deposit and for which there are few, if any, samples or measurements. The estimates are based on an assumed continuity or repetition, of which there is geologic evidence; this evidence may include comparison with deposits of similar type. Bodies that are completely concealed may be included if there is specific geologic evidence of their presence. Estimates of inferred reserves should include a statement of the specific limits within which the inferred material may lie.

As may be noted, these categories are analogous to *proved, probable* and *possible,* but they are too general and have too large a margin of error to satisfy the practical business needs of the miner.

These new definitions are being found so useful that they are creeping into worldwide acceptance. We propose to accept the concept, modifying it in one way.

As mentioned above, estimators of national reserves or resources commonly do not have access to the estimates of ore reserves of individual companies. Such information is jealously guarded. Accordingly, in practice the estimator cannot separately assess "measured" and "indicated" reserves but must combine them into a single figure, and indeed he finds it no detriment to his purpose to do so. The "measured" reserves do not constitute or express a characteristic of the deposits of a district or of a nation but represent only the bulk sentiment of mine management as to how thoroughly a deposit should be developed before mining is started—indeed, they commonly represent only the bulk sentiment of the first entrepreneurs, grown into a tradition through the practice of attempting to find each year as much ore as was mined out in that year, as discussed in more detail farther on.

We suggest, therefore, that the terms "measured" and "indicated" be abandoned and be replaced by a single term that will include them both. For want of a better we suggest "demonstrated" reserves, in contradistinction to "inferred" reserves. Thus, we may say we have "demonstrated" the existence of a hundred million tons of reserves, for example, without having "measured" them. The proposed definition of demonstrated reserves is as follows:

Demonstrated reserves are those for which tonnage and grade are computed partly from specific measurements, samples, or production data and partly from projection for a reasonable distance on geologic evidence. They may include some orebodies concerning which the sites for inspection, sampling, and measurement are spaced so closely and the geologic character so well defined that size, shape, and mineral content are well established, and other orebodies concerning which the sites available for inspection, sampling, and measurement are too widely or otherwise in appropriately spaced to permit the bodies to be outlined completely or the grade established throughout.

PERIODIC REVIEW

Clearly, estimates of reserves and resources, and especially their division into categories, are never final. With changes in economic conditions, some resources will go from a less favorable to a more favorable category, or vice versa. A growing part of the reserves will disappear through exploitation. Within the "reserves," better knowledge acquired through new mining works will allow a "possible" or "inferred" part to pass on to a more certain category. This may happen even through nothing more portentous than increased availability of information already in existence. Finally, entire estimates may be revised through a better geologic or mining or economic appreciation.

The estimates are entirely valid only for the date on which they are made and must be reviewed periodically. This is common practice in mining circles in estimating reserves; it now happens only incidently in appraising resources.

FURTHER CONSIDERATIONS

In order not to complicate the discussion, we have reasoned up to now that the miner and the economist are interested in knowing that such and such a deposit, or that such and such a region, contains X million tons of ore. That is true, but he would like, if possible, to know much more. He would like, in fact, to be able to say: This deposit—or this region—is likely to produce Y million tons annually over a period of Z years.

The miner is fortunate in this regard, for he himself ordinarily decides the annual production of the deposit with which he is concerned. Both the life and the annual production are conditioned not only by the technical characteristics of the deposit, but also by the market possibilities, by the profit per ton or other unit and the necessity of a sufficient annual production to meet the financial burden, by the level of income the miner desires, and by the number of years over which he wishes to amortize the investment. At the outset, he will "prove" the reserves only to the extent necessary to support production at the desired rate for the number of years desired for amortization; afterwards, each year he will try to "prove" new reserves equal to the tonnage extracted during the year, rarely at a greater rate unless he can do so in quantity sufficient to expand the annual rate of production profitably. What he does not and cannot know is how long he can perpetuate his operations in this fashion.

The economist is in a similar quandary. He may know, for example, that a country contains ten billion tons of iron ore, which at present rates of production would last, say, a thousand years, and he contents himself with this information because he has no other. But what he would really like to know is how much the country would be able to furnish annually under the various economic conditions that he cares to postulate.

The statement that a country has a specified number of years of reserves of a metal at present rates of production seems precise, but in reality it is meaningless as a measure of reserves or resources because it makes no allowance for the fact that production rates do not remain constant—the production for any particular year or period is likely to be a unique quantity—or for the effect of depletion on the one hand or new discovery on the other.

CONCLUSION

Simple knowledge of the mineral resources of a mine or region, however categorized, is only partial information on what could be called the ultimate possibilities of the mine or region. Yet such information is about all that we, as geologists and engineers, are in a position to furnish. All the more reason to try to clear up the ambiguities surrounding it. Such is the aim of this report.

In so aiming, we have indicated the different, yet supplemental, ways in which the mining man and the economist approach the matter of mineral resource appraisal, and have indicated the differing needs of each; we have made a distinction between reserves and resources and have shown the relation of one to the other, offering definitions for each; we have recommended common acceptance of specific definitions of *proved, probable,* and *possible* ore as applicable to the needs of the miner, and have recommended definitions of *demonstrated* and *inferred* reserves as meeting the needs of the man who would assess the reserve or resource position of a whole industry or of a region or a nation. And finally, we have pointed out the necessity of periodic review of any and all estimates of reserves and resources because of the unique changeability of the factors that condition such estimates. We cannot emphasize too strongly the necessity of specifying the limiting factors or assumptions in any estimate of reserves or resources, such as cut-off value, for example, or some particular technology. The benefit of more definite concepts and of commonly accepted definitions will be partly offset if ambiguity persists through failure to state the assumptions controlling the estimate.

12 RUE DE BOURGONNE, PARIS,
OFFICE OF THE SECRETARY,
U. S. DEPT. OF THE INTERIOR,
WASHINGTON, D. C.,
May 16, 1956

PARTIAL BIBLIOGRAPHY

1. Blondel, F. and Lasky, S. G., 1955, Concepts of mineral reserves and resources: Survey of World Iron Ore Resources, p. 169–174, United Nations, New York.
2. Einecke, G., 1950, Die Eisenerzvoräte der Welt, p. 2: Dusseldorf.
3. Fennel, J. H., 1939, Ore reserves: Trans. Inst. of Mining and Metallurgy, v. 49, p. 315, London.
4. ——, 1951, Ore reserves: Mining Magazine, p. 217–218, London, October.
5. Kendall, J. D., 1901–1902, Ore in sight: Trans., Inst. of Mining and Metallurgy, v. 10, p. 143, London.
6. Lasky, S. G., 1945, The concept of ore reserves: Mining and Metallurgy, v. 26, p. 471–474, New York. Translation in French in La Chronique des Mines coloniales, Sept.-Oct. 1953, Paris.
7. ——, 1949, National mineral resource appraisal: Mining Congress Journal, January, Washington, D. C.
8. ——, 1950, Mineral resources appraisal by the United States Geological Survey: Quarterly, Colorado School of Mines, v. 45, no. 10.
9. Leith, C. K., 1938, Mineral valuations of the future: American Institute of Mining and Metallurgical Engineers, p. 47–48.
10. Mikami, H. M., 1944, World iron ore map: ECON. GEOL., v. 39, no. 1, p. 1–2.
11. Proceedings of the United Nations Scientific Conference on the Conservation and Utilization of Resources (UNSCCUR), 17 August to 6 September 1949. Vol. II, Mineral Resources, p. 1–40.
12. 1952, Outlook for key commodities: Resources for Freedom, v. 2, Rept. President's Materials Policy Comm., Washington, D. C.
13. 1947, Mineral position of the United States: U. S. Bureau of Mines and U. S. Geol. Survey, Washington, D. C.

16

Reprinted from *Mining Mag.* **130:**183, 185 (1974)

CLASSIFICATION OF MINERAL AND ENERGY RESOURCES

The U.S. Bureau of Mines and the U.S. Geological Survey have recently redefined ore reserve and resource categories following their original recommendations of 1944

AS highlighted in a recent leading article in *Mining Magazine,* the consistency of classification of both ore reserves and total mineral resources throughout the world leaves much to be desired. Currently, the most commonly used definitions of 'ore reserves' are those originally recommended by the U.S. Bureau of Mines and the U.S. Geological Survey, back in 1944. While a great number of mines do base their own ore reserve figures on these recommendations, there are still some major discrepancies—and where potential resources of a metal or mineral are described, the worldwide variations in the basis of estimation are considerable.

Redefined categories

Now the USBM and the USGS have redefined ore reserve and resource categories to cover all possibilities for extraction now and in the future. In geologic terms, a mineral or energy resource is a concentration of naturally occurring solid, liquid or gaseous material in or on the earth's crust in such form that economic extraction is currently or potentially feasible. This is the basis for the USBM/USGS classification of a resource, while their classification of a reserve is the portion of an identified resource which can be produced at a profit at the time of classification (an identified resource being one estimated from geological and engineering evidence).

The full category covering all materials that have present or future value and comprise identified or known materials plus those not yet identified, but which, on the basis of geological evidence are presumed to exist, is classified as 'Total resources'.

The philosophy behind the new definitions from the USBM/USGS is that long-term public and commercial planning must be based on the probability of discovering mineral resources in as yet undiscovered deposits and must also make allowances for such factors as the technological development of economic extraction methods for mineral deposits that are currently unworkable on cost grounds. Possible substantial price movements relative to the overall trend must also be taken into account by those assessing resources. It is felt that all the components leading to the estimation of 'total resources' must be continuously re-assessed in the light of new geological and technological knowledge and of shifts in economic and political conditions. It is also important that the classification system must be uniformly applicable to all commodities so that data for possible alternatives or substitutes can also be compared.

'Total resources'

Because of these necessities, 'total resources' are classified both in terms of economic feasibility (i.e., as economic and sub-economic, with the latter sub-divided into paramarginal and marginal) and also in terms of the degree of geological assurance (i.e., identified and undiscovered, each sub-divided into further categories). A visual representation of the USBM/USGS recommended classifications is shown in Fig. 1 below.

General guides for the use of this classification system are as follows:

1. Resource categories and definitions in the classification should be applicable to all naturally occurring concentrations of metals, non-metals and fossil fuels. The categories may be sub-divided for special purposes. (Definitions as recommended are set out below.)
2. Definitions may be amplified, where necessary, to make them more precise and comfortable with accepted usage for particular commodities or types of resource evaluations.
3. Quantities and qualities may be expressed in a variety of terms and units to suit different purposes, but must be clearly stated and defined.

Recommended definitions

Resource: A concentration of naturally occurring solid, liquid, or gaseous materials in or on the earth's crust in such form that economic extraction of a commodity is currently or potentially feasible.

Identified resources: Specific bodies of mineral-bearing material whose location, quality and quantity are known from geologic evidence supported by

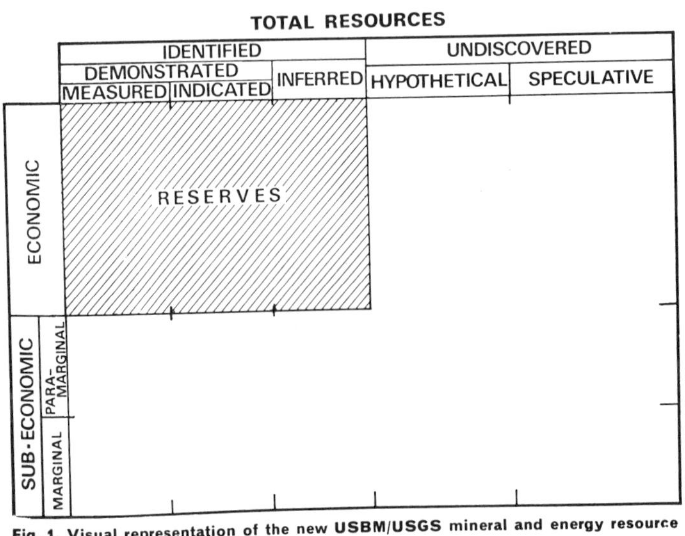

Fig. 1. Visual representation of the new USBM/USGS mineral and energy resource and reserve category proposals

engineering measurements with respect to the demonstrated category.

Undiscovered resources: Unspecified bodies of mineral-bearing material surmised to exist on the basis of broad geologic knowledge and theory.

Reserve: That portion of the identified resource from which a usable mineral and energy commodity can be economically and legally extracted at the time of determination. The term *ore* is used for reserves of some minerals.

The following definitions for measured, indicated, and inferred are applicable to both the Reserve and Identified-Subeconomic resource components.

Measured (proved): Material for which estimates of the quality and quantity have been computed, within a margin of error of less than 20 per cent, from sample analyses and measurements from closely spaced and geologically well-known sample sites.

Indicated (probable): Material for which estimates of the quality and quantity have been computed partly from sample analyses and measurements and partly from reasonable geologic projections.

Demonstrated: A collective term for the sum of materials in both measured and indicated resources.

Inferred (possible): Material in unexplored extensions of Demonstrated resources for which estimates of the quality and size are based on geologic evidence and projection.

Identified-Subeconomic resources: Materials that are not Reserves, but may become so as a result of changes in economic and legal conditions.

Paramarginal: The portion of Subeconomic Resources that (a) borders on being capable of production economically or (b) is not commercially available solely because of legal or political circumstances.

Submarginal: The portion of Subeconomic Resources which would require a substantially higher price (more than 1·5 times the price at the time of determination) or a major cost reducing advance in technology.

Hypothetical resources: Undiscovered materials that may reasonably be expected to exist in a known mining district under known geologic conditions. Exploration that confirms their existence and reveals quantity and quality will permit their reclassification as a Reserve or Identified-Subeconomic resource.

Speculative resources: Undiscovered materials that may occur either in known types of deposits in a favourable geologic setting where no discoveries have been made, or in as yet unknown types of deposits that remain to be recognised. Exploration that confirms their existence and reveals quantity and quality will permit their reclassification as Reserves or Identified-Subeconomic resources.

Guidelines

These guidelines are laid down principally for the benefit of the USBM and USGS, with the former's area of responsibility covering reserve estimates and the latter's estimations of total resources. Every year the USBM and USGS will meet to confer and agree on estimates in all the resource categories defined above. Agreed data will be used in Bureau or Survey publications.

Obviously a great deal of thought and time has been put into the new recommendations and it would seem sensible for the USBM and USGS counterparts in other countries to follow this lead. This is not to suggest that they should draw up their own definitions, but should use the U.S. ones as their own guidelines. Mining companies everywhere should also consider using the reserve definitions for their own estimations. The sooner there is international agreement on the estimation of metal and mineral resources and reserves, the better for all involved in the mining industry worldwide.

17

Reprinted from *Eng. and Mining Jour.* **160:**111, 112–115 (1959)

Estimating Ore Reserves Follows Logical Steps

JOHN A. PATTERSON

A RECENT REVISION of Noah Webster's dictionary defined a reserve as "that which is reserved; a store; stock; extra supply." Ore might be defined as mineral material that occurs in nature in such quantity, grade, and configuration as to allow profitable removal from the earth. An ore reserve then, should be the stock of economically minable material in a mineral deposit.

The determination of an ore reserve must be based on information varying from specific assays and measurements to geologic knowledge on the setting and habit of the orebody. The reliability, and hence worth, of an ore reserve estimate depends largely on the quantity and quality of factual data available and the location of the points that the data represent. As we move from an area in an orebody where there is ample evidence of the dimensions and grade of the ore to unknown or less well-explored parts of the orebody, we similarly move from a reserve estimate that is highly reliable to one that is less reliable. Geologic knowledge and experience with the orebodies in an area become more important as the sample data become sparse.

The geologic knowledge and experience used in an ore reserve estimate should be specific and not based on speculation. We in the mineral industry, needing at least a modicum of optimism to survive, are often prone to draw conclusions for which there actually is little evidence. We are often in the same situation as the blind men who encountered the elephant. The blind man who felt the side of the elephant knew he was dealing with something big: he thought it was a house. The blind man who examined the leg thought it was a tree; and the one who examined the tail and thought it was a snake, and did not have much basis for deciding the object was very large. Each man needed to do a little more exploring before he could draw a valid conclusion.

[*Editor's Note:* Material has been omitted at this point.]

Calculating the Reserve

The actual process of calculating the ore reserve involves the following steps: (1) Analyzing the basic data to determine the sample points at which we know ore exists, (2) Outlining the ore blocks as we interpret them, giving consideration to the assignment of the reserve into a measured, indicated or inferred category depending on the criteria establishing the block and (3) Computing the dimensions of the block, the volume of ore, the tonnage, and the average grade.

Analyzing the Data

Analyzing the Data, the first step, involves two decisions: one, the determination of the accuracy of the data and how it should be used: and two, what is ore.

The most reliable data available is probably provided by assays and measurements on core, where the core recovery is very high, and by carefully cut channel samples. Where core recovery is low or when the channel sample is really a carelessly cut chip sample, the data might be very misleading. In certain cases of low core recovery, the material lost is the higher grade fractions and the core assays are lower than they should be. In other cases, the high grade material is retained, while lower grade host rock is lost, resulting in core assays which are higher than they should be. Careful recovery of the sludge could provide a basis for a dependable correction. It is easy when cutting chip samples to get a larger proportion of the softer materials. If these soft materials are the high grade fraction, the sample will produce a higher grade than actually exists. Careful sampling of the individual ore streaks will help solve this problem. Similar types of problems exist when dry-cutting or wet-sludge samples are taken. Grab samples, because they give no dimensions, usually are of little value in ore estimation.

In uranium exploration, gamma ray logging is the commonest method now used to determine thicknesses and grades of mineralization; it is also

Fig. 2 Classification Leans Heavily on Geology

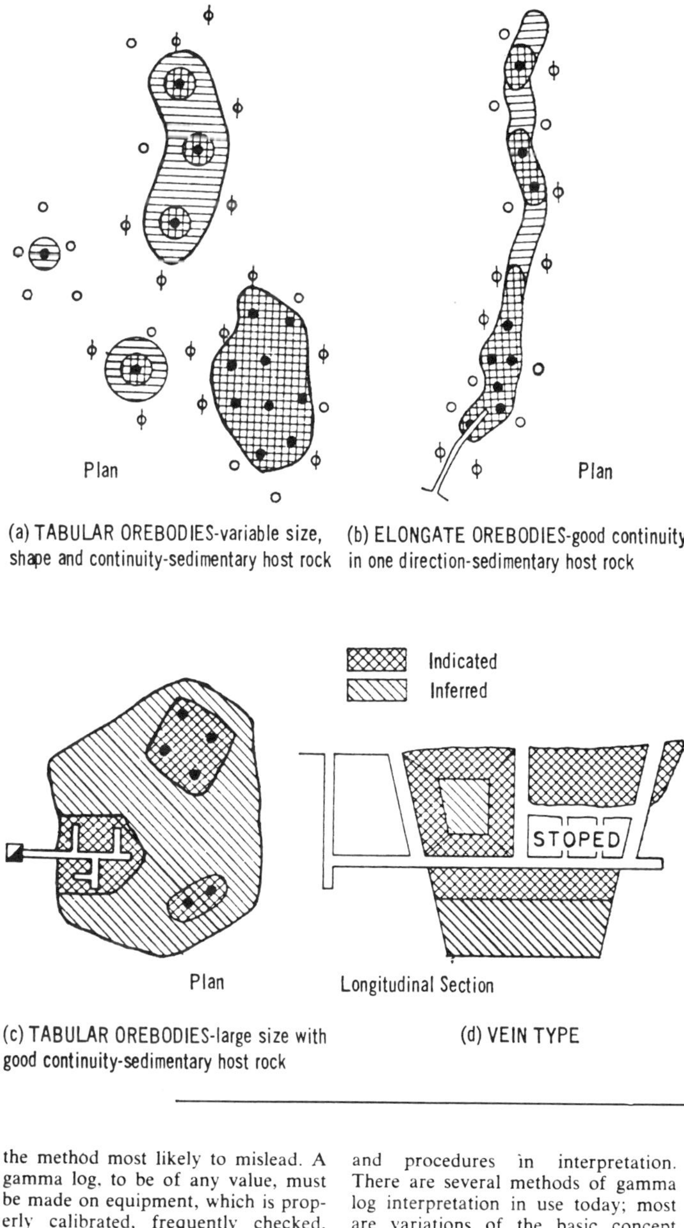

(a) TABULAR OREBODIES-variable size, shape and continuity-sedimentary host rock

(b) ELONGATE OREBODIES-good continuity in one direction-sedimentary host rock

(c) TABULAR OREBODIES-large size with good continuity-sedimentary host rock

(d) VEIN TYPE

The "habit" of ore in an area is one of the major starting points for ore reserve classification. Every effort must be made to determine it as accurately as possible. In a new area this determination is highly critical and must be a major exploration effort if "reserves" is to be a meaningful term. Once the habit is known, classification can proceed.

Ore of variable continuity in tabular pods of varying size and shape is a common situation for uranium deposits. It is illustrated in (a) (*left*). One pod appears from the drilling to have continuity between a number of holes. Knowledge of the area would help one to decide if this is a reasonable expectation. Another pod has three ore holes on it, but they are spaced too far apart to allow indicated ore between them. Here an indicated block is made for each hole and an inferred block between them and around the indicated blocks. A single ore hole indicated block has been given an inferred annular ring. This is warranted by the adjacent mineralized holes and the location of the ore hole with respect to the other orebodies and the general ore trend. The other single hole block has been made entirely inferred as it is off the main ore trend and the results of the surrounding drilling limit the possibilities.

For a very strong ore trend, perhaps controlled by an ancient stream channel cut into an underlying formation, sketch (b) illustrates the approach. Indicated ore has been extended along the channel direction for some distance, but only a short distance across. Ore has been inferred along but not across the trend.

Large, tabular orebodies with very good continuity are handled as in (c) (*left*). Ore has been indicated around mine workings and groups of ore holes, and inferred between the indicated blocks.

A vein type deposit is shown in (d) (*left*). Ore is indicated where exposed in mine workings and inferred at the center of large blocks of indicated ore and in projections into unexplored ground at depth. The block above the stoped area might be considered as measured if the distances were appropriate.

the method most likely to mislead. A gamma log, to be of any value, must be made on equipment, which is properly calibrated, frequently checked, and consistently used. To properly interpret the log, once it has been obtained, one must know the characteristics of the equipment and the character of the ground being logged. Correlation with chemical assays from a number of cored holes with high core recovery is a must for proper interpretation. Only in this manner can one compensate for the vagaries of ore equilibrium, hole conditions, instrument and operator individuality and procedures in interpretation. There are several methods of gamma log interpretation in use today; most are variations of the basic concept that the total counts registered through a radioactive zone are proportional to the thickness of that zone multiplied by its average grade. Because of the variation of equilibrium within orebodies, varying hole conditions and equipment aberrations, the results interpreted from any one hole can be considerably in error; the average results from interpreting a number of holes in an orebody, however, can approach the actual average for the orebody and thus give useable data.

As to what constitutes ore for the deposit being considered, many considerations are involved. However, the problem resolves into this: What is the minimum thickness that can be mined with the mining system that we envisage for the orebody, and what is the minimum grade that we can mine at this thickness that will repay the cost of mining? The answers can only be obtained by a study of the mining methods and costs involved. The thickness and grade combination decided upon, often called the cutoff, is the minimum that we will call ore.

Fig. 3 Random Sample Scheme Is Simplest

The general outline method of constructing ore blocks and computing tonnage and grade is a simple system used by the AEC.

Let us look at the Block I, the drilled out ore block. The sample data has been examined and the holes that have been identified as ore intercepts are marked as such on the map; similarly mineralized and barren holes are identified. It is usually handiest to record the thickness and grade combination for each hole right on the map. Since the drill hole spacing is close enough to allow us to consider the area as indicated ore, we outline the ore block as the data requires. A line is drawn around the periphery of the block, paying attention to the values in the adjacent mineralized holes, and the vertical location of the ore—it must all be in the same horizon. On completing our line we have an ore block with a number of samples within it.

Similar analysis results in the measured blocks of ore in the mine pillars, the indicated ore blocks exposed in the workings and on the outcrop, the inferred ore block projected between the ore exposed on the outcrop and underground, and the inferred block projected between the drilled ore and ore exposure at the limits of the underground workings. Each of these blocks is drawn keeping in mind the definitions and criteria for the various classes of ore.

Once we have outlined the ore blocks, we are ready to compute the tonnage for each block. As illustrated in Fig. 3 for the drilled block, the average thickness is computed by averaging the thicknesses of the sample points within the block. This is simply the sum of the thicknesses divided by the number of samples. The area is easily obtained from the map by using a planimeter, which measures the actual map area in square inches; using the map scale, we can compute the area in square feet in the block. This area multiplied by thickness produces the block volume. The volume of ore must be converted to tonnage by dividing by the tonnage factor, a number based on specific gravity measurements that tells us how many cubic feet of rock in the ground are needed to make a ton of ore. These items can be combined into the formula shown in the table.

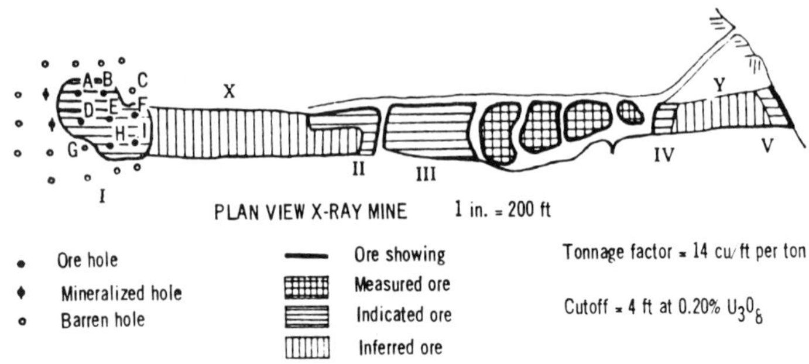

Calculations Drilled Orebody, Block I

Hole	Thickness feet		$\%U_3O_8$		Product
A	5	×	0.22	=	1.1
B	4	×	0.30	=	1.2
D	8	×	0.25	=	2.0
E	6	×	0.30	=	1.8
F	5	×	0.24	=	1.2
H	7	×	0.22	=	1.54
I	4	×	0.28	=	1.12
Totals:	39				9.96

$$\text{Avg. thickness} = \frac{39}{7} = 5.6 \text{ ft}$$

$$\text{Avg. grade} = \frac{9.96}{39} = 0.25\%$$

Area (planimetered) = 0.26 sq in.

$$\text{Tonnage} = \frac{\text{Avg. thickness} \times (\text{map scale})^2 \times \text{map area}}{\text{Tonnage factor}}$$

$$\frac{5.6 \times (200)^2 \times .26}{14} = 4160 \text{ tons}$$

The grade is determined by combining the sample data for the block according to the length of samples as shown. The products of the thickness and grades are summed up and divided by the sum of the thicknesses. The resulting average grade in this case is 0.25%.

A similar procedure is followed for the other measured and indicated blocks, using channel sample data for the ore exposed by underground workings. For the two inferred blocks the thicknesses and grades of the adjacent indicated blocks are used to determine the thickness and grade for the inferred block. The inferred block "x" in Figure 3 is given the average thickness and grade of the adjacent indicated Blocks I and II.

When combining the various blocks, the average grade is determined by weighting according to tonnage as illustrated for the indicated blocks.

Other schemes are shown on next page.

Combining Indicated Blocks

Block	Tons		Grade		Product
I	4160	×	0.25	=	1040
II	600	×	0.28	=	168
III	2100	×	0.24	=	504
IV	240	×	0.20	=	48
V	350	×	0.25	=	87.5
Totals	7450				1847.5

$$\frac{1847.5}{7450} = 0.248\%\ U_3O_8 \text{ Average grade for Indicated Ore}$$

Calculations for Inferred Ore Block "x"

Block	Avg. thickness	Average grade
I	5.6	0.25 (One side of block)
II	4.8	0.28 (Other side of block)
Block "x"	5.2	0.26

$$\text{Tonnage} = \frac{5.2 \times \$40,000 \times .42}{14} = 6240 \text{ tons}$$

Lower grade material will not be included regardless of thickness; thinner material is included only where the grade is the high enough to allow dilution to something better than the cut-off thickness and grade. Thus, using a cut-off of 4 ft @ 0.15%, a 10-ft 0.06% sample would not be acceptable, but a 2-ft @ 0.30% one would be: Consideration of the detailed data for a sample point enables one to arrive at the optimum thickness and grade combination appropriate for the cut-off and mining method.

While the cut-off tells us what should be mineable as far as operating costs are concerned, the final tonnage and grade of the orebody must provide sufficient margin of dollar return after mining to pay for the cost of developing the property. If there is not enough margin to repay this cost, it is not a mineable ore reserve under existing economics, regardless of how cheaply you could have mined a ton of ore once the property was developed and the development cost amortized.

Constructing the Ore Blocks

After we decide what our data is telling us and what is ore, we can proceed with the next step; constructing the ore blocks.

The General Outline Method, a system used in the AEC, is explained and illustrated in Fig. 3. It uses an approach in drawing the ore blocks and in the calculations that considers the sample points as random samples in the ore body. Each sample in a block has equal influence in determin-

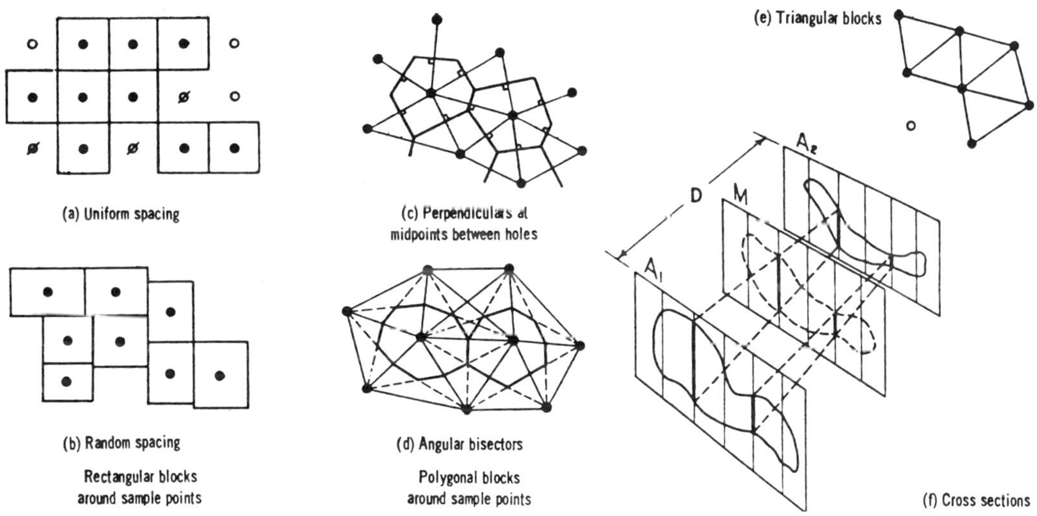

Fig. 4 Other Systems Fit Specific Needs

SKETCH "a" shows a system using squares. This is very convenient if the drill holes are on a uniform grid. If the grid varies, we end up with the pattern in Sketch "b". Note that the influence from a hole is extended halfway toward the adjacent holes. This system is very useful to keep reserve calculations current with drilling progress.

More complicated systems are shown in the center of the illustration above. Sketch "c" uses perpendiculars erected at the mid-points between holes to form the boundaries of the block. For a square drill hole grid, the blocks would end up as the squares of Sketch "a". The system illustrated in Sketch "d" uses bisectors of the angles formed on the outer corners of the triangles formed by lines connecting a hole and its adjacent holes. The bisector lines are connected as shown to form the blocks.

Two systems in use that do not form a block for each sample point are illustrated on the right. Sketch "e" illustrates a system that forms triangular blocks by connecting the holes as shown. (See "How to Estimate Ore Reserves" by H. J. Wolf in E&MJ December 1958, p 92). The triangles should be of uniform shape and ore thickness for best results. This system is most accurate for blocks that form equilateral triangles; where the triangles vary from the equilateral condition, a correction should be made by adjusting sample weights proportionately to the angular variation from the 60 deg that prevails for the equilateral triangle. The similarity of this system to the polygon system in Sketch "c" is apparent when we notice that the polygons are really constructed of pieces of the triangles connecting the holes.

Sketch "f" illustrates the cross section or fence line system. This is often used where rows of holes have been drilled across a deposit, the distance between the rows usually being considerably larger than the distance between holes within a row. In this system, the cross sectional area is determined for two places where the orebody has actually been measured and in an interpolated section. The data are used in this formula:

$$\text{Volume} = \frac{D}{6}(A_1 + 4\ M + A_2)$$

Where: D is the distance between end sections; A_1 is the area of section at A_1; A_2 is the area of section at A_2; and M is the area of interpolated section at M. The middle cross section is given more influence than the end sections. Dividing by 6 produces the average cross sectional area. This system can be used for vertical orebodies also, where the sections become level plans.

---

ing the average grade and thickness of the block.

There are other systems commonly used to organize ore blocks. Most of them construct an ore block for each hole or sample; the thickness and grade of the sample are the thickness and grade of the ore block. The boundaries are drawn around the block on the basis that the influence of one sample should extend until the influence of another sample takes over, which would be at the mid-distance between two samples. In these systems the influence of a hole is related to the area of its block, a factor often related more to the problems of drill site selection than the value of a sample. The most important of these systems are illustrated in Fig. 4.

It can be seen that all of these methods involve more work than the General Outline Method; either in the construction of the blocks or in the subsequent calculations. The more blocks you create, the greater will be the calculation effort. Systems detailing blocks for various parts of an orebody are often misleading for uranium orebodies as one can rarely find the actual tonnage or grade calculated for a detailed block, though the average for the orebody obtained by combining all the blocks probably will be a reasonable estimate.

Computing the Tons and Grade

The computation of tons and grade is the final step in ore reserve estimation and is similar for all methods. In general, each block is computed separately and then totaled in appropriate combinations of classes, as explained in the General Outline Method. More details are available in USBM Bulletin 356 "Sampling and Estimation of Ore Deposits" by C. F. Jackson and J. B. Knaebel, 1932, as well as most texts on field geology and mine examination and valuation.

In conclusion, the estimation of an ore reserve involves consideration of samples and measurements together with specific information on the geologic setting and habit of the orebody. Analysis of the data available enables recognition of ore, and construction of measured, indicated, or inferred ore blocks. Computation of the volume contained in each block, conversion to tonnage, and grade computation completes the process.

18

Reprinted from *Mining Eng.* **8**:401-414 (1956)

Methods Used to Determine Grade And Reserves of Pegmatites

by James J. Norton and Lincoln R. Page

Recent work on pegmatites, especially in the Black Hills, S. D., has shown that reserves of unmined pegmatite can be computed by projections down the structure. The grade of a pegmatite deposit can be satisfactorily determined by measurements of grains on representative exposures, by production records, or by assays.

EFFECTIVE methods for determining grade and reserves of pegmatites in advance of mining have been developed in recent years. When intensive work began on the economic geology of pegmatites during the early years of World War II, it was widely believed that grade and reserves could be estimated only in a most general way. The literature indicated that not all pegmatite deposits are unpredictable, but statements emphasizing unpredictability and irregularity seem to have been more widely accepted.

Many standard references reflect the prevailing opinion that minerals are unsystematic in their occurrence in pegmatites. Pegmatites have been described as "notorious for their irregular structure," and minerals within pegmatites as "irregularly distributed," "irregularly disseminated," of "erratic occurrence," "proverbially erratic in distribution," "characterized by extreme irregularity in . . . size and distribution of the essential minerals, and variety and proportion of the accessory minerals," and "spotty."[1-6]

Over many years, however, geologic techniques have been applied with increasing success to economic analysis of pegmatites. Investigations, chiefly since 1939, have shown that mineral distribution and structure of pegmatites are largely systematic and that detailed geologic study can be used to predict the size and shape of mineral deposits in pegmatites.

Because the writers are most familiar with the pegmatites of the Black Hills, S. D., examples cited are from that area.* Pegmatites mined in other districts are similar to those in the Black Hills,[8] and the techniques described here have been applied by geologists and engineers working in other districts.

* All examples are described more completely elsewhere.[7]

J. J. NORTON, Member AIME, is a Geologist with the U. S. Geological Survey in Denver. L. R. PAGE is Assistant Chief Geologist, U. S. Geological Survey, Washington, D. C.

TP 4225I. Manuscript, Feb. 1, 1955. New York Meeting, February 1950.

In actual practice few pegmatite mine operators attempt to make accurate estimates of grade and reserves in advance of mining. Many deposits are too small for mining except on a very small scale; few pegmatite mines have more than ten employes. Hence there is little systematic preparation for mining, and the capital outlay may be so small that only a general knowledge of reserves is necessary. Technical personnel are rarely employed by pegmatite mine operators to make such estimates. In the future, however, as capital outlay becomes larger, estimates of grade and reserves will have to be more accurate.

Geologic Aspects of Determining Grade and Reserves

Estimating grade and reserves requires that the type of pegmatite, the size and shape, and the internal structure be determined by detailed geologic mapping and careful mineralogic study. The wide variations in the texture and distribution of minerals within many pegmatite deposits make a thorough geologic study of this kind necessary before it is possible to decide what sampling methods should be used to determine mineral content or grade.

Types of Pegmatites: The first step in economic appraisal of a pegmatite is to determine the mineralogical and structural type. In general, pegmatites may be classified into two groups: 1) those in which component minerals are more or less evenly distributed throughout the entire body and show relatively minor variations in texture, and 2) those in which component minerals have been segregated into units with recognizable differences in texture, mineralogy, or both. Johnston[9] has classified these types of pegmatites as homogeneous and heterogeneous, respectively. Cameron et al.[10] have used the terms *unzoned* and *zoned* for similar types.

The difference between the homogeneous and heterogeneous types is of great commercial importance. Homogeneous pegmatites can seldom be mined where the products are concentrated by the hand methods now commonly used, but they have been mined in a few places where a milling process has been perfected for recovery of the useful minerals, and perhaps many more of them will be mined in the future by large-scale operations. In North Carolina and in Maine feldspar is recovered from fresh homogeneous pegmatite, and weathered homogeneous pegmatite is mined for kaolin and scrap mica in southeastern U.S. and for cassiterite and tantalite in Africa.[11]

Heterogeneous pegmatites are the source of most of the pegmatite minerals produced in the U.S. Pegmatites of this type that have been profitably mined contain units rich in minerals that can be extracted by selective mining: mica, spodumene, amblygonite, lepidolite, feldspar, beryl, and columbium and tantalum minerals.

A mineable concentration of industrial minerals (hereafter called a *deposit*) may consist of an entire homogeneous pegmatite, or it may be only part of a heterogeneous pegmatite. Detailed geologic mapping is necessary to determine the size and shape both of the entire pegmatite body and of the component internal units. In homogeneous pegmatites the tonnage of rock containing any specific industrial mineral is the total tonnage of the pegmatite body; it can be found by projecting all local variations of strike, dip, and plunge to determine size and shape of the deposit where it is not exposed. In heterogeneous pegmatites the location and attitude of the contacts of the internal structural units must be mapped to determine the size and shape of each unit that contains a concentration of an industrial mineral.

Size and Shape of the Entire Pegmatite: The structural relation between pegmatite and wall rock is of greatest importance in predicting the size and shape of pegmatite bodies in depth. Many pegmatites are, for the most part, concordant with foliation of the country rock, though locally the contact may be crosscutting. The plunge of such concordant pegmatites is essentially parallel to the average plunge of linear structures in the schist and the average plunge of the principal irregularities or *rolls* of the contact. Rolls ordinarily plunge parallel to the nearest crest or keel of the pegmatite, but in some places they merely reflect local variations in shape. Concordant pegmatites have a variety of shapes, but most are irregularly lenticular. The thickest part is commonly near the top or crest, and beneath the thickest part, or *bulge*, the pegmatite thins gradually down dip. The isopachous map of the two pegmatites mined at the Victory mine, Custer County, S. D., Fig. 1, shows this change of thickness with depth. Some concordant pegmatites such as the Victory (Figs. 1, 3, and 4) and the Edison (Fig. 5) are multiple intrusives consisting of two or more individual lenses separated by discontinuous schist partings. Other concordant pegmatites are arcuate, sinuous, or tabular. Recognition of these irregular shapes and their accurate projection along the plunge are necessary before reserves can be adequately estimated.

Size and shape of discordant pegmatites are predicted less readily than those of concordant pegmatites, because they are controlled by fault planes, joints, or other structures that may not be well understood. Discordant pegmatites tend to be tabular but are commonly branching. Pegmatites of this kind have formed in granite, quartzite, gneiss, or other rock that was competent at the time of intrusion. Extensions of such deposits beyond known areas may be predicted in the same fashion as extensions of metal veins that are controlled by similar structures. Projections in depth must be based on the strike and dip of the host fracture and the plunge of fracture intersections.

Size and Shape of the Internal Units of Heterogeneous Pegmatites: The internal structure of a heterogeneous pegmatite must be understood in all dimensions before it is possible to determine the feasibility of selective mining for a particular pegmatite mineral. In recent work the internal units have been classified as zones, fracture-filling units, and replacement units.[3] Contacts between pegmatite units are for the most part gradational. Where gradation takes place over a distance measurable in feet, the contact must be defined by the occurrence of a diagnostic mineral or group of minerals, by the percentage of an individual mineral, or by differences in texture.

Most zones are continuous throughout the pegmatite and reflect to some degree the shape of the pegmatite body. Outer zones are shaped more like the pegmatite as a whole than are inner zones and are more regular in thickness. Thickness of a zone may be uniform or may change gradually, rather than abruptly or irregularly. Some zones thicken or thin to form a series of discontinuous lenses, the long axes of which parallel the plunge of the irregularities in the pegmatite contact. A few zones are hood-shaped or trough-shaped and are approximately parallel to the crest or keel, respectively, of the pegmatite. Incomplete or irregular zones plunge parallel to the nearest keel or crest of a conformable pegmatite or to the intersection of the different segments of a branching pegmatite. Shapes of various types of zones are illustrated in Figs. 2 through 5.

Zones are more abundant and larger than fracture-filling and replacement units and are economically the most important units in a pegmatite. Thus it is essential that they be recognized and that their shape and attitude be determined in detail before calculations of tonnage are made for reserve estimates. Fracture-filling and replacement units are rarely large enough to be mined. Depth extensions of some fracture-filling units can be projected along the strike and dip of the controlling fracture, but many of these bodies follow irregular, intersecting fractures. These units may be zoned; if so, the zones reflect the shape of the fracture-filling unit. Replacement units, on the other hand, have an extremely irregular and unpredictable shape, except where controlled by such structures as fractures or gneissoid banding. The size of a replacement body can be predicted if the structural control can be recognized and is thoroughly understood, but predictions are less accurate than for other units. In general, prediction of size and shape of replacement units in pegmatites is similar to prediction of the size and shape of contact-metamorphic deposits. Few replacement units are large enough to be mined profitably. The spodumene deposits at Tinton, S. D.,[13, 14] are the best examples of large deposits in the U.S.

Mineralogy and Texture of Deposits: Industrial minerals produced from heterogeneous pegmatites are concentrated in definite zones or in shoots or other readily defined parts of these zones. The content of any reasonably large block of rock, at least several hundred tons in size, is normally about the same as the average content of the entire deposit. Irregularities of mineral distribution and variations in the size of crystals or aggregates may cause small

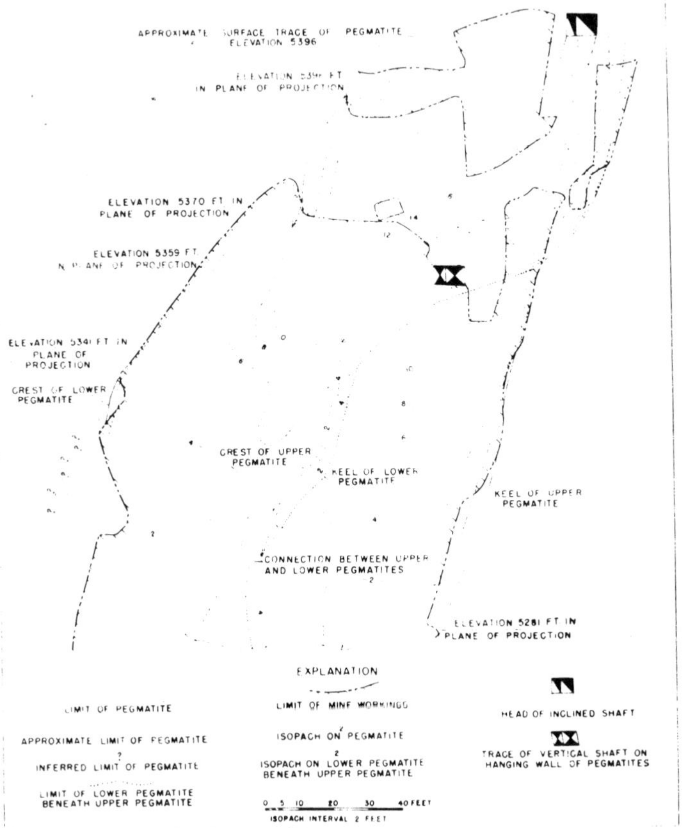

Fig. 1—Isopachous map of the Victory pegmatite, Custer County, S.D., projected to a plane striking N 88° W and dipping 48° SW.

parts of the zone to be much richer than the zone as a whole.

In general, the texture of a pegmatite is more nearly that of a porphyritic than an equigranular rock. Differences in grain size of minerals from one deposit to another may be extremely great. In some deposits all minerals or aggregates occur in masses of about the same size, but in others individual crystals or aggregates range from a fraction of an inch to more than 40 ft long. The matrix commonly consists of grains only a few inches in diameter.

Distribution of minerals within most zones is fairly uniform, though small blocks of rock may be very rich or very lean in certain minerals. On the other hand, mineral proportions in a zone may change gradually along the structure, as at the Hugo pegmatite, Fig. 2, where potash feldspar is abundant in the top of a perthite-quartz-albite pegmatite unit but decreases to zero with depth. Such zones can be subdivided into smaller units on the basis of the percentage of perthite, and the content of these units can be satisfactorily estimated. This type of gradation in the potash feldspar content is characteristic of pegmatites that contain perthite-rich hoods. Similarly, the proportion of muscovite in wall zones that consist essentially of quartz, albite, and muscovite may show a gradual change with depth. Muscovite-bearing zones of this type commonly are lean in mica adjacent to a perthite-bearing hood, but the quantity of muscovite increases downward until, below the perthite-bearing hood, mica is very abundant. Some of the richest mica mines in the Black Hills, such as the White Spar and Buster Dike, were discovered beneath a perthite-rich hood. At depth a few pegmatites may become leaner in large muscovite and possibly in total muscovite. At one end of the surface exposure the New York pegmatite has what is probably a remnant of a perthite-rich hood surrounded by a narrow quartz-albite-muscovite zone that contains far too small a quantity of large muscovite to be mined for sheet mica. In the upper levels of the mine the quartz-albite-muscovite zone is said by Sterrett[15] to have contained 6.6 pct recoverable crude mica in rock that was being mined at the time of his visit in 1908. During mining on the 300 level in 1943 only 1.6 pct mica was recovered from the rock.

The richest parts of some zones form shoots that plunge in the same direction as the pegmatite. The quartz-albite-muscovite wall zone at the New York mine, for example, contains shoots that are richer in mica and contain more large mica crystals than the rest of the zone. Much of the mica produced in North Carolina has been from rich shoots in pegmatite otherwise too lean to be mined.

Another type of change in mineral proportions from one part of a pegmatite to another is illustrated by certain spodumene-bearing pegmatites, such as the Mateen and the Edison, Pennington County, S. D. The spodumene is pseudomorphically altered, either partly or wholly, to micaceous and clay minerals. Where the alteration is complete, the shape and cleavage of the spodumene crystal can still be recognized, even though virtually no true spodumene remains. Degree of alteration increases from the inner part of the pegmatite outward and from the top downward. The proportion of spodumene crystals, including material that has been pseudomorphically altered, may be essentially the same from one part of the spodumene-bearing unit to another, but the proportion of lithia and unaltered spodumene changes gradually from one place to another. Changes may be so systematic that lines of equal lithia content can be drawn on maps or sections.[16]

Changes in mineral content across zones are normally greater than changes parallel to the structure. Most mica-bearing wall zones, for example, are richest in the outer part, and the mica content gradually decreases inward. The mica and beryl content of intermediate zones commonly increases from the outside to the center and decreases from the center to the inside of the zone. Similar variations occur in zones that contain other industrial minerals, and as a result the thickness that can be mined varies with changes of economic conditions.

Mineral distribution of replacement units is less constant and not so well understood as in zones and fracture-filling units. The proportion of relic minerals and of introduced minerals in the replacement units of the Hugo pegmatite varies greatly and unpredictably, Fig. 2.

Processing and Marketing Factors Affecting Reserve Estimates

Pegmatite mining operations range in size from one-man enterprises to companies with a capital investment of several million dollars. In general, however, the mines are small in relation to mines in most other parts of the mineral industry. Operations differ greatly in efficiency, and a deposit that may be very profitably operated by one may be unprofitable for another. Many operations that are relatively large, by the standards of the pegmatite mining industry, make no use of technical personnel.

A small pegmatite mine may be worked by only one or two men, using little mechanical equipment, and the mining method may be a simple process of following the ore in a deposit that is small and has a simple structure. Hand-cobbing is the standard concentrating method, and it is generally the only processing that a small miner provides before selling to a local buyer. An operator of this sort can move readily from one deposit to another that appears more promising. He can change his mining from one pegmatite mineral to another as market conditions change. Such mines may operate on a week to week basis, and there is ordinarily no real necessity for making careful reserve estimates using controlled techniques.

Larger operators, on the other hand, may work many properties and do the work on a much larger scale, using mechanical equipment and milling methods that add greatly to the capital costs. To justify the necessary costs of such operations, they must make reserve estimates.

Evaluating the probable financial rewards from such an enterprise requires detailed knowledge of many problems in processing and marketing pegmatite minerals. Each mining company is engaged primarily in the feldspar, mica, or lithium business. A feldspar company concentrates its mining on the feldspar of a pegmatite, although other minerals that are recovered are sold as byproducts. If the pegmatite contains a unit lean in feldspar but rich in other industrial minerals, the feldspar company may not mine it. A single pegmatite may be mined successively by different companies as a lithium pegmatite, a mica pegmatite, and a feldspar pegmatite.

A feldspar company sells its feldspar in a ground state at about twice the price of hand-cobbed feldspar produced by small mine operators, who ordinarily sell their product to a feldspar company for grinding and marketing in the regular trade channels. Evaluation of the feldspar company operation must be in terms of its ability to produce and sell ground feldspar, unless the economics of mining can be completely separated from milling.

Similarly, companies that are primarily mica miners may grind their own scrap mica and rift and trim their own sheet mica. Value of the sheet mica recovered from crude mica depends very largely on efficiency of the rifting and trimming shops, which varies greatly from time to time. At the beginning of World War II there were very few experienced mica workers in the Black Hills, and recovery of sheet from crude mica was low. As the mica workers became more experienced, recovery became much greater, and the dollar value of mica produced from each mine was higher.

Companies mining lithium minerals ordinarily process the material themselves and sell it eventually as lithium chemicals. Unprocessed lithium minerals are sold from only a few domestic mines, and the price is a nominal one that may not at all times be the same as it would be if all the lithium minerals changed hands in the open market before going to chemical plants.

Adequate evaluation of a deposit must include consideration of all possible products. Recovery of 5 lb of byproduct beryl per ton of rock would, at a price of $600 per ton, increase the value of the rock mined by $1.50 per ton. One pound of tantalite per ton might increase the value by twice this amount.

Appraisal of a pegmatite deposit must be not only in terms of the overall mineral content, but also in terms of the recovery in the process to be used, whether it be hand-cobbing or one of the milling methods of varying degrees of efficiency that have been used for pegmatite minerals. Most of the muscovite in mica-bearing zones is too small to be recovered by hand sorting or to be cut into sheet mica. The productive zone of the Victory pegmatite, for example, was estimated to contain about 24 pct muscovite, but only one third was recovered as crude mica suitable for rifting and trimming into sheet mica. It would be of no interest to the operator of this mine that his mica-bearing zone contained 24 pct muscovite, because only part of the muscovite could be recovered and used for the purpose for which he was mining the pegmatite.

Some of the industrial minerals of pegmatites may be intergrown with deleterious minerals seriously affecting the value of the deposit. Potash feldspar containing more than a negligible quantity of intergrown tourmaline, garnet, or biotite is valueless for ceramic uses. In many deposits potash feldspar is intergrown with quartz to form graphic granite. The quartz cannot be separated in mines where the concentrating method is hand cobbing; consequently the aggregate must be sold as a low-grade product at a low price. Intergrowths of plagioclase with potash feldspar may also lower the

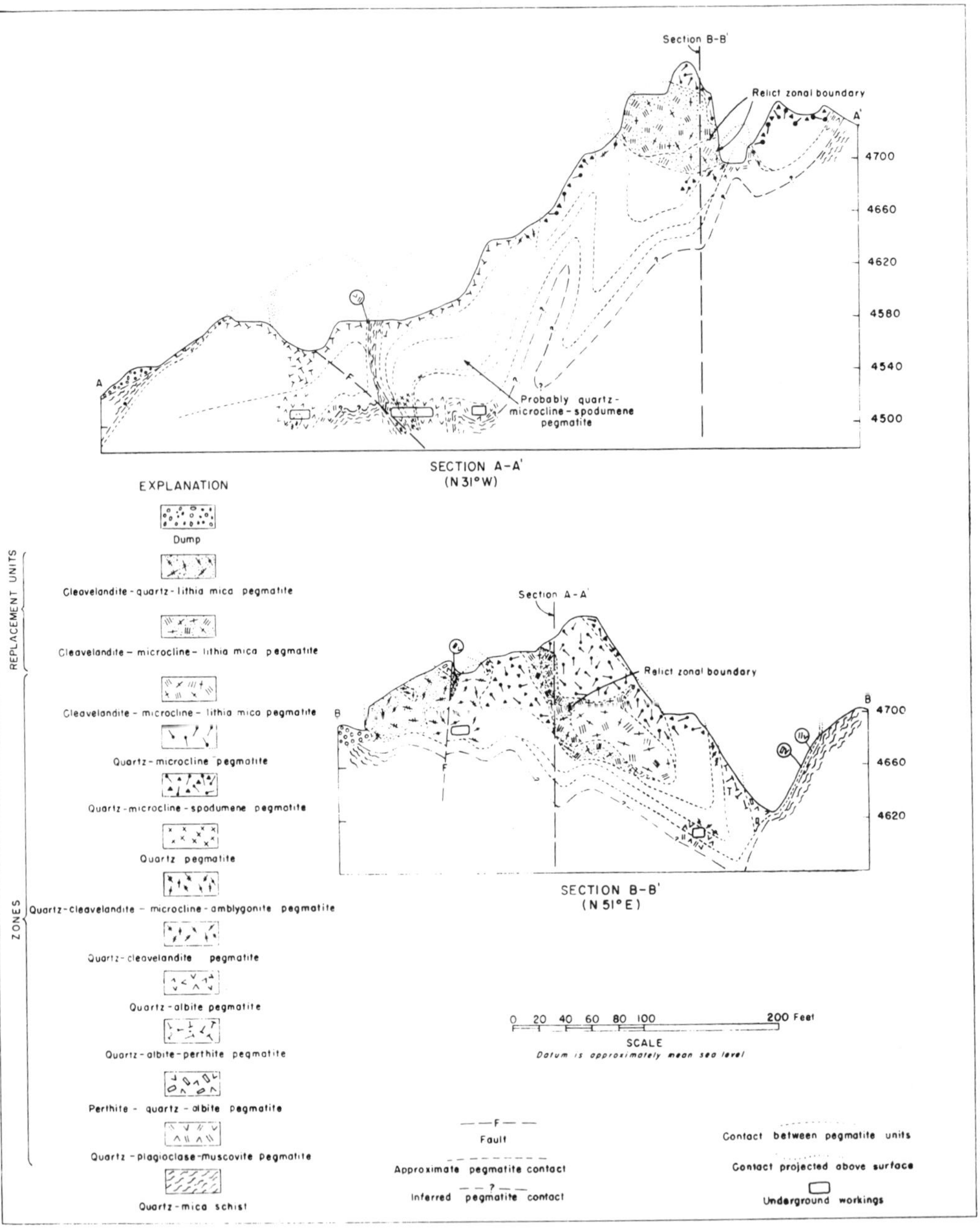

Fig. 2—Geologic sections, Hugo pegmatite, Keystone, S. D.

value of the feldspar product. Where this type of feldspar-bearing rock is beneficiated by flotation, the resulting product normally contains considerable quantities of soda feldspar and has a lower price than clean potash feldspar. Similar intergrowths of amblygonite with quartz or plagioclase, muscovite with quartz, beryl with quartz and plagioclase, and columbite-tantalite with quartz or plagioclase make hand methods ineffective for recovering the minerals sought. Most plagioclase of pegmatites is intergrown with quartz and cannot be marketed as soda feldspar. The chief reason that many pegmatite deposits cannot be mined at the present time is that intergrown minerals cannot be separated profitably at prevailing prices.

Because individual minerals range widely in chemical composition it is necessary to know the composition of the industrial minerals of each de-

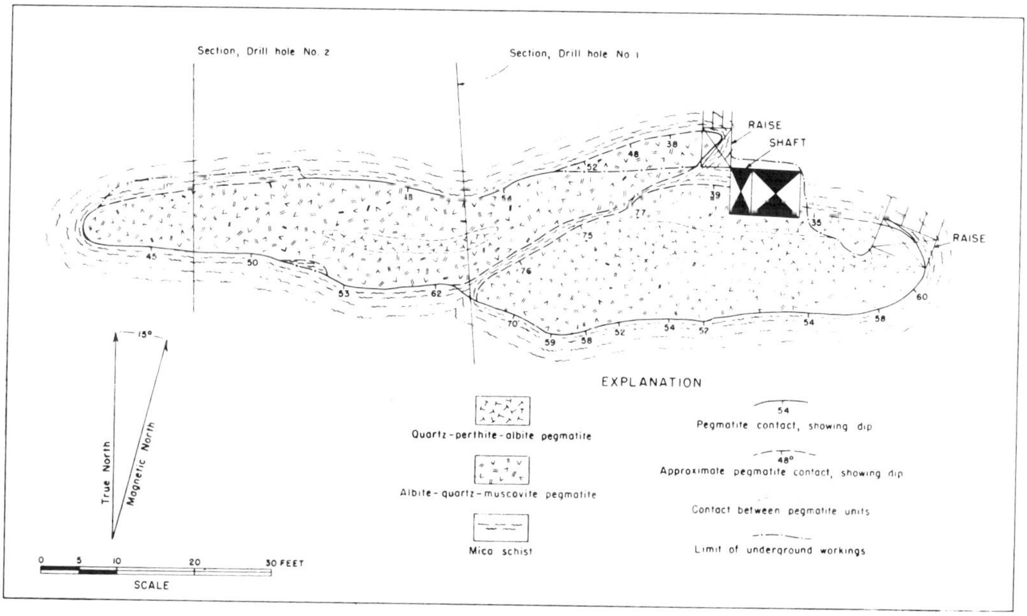

Fig. 3—Geologic map of the 60 level, Victory mine, Custer County, S. D.

posit within a single pegmatite. The value of spodumene, amblygonite, and lepidolite depends directly on the lithia content. The idealized formula for spodumene indicates a lithia content of 8.0 pct, but nearly all spodumene concentrates contain 4 to 7 pct lithia, largely because sodium and other elements occupy the lithium positions in the lattice. Likewise, in beryl the BeO content, on which the price depends, may range from 10 to 14 pct. Iron in potash feldspar may subject the product to penalties. On the other hand, high aluminum content in potash feldspar is considered desirable.

Variations in the properties of muscovite mica having influence on the market value have been discussed many times, most recently by Jahns and Lancaster.[17] The physical properties of muscovite determine whether or not it can be used as sheet mica or sold only as scrap. Value of trimmed sheet mica depends largely on the size of the sheets. Physical defects such as waviness or inclusions also affect the value of sheet mica. During World War II when the Federal Government purchased the entire sheet mica output of all mica mines—except those that produced chiefly stained mica—most of the sheet mica produced in the U.S. was purchased at a fixed price per pound without regard to size or quality. In normal times, however, the better the quality and the greater the size of sheet mica, the higher its value.

Determination of Grade

Because the term *grade* is used with several meanings in pegmatite mining, its common use in reference to the metal content of an ore deposit may be confusing. *Grade* is used in commerce to indicate the size of sheet mica and also the quality of potash feldspar.

Statements about grade should be carefully qualified. For example, a figure for the overall metal content is misleading if the metal occurs in two minerals and only one of these minerals can be used commercially. Statements expressing the total quantity of a mineral in a deposit do not convey sufficient information if, as in sheet mica deposits, the crystals recovered must be greater than a certain size.

The grade, or mineral content, of a pegmatite deposit may be determined by assays, grain measurements of representative faces, or production records. It may be expressed as: 1) percent of mineral, 2) percent of recoverable mineral, 3) percent of oxide or element, 4) dollar value of the industrial minerals per ton of rock, or 5) dollar value of recoverable minerals per ton of rock. The choice will depend on the purpose for which the grade estimate is to be used.

Grade expressed in terms of percent of mineral may not be useful because most pegmatite mines use hand sorting, and in these mines the total percentage of mineral is of no practical interest. If, on the other hand, mechanical methods of beneficiation are to be used, it is necessary to know the total percentage of mineral. The total percent of mineral is also a convenient way of expressing the grade of potential reserves.

The method of expressing grade in terms of percent recoverable mineral unfortunately makes the grade figure depend in part on the mine efficiency. A better figure can be obtained by assuming that all mineral grains greater than a certain minimum size can be recovered. Minimum size for mica books that will give sheet or punch mica is about 1½x1½x⅛ in. Feldspar, beryl, spodumene, amblygonite, or lepidolite seldom can be sorted economically in pieces less than 1½ in. average dimension.

Where more than one mineral is produced from a pegmatite deposit, it may be desirable to express grade by a single figure in terms of dollar value per ton of rock. Such a figure must be used in conjunction with percentage figures, which remain constant, whereas grade expressed in terms of dollar value varies with prices change.

Assays of Channel, Bulk, or Core Samples: Chemi-

Table I. Results of Assays by Various Methods, Metals Reserve Samples from the Helen Beryl Pegmatite, Custer County, S. D.*

Sample Cut No.	Weight of Sample, Lb	BeO Content Determined by Measuring Beryl on Exposed Faces: Area Measured, Sq Ft	Beryl, Pct	BeO, Pct**	Spectrographic Assays: Geological Survey	Analyst *A*	Chemical Assays: Geological Survey	Analyst *A*	Analyst *B*	Analyst *C*: Average of Three Determinations	"Best Shift"†
1	6,400	80.5	0.25	0.03	0.03	0.09	0.03	0.19	0.11	0.18	0.14
2	5,150	48.6	0.60	0.08	0.03	0.09	0.07	0.06	0.15	0.13	0.15
3	6,620	64.7	0.23	0.03	0.03	0.05	0.04	0.16	0.09	0.12	0.12
4	4,000	20.0	0.76	0.10	0.08	0.07	0.03	0.13	0.16	0.08	0.15
5	9,350	94.0	0.23	0.03	0.006	0.02	—	0.12	0.07	0.03	0.04
6	7,400	36.0	0.60	0.08	0.08	0.09	0.10	0.16	0.14	0.34	0.33
Composite of samples 1 to 6					0.04	—	0.04	—	0.12	0.12	0.14
Weighted average‡ of samples 1 to 6				0.05	0.04	0.06	0.04	0.14	0.11	0.15	0.15
Average BeO content from each method				0.05	0.05				0.11§		

* After L. R. Page et al., Ref. 7, p. 130.
** Refractive indices show that average BeO content of the beryl is 12.5 pct.
† The meaning of the phrase *best shift* was not explained by the analyst.
‡ Weighted according to number of pounds in each sample.
§ Column headed *best shift* disregarded.

cal and spectrographic assays of channel, bulk, or core samples have limited value unless the mineralogy of the deposits is well understood. They have been used chiefly to determine the BeO, Li_2O, Ta_2O_5, and Sn content of pegmatite deposits, but results may be misleading. The BeO assay will include not only BeO in recoverable beryl but also in minerals not saleable at present or too fine grained to be recovered by hand methods. Spodumene is the pegmatite mineral most commonly used in the U.S. for production of lithium metal and salts, but assays for lithia will also include the lithia in amblygonite, triphylite-lithiophillite, and lithia micas. Derry[18] has evidence that appreciable quantities of lithia may occur in feldspar from spodumene pegmatites in Quebec. Similarly Ta_2O_5 and Sn may be in more than one mineral.

To determine how large a sample must be in order to be representative, it is necessary to consider: 1) average grain size, 2) range in grain size, and 3) evenness or unevenness of the distribution of the mineral sought. It is apparent that a fine grained, evenly distributed mineral can be adequately sampled by small, widely spaced channel samples or drill core samples. A mineral showing wide variation in grain size or very uneven distribution can be sampled only by taking a large volume of rock. Beryl deposits in the Bob Ingersoll No. 1 and No. 2 pegmatites, Keystone, S. D., would require bulk samples so large that the sampling operation would be similar in magnitude to the actual mining operation. Of the 150 tons of beryl produced from the Bob Ingersoll No. 1 pegmatite, about 100 tons have been from crystals and aggregates of crystals ranging from 11 to 61 tons. Similarly, 117 tons of the total production of about 300 tons of beryl from the Bob Ingersoll No. 2 pegmatite was from crystals and aggregates ranging from 6 to 34 tons.[19] Average beryl content of each of these deposits is probably less than 1 pct. It is apparent that a bulk sample would have to be exceedingly large, and other methods of determining beryl content may be more suitable. Many other beryl deposits, however, are amenable to normal sampling procedures because they are evenly grained and evenly distributed.

Grade Determined from Grain Measurements: Measurements of the area of a mineral on representative exposures can be used satisfactorily to determine grade except where the pegmatite is very fine grained. This method is essentially a modification of areal methods of determining rock composition that date back to Delesse in 1847.[20] Total area of the rock exposure and area of the mineral can be used to obtain volume percent, or corrections can be made for differences in specific gravity, and the percentage of mineral by weight can be determined. As in other types of sampling it is necessary to consider grain size and evenness of distribution in determining how large an area should be measured. Experience has shown that for rocks with an average grain size of 2 in. or less, 100 sq ft normally is adequate; for an average grain size of 6 in., 1000 sq ft should be measured; for 12 in., 5000 sq ft.

Care must be exercised to insure that the faces are representative. Accessible exposures in many abandoned mines are virtually barren because they were stripped of all valuable minerals when the mine was closed, and selection of representative faces is difficult or impossible.

The grade of beryl determined by grain measurements on the walls of six sample cuts in the Helen Beryl pegmatite, Custer County, S. D., is compared with spectrographic and chemical assays in Table I. Chemical assays by four different analysts failed to check each other satisfactorily. Two spectrographic analysts obtained approximately the same results. The spectrographic assays and grain measurements each indicate a grade of 0.05 pct BeO, but they are not in agreement with the 0.11 pct BeO determined chemically.

Grain measurements have been used successfully to predict the beryl content, before mining, of a block of beryl-bearing pegmatite at the Wood Tin mine, Pennington County, S. D.[21]

Large-scale maps and sections made in 1943 to show the size and distribution of large muscovite books in the underground workings at the White Bear pegmatite, Custer County, S. D., show that 66 sq ft, or 4.2 pct recoverable mica, were exposed in an area of 1560 sq ft of pegmatite. Actually, from 900 tons of rock mined, about 37 tons, or 4.1 pct mica, were recovered.

Mineral measurements show that the average beryl content of the drill core in the Helen Beryl

Table II. Muscovite Content of Plagioclase-Quartz-Muscovite Zone, Hot Shot Pegmatite, Custer County, S. D.*

Location	Muscovite in Drill Core, Estimated Percent				Muscovite, Estimated Content**		Muscovite, Measured Content**	
	Hole 1	Hole 2	Hole 3	Holes 1, 2, and 3	Area Examined, Sq Ft	Muscovite, Pct	Area, Sq Ft	Muscovite, Pct
Hanging wall	12	4	28†	19†	125	16	27	12
Footwall	10	7	17	13†	30	8	17	6.5
Both foot and hanging wall	12	6	23	16	155	14	44	10

* Determined from grain measurements on drill cores and from visual estimates and grain measurements in underground workings.
** Ref. 23.
† Erroneously reported in Ref. 21.

pegmatite is 0.19 pct. Spectrographic assays of the core samples, converted from percent BeO to percent beryl, indicate that the drill core contained 0.21 pct beryl.[22] Thus, for all practical purposes, the mineral measurements were as good as the spectrographic assays and less expensive. Very little additional work was required by the geologist, M. H. Staatz, who logged the core, to measure the observed beryl crystals. It is doubtful that the cost of measuring all beryl crystals found in the core exceeded $100. The cost of splitting and grinding the core, transporting the samples, and having the samples assayed was more than $1000.

The muscovite content of the plagioclase-quartz-muscovite wall zone of the Hot Shot pegmatite was determined,[23] as shown in Table II, by visual estimate of the muscovite content of three drill cores, by visual estimate of the quantity of muscovite exposed in this same zone in the underground workings, and by grain measurements in the underground workings. Differences in the figures for mica content in the drill cores are typical of the differences to be expected in small samples. The average value of these figures, weighted according to the length of drill core, shows the deposit contains about 16 pct muscovite. All values for mica content, whether high or low, should be used in obtaining weighted averages. The not uncommon practice of arbitrarily discarding all high values is not justified. In pegmatite containing very coarse crystals some small samples are expected to be very rich in muscovite where one or more large crystals are cut by the drillhole. These high values may compensate for other samples containing no muscovite. The largest samples in Table II are areas in the underground workings where the muscovite content was visually estimated as 16 pct for the hanging wall part of the zone, 8 pct for the footwall part of the zone, and 14 pct for both areas. These figures differ somewhat from those obtained from drill core 1, which cut the pegmatite near the underground workings, and from those obtained by measurements of mica crystals in smaller areas. The drillhole samples, however, provide an estimate of mica content that is adequate for most purposes.

A method similar to the Rosiwal technique can be used to determine grade by measuring the intercepts of grain boundaries on lines at fixed intervals on an exposure. This method is especially useful for coarse grained pegmatite. Jahns[24] has made measurements of this kind with a mechanical device consisting of a wheel and attached counters to record revolutions.

Microscopic grain counts of crushed samples have been used successfully by the U. S. Bureau of Mines to determine grade.[25] This method is most accurate if the size distribution of the crushed fragments of the various minerals is determined. In a crushed sample of quartz-spodumene pegmatite, for example, average size of the spodumene fragments may be smaller than the average size of the quartz fragments, and grain counts would indicate that the spodumene content is higher than it actually is. If each sample is sieved and spodumene grains are counted in each sieve fraction, the spodumene content of each sample can be readily determined by using the percent spodumene and the total quantity of material in each sieve fraction to compute a weighted average.

Microscopic grain counts may be impracticable for low grade deposits, especially if the minerals are not readily distinguishable by optical methods. An ore that contains 0.5 pct beryl has only 5 grains in 1000, and an accurate microscopic grain count would be long and tedious because the optical properties of beryl are somewhat similar to the optical properties of quartz, plagioclase, and muscovite. Stevens[26] has successfully used microscopic grain counts to determine the beryl content of products of a USBM pilot mill at Rapid City, S. D.

Minerals such as microlite that have a high specific gravity can be concentrated by means of heavy liquids before microscopic grain counts are made.

Grade Determined from Production Records: Production records show how much of an individual mineral can be recovered by any particular method of concentration. If such records are not available, a useful substitute method is to take large samples and concentrate them under conditions simulating actual mining. Production records were used with success during World War II in determining the tenor of sheet mica deposits in pegmatites. The tenor of such deposits can also be obtained by grain

Table III. Mica Recovery, New York Mine, Custer County, S. D.

Date	Rock Mined, Tons*	Mica Recovered	
		Weight, Tons*	Pct
June 1 to July 22, 1943	580	10.5	1.8
June 1 to Sept. 17, 1943	1800	28	1.6
June 1 to Nov. 17, 1943	2000	32	1.6

* Based on foreman's records and measurements of volume of rock mined. Published with permission of the owners.

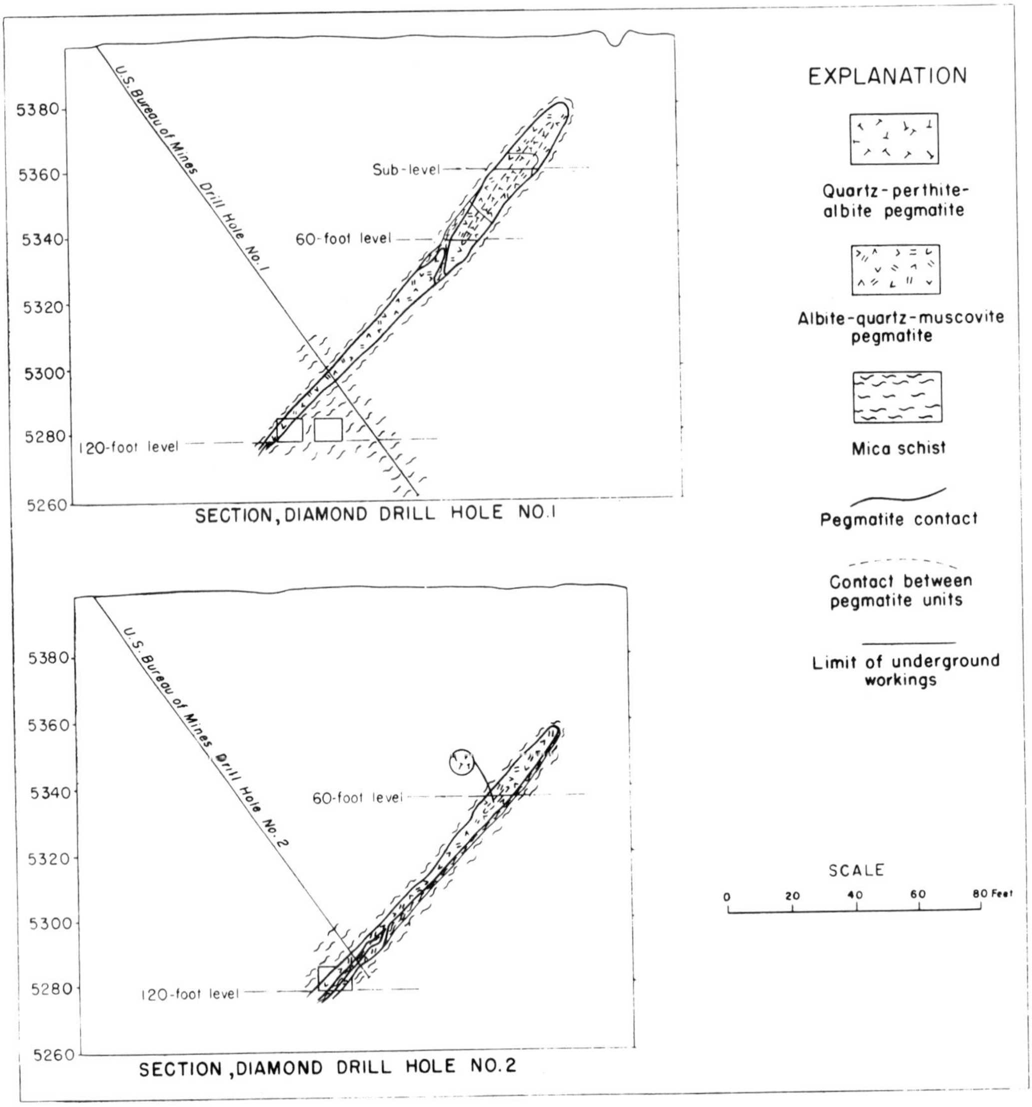

Fig. 4—Geologic sections, Victory pegmatite, Custer County, S. D.

measurements of mica books that are larger than the smallest marketable cut mica.

The percent recoverable mica in the deposit worked at the New York mine during 1943 was determined at three different times while the mine was in operation. Table III shows that the percent mica recovered from June 1 to July 22, 1943 was 1.8 pct, and that by September 17, 1943 the figure for recovery was reduced to 1.6 pct. The recovery figure was still 1.6 pct when the mine closed November 17, 1943. The consistency of the recovery figures seem to indicate that the tenor of any one large block of rock is essentially the same as that of any other large block in the mica zone that was mined.

Careful estimates and measurements of the mineral distribution on exposures in the lower 42 ft of the vertical shaft and in crosscuts of the Hot Shot pegmatite were compared with the results obtained by the USBM[27] in hand sorting perthite, muscovite, and phosphate minerals from the rock obtained from these workings. Table IV shows the quantity of each of these minerals estimated to be in each round, and also the quantity of each of these minerals that was actually recovered. Thus, this table shows approximately how much of each mineral is lost during hand sorting of pegmatite. Rounds estimated to contain relatively small quantities of muscovite were also, in general, those in which small proportion of the total estimated quantity of muscovite was recovered. In only a few of the rounds estimated to contain less than 500 lb of muscovite was more than 10 pct of the total estimated quantity of the muscovite recovered. On the other

Table IV. Recovery by Hand Sorting of Perthite, Muscovite, and Phosphates from the Hot Shot Pegmatite, Custer County, S. D.*

Pegmatite Units	Round	Perthite			Muscovite			Phosphates (Except Amblygonite)		
		Estimated Weight, Lb	Recovered Weight, Lb	Estimated Weight Recovered, Pct	Estimated Weight, Lb	Recovered Weight, Lb	Estimated Weight Recovered, Pct	Estimated Weight, Lb	Recovered Weight, Lb	Estimated Weight Recovered, Pct
Perthite-quartz-plagioclase pegmatite.	Shaft 1	19,800	6,400	32.3	860	103	12.0			
	Shaft 2	37,800	10,400	27.5	1,140	33	2.9			
	Shaft 3	33,900	15,200	44.8	1,380	16	1.2	284	None	0.0
	Shaft 4	16,900	12,800	75.7	580	21	3.6			
	Shaft 5	28,500	17,600	61.8	385	22	5.7			
	Shaft 6	24,900	17,600	70.6	455	13	2.9			
	Shaft 7	32,400	19,200	59.3	1,770	40	2.3			
Perthite-quartz-plagioclase, quartz-plagioclase-muscovite, and cleavelandite-lithia mica-quartz pegmatite.	Shaft 8	21,600	17,600	81.5	580	9	1.6	120	None	0.0
Perthite-quartz-plagioclase, quartz-plagioclase-muscovite, quartz-spodumene-plagioclase, and cleavelandite-lithia mica-quartz pegmatite.	Shaft 9	8,650	8,000	92.4	1,100	53	4.8	None	64	—
Perthite-quartz-plagioclase, quartz-plagioclase-muscovite, quartz-plagioclase-perthite, quartz-spodumene-plagioclase, and cleavelandite-lithia mica-quartz pegmatite.	Shaft 10	1,930	1,600	82.9	3,120	92	3.0	500	220	44.0
Quartz-plagioclase-muscovite, perthite-quartz-plagioclase, quartz-spodumene-plagioclase, and quartz-plagioclase-perthite pegmatite.	Shaft 11	1,610	800	49.7	1,810	80	4.4	Tr.**	30	—
Quartz-plagioclase-muscovite, perthite-quartz-plagioclase, and quartz-spodumene-plagioclase pegmatite.	Shaft 12	1,610	1,600	99.4	1,010	349	34.6	Tr.**	18	—
Quartz-plagioclase-muscovite, perthite-quartz-plagioclase, cleavelandite-lithia mica-quartz, and quartz-plagioclase-perthite pegmatite.	South crosscut 1 and North crosscut 1	18,430	7,200	39.1	1,770	130	7.3	1 650	122	7.4
Perthite-quartz-plagioclase, cleavelandite-lithia mica-quartz, and quartz-plagioclase-perthite pegmatite.	North crosscut 2	8,260	1,200	14.5	Tr.**	27	—	1 330	87	6.5
	North crosscut 3	7,060	800	11.3	Tr.**	37	—	2 340	11	0.5
Cleavelandite-lithia mica-quartz, perthite-quartz-plagioclase, quartz-plagioclase-perthite, and plagioclase-quartz pegmatite.	North crosscut 4	17,800	2,400	13.5	270	35	13.0	590	14	2.4
Plagioclase-quartz and perthite-quartz-plagioclase pegmatite.	North crosscut 5	10,500	1,600	15.2	750	296	39.5			
Plagioclase-quartz pegmatite.	North crosscut 6	None	800	—	2,340	195	8.3			
Plagioclase-quartz and plagioclase-quartz-muscovite pegmatite.	North crosscut 7	None	None		675	71	10.5			
	North crosscut 8	None	None		3,740	520	13.9			
	North crosscut 9	None	None		1,940	355	18.3			
Perthite-quartz-plagioclase and plagioclase-quartz pegmatite.	North crosscut 10	7,820	800	10.2	750	215	28.7			
Perthite-quartz-plagioclase pegmatite.	North crosscut 11	32,800	3,200	9.8	350	34	9.7			
Plagioclase-quartz, perthite-quartz-plagioclase, and plagioclase-quartz-muscovite pegmatite.	North crosscut 12	2,560	1,600	62.5	650	152	23.4			
Plagioclase-quartz-muscovite and plagioclase-quartz pegmatite.	North crosscut 13	None	None		610	246	40.3			
	South crosscut 1	**See North Crosscut 1. Both rounds sorted together.**								
Perthite-quartz-plagioclase and quartz-plagioclase-muscovite pegmatite.	South crosscut 2	2,660	400	15.0	820	123	15.0	None	30	—
	South crosscut 3	20,660	1,600	7.7	1,130	72	6.4			
	South crosscut 4	23,200	1,600	6.9	870	77	8.8			
Perthite-quartz-plagioclase and plagioclase-quartz pegmatite.	South crosscut 5	23,800	1,600	6.7	131	40	30.5			
	South crosscut 6	4,910	400	8.1	290	24	8.3			
Plagioclase-quartz, plagioclase-quartz-muscovite, and perthite-quartz-plagioclase pegmatite.	South crosscut 7	430	None	0.0	1,280	23	1.8			

* After H. G. Stephens and D. B. Stewart: Ref. 23. Weight of minerals recovered from each round by hand-sorting is from A. B. Needham, Ref. 27, Table V. Estimated quantity of the minerals in each round was determined from the estimated mineral distribution on exposed faces.
** Tr. means less than 50 lb was estimated to be in round.

hand, in rounds estimated to contain more than 500 lb of muscovite, much of which was in large books, recovery was somewhat higher. The recovery of perthite shows a similar tendency.

Determination of Reserves

Reserves of pegmatite minerals can be calculated after the pegmatite has been mapped, geologic sections constructed, and grade of zones that attract interest determined. For computing reserves of any particular category, the pegmatite can be divided on the basis of grade into suitable blocks of ore. Geologic sections, properly spaced to show variations in shape and size, can be used to determine the average cross-sectional area of an ore block, and the volume of the block can be obtained by multiplying by the distance between geologic sections. At some properties, as at the Edison, Fig. 5, geologic plans at fixed vertical intervals are readily used to determine the volume of a block of ore. Volume determinations calculated from geologic sections should not differ significantly from those obtained from plans. Specific gravity of the pegmatite can be estimated and used to convert volume to tonnage. One ton of pegmatite in place normally occupies about 12 cu ft.

The problem of determining the tonnage of the deposit is the same regardless of the mineral sought. In determining grade, however, each pegmatite mineral presents certain problems that are peculiar to that mineral:

Mica: Evaluation of a mica deposit involves determination of 1) total mica content, 2) content of recoverable mica, 3) content of sheet in mica recovered, 4) content of various sizes of sheet mica, and 5) content of various qualities of sheet mica. These can be expressed in percentage terms, but the only simple way to express the value of the deposit is in terms of dollar value of the recoverable mica per ton of rock. Value of byproducts and cost of rifting and trimming the mica can also be incorporated in this figure. The term *grade* may be confusing if used in reference to the amount of mica in a deposit, as it is used extensively in the sheet mica industry to indicate the size of trimmed sheets.

The Victory and the Tin Queen pegmatites, near Custer, S. D., show how a figure expressing dollar value per ton of rock can be useful for mica deposits. The Victory pegmatite, from which about 8.8 pct crude mica was recovered, produced an average of 19 lb of sheet and 65 lb of washer punch and other byproduct mica per ton of rock mined in 1943 and 1944. On the other hand, only 1.6 pct crude mica was recovered from the Tin Queen pegmatite and only 1.3 lb of sheet and 3.8 lb of washer punch were produced per ton of rock mined. Thus the Victory pegmatite, which produced 5½ times as much crude mica per ton of rock as the Tin Queen, produced about 15 times as much sheet mica and 17 times as much washer punch and other byproduct mica as the Tin Queen. Furthermore, rifting and trimming costs for the Tin Queen mica were probably greater than for the Victory mica, and to compare these two mines it is necessary to subtract the rifting and trimming costs from the value of the mica. The Tin Queen pegmatite was a source of considerably greater quantities of byproduct minerals, such as feldspar, amblygonite, and beryl, than the Victory pegmatite. If the dollar value of these minerals per ton of rock mined is taken into consideration, then the value of the products of the Tin Queen pegmatite is increased by about 50 pct. The only way to compare these two mines in simple terms is to determine the dollar value of the minerals produced per ton of rock.

Of all the mica mines in the Black Hills, the Victory is the one for which the most data are available on predicted reserves compared with actual production. The history of the authors' reserve calculations at this mine, therefore, will be given in detail to illustrate the degree of accuracy obtainable in estimating reserves of mica. The Victory program pegmatite was carefully studied by the USGS in 1943 and 1944[28] and was core drilled by the USBM in February 1944.[29] The pegmatite is a multiple intrusive consisting of two structurally independent, but connected, bodies that continued down the plunge as far as the 120 level, Figs. 1, 3, and 4.

Reserves in unmined rock above the 120 level were estimated as of March 1, 1944. The crude mica recoverable was 7.6 pct of the rock mined on the 60 level.** The waste rock from mining in the upper levels contained 17.1 pct total mica according to the USBM.[30] Average muscovite content in two drillholes that cut the pegmatite near the 120 level was 32 pct, suggesting that the muscovite content of the pegmatite might increase downward.

On the basis of information from geologic mapping and diamond drilling, the quantity of unmined pegmatite estimated to be above the 120 level on March 1, 1944 was 4450 tons. If the content of recoverable mica were 7.6 pct, a total of 338 tons of mica could be produced. When the mine closed at the end of 1944, the operators' records showed that 285 tons of mica had actually been produced from this block. Computations based on the isopachous map, Fig. 1, showed that 3250 tons of pegmatite were mined. These figures indicate that the pegmatite contained 8.8 pct recoverable mica.

** Actually the figure originally was 5.77 pct, but it was based on erroneous data. The chief error was an incorrect assumption by the mine operator on the capacity of the mine cars.

Table V. Percent of Lithia Found in Analyses of Spodumene-Bearing Samples from the Edison Mine, Pennington County, S. D.*

Sample Selected As	U. S. Bureau of Mines	Analyst *A*	Analyst *B*	Analyst *C*	Analyst *D*	Analyst *E*	Analyst *F*
Low	0.79	0.94	Trace	1.10	1.24	1.38	1.48
Medium	2.80	3.12	2.88 3.10	3.50	3.50	3.52	3.55
High	4.63	5.54	5.54 5.50	6.05	6.00	5.41	5.74

* See Ref. 31. Table VII.

The isopachous map also showed that about 2550 tons of pegmatite remained unmined and that on March 1, 1944, when 4450 tons of pegmatite were predicted on the basis of geologic sections, the actual quantity was about 5800 tons, indicating an error of 23 pct. Similarly there was a 14 pct error between the mica content of 7.6 pct indicated by the original data and the 8.8 pct actually obtained. The discrepancy between the original and final figures for mica content can be explained partly by the fact that during the later months of the operation the broken rock was hand sorted for mica twice, instead of once.

Lithium Minerals: The principal lithium minerals produced from pegmatites are spodumene, lepido-

Table VI. Predicted Grade and Reserves Compared with Production Figures, Edison Spodumene Mine, Pennington County, S. D.

Reserves	Above Adit Level			Between Adit Level and 50-Ft Level		
	Spodumene-Bearing Pegmatite, Tons	Li_2O, Pct	Li_2O, Units	Spodumene-Bearing Pegmatite, Tons	Li_2O, Pct	Li_2O, Units
Predicted: Based on geologic sections	40,000	1.28	51,000	64,000	1.03	65,900
Predicted: Based on geologic plans				85,000	1.04	88,400
Mined 1945 to 1949*	34,000	1.35	45,900	30,000	1.02	30,600
Estimated reserves, Jan. 1, 1950*	Several thousand			40,000 to 50,000		

* From F. F. Clarke, Lithium Corp of America: Personal communication. Published with his permission.

lite, and amblygonite. Because some deposits include two or more lithium minerals, not only the lithia content of the pegmatite must be determined, but also the grade for each of the lithium minerals present. Grain measurements can be used to determine the grade of these minerals, and if only one of them is present, chemical analyses can be used. Production records may be used to give the grade in terms of recoverable mineral.

Chemical assays are useful to determine the grade of such spodumene deposits as the Edison pegmatite, Pennington County, S. D., where the spodumene is fine grained and uniformly distributed. The USBM used chemical methods to assay channel, bulk, and drill core samples from this deposit.[31] To check the assays, each of three samples analyzed by the USBM was also analyzed by six private laboratories. Results are shown in Table V. It is noteworthy that the greatest errors are in the sample that was selected as having a low content of lithia; results for this sample range from a trace to 1.48 pct Li_2O. Only one determination shown on the table was lower than the USBM determination for the same sample. Actually it was found that reserves computed for the Edison pegmatite, based on USBM assays, were very nearly correct, as shown in Table VI. The inadvertent mining of waste rock with ore probably compensated for any errors in assays.[32] Recent experience has shown that the flame photometer gives better results than the more difficult chemical methods to determine lithium.

In estimating spodumene reserves in the Edison mine the tonnage of unmined pegmatite was determined both from geologic plans and from sections. Fig. 5 consists of geologic plans that illustrate the structure of this pegmatite, showing the multiple nature of the intrusive and the zoning. The core of quartz-spodumene and quartz-spodumene-albite pegmatite is the orebody. Reserves in various blocks of ore above the adit level, between the adit level and the 50 level, between the 50 level and the 100 level, and below the 100 level were calculated in advance of mining. In the deeper parts of the pegmatite the grade is lower than at higher levels because more of the spodumene is altered to micaceous and clay minerals.

In five years of mining, from 1945 through 1949, most of the spodumene-bearing pegmatite above the adit level and much of the pegmatite between the adit level and the 50 level was mined. Table VI compares predicted reserves and grade with what was actually mined. Of the 40,000 tons of spodumene-bearing rock predicted above the adit level, 34,000 tons were mined and "several thousand" tons[32] still remained in place in 1950. The predicted grade was 1.28 pct Li_2O, and the grade of rock that was mined was 1.35 pct Li_2O. Table VI also shows that probably more than half the spodumene-bearing pegmatite between the adit level and the 50 level had not yet been mined in 1950. The grade of rock mined in this block has been 1.02 pct Li_2O, comparing remarkably well with the 1.03 and 1.04 pct Li_2O predicted. In view of the very complex nature of the Edison pegmatite as shown in Fig. 5, and the inability of analysts to check each other, production records have verified predictions more closely than expected.

Most amblygonite that has been mined in the U.S. is from very large crystals, some of them probably weighing hundreds of tons. The extremely large size of amblygonite in many deposits makes it necessary to measure crystals on a very large area of pegmatite to determine grade. On the other hand, amblygonite in some deposits is relatively fine grained, and much of it cannot be recovered by hand sorting. The grain size of amblygonite is as important a factor as grade in determining whether or not a deposit can be mined.

Lepidolite usually occurs in large aggregates associated with cleavelandite, quartz, and other minerals. The lepidolite must be assayed to determine whether or not it carries the minimum percent Li_2O acceptable in the market. Grade is best expressed in terms of percent lepidolite rather than units of lithia because the product is sold as tons of lepidolite rather than units of lithia.

Beryl: Estimation of reserves of beryl can be most readily accomplished if grain measurements are used to determine grade. Beryl is unevenly distributed and varies widely in grain size in many deposits, and consequently a large quantity of rock must be examined to determine grade.

Chemical and spectographic analyses of channel samples, drill samples, or small bulk samples can also be used in determining the grade of deposits in which beryl is uniformly distributed and is the only beryllium mineral present, but it is best to confirm the analyses by grain measurements. Experience with chemical assays has frequently been unsatisfactory because beryllium is difficult to determine and normally forms such a very small percentage of a deposit that small absolute errors in assays can cause errors of 200 pct or more in computing grade. Spectographic analyses have proved more satisfactory. Table I shows results of BeO de-

Fig. 5 (right)—Plans showing geology projected to the adit, 50 and 100 levels, Edison spodumene mine, Pennington County, S. D.

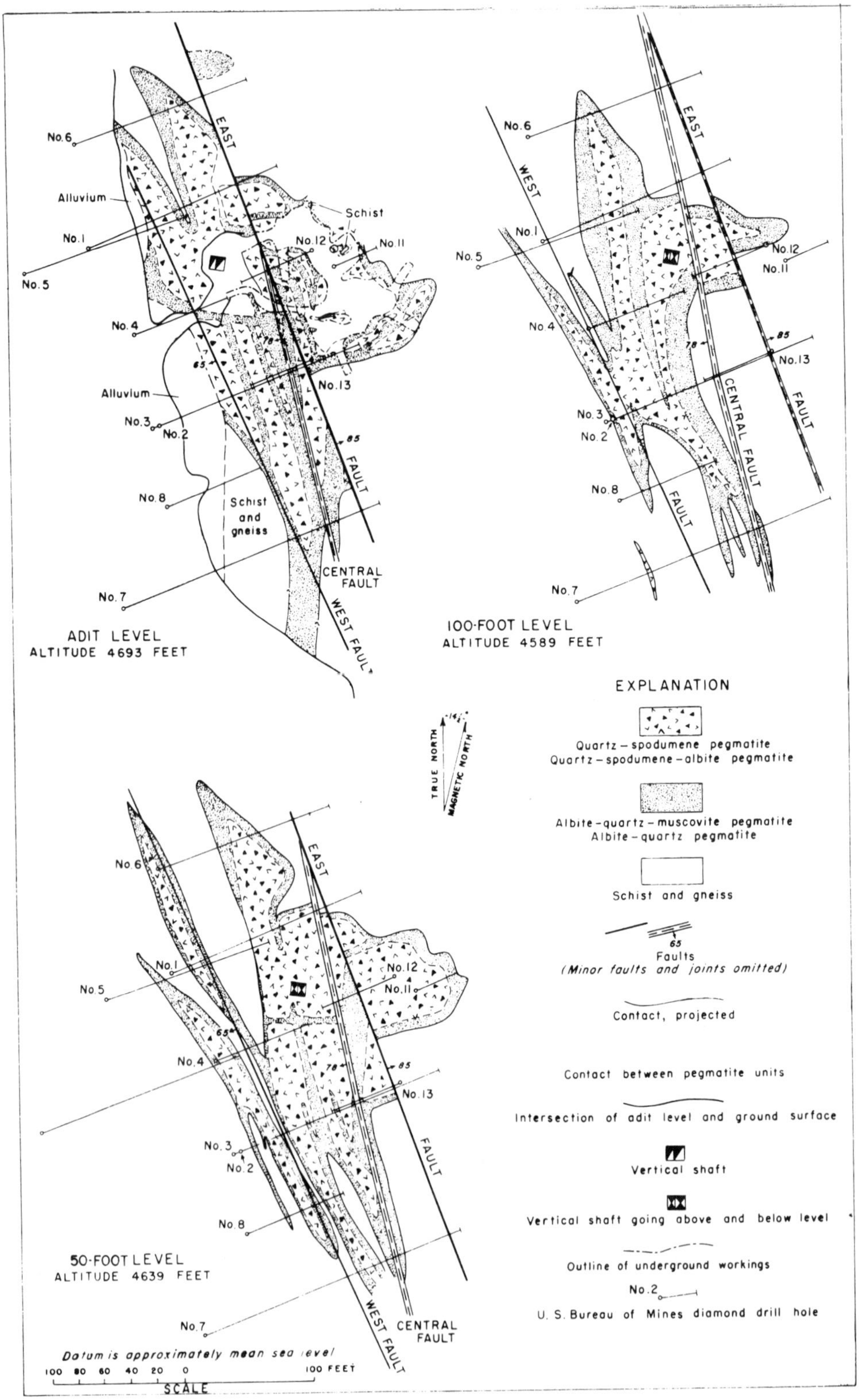
ADIT LEVEL
ALTITUDE 4693 FEET
100-FOOT LEVEL
ALTITUDE 4589 FEET
50-FOOT LEVEL
ALTITUDE 4639 FEET
EAST
FAULT
CENTRAL FAULT
WEST FAULT
Alluvium
Schist
Schist and gneiss
No. 1
No. 2
No. 3
No. 4
No. 5
No. 6
No. 7
No. 8
No. 11
No. 12
No. 13
TRUE NORTH
MAGNETIC NORTH
EXPLANATION
Quartz – spodumene pegmatite
Quartz – spodumene – albite pegmatite
Albite – quartz – muscovite pegmatite
Albite – quartz pegmatite
Schist and gneiss
Faults
(Minor faults and joints omitted)
Contact, projected
Contact between pegmatite units
Intersection of adit level and ground surface
Vertical shaft
Vertical shaft going above and below level
Outline of underground workings
U. S. Bureau of Mines diamond drill hole
Datum is approximately mean sea level
100 80 60 40 20 0 100 FEET
SCALE

154

terminations for samples from the Helen Beryl pegmatite by grain measurements, spectographic assays, and chemical assays. Grain measurements and spectographic assays by two different analysts gave about the same results. Chemical assays by four different analysts, however, differed substantially. It is noteworthy that samples that apparently contain the smallest amount of BeO are those in which the largest errors are made. For example, Table I shows that for sample 5 one of the chemical analysts obtained a result nearly three times as much as the average of all the other determinations of this sample.

These difficulties of analysis prevail for all but very high grade ores, and the use of grain measurements is thus nearly always necessary. For the purpose of determining the BeO content of the beryl itself, a graph such as Schaller's[33] can be constructed, showing the omega index of refraction plotted against percent BeO.

Page[21] has shown that in the Wood Tin pegmatite, Pennington County, S. D., 0.9 pct beryl was recovered in a 4400-ton block of rock from which a recovery of 1.0 pct was expected on the basis of grain measurements.

Feldspar: Many feldspar mines are small hand sorting operations for which reserve estimates are seldom necessary, but relatively large mining and milling operations that should use reserve estimating techniques are becoming more and more common. In either type of operation the percent recoverable feldspar is a figure of importance in expressing the value of a deposit, and for milling operations the total feldspar content must also be determined.

The term *grade* is used in the potash feldspar industry for quality of feldspar, and consequently the use of this word to express the amount of feldspar in a deposit can be ambiguous. Much of the feldspar of pegmatites, especially plagioclase, occurs in grains that are too small or too much intergrown with other minerals to be recovered by hand sorting. For mines that use milling methods, or where for some other reason the total feldspar content must be determined, grain measurements may be used to determine the percent feldspar. In very coarse grained deposits the intercepts of feldspar crystals on lines separated by fixed intervals can be used to determine the feldspar content much as the Rosiwal method is used in thin section work.

Much of the better quality potash feldspar has been mined from very coarse grained, hood shaped, incomplete zones in the upper parts of pegmatites. Perthite is most abundant at the top of these units, and as the deposit becomes leaner downward and inward the groundmass minerals, chiefly quartz and plagioclase, increase in quantity. In blocking out reserves these gradual changes must be considered.

Other Minerals: The other industrial minerals of pegmatites, including cassiterite, columbium-tantalum minerals, pollucite, and uranium minerals, are distributed through pegmatite units in much the same way as the minerals that have already been discussed. Normally these minerals form only a small quantity of the rock, even in the richest deposits. Grade may be determined by grain measurements, by visual estimates on representative faces, or by assays.

The variation in metal content in most of these minerals requires that both grade and the metal content of the mineral itself be determined. Columbite-tantalite, for example, is sold as columbium ore if it contains less than 35 pct Ta_2O_5 and as tantalum ore if above that figure. Other requirements for total Cb_2O_5 plus Ta_2O_5 content, percent SnO_2, and percent TiO_2 would have to be considered in blocking out reserves. Cost and difficulty of analyses for Cb_2O_5 and Ta_2O_5 have led to the practice of purchasing columbite-tantalite on the basis of specific gravity. Since the mineral may have a different composition in each zone, reserves must be stated in terms of tonnage containing columbite-tantalite of some specified Ta_2O_5 content or specific gravity.

Acknowledgments

All the geologists of the U. S. Geological Survey who worked on South Dakota pegmatites between 1939 and 1950 and many who worked in other districts have contributed to the formulation of the ideas presented in this paper. It is impossible to give specific credit to each of these men. Those who worked on the pegmatites mentioned in this paper include, in addition to the authors, J. W. Adams, L. C. Pray, T. A. Steven, M. H. Staatz, Peter Joralemon, D. B. Stewart, H. G. Stephens, M. P. Erickson, and R. E. Roadifer. Aid was also given by mine operators, engineers of the U. S. Bureau of Mines, and officials of the Colonial Mica Corp., a Metals Reserve Co. subsidiary.

References

1 H. E. McKinstry: *Mining Geology*, pp. 279, 390. Prentice-Hall, New York, 1948.
2 A. M. Bateman: *Economic Mineral Deposits*, p. 745. John Wiley & Sons, New York, 1942.
3 W. Lindgren: *Minerals Deposits*, 4th ed., p. 758. McGraw-Hill Book Co., New York, 1933.
4 R. R. Chowdhury: *Handbook of Mica*, pp. 6-7. Chemical Publishing Co., Brooklyn, N. Y., 1941.
5 B. C. Burgess: Feldspar. In *Industrial Minerals and Rocks* by S. H. Dolbear et al., 2nd ed., p. 347. *AIME*, New York, 1949.
6 E. H. Dawson: Mica. See Ref. 5, p. 564.
7 L. R. Page et al.: Pegmatite Investigations 1942-1945, Black Hills, S. Dak. USGS *Prof. Paper 247*, 1953. (This is the principal reference. Other references are cited separately.)
8 E. N. Cameron, R. H. Jahns, A. H. McNair, and L. R. Page: Internal Structure of Granitic Pegmatites. *Economic Geology* Monograph 2, 1949.
9 W. D. Johnston, Jr.: Beryl-Tantalite Pegmatites of Northeastern Brazil. Geol. Soc. of America *Bull. 56*, 1945, pp. 1024-1025.
10 E. N. Cameron, D. M. Larrabee, A. H. McNair, J. J. Page, V. E. Shainin, and G. W. Stewart: Structural and Economic Characteristics of New England Mica Deposits. *Economic Geology*, 1945, vol. 40, pp. 372-373.
11 J. S. Baker: World Survey of Tantalum Ore. USBM *Inf. Circ. 7319*, 1945, p. 13.
12 Ref. 8, pp. 13-97.
13 W. C. Smith and L. R. Page: Tin-Bearing Pegmatites of the Tinton District, Lawrence County, S. Dak. USGS *Bull. 922-T*, 1941, p. 609.
14 Ref. 7, p. 204.
15 D. B. Sterrett: Mica Deposits of the United States. USGS *Bull. 740*, 1923, p. 295.
16 M. P. Erickson and T. A. Steven: Mateen Spodumene Prospect. In Ref. 7, Plate 27.
17 R. H. Jahns and F. W. Lancaster: Phyiscal Characteristics of Commercial Sheet Muscovite in the Southeastern United States. USGS *Prof. Paper 225*, 1950.
18 D. R. Derry: Lithium-Bearing Pegmatites in Northern Quebec. *Economic Geology*, 1950, vol. 45, p. 103.
19 A. I. Johnson: Personal communication.
20 A. Holmes: *Petrographic Methods and Calculations*, p. 312. Thomas Murby and Co., London, 1930.
21 L. R. Page: Uranium in Pegmatites. *Economic Geology*, 1950, vol. 45, p. 26.
22 M. H. Staatz et al.: USGS report in preparation.
23 H. G. Stephens and D. B. Stewart: Mineral Distribution in the Hot Shot Pegmatite, Custer County, S. Dak. USGS unpublished report.
24 R. H. Jahns: The Genesis of Pegmatites. II. Quantitative Analysis of Lithium-Bearing Pegmatite, Mora County, N. Mex. *American Mineralogist*, 1953, vol. 38, p. 1096.
25 J. P. Gries: Investigation of the Beecher No. 2 Lithium-Bearing Pegmatite, Custer County, S. Dak. USBM *R.I. 4632*, 1950, pp. 0-13.
26 E. H. Stevens: Oral communication.
27 A. B. Needham: Investigation of Mica Deposits at the White Bear, Silver Dollar, Buster Dike, and Hot Shot Mines, Custer County, S. Dak. USBM *R.I. 4693*, 1950, pp. 52-53, Table V.
28 Ref. 7, pp. 206-210.
29 A. B. Needham: Investigation of Mica Deposits at the Victory, Jack Rabbit, Rainbow, and Midas Mines, Custer County, S. Dak. USBM *R.I. 4507*, 1949.
30 G. A. Munson: USBM unpublished report.
31 F. F. Clark, P. Zinner, et al.: Edison Spodumene Mine, Pennington County, S. Dak. USBM *R.I. 4234*, 1948, pp. 9-23.
32 F. F. Clarke: Personal communication.
33 W. T. Schaller: Beryl—Preliminary Statement. USGS unpublished memorandum, 1948.

19

Reprinted from *Eng. and Mining Jour.* **163:**80–88 (July 1962)

VALUATION OF LARGE, GOLD-BEARING PLACERS

A. Daily

SAMPLES from large gold-bearing deposits are obtained with placer drills such as the above BE 22-T. Volume and weight estimates are keyed to drilling procedure.

In this article, Arthur Daily describes a reliable procedure for estimating the amount of gold in a soil. According to the author, final gold recoveries have been within 5% of those predicted. Although the procedure requires careful study, it can be performed simply and rapidly.

THE SUBJECT OF VALUATION of potentially economic mineral bearing alluvial deposits has been given a great deal of study by placer mining engineers, particularly since the introduction of bucket-line dredges in the U.S. shortly before the turn of the century. Although several articles have been published on the valuation of alluvial deposits, particularly those containing tin, very few that are easily available to the engineering profession have appeared on valuation of large auriferous stream valley deposits minable with bucket-line dredges, or with a combination of dragline and floating washing plant as used during recent years. One reason for the latter situation is that, historically, much work has been done by specialists who developed procedures they considered personal property and confidential. Understandably, most dredging companies also consider valuation procedures confidential, as well as the relation between recovery and estimate.

A valuation is an estimate of the weight and usually the value of commercial quality minerals that can be recovered, by use of appropriate equipment and methods, by mining and concentrating part of an alluvial deposit lying within defined areal and vertical limits.

This article presents principles and procedures proved reliable by close agreement between actual recovery and valuation at various properties over long periods of time. Two principles for valuation are:

- Basic valuations should be calculated by a placer engineer using procedures that involve minimum exercise of discretionary judgement anu personal opinion.
- Management, with advice from a qualified placer mining engineer, should decide what over-all recovery factors should be applied to a basic valuation for technical and policy reasons.

Procedures are treated in three parts: establishment from prospect drill log data of depth of mining section and unit gold content, areal valuation and application of management and experience factors.

Scope of Article

The scope of this article is unit (drill hole) and areal valuation of large thawed or frozen auriferous gravel deposits using data obtained by prospecting with a churn-type placer drill, resulting in a basic valuation, and preparation of estimates of recovery before and after extractive mining operations are under way, The data from drilling that are required for valuation are described in detail. Knowledge of the appropriate techniques for obtaining such data in the field, including selection of drilling pattern, is presupposed.

The procedures described are equally applicable to valuation of alluvial platinum bearing deposits, with modifications to tin bearing deposits, and under certain conditions to so-called heavy mineral (other than tin) bearing deposits. Also, the system of areal valuation is applicable in some situations for use with data obtained by methods other than drilling. Well known American designs of drilling machines include engine-powered drills manufactured by Buffalo-Springfield Co., successor to Keystone Driller Co., Bucyrus-Erie Co. and C. Kirk Hillman Co.; and the Empire and Ward hand-powered drills manu-

Mr. Daily is a consultant residing at 488 Fairbanks Ave., Oakland 10, Calif. For fifteen years he held engineering and management positions at large dredging operations. For the past fifteen years, he has acted as consultant on a number of projects in several countries ranging from the Arctic to the South Pacific.

factured respectively by New York Engineering Co. and Mills Iron Works Inc.

Usually open pits are preferable to placer prospect drills for prospecting shallow alluvial deposits, say less than ten feet in depth, which are dry or can be dewatered. Jet drilling usually is preferable for prospecting beach sand deposits for heavy minerals. The samples required for prospecting diamondiferous alluvial deposits are much larger than can be obtained with a placer drill because of the dilute dispersion of valuable mineral, wide range of particle sizes, and the high unit value of diamonds.

Separation of Functions

With respect to valuation:

A The initial functions of management (or representative of a venture group) are to obtain rights to prospect in a designated area, engage or assign a placer engineer to direct prospecting, authorize the general program to be followed, establish the criteria to be used, and provide funds. In the absence of action by management in whole or in part in respect of technical matters, it devolves upon the placer engineer as either consultant or employee to make the necessary decisions.

B The functions of the placer engineer are to direct and closely supervise the prospecting program, and to prepare a basic valuation based on the prospect drilling data and the plan of the area, i.e., an estimate on which is shown average depth, volume of material, quantity of valuable mineral and usually its value, that is present in a designated area.

C It is strictly the prerogative of management to apply to a basic valuation such over-all discretionary factors as are deemed advisable for technical or policy reasons, or for budget of other estimates. The advice of a placer mining engineer who has had broad experience with applicable types of mining operations is essential for competent evaluation of technical factors by management.

Field Drilling—Duties of Principal Personnel

The placer engineer engaged or assigned to direct a prospect drilling program is responsible for the conduct of the work, use of appropriate techniques and procedures, and calculation of the basic areal valuation. The related activities of the engineer are divided into two distinct phases: accomplishment of field work ending with completion of drill logs showing depth of mining section and unit gold content (and value) for each hole, and calculation of the basic areal valuation.

The engineer lays out the drilling plan, establishes the sequence in which the holes are to be drilled, prescribes the general procedures to be followed by the driller and panner, specifies the data to be recorded and provides the drill log forms. Also, he assures security of sampling and samples, generally supervises the work and arranges a plan on which the location and surface elevation of each hole are shown.

The principal duties of the driller are to operate the drill, supervise the crew, accurately measure drilling depths and cores, obtain data on formation and maintain the drill and accessory items. The panner, in conformance with instructions by the engineer, measures the volume of, and concentrates (pans), each increment of core; promptly provides the driller with sufficient information on panning results for guidance in selecting length of drive; and performs general work under direction of the driller. Also he places the final samples of concentrate in suitable containers bearing identification marks, and personally hands them to the engineer. The gold in each sample is separated from residual gangue material and is weighed in the project office, either by the engineer or, under his close supervision, by a qualified person.

All of the data resulting from these activities are recorded on the log for the hole. The panner may record the data he obtains, and that reported by the driller or, depending on circumstances, the driller, a drill foreman, or the engineer may record the data. The engineer completes entries of data not obtained in the field.

The engineer then analyzes the data recorded on the log for each hole, calculates the corrected weight of gold, fixes the expected depth of mining section (and overburden separately if necessary) and calculates the gold content by weight per unit of area and of volume. The fine gold and silver content of a sample of raw gold representative of that occurring in the deposit is determined by fire assay, and the value in money per unit of area and of volume usually is calculated and entered on each log. The engineer, when data for a sufficient number of holes are available, completes the prospect drilling plan and calculates a basic valuation showing estimated volume of material and the gold content (and value) for the area or for a portion(s) thereof which qualifies within selected criteria for mining.

Drive Shoe Factor

Steel pipe casing with a bevel edged drive shoe is used for practically all placer prospect drilling in thawed ground, to assure obtaining as representative a sample as possible. The commonly used sizes of casing have nominal inside diameters of six, five or four inches, and the corresponding outside diameters of standard drive shoes are 7½, 6½ and 5¼ in. The typical relationship between casing and shoe is shown in Fig. 2. For hand-powered drilling with light equipment such as the Empire drill, a serrated end on casing may be used rather than a separate shoe.

As the casing with shoe is driven into the formation, theoretically the bottom edge of the shoe cuts a cylindrical sample of material which is forced inward and upward into the casing. As the inside diameter of the casing is smaller than the cutting edge diameter of the shoe, the theoretical linear rise of core in the casing per unit of drive is in proportion to: area of cutting edge of shoe/area of inside of casing.

This relationship commonly is called the *drive shoe factor* and, for a new 7½-in.-dia shoe and American 6-in. heavy pipe casing having an ID of 5.761 in., is 0.3068/0.1803 = 1.695, which in the English system of measures is equivalent to 20.3 in. of rise per foot of drive.

This theoretical relationship assumes that the shoe cuts a perfect cylinder of material that occupies the same volume inside the casing as in place in the ground. Actually, this theoretical condition seldom is realized in practice. Rocks encountered by the cutting edge of the shoe may not be cut cleanly and may be forced into or away from the casing together with more or less fine material. The character of formation may be such that a compacted zone forms within the shoe, forcing some material outward into the surrounding formation. As material is forced into the casing it may be compacted or it may expand, resulting in an apparently deficient or excessive rise of core. Loose sandy formation may be encountered, often under artesian pressure, resulting in a "run-in" of material into the casing, sometimes of major propor-

XYZ Company Prospect Drilling Field Log

Project **Line No.** **Hole No.**

Elev. Surface ______ ft
Elev. Bottom MS____ft

Coordinates ______________________ Date Started ______________ 19______
Date Finished 19______

DRILL DATA
Make ______________________
Casing ID ______________________ in
Casing Area ______________ sq ft
Drive Shoe Dia { Outside ______________ in
Start ______________ in
Finish ______________ in
Effective ______________ in
Effective Area ______________ sq ft

DEPTH
Water Level ______________ ft
Overburden ______________ ft
Gravel ______________ ft
Top Bedrock ______________ ft
In Bedrock ______________ ft
Total drilled ______________ ft

MINING SECTION ________ ft

TIME LOG
Moving ______________ hrs
Drilling ______________ hrs
Pulling ______________ hrs
Delays ______________ hrs
Total ______________ hrs

CALCULATED VALUES

FACTORS
Drive Shoe ______________
Casing ______________
Theor. Rise of Core ______________ in. per ft of drive

GOLD
Au Wt. Actual ______________ mg
Au Wt. Corrected ______________ mg
Au Fineness ______________ PPM
Au Value ______________ $U.S./fine oz T

Au Wt. Aver. ______________ mg per sq ft
Au Wt. Aver. ______________ mg per cu yd
Raw Gold Value ______________ ¢ U.S. per mg
Au Aver. Value ______________ ¢ per sq ft
Au Aver. Value ______________ ¢ per cu yd

Time Hr Min	Depth Drive Ft	In.	Core Before Pumping Ft	Core After Pumping Ft	Meas. Vol. *In.	Colors 1	Colors 2	Colors 3	Est. Field	Wt. mg Eng	Au Corr	Remarks Formation

Subsequent pages:
Omit detail section above and Notes below.

Add legend of abbreviations for describing formation if desired

Notes: Drive Shoe Factor = area within effective cutting edge of shoe/area inside of casing
Casing Factor = unit of volume/area within effective cutting edge of shoe

* Record measured volume in same linear units as core measurements

Driller ______________ Panner ______________ Engr. ______________
Calculated by ______________ Date ______________ LINE NO. ________ HOLE NO. ______

FIG. 1 On the sample form for recording data from prospect drilling, adequate space is provided for items which should be entered. The units used on the form can be in any desired system of measurement and currency.

tions, say ten to one. A boulder may be encountered and the severe pounding action of driving and drilling through or by it may loosen material in the vicinity which reports in the core, or may force material away from the hole.

Competent drillers employ various techniques in an endeavor to obtain a representative sample having a volume as close as possible to theoretical, and especially to prevent excessive run-ins. However, despite experience and the use of tested techniques, core rises frequently are erratic, low or high, for reasons unknown, and in many formations are consistently low.

Various theories and opinions have been expressed regarding what actually happens under different conditions, whether material expands or contracts in volume when forced into the casing, and what the drive shoe factor should be. The most important point in relation to basic valuation is *what* occurred, rather than why.

For each cycle of the drilling operation the driller measures the depth to cutting edge of shoe immediately before and after the drive, the depth to top of material in the casing after the drive and after pumping out the core. For convenience, the core measurements usually are recorded as the distance above (occasionally below) the cutting edge of the shoe.

Good practice where run-ins are not occurring is to leave a short plug of core, say one to four inches, above the cutting edge of shoe at completion of each cycle of drive and pumping. Much longer plugs, up to several feet in length, may be necessary when drilling through very loose formations. Drilling and pumping below the shoe should never be done except on specific instructions by the engineer; occasionally it is justified to break boulders or to obtain data on formation, particularly to confirm that suspected bedrock is true bedrock.

The sharp cutting edge of a new shoe gradually becomes worn to a rounded section such as shown in the right half of Fig. 2. The rate of wear depends on the character of formation and the rate of drilling decreases as the cutting edge becomes rounded. In some formations the rate of wear is so slow that a satisfactory drilling rate is realized for several hundred feet of hole, whereas in other formations the cutting edge is fully rounded after drilling 40 feet.

As the shoe becomes worn, the cutting edge moves inward and the effective cutting diameter decreases. This results in a corresponding decrease in drive shoe factor and theoretical rise of core per unit of drive. To illustrate the important degree of magnitude of this change, assume that a 7½-in. dia shoe is worn to an effective cutting diameter (approximately mean) of 6.75 in., then the corresponding drive shoe factor is 0.2398/0.1808 = 1.37 as compared with the new shoe factor of 1.695, and the corresponding theoretical rise of core is 16.5 instead of 20.3 in. per ft of drive.

When wear conditions are severe the effective diameter often becomes less than the mean. Because, other things being equal, the volume of material deflected into the casing is determined by the effective cutting diameter of the shoe, the diameter before and after drilling a hole should be recorded on the log to provide the engineer with data essential for calculating unit gold content. If a machined shoe is not used, as often with an Empire drill, corresponding measurements are made of the drive end of the casing.

Core Volume Measurement

Upon completion of a drive and determination of the position of the top of core, the core is pumped out, using the drill bit as necessary to loosen or break up material. The core removed by pumping is delivered into the inclined dump trough, then washed into a measuring bucket resting a tub or similar container. The material is completely disintegrated, by hand if necessary, and the slime overflowed into the tub. The sand and coarser material remaining in the measuring bucket are compacted by decanting the water and jarring the bucket, and the volume determined by measurement. If the quantity of slime is significant, the volume for each drive is measured, and if not significant it is accumulated and volume measured for the hole.

The volume of washed core (and slime) should be measured and recorded as linear inches of a cylinder having the same inside diameter as the casing being used, to facilitate making calculations as described later. A bucket having an inside area two times that of the casing and a height of, say, 16 in. or 40 cm, is very convenient for this purpose, as the measured depth in the bucket simply has to be doubled to obtain the equivalent length of casing.

If the concentrate resulting from panning a sample contains gold, the panner counts the number of particles, or "colors" separately as nuggets and, usually, for three particle size or estimated weight ranges, and estimates the weight in each increment of sample and of total weight for the hole.

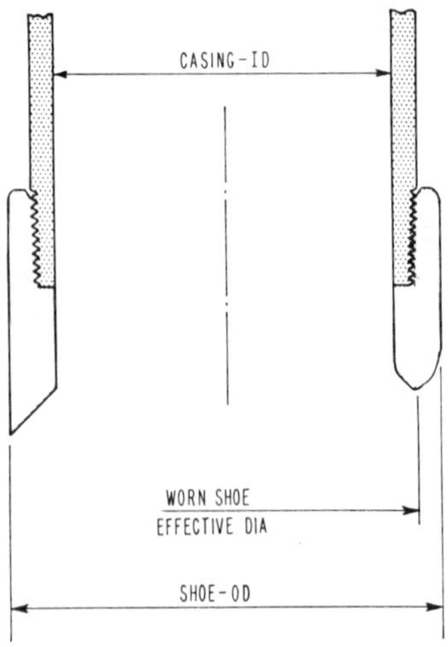

FIG. 2 Effective diameter of bevel-edged drive shoe changes as shoe wears.

Accurate determination and recording of each of these items, depth to cutting edge of shoe, core height and volume measurements, color count and estimated weight of gold, are the essential minimum requirements for preparing sound valuations from drilling data.

Prospect Drilling Log Form

Fig. 1 is a sample form for recording data from prospect drilling, and the related calculations for unit value made by the engineer. This form is more elaborate than those commonly used. However, it has been found from experience that all of the items for which spaces are provided are (or should be) entered on the drill log. Therefore it is better to provide a space for each item and thus tend to prevent inadvertent omissions. The units used on the form can be in any desired system of measurement and currency.

For individual cycles of drive and core recovery it is often found during drilling that: (a) the rise of core is greater than the theoretical rise calculated from the drive shoe factor and/or: (b) the measured volume of washed core exceeds the theoretical volume pumped from the hole. Frequently, excessive rise or volume or both occur when drilling the principal gold bearing horizons in a formation.

Whenever either of these situations occurs and gold is recovered from the sample, the volume of core must be assumed to be larger than should have been obtained, and the weight of recovered gold must be reduced or "corrected" proportionally to correspond to a theoretically normal core volume. On the other hand, if rise of core and/or measured volume are less than theoretically normal, no upward correction of weight should be made in any circumstances. This practice has been proved sound at many properties.

Estimated Average Weight

If the data on gold recovered consists only of the panner's color counts and estimated weights by sample increments and/or total for the hole, and the actual total weight, the engineer is obliged to assign estimated weights for one color of each size. This is not an engineering method and is not reliable, but it is all that can be done in such circumstances. At some projects the gold recovered from each drill hole has been separated by particle sizes into fractions, each containing the number of colors reported by the panner for each size, and each fraction weighed, from which the average weight of a color of each size is calculated. For example, assume that 2 nuggets, 5 No. 1 colors, 27 No. 2 colors, and 260 No. 3 colors were recovered from a hole. Using a low-power binocular microscope, the 2 nuggets, the 5 largest, the 27 next largest, and the remaining colors without counting, are separated and placed in individual piles. Each nugget and the contents of each pile are weighed, and the individual particle weights calculated, from which a fairly accurate estimate of weight for any sample increment can be made. The resulting data are very useful data for valuation as well as for other purposes, particularly for studies on recovery methods. This procedure is not needed for purposes of valuation if few or no corrections are necessary for a hole.

Correction for Excess Rise

The data required for calculation of a correction for excess rise of core are: the theoretical rise of core per unit of drive (corresponding to the drive shoe factor established for the hole); the length of drive; the actual rise; and the estimated weight of gold recovered from the sample. Usually the engineer can select a diameter that satisfactorily represents the condition of the shoe for the entire hole, by study of the effective cutting diameter of the shoe before and after drilling, and the position of the horizon(s) where excess gold bearing cores were

Table I. Prospect Drilling Factors American Extra Heavy Pipe

Drive Shoe Effective Diameter (in.)	Drive Shoe Area (sq ft)	Drive Shoe Factor	Theor. Rise Core (in./ft drive)	Casing Factor (ft/cu yd)	Casing Factor (m/m³)
A. Inside diameter—nominal 6 in., actual 5.761 in.; area 0.1809 sq ft					
7.500	0.3068	1.696	20.3	88.0	35.09
.325	.2967	.640	19.7	91.0	36.28
.250	.2867	.585	19.0	94.2	37.55
.125	.2769	.531	18.4	97.5	38.87
7.000	.2673	.478	17.7	101.0	40.27
6.875	.2578	.425	17.1	104.7	41.75
.750	.2485	.374	16.5	108.7	43.32
.625	.2394	.323	15.9	112.8	44.96
.500	.2304	.274	15.3	117.2	46.72
B. Inside diameter—nominal 5 in., actual 4.813 in.; area 0.1263 sq ft					
6.500	0.2304	1.824	21.9	117.2	46.72
.375	.2217	.755	21.1	121.8	48.55
.250	.2131	.687	20.2	126.7	50.51
.125	.2046	.620	19.4	132.0	52.61
6.000	.1963	.554	18.6	137.5	54.84
5.875	.1883	.491	17.9	143.4	57.17
.750	.1803	.428	17.1	149.8	59.70
.625	.1726	.367	16.4	156.4	62.37
.500	.1650	.306	15.7	163.6	65.24
.375	.1576	.248	15.0	171.3	68.30
C. Inside diameter—nominal 4 in., actual 3.826 in.; area 0.1007 sq ft					
5.250	0.1503	1.493	17.9	179.6	71.62
.125	.1433	.423	17.1	188.4	75.12
5.000	.1364	.355	16.3	197.9	78.92
4.875	.1296	.287	15.4	208.3	83.06
.750	.1231	.222	14.7	219.3	87.44
.625	.1167	.159	13.9	231.4	92.24
.500	.1104	.096	13.2	244.6	97.50
.375	.1044	1.037	12.4	258.6	103.11
.250	.0985	0.978	11.7	274.1	109.28

reported. An extreme combination of conditions occurs in some places, of deep ground, very rapid rate of shoe wear, and two widely separated important gold bearing horizons with frequent excess core rises in both. Use of two drive shoe factors for such a hole may be justified.

The corrected weight of gold for any individual sample for which a correction is necessary because of excess rise of core is calculated as follows: theoretical rise for the drive/actual rise for the drive (always less than unity) times the estimated weight of gold recovered from that sample. The amount of correction, obtained either by difference or by direct calculation, rather than the corrected weight, is entered on the log, because for many holes the number of samples for which corrections are required is much less than the total number that contained gold.

Correction for Excess Volume

The data required for calculation of a corrrection for excess measured volume are: the length of core that was pumped out as determined by measurement, and the length of core equivalent to the measured volume of washed material. If the volume of slime washed out of the sample was greater than reasonably expected to have occupied the interstices of the material in place, the estimated or measured amount of excess should be added to the washed volume. A correction is necessary if the measured volume exceeds the theoretical volume.

The corrected weight of gold is: length of core pumped/equivalent length of washed and measured core, times the estimated weight of gold recovered from that sample. The amount of the correction, rather than the corrected weight, is recorded on the log, as for excess rise of core. The measured volume may exceed theoretical where the core rise for the corresponding drive was less or more than theoretical. In the latter case, the two corrections for that increment of sample are additive. The volume of core samples recovered from drilling in many deposits is consistently less than theoretical. Therefore the degree of corrections resulting from calculation as described above is the logical minimum that is justified.

Total Correction

The foregoing procedures may appear complicated and cumbersome. Actually, the corrections for a deep hole for which many excess core rises and measured volumes were reported can be calculated by slide rule in a few mintues, provided all pertinent data are recorded on the log.

The corrected weight of gold from a drill hole is the actual weight less the sum of all corrections made by reason of excess core rise and measured volume. It will be noted that nuggets are considered to be colors and that no special correction is made for them. The relationships between drives and core rises, and of pumped samples to core in the pipe, often are not as simple and easily correlated as stated above for descriptive purposes, because of the great differences in the character of formation in various horizons in alluvial deposits, and the techniques employed during drilling to obtain as representative samples as possible. Where a complex situation is recorded for auriferous material, the data have to be analyzed and the correlations of position decided by the engineer.

The degree of magnitude of corrections varies through a wide range. At one property where over five hundred holes had been drilled, the total of corrected weights was 77% of actual weight, and at another for over one hundred holes it was 68%. A figure close to 80% is not uncommon.

Depth of Mining Section

A common practice for establishing depth of material that is expected will have to be mined to recover the gold is to add about one foot to the depth to the bottom of the deepest economic concentration as reported on the log of each drill hole, and always to establish depth as units, i.e., whole feet or even decimeters. This practice is satisfactory when the bottom of economic formation is sharply defined, such as a clay-like formation. At several dredging properties where the limit is on or in a jointed, fractured or decomposed bedrock it has been determined from experience that, in order to mine all of the ground that will pay to mine, the average digging depth was at least one foot more than originally estimated by the practice described above, and in some places was three or more feet greater. Where this occurs it does not necessarily follow that the drilling data were not accurate, or that the dredging operation was not sufficiently controlled to avoid mining of uneconomic ground. Therefore the depth of mining section established for each drill hole should be ample, and compatible with the character and topography of bedrock.

In some places a very hard irregular bedrock is present which cannot be dug effectively by a dredge and on which a concentration of valuable mineral exists, e.g., the limestone bedrock with pinnacles found in some properties in Malaya. The depth of mining section established for valuation at such places should be less than that indicated by drilling, i.e., compatible with the type of bedrock, and the mineral content below the selected depth in the deposit must not be included in the valuation.

Casing Factor

The calculation of unit gold content from drill hole data necessitates establishment of the theoretical length of drive that would produce in the casing a core having the unit volume (e.g., cubic yard or cubic meter) selected for the valuation. This is commonly called the *casing factor* and is: the unit of volume/area within the effective cutting edge of the drive shoe. Using English measures, for 6 in. nominal diameter casing and a new 7½-in.-dia shoe, the factor is: 27 (cu ft)/0.3068 (sq ft) = 88.0 (ft). For a shoe worn to an effective diameter of 6.75 in., having an area of 0.2485 sq ft, the factor is 108.7.

Because it is determined by the diameter of the effective cutting edge of the shoe, the casing factor for each hole must correspond to the actual shoe diameter for that hole. Drive shoe and casing factors are listed in Table 1 for new and worn shoe diameters for three common sizes of American extra heavy drill pipe.

Radford Factor

Theoretically, for a 7½-in. OD drive shoe the volume of sample produced by a drive of 1.00 ft is 0.3068 cu ft. From experience and tests conducted many years ago, the well-known placer engineer, Mr. Radford, concluded that, for this size of shoe, a constant theoretical volume of 0.27 cu ft per ft of drive should be adopted, which is 88.0% of theoretical. This relationship became known as the "Radford Factor." Use of this same relationship also was recommended by the Keystone Driller Co., which called it the "Keystone Constant."

The Radford Factor of 0.27 under the conditions defined above corresponds to an effective cutting edge of shoe diameter of 7.036 in. Use of the Radford Factor or Keystone Constant 0.27 therefore can be considered as assuming that the effective cutting diameter of a 7½-in. shoe always is 7.036 in., which probably is approximately true during drilling of many holes. However, adoption of any figure as a constant does not allow for the fact that the actual effective diameter of a shoe varies through a wide range. Therefore, the factors used for calculating corrections should correspond to and accurately reflect the conditions which existed during drilling.

Starting with the Radford Factor, the corresponding theoretical rise of core per unit of drive is calculated by the same formula as given above. Years ago a standard weight pipe was widely used as casing which had an inside diameter close to and assumed to be 6.00 in. The theoretical drive shoe factor for this casing and the standard 7½-in. shoe is 1.57, and the corresponding rise of core is 18.8 in. per ft of drive for a sharp shoe and 16.5 in. for the Radford Factor equivalent shoe.

The theoretical casing factor for a shoe having an effective cutting edge diameter of 7½ in. is 88.0 as given above; applying the Radford Factor or Keystone Constant the factor is 100.0. This latter factor, 100 linear ft of drive per cu yd of material in place, was very convenient and was widely used for calculating corrections using core volumes measured in decimal fractions of a cubic yard, 0.01 cu yd being equivalent to 1.00 ft of drive. Tables similar to Table 1 are available which show factors for various shoe and casing diameters, some showing the relationship between Radford Factor and theoretical figures.

Calculation of Unit Weight

The average gold content by weight for a drillhole, expressed in English system measures, is calculated as follows:

$$\text{Au (mg per sq ft)} = \frac{\text{Corrected weight of Au (mg)}}{\text{Effective area of shoe (sq ft)}}$$

$$\text{Au (mg per cu yd)} = \frac{\text{Corrected weight of Au (mg)} \times \text{casing factor (ft)}}{\text{Depth of mining section (ft)}}$$

For convenience in calculation usually the value per milligram of raw gold is calculated from the fineness and for the desired currency and price of gold. The value of one milligram of fine gold in U.S. funds, Au at $35.00 per fine ounce Troy, is 0.1125¢. Unit values for a hole are calculated simply by multiplying unit weights by value per milligram. The value of silver associated with the gold is disregarded for drill hole valuation, unless it is exceptionally high.

It is desirable, in some circumstances, to use weight of gold for calculating drillhole and areal valuation, and to convert only total areal valuation to a currency value.

A valuation may be made in terms of value per unit of volume only, i.e., without calculation of corrected weight of gold per unit of area or of volume. The formula for calculating the value per unit of volume becomes—English measures:

$$\frac{\text{Corrected weight of Au (mg)} \times \text{Factor}}{\text{Depth of Mining Section (ft)}}$$

in which the Factor is the constant:

$$\frac{\text{Value of raw gold (¢/mg)}}{\text{Effective area of shoe (sq ft)}}$$

Use of such a factor and formula saves a great deal of time when valuing a deposit in which a large number of holes have been drilled, an average fineness figure can be used, and only value per unit of volume is desired.

Calculation of average unit weights and values completes the valuation for a drillhole. Most engineers separate the gold from a sample, ascertain the weight, and calculate unit weights as soon as possible after completion of a hole. Average unit values can be added later when assay data are available.

Valuation in Frozen Ground

In frozen ground a short section of casing is set in the surface zone and the remainder of the hole drilled without casing. Water is poured into the hole only in sufficient quantity to permit drilling and pumping, and ice is added to keep the sides of the hole frozen above the drill bit. The diameter of a hole is variable because of differences in character of formation and the quantity of water that has to be introduced for drilling. The washed volume of each increment of sample is measured by the same procedure as described above for thawed ground, but recorded as an absolute value in suitable units. All water is removed from the hole with a pump after drilling is completed.

Immediately thereafter the hole is filled with water by stages which, in the auriferous horizons, correspond as closely as possible to the depths for each cycle of drilling and pumping. The rate of seepage of water into the hole is determined before and after filling. Depth to water surface is measured with a float, and the volume introduced for each stage is carefully measured in the same units as used for the samples. Columns for recording water measurements and time of observation are added on the log form for drilling in frozen ground. The average area and diameter of each section of the hole from which gold was recovered is calculated from the water measurement data including allowance for seepage.

The diameter (and volume) and the

weight of gold recovered from the sample is arbitrarily corrected upward to correspond to the bit diameter plus one-fourth inch for any section of hole where the water measurement volume shows a smaller average diameter because of ice buildup on the wall of the hole. The diameter of the bit should be measured and recorded frequently when drilling in auriferous horizons to produce a basis for this type of correction. If the measured volume of a washed sample that contained gold exceeds the water measurement volume for the corresponding section of the hole, the weight of gold recovered is corrected downward proportionately.

The weight (or value or both) of gold per unit of area or volume is calculated for each section of the hole from which gold was recovered, using corrected weight where applicable. The desired average unit figure(s) for the entire hole is calculated as the weighted mean of the several sections, to compensate for the variations in diameter of hole and volume of sample from place to place. Drive shoe and casing factors do not apply to holes drilled in frozen ground.

Areal Valuation

The areal valuation procedures described below are designed to meet the following criteria: (a) elimination, as far as possible, of the exercise of discretionary judgment; (b) a sound, simple and rapid method of calculation for large areas; (c) simple, fast calculation of the proportional part of the valuation of a large area that applies to any small, irregular area therein. As the basic unit of area used is a triangle, it is called the "triangle system" of valuation. Calculation of volume and total gold content by triangular areas is a method of valuation that has been used to some extent since the early days of dredging, but frequently without application of limiting provisions as described herein.

The basic data required are: (a) the completed drill log for each hole pertinent to the area; (b) an accurate plan to be used for valuation purposes only, on a scale not smaller than 1:2400 (or 1 in. = 200 ft). On the plan there should be shown for each drillhole the position, identifying number, surface elevation, depth of mining section (overburdens separate if desired); and unit weight or value of gold, using a suitable legend; also, selected coordinate lines if a coordinate system has been established, and the boundary line of any area which for any reason cannot be mined, or which is subject to royalty or other condition related to mineral content. In some cases the principal drainage channels and the toe of steep hillside slopes may be desirable additions, but other culture and topography should not be shown.

The depth of mining section and unit weight or value of gold as calculated and recorded on the log for each drill hole are deemed to be frozen and not subject to change, with the single exception of extremely high valued holes.

When such a hole was found in the field, customarily another hole was drilled nearby. If so, the average depth and value of the two should be assigned to a point located midway between them, without arbitrary adjustment of value. A discretionary adjustment of the value of an occasional high valued hole is justified, say to twice the value of the nearest two holes. No adjustments of the values of high holes should be made if a number were reported.

The area within the perimeter drillholes as shown on the plan is marked off into triangles having one drillhole at each corner. The number of triangles that radiate from one hole may vary considerably unless the drilling pattern was very regular. If the drilling pattern was regular, i.e., squares or equilateral triangles, the area influenced by an unusually high valued hole will be the same for any arrangement of triangles. If the pattern was irregular, in some places the area influenced by such a hole can be kept to a minimum by arranging the layout of triangles so that the high valued hole influences the least area. Discretionary judgement should be exercised in such cases on the side of conservatism.

Calculation Theory

The theory of calculation is: (1) along a straight line between any two adjacent drillholes the depth of mining section and the gold content per unit of area change at a uniform rate; for the area of a triangle; (2) average depth is the arithmetical mean depth of the three holes; (3) average gold content and value are the arithmetical means of those of the three drill holes; (4) total volume is area times mean depth; and (5) total gold content or value is total area times average gold content or value per unit of area. Calculation of total gold content or value from units of area produces the same totals as calculation by volume factors because the area factors include weighting for depth.

Average gold content or value for units of either area or volume can be used for calculating total gold content or value for each triangle. Use of the area factor involves less calculation because a weighted figure for each hole does not have to be calculated. Sample forms and calculations for both procedures are shown on p 87, in the English system of measures and in units of value in U.S. funds rather than weight of gold.

These calculations determine total and average figures for the life of a property, so it is desirable to record them in a bound book, a style with quadrille ruling being convenient. An index schedule with one line for each triangle is useful for quick reference.

The next steps are establishment of the limits of potentially mineable area on the basis of economic and other considerations, and valuation of the selected area. Usually the dredging limit passes through few drill holes and therefore only a portion of many triangles lies within the area to be valued. Average depth and unit value for a triangle are used for a partial area. A form heading containing the following can be used for listing the data: (1) Triangle No.; (2) Part or All; (3) Area (sq ft); (4) M.S. (ft); (5) Vol. (cu yd); (6) Average Value (¢/sq ft) (¢/cu yd); (7) Gross ($).

The figures for triangles which lie entirely within the area to be valued are posted from the index. The average depth and unit value figures for triangles which are partially within the area to be valued are posted from the index, the partial area measured, and volume and gross value calculated from the factors. When many partial area calculations are to be made using volume, not area, factors, it is helpful to calculate and enter on the index a factor for each triangle to facilitate computing gross value directly from area.

The same form can be used for preparing valuations of small irregular areas such as those dredged in two cleanup periods. An engineer experienced in the use of a planimeter can obtain a sum of areas of parts of triangles that will differ from the measurement of total area by less than 0.5%, which is an acceptable degree of accuracy in relation to that of other figures.

Other Procedures

A variation of the triangle system as described above may be desirable for valuation of stream valley deposits where the drilling pattern was holes relatively close together on lines relatively far apart, e.g., holes 50 ft apart on lines 500 ft apart. A high or low valued hole at the acute angle point of a long narrow triangle has a significant effect on the valuation of the area near the short base of the triangle, which may be several months distant in dredging time. More satisfactory estimates of recovery under such conditions can be made for small areas, particularly where the rate of advance of cut is slow, if only the two holes at the ends of the short base of the triangle are used for calculating the average figures. The arithmetical mean of the two depths will be the average depth and the weighted mean of the unit values will be the average values for the entire triangle. The three-hole and two-hole methods of calculation can be used in combination for stream valley areas where the basic pattern of drillhole layout was that described above, and where holes at intermediate points between the lines have also been drilled as often done to closely define the irregular lines of economic limits for mining along the sides of the deposit, or irregularly in the center of the valley for data on depth.

The initial plan of prospect drilling usually is laid out on a pattern of widely spaced holes. If the results are satisfactory additional holes are drilled on closer spacings until sufficient data are available for preparation of an acceptably reliable valuation. Preliminary estimates for relatively long and narrow deposits can be made by the block method of valuation, and the triangle system used for final valuation after drilling is completed. The block estimates usually are satisfactory until all field data are available and the time and effort required for calculating a triangle valuation are justified.

A. Calculation from drillhole gold value per unit of area:

1 Triangle No.	2 Hole No.	3 Area sq ft	4 Depth M. S. ft	5 Vol. cu yd	6 Av. Value ¢/sq ft	7 Total Value $	8 Av. Value ¢/cu yd
	17–2		38		27.1		
37	17–3	$\frac{380 \times 90}{2}$	42		47.3		
	18–3		49		42.8		
Total		17,100	129	27,200	117.2	6,691	—
Aver.		—	43.0	—	39.1	—	24.6

Notes: Cols. 1, 2, 4, and 6 posted from logs, Col. 3 by measurement.

Cols. 4 and 6: Aver. $= \dfrac{\text{Total}}{3}$

Col. 5 Total $= \dfrac{\text{Total Col. 3} \times \text{Aver. Col. 4}}{27}$

Col. 7 Total = Total Col. 3 × Aver. Col. 6.

Col. 8 Aver. $= \dfrac{\text{Total Col. 7}}{\text{Total Col. 5.}}$

B. Calculation from drillhole gold value per unit of volume:

1 Triangle No.	2 Hole No.	3 Area sq ft	4 Depth M. S. ft	5 Vol. cu yd	6 Av. Value ¢/cu yd	7 M. S. ¢/cu yd	8 Total Value $	9 Av. Value ¢/sq ft
	17–2		38		19.2	730		
37	17–3	$\frac{380 \times 90}{2}$	42		30.4	1277		
	18–3		49		23.6	1156		
Total		17,100	129	27,200	—	3163	6,691	
Aver.		—	43.0	—	24.6	—	—	39.1

Notes: Cols. 1, 2, 4, and 6 posted from logs. Col. 3 by measurement.

Col. 4 Aver. $= \dfrac{\text{Total}}{3}$

Col. 6 Aver. $= \dfrac{\text{Total Col. 7}}{\text{Total Col. 4}}$

Col. 9 Aver. $= \dfrac{\text{Total Col. 8}}{\text{Total Col. 3}}$

Col. 5 Total $= \dfrac{\text{Total Col. 3} \times \text{Aver. Col. 4}}{27}$

Col. 8 Total = Total Col. 5 × Aver. Col. 6.

Management Factors

A basic valuation calculated from reliable drilling data by the procedures described above constitutes a starting point for preparing estimates of recovery from mining. Technical, business and policy problems have to be carefully examined which usually results in application to the basic valuation of over-all factors, herein called management factors because the decisions are properly a function of management. This term includes the holder of mineral rights during the period prior to undertaking an extractive operation.

Management factors pertaining to valuation are divided into two phases: (a) those applied before mining is started, for technical and policy reasons; (b) those applied after mining is in progress, for the purpose of estimating future progress and production as closely as possible.

Pre-Operating Factors

The first activity to be undertaken after completion of a basic valuation is a detailed technical study of conditions and things that limit or affect mining and recovery. The study will include the range of depth of ground, surface and bedrock topography, character of material and of bedrock, degree of induration of the formation, water supply and natural water levels, vegetation, and climate conditions. A study will also be made of the size of the property, transportation facilities, types and sources of power, availability of labor by classifications, and applicable laws and regulations.

Technical studies of conditions, concentrating devices and plant design that influence efficiency of recovery are next in order. The shape, size range and distribution of the particles of gold, and its susceptibility to amalgamation have to be evaluated in relation to the efficiency of concentrating devices. The studies will include the effect on recovery of the characteristics of the material in relation to washing and disintegration, the amount of slime size particles in the deposit and the probable turbidity of pond water, the quantity, particle size and shape of heavy minerals, and the presence of minerals that adversely affect amalgamation.

From these studies an over-all recovery factor applicable to the basic valuation is developed. This factor may be unity for many properties where conditions that adversely affect recovery are minimal and adequate funds are available for installation of efficient equipment of suitable size. The valuation resulting from the procedures described above should be within plus or minus 5% of recovery that will be realized from mining the selected area as a whole, provided that the drilling and technical evaluation were done by competent personnel.

A preliminary selection of type and size of mining machinery and auxiliary facilities now can be made, and a preliminary estimate of capital cost and operating expense prepared. The estimate may show that the area under consideration cannot be mined profitably, or that only a selected portion can be mined profitably in which case a revised plan and estimate are made. The amount of funds available for capital investment may impose limitations, and compromises between maximum attainable efficiency of recovery and of initial and operating cost may be necessary.

Competent technical studies of these types and scope can be made only by a placer mining engineer qualified by long and broad experience. The decisions, however, rest with management as a proposed project must be evaluated as a business venture.

An over-all recovery factor of less than unity often is applied to the valuation by management when preparing the estimate of recovery, cost and profit that is used to justify undertaking a mining project. This is done for policy reasons and as a prudent precaution to allow for unforeseen contingencies, and does not constitute a change of the valuation itself.

Experience Factors Applied

R/E (Recovery/Estimate) statements are prepared at selected intervals after mining is under way, usually for periods of two or three cleanups, for period to date, and at annual intervals for years to date. The absolute and percentage relationships between valuation and the results of dredging are shown on these statements for identical areas. The valuation used is the basic valuation modified by application of any management factors established as a result of technical studies as described in the preceding section.

The boundary of a dredged area, as used for preparing R/E statements, should be the mean between the true lines of actual mining on the surface and at the bottom of the pond. If only the surface of the bottom line is used, the calculated volume will be too high or too low respectively. Both of these lines sould be shown on the operating plan, usually drawn on a scale of 1:600 or 1:500. A varying volume of tailing is mined whenever a cut is dredged that adjoins a previously dredged area, and the volume of tailing properly is included in that credited to the dredge, but must be excluded from that used in the R/E statement. When a production estimate is being prepared for such an area that is to be dredged, due allowance must be made for dilution that will occur from redredging tailing. In many places the volume of tailing may amount to five or more percent of total to be dredged.

Periodic R/E statements may show that the depth being mined is consistently more or less than estimated. Thereupon a study should be made to ascertain whether or not efficient control of digging depth in relation to gold content is being exercised. If the control is not efficient, appropriate improvements of practices should be made. If efficient, an experience factor applying to depth is developed which is added to or subtracted from the average depth as calculated from the valuation, and is used thereafter until conditions change in the field. Application of a depth factor does not change estimated gold content per unit of area or the total for a given area, but does change the average gold content per unit of volume and affects estimates of advance of face of cut.

Appraisal of Causes

A study and appraisal of probable causes should be made wherever the quantity of gold recovered from a substantial area is appreciably more or less than estimated. Application of a plus factor for areas to be mined should be done with caution until experience is adequate. A thorough study may show that improvements in mining practices or recovery methods can be made where recovery is less than estimated, and one of the important duties of management is to prove beyond reasonable doubt that the efficiency of mining and recovery are as good as possible. If recovery is proved to be efficient, a minus recovery factor is applied to valuation figures until conditions change in the field. An R/E statement provides management with useful data for appraising the reliability of the basic valuation and the efficiency of operation, and for applying depth and recovery factors so that estimates of future production will be as realistic as possible.

The recovery factor developed from long experience has been, at some properties, less than 80% of the original valuation based on prospecting data and a satisfactory explanation has not been found. In all such cases with which the author is familiar the valuations were not made in conformance with the principles and procedures described above, but experience factors were used after mining was started for making estimates of future recovery. Average recovery for large areas at several large properties where the system of areal valuation described above was used has been of the order of 100 to 105% of original valuation, though at times rather wide variations for one or more dredges have occurred for periods of upwards of a year, resulting in application of experience factors for preparation of budget estimates.

Comments

Arguments can be advanced against the validity of various assumptions represented by the procedures described, and doubtless some have merit. However, what really happens at the bottom of a drill pipe in a great many places, and why, remains a matter of conjecture. Therefore there is much to recommend the simple, conservative and uniform practices described for calculating the unit gold content of the formation at a drill-hole, and the valuation of large areas, with minimum exercise of discretionary judgment representing great differences of personal opinion, and to place over-all appraisal of valuation with management where it belongs. The importance of competent technical evaluation of things relating to mining and recovery is emphasized. The soundness of these principles and procedures has been proved by that most important criterion, recovery very close to estimate.

Over one hundred prospect holes were drilled in a certain auriferous gravel deposit. A valuation was made by an experienced engineer who applied discretionary judgement in appraising and interpreting the data shown on the drill logs, and arriving at this estimate of depth and gold content. The estimate of recovery, cost and profit being satisfactory, a dredge was installed. Recovery was disappointingly low during several years of operation, about 68% of estimate.

Subsequently an entirely independent revaluation was made, using the same logs but conforming to the procedures described above, without exercise of discretionary judgement. Many corrections were made because of excess core rise and measured volume reported on the logs. the new R/E statement for the total area mined showed that actual recovery had been 97% of the revised estimate, proving that a reliable prediction of recovery was possible from the data available for the original estimate.

Acknowledgment

I am very grateful to Dr. Charles F Park, Dean of the School of Mineral Sciences, Stanford University, for many helpful recommendations on composition and arrangement, and to Mr. James D. Crawford, Vice President and General Manager of Alaska Operations, United States Smelting, Refining and Mining Company, for helpful technical information on procedures.

20

Reprinted from pages 429–431, 432, 439, and 452 of *South African Inst. Mining and Metallurgy Jour.* **63:**429–452 (1963)

VALUE CONTOURS AND IMPROVED REGRESSION TECHNIQUES FOR ORE RESERVE VALUATIONS

By **D. G. Krige,** M.Sc.(Eng.) (Member) and **H. J. Ueckermann,** B.Sc.(Eng.) (Affiliate)

SYNOPSIS

A new type of value contour plan based on a two-dimensional moving average of underground gold and uranium development values is suggested. The practical advantages of such plans are discussed briefly in relation to mine planning, the determination of value trends and correlations, and certain geological interpretations. Based on such plans improved statistical regression techniques are developed for eliminating certain bias errors and for increasing the efficiency of the valuation of ore reserves in the South African gold mines.

SECTION I

VALUE CONTOUR PLANS

GENERAL

The variability in the distribution of gold values in the reefs of the Witwatersrand system is well known. Not only do appreciable variations in gold content occur between adjacent sampling sections even where taken in the same groove across the reef, but also when viewed on a regional basis significant variations in average value are found between different portions of a mine. On some mines the higher and lower grade areas form a rather complex distribution pattern with no apparent directional value trends, whereas on other mines, notably on the East Rand, the gold is clearly concentrated along well defined 'pay shoots' trending roughly in the same direction. In spite of all this variability, there is ample evidence of some degree of local association or correlation of values on all gold mines without which the blocking and selective mining of ore would have been impracticable.

A knowledge of any regional distribution patterns can be of invaluable assistance in the planning and general layout of a mining property, especially where payability is low and selective mining has to be practised. To assist in the definition of any such patterns, contour assay plans, often referred to as 'spotted dog' plans, have been in general use for many years on a number of mines. The method of construction of these contour assay plans was described in a paper to the South African Institute of Mine Surveyors[8], and is essentially the systematic grouping in grade categories of sample values obtained from stoped-out areas. A portion of such a 'spotted dog' plan for the Virginia mine is shown in Fig. 1 in a somewhat simplified form, and although there are no well defined major pay shoots on this mine, the plan clearly provides evidence of some minor value trends in the area covered by the plan. (The value contours shown have been transferred from Fig. 3.)

DISADVANTAGES OF CONVENTIONAL CONTOUR ASSAY PLANS

In the construction of such contour assay plans, the human element plays a large part in deciding on the grouping of sample values and any preconceived ideas as to the direction of value trends may materially influence the system of grouping. These plans are therefore subject to human bias and as such they are of doubtful scientific value. Furthermore, these plans can only be constructed after stoping has commenced, because during the initial development period, reef development is done on too large a grid to make effective grouping possible. These plans may therefore contribute to improved stoping efficiency at a later stage but are impractical for the advanced planning of the development layout and for estimating tonnages and grades of ore to be selected for mining in newly developed areas.

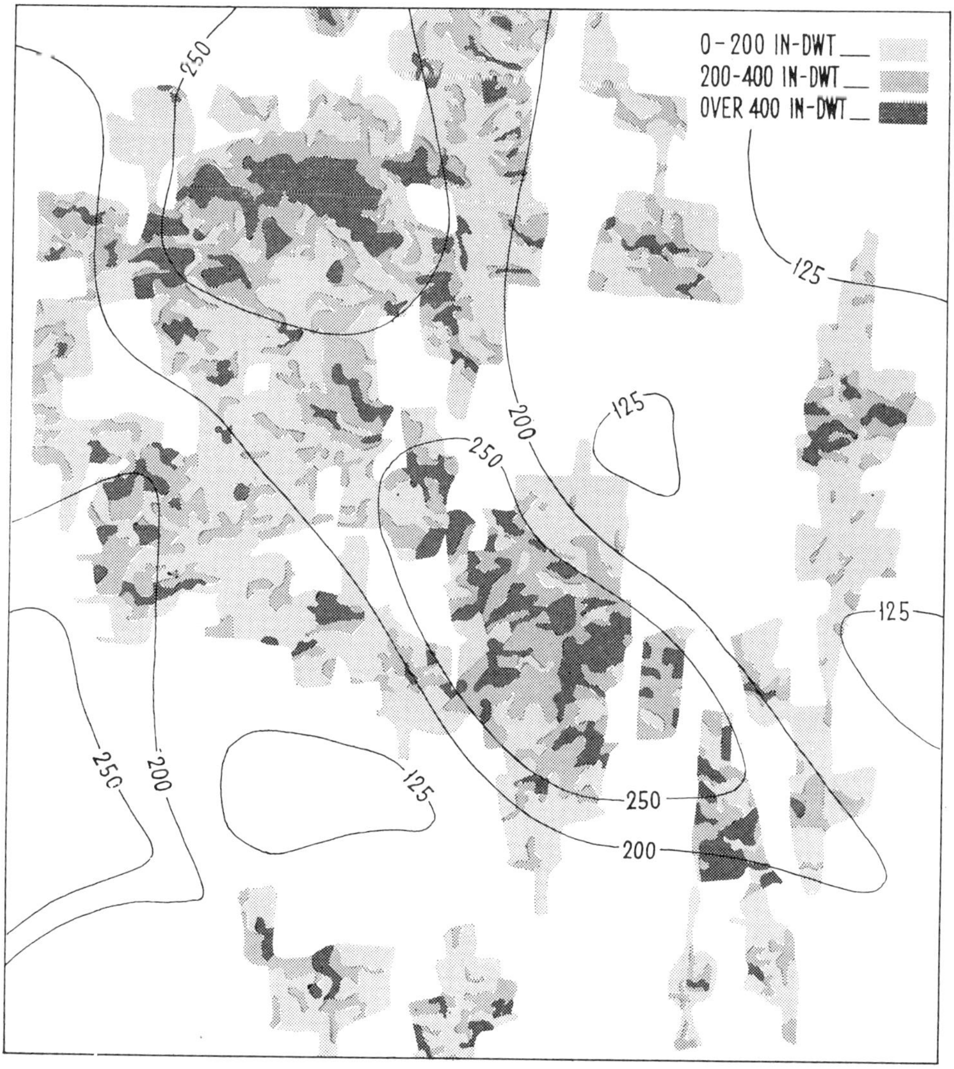

SHOWING PORTION OF 'SPOTTED DOG' PLAN

0 500 1000

ENG. FT.

Fig. 1

VALUE CONTOUR PLANS BASED ON A MOVING AVERAGE OF DEVELOPMENT VALUES

To eliminate the major disadvantages of the 'spotted dog' type of contour plan, and also to provide a plan suitable for use with the new ore reserve regression technique described in Part II of this paper, a new approach based on development values only, was developed. It was obvious that the overall pattern of regional trends could only be obtained by using a method which would even out the high local variation in development values and also provide a reliable estimate of the average regional value at any point in the area covered by development. The well known principle of a moving average seemed admirably suitable for these requirements and its application in two dimensions formed the basis of the method used.

The principle of a moving average for a series of sample values taken along a drive is illustrated graphically in Fig. 2, where the in.-dwt value of each sample section has been plotted against the corresponding distance from the start of the drive. Curve A shows the moving average obtained by plotting at the centre of all 50 ft stretches, the average values of all samples taken within those stretches. Similarly curve B represents the moving average obtained by averaging 75 ft stretches at a time. Curves A and B clearly show the main value trends along the drive which are not as evident in the original data because of the variability of the individual sample values.

The same principle can now be applied in two dimensions by plotting at the centre of a circle of specified radius, the average value of all development sampling within that circle. By moving the circle to cover a sufficient number of selected points within the developed area of the mine, a series of spot values can be obtained, which can be used to construct a value contour plan of the area by interpolation between spot values.

A circle is used in preference to a square or other shaped area because of ease of construction and also to avoid abrupt changes in average value as a slight movement of an area with straight sides could suddenly include or exclude a whole line of raise or drive stretches. The radius selected for the circle will depend primarily on the nature of the development grid. In general, the further apart the raises and drives are spaced, the larger will be the radius required, since sufficient development values must be included in each circle to give a reasonably reliable average contour value. The Virginia mine contour plans were based on a circle of 500 ft radius and although the radius chosen was somewhat arbitrary, the overall pattern of contour values is evidently not very sensitive to minor variations in radius. This principle is illustrated in one dimension by the similarity between curves A and B in Fig. 2. Between 1,000 ft and 4,500 ft of development sampling were included in each circle with an average of about 2,000 ft per circle.

Contour plans for the Virginia mine were prepared on the above basis for the gold and uranium value distributions and for the corresponding reef channel widths. In.-dwt and in.-lb values were used in preference to dwt per ton and lb per ton units since the latter are dependent on stoping width, which in turn is subject to changes in mining policy and hanging wall control. Furthermore, by superimposing the necessary allowance for exterior waste on the channel width contours, a theoretical stoping width contour plan can conveniently be arrived at which can then be used in conjunction with the in.-dwt and in.-lb value contour plans for arriving at the corresponding dwt per ton and lb per ton contour plans where required.

[*Editor's Note:* Material has been omitted at this point.]

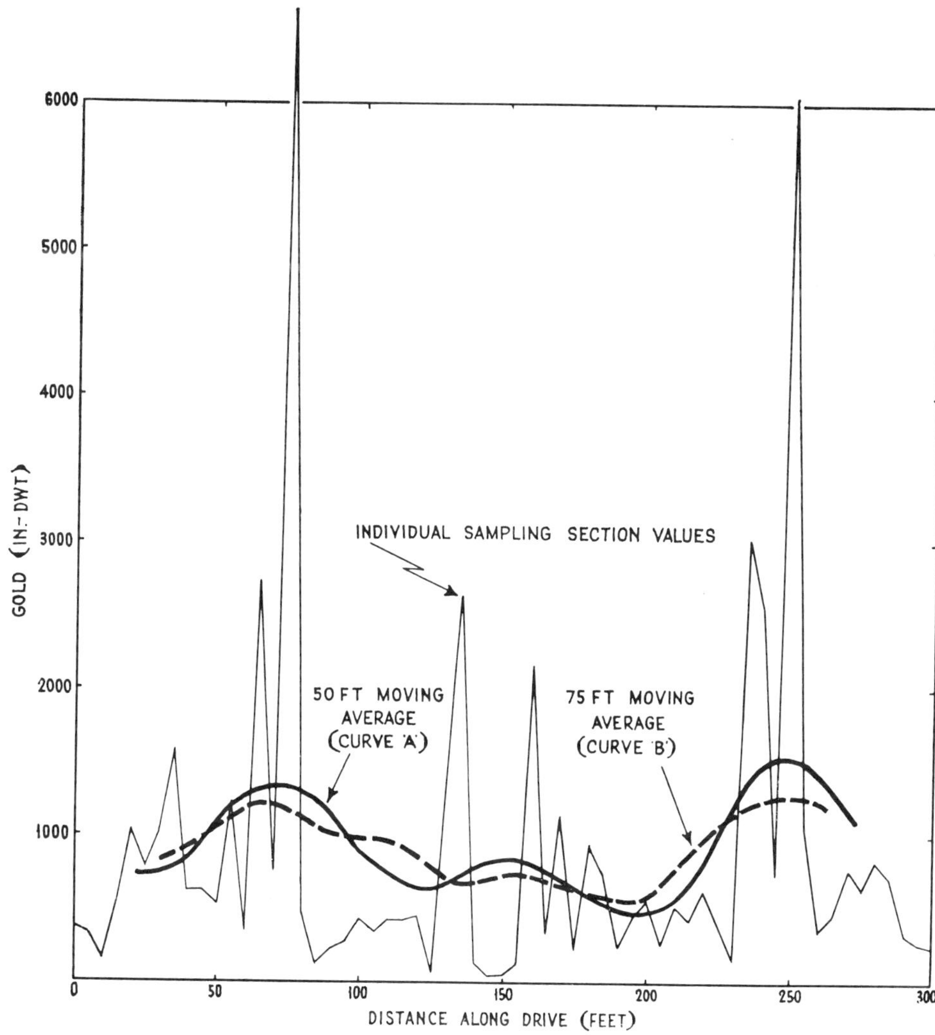

Fig. 2—Showing 50 ft and 75 ft moving average of development values along a drive

[*Editor's Note:* Material has been omitted at this point.]

ADVANTAGES AND APPLICATION OF VALUE CONTOUR PLANS BASED ON A MOVING AVERAGE OF DEVELOPMENT VALUES

Value contour plans prepared on this basis have the following advantages and applications:

(i) The method of construction eliminates the human element in estimating contours.

(ii) Once the plan has been prepared, little effort is required in keeping the plan up to date. It is suggested that the plans be brought up to date at the end of each quarter, after the development for that quarter has been plotted on the assay tracings. As the major portion of the new development will be concentrated around the periphery of the developed area and as stope values are not taken into account it may only be necessary to add a few additional circles each quarter and the small amount of internal development can easily be added to the totals and averages of existing circles.

(iii) Such plans show the overall value trends on a mine and since they are based exclusively on development values, they can be prepared at the stage when no stoping data is available but when a visual indication of trends, normally not evident from a study of the assay plans, will assist in the planning and development layout of the mine. Where a clear pattern of trends is indicated, the development programme will obviously be designed to suit the pattern disclosed and thereby make maximum use of the footage advanced.

(iv) Reliable regional values are indicated for the developed areas of a mine, thus enabling visual comparisons to be made between various sections of a mine and estimates of the likely tonnages and grades to be expected from those sections.

(v) Forecast estimates of tonnage and grade from areas still to be exploited can be made from the value contour plans by extrapolating the contours beyond the developed area. The pattern of trends disclosed in the developed area will serve as a guide in extending the contours and increasing or decreasing trends of payability will thus be taken into account.

(vi) The direction and extent of payable areas as disclosed by the value contours may assist in determining the most suitable system to be employed for the blocking of ore reserves.

(vii) These contours plans are ideal for determining regional correlations between gold and uranium values, gold content and channel width, etc.

(viii) Such plans can provide the basis for certain geological interpretations. In the case of the Vaal reef on the Hartebeestfontein mine, for example, such plans have served the purpose of providing an indication of the extent of the horizontal displacements along certain of the major faults. The plans for Virginia show a definite change in the distribution patterns for Basal reef proper and for the Leader-Basal reef beyond the line of coalescence.

(ix) Such contour plans can be used in the application of an improved ore reserve regression technique, described in Part II of this paper.

[*Editor's Note:* Material has been omitted at this point.]

BIBLIOGRAPHY

1. COXON, C. H. and SICHEL, H. S. 'Quality control of routine mine assaying and its influence on underground valuation.' *J. S.Afr. Instn. Min. Metall.*, May, 1959.
2. DE WIJS, H. J. 'Statistics of ore distribution.' *Geologie en Mijnbouw*, November, 1951, January, 1953.
3. JOSEPH, M. E. 'Statistical methods in sampling'. *Proc. Aust. Inst. Min. Metall.*, June, 1962.
4. KRIGE, D. G. 'A statistical approach to some mine valuation and allied problems on the Witwatersrand.' Unpublished Master's thesis, University of the Witwatersrand, 1951.
5. KRIGE, D. G. 'Effective pay limits for selective mining'. 'Economic aspects of stoping through unpayable ore.' *J. S.Afr. Instn. Min. Metall.*, January, 1962.
6. KRIGE, D. G. 'Statistical applications in mine valuation.' *J. S.Afr. Instn. Min. Surv.*, June and September, 1962.
7. MATHERON, G. 'Lognormal theory of systematic sampling of deposits' (French text). *Annales des Mines*, September, 1957.
8. MORTON, D. S. 'The construction and application of contour assay plans.' *J. S.Afr. Inst. Min. Surv.*, June, 1953.
9. ROSS, F. W. J. 'The development and some practical applications of a statistical value distribution theory for the Witwatersrand auriferous deposits.' Unpublished Master's thesis, University of the Witwatersrand, 1950.
10. ROWLAND, R. ST. J. and SICHEL, H. S. 'Statistical quality control of routine underground sampling.' *J. S.Afr. Inst. Min. Metall.*, January, 1960; also March and May, 1960.
11. SICHEL, H. S. and ROWLAND, R. ST. J. 'Recent advances in mine sampling and underground valuation practice in South African goldfields.' *7th Com. Min. Metall. Congr.*, Johannesburg, 1961.

21

Reprinted from *Canadian Inst. Mining and Metallurgy Bull.* **67**:131–142 (1974)

A Geostatistical Study of the Eagle Copper Vein, Northern British Columbia

A. J. Sinclair and J. Deraisme

Abstract

Geostatistical methods developed by Matheron are ideally suited for ore reserve estimations of the Eagle copper vein, owned by Davis-Keays Mining Co., Ltd., and located in northern British Columbia. A geostatistical study of this deposit is described in considerable detail to illustrate the application of the methods to vein-type deposits. Apart from providing global and local estimates of ore reserves, the study points to several practical applications of geostatistics in guiding future work.

Global estimates for the entire orebody give a mean vein thickness of 4.02 feet and a mean accumulation (grade x thickness) of 17.0 feet per cent, with associated variances of 0.038 and 1.33 respectively. From these figures, a mean grade of 4.23% Cu and a variance of 0.0716 are calculated. The relative standard deviations of vein thickness, accumulation and grade are 4.8%, 6.8% and 6.3% respectively.

Kriging, based on several plans suited to the disposition of available data, is used to obtain local estimates for vein thickness and accumulation, from which local grade estimates are obtained. Corresponding variances are also estimated.

The study has been based on an assumption of geostatistical isotropy that must be tested by the preparation of variograms in directions other than parallel to existing development drifts. If the assumption proves valid, some practical results of the study are as follows.

(1) — A sample spacing of 10 feet should be maintained for local ore reserve estimations of the quality indicated here.

(2) — The location of additional samples should be guided by the knowledge of their range of influence, approximately 100 feet, as shown by experimental variograms. Because samples have no influence beyond 100 feet from their locations, the best possible reserve estimates currently available beyond this range are global mean values. By the same token, global values represent the best available estimates for parts of the vein where continuity is known but for which samples are not available.

(3) — Kriging provides the best possible reserve estimation for ore control during mining. A practical kriging procedure is described that can be applied rather simply at the mine site to new assay data as it becomes available, without recourse to the more laborious calculations essential in setting up the procedure.

Alastair J. Sinclair is an Associate Professor in the Department of Geological Sciences, University of British Columbia. He obtained his B.A.Sc. and M.A.Sc. degrees from the University of Toronto and his Ph.D. from the University of British Columbia, all in economic geology and related fields. From 1962 to 1964, Dr. Sinclair taught in the Department of Geology, University of Washington, following which he moved to his present department. His research interests center on mineral deposits, with particular emphasis on mineral exploration and data interpretation procedures, including geostatistics.

Jacques Deraisme is a mining enginering graduate of the University of Nancy, France. Since 1971, he has been employed as a research engineer at the Centre de Morphologie Mathématique (Paris School of Mines), Fontainebleau, France, where he has worked on a wide variety of geostatistical problems in the field of ore reserve estimation.

Keywords: Geostatistics, Eagle vein, Copper, Vein deposits, Kriging, Variograms, Global estimators, Ore reserves.

Introduction

IN SPITE of the recent surge in geostatistical studies of bedded, residual and porphyry-copper types of mineral deposits, published accounts of comparable studies for vein deposits are virtually non-existent, particularly in English-language journals. In fact, because of the common, local, apparently erratic nature of assay values in many vein deposits, a general feeling has arisen that geostatistical methods of ore reserve estimation are not as applicable to veins as they are to larger, more uniform bodies. The Eagle vein complex seemed a useful example to offset both the deficiency and the misconception, because abundant, high-quality assay information was publicly available (see Trimble, 1972). It should be made clear at the outset, however, that considerably more workings and assay data exist for the deposit than are considered here. Consequently, this study, although based on actual data, represents ore reserve estimations at an early stage of development that has since been surpassed at the mine site.

The Eagle vein complex is owned by Davis-Keays Mining Co., Ltd., of Vancouver, B.C. It is on the company's Yedhe Creek property, Liard Mining Division, northern British Columbia, centered approximately at latitude 58°33′N and longitude 125°29′W (Fig. 1). The area is rugged, with elevations ranging from about 4000 to 8000 feet a.m.s.l. Development workings on the vein are between 5900 and 7400 feet a.m.s.l.

The vein occurs in a tightly folded sequence of shale, limestone and dolomite of Late Precambrian age. Adjacent to the vein, these wall rocks have undergone various degrees of silicification and decalcification. Contacts between vein and altered wall rock are sharp.

Sulphides occur in both massive and disseminated form. Chalcopyrite is by far the most abundant (80 to 95% of total sulphides), with pyrite and negligible quantities of covellite being the only others recognized to date. Chalcopyrite is commonly present as massive patches of large anhedral grains. Less commonly, it forms a fine, fracture filling in the earlier-deposited pyrite. Pyrite forms small irregular masses or individual grains distributed unevenly throughout the vein. Covellite occurs sporadically throughout the mine in very minor amounts, principally as a thin rim around some chalcopyrite grains and along thin irregular fractures within pyrite grains. Small amounts of limonite are associated spatially with covellite. Oxidation at and near the surface, and along some fracture

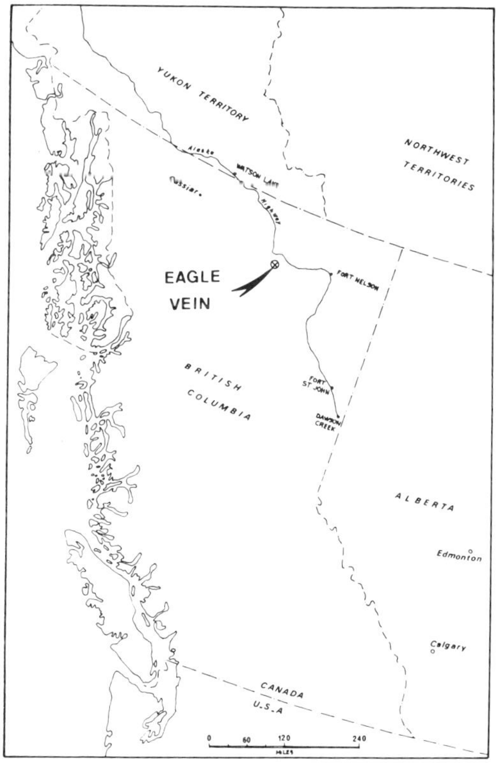

FIGURE 1 — Location of the Eagle vein.

zones, has produced abundant limonite, malachite and azurite locally. The vast proportion of the vein, however, is essentially unoxidized. Gangue is principally quartz with lesser but variable amounts of carbonate.

The vein strikes approximately N36°E and dips steeply to the northwest. Dips are variable, averaging about 80 degrees in the upper part of the vein and about vertical in the lower levels. Local reversals of dip have been observed in the northern section of the lower workings.

Average vein width is about 4 feet, but considerable variation exists, mostly in the range of 1 to 9 feet, giving the vein a complex lenticular form. Despite abundant post-ore faults, the vein is essentially continuous. Only two faults with significant lateral offsets (10 to 20 ft) are known.

Description of Data

Data for this study, taken from Trimble (1972), consist of vein width and copper assay values for 693 locations on the three main development levels of the Eagle vein; i.e. 7325, 6930 and 6360 levels (Fig. 2). These data were obtained originally in late 1969 and early 1970 as part of a re-evaluation of the Eagle deposit following an earlier preliminary feasibility study. During this re-evaluation, the vein was resampled at somewhat irregular intervals (average interval of about 8.5 ft) along the three main development levels. At each sample site, a sample was taken across the vein, generally from the backs, using a "wasp" air hammer. The samples can therefore be considered as intermediate in character between those obtained from the widely recognized procedures of channel and chip sampling. Average sample weight

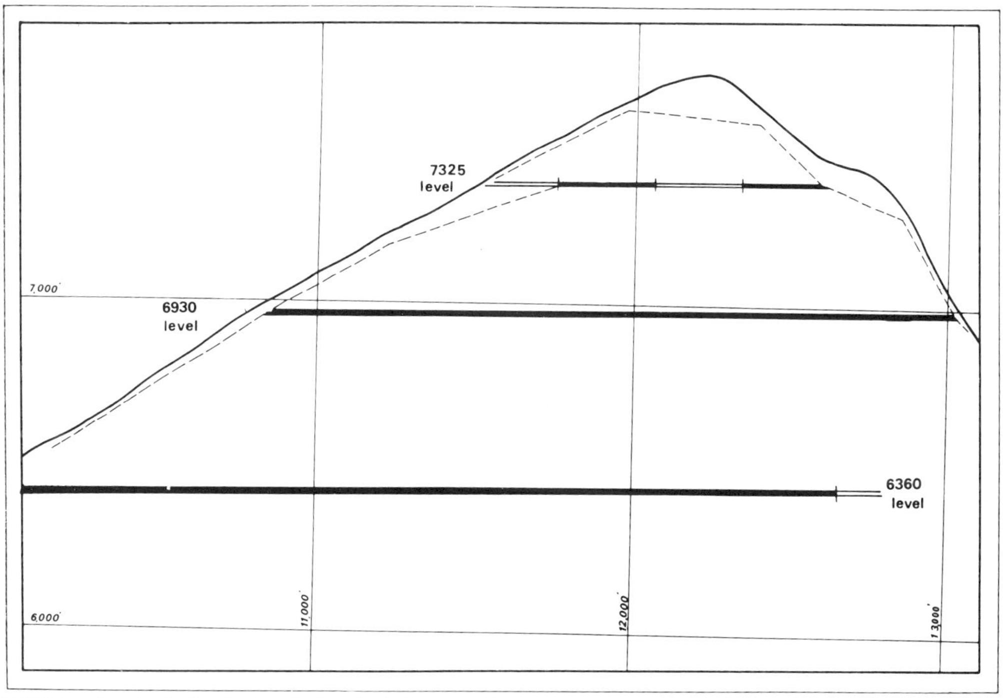

FIGURE 2 — Longitudinal projection, Eagle vein. Solid parts of levels show disposition of data available for this study.

was about 4 lbs. Vein widths were also recorded. Samples were crushed, split and sent to a registered B.C. assay office for Cu analysis.

Sample locations were fixed by noting their three-dimensional coordinates relative to the mine coordinate system. Locations listed by Trimble (1972) and used in this study were taken from level plans and refer to the center of each sample.

Geostatistics Procedure

In general, it is not sufficient to calculate the average grade of an orebody or parts of it without having some appreciation of the accuracy with which such estimates are known. Attesting to this are numerous examples of ore reserve calculations that led to huge pre-production expenditures only to find, as development progressed, that ore did not exist in the grades or amounts forecast. Geostatistical procedures developed by Matheron (e.g. 1971) provide a formal means of dealing with this difficulty.

Geostatistics is concerned with regionalized variables — those with values which show some relationship to adjacent values — including ore grade, vein thickness, accumulation (grade × thickness) and many others. In our study of the Eagle vein we will be concerned with two regionalized variables — vein thickness and accumulation. Assay values in this case do not represent a single regionalized variable, because they are based on different supports; that is, sample volumes vary. We will make various estimates of mean values and variances for vein thickness and accumulation, and from these calculate corresponding parameters for ore grade.

A formal geostatistical study of this nature involves a series of well-defined steps:

(1) critical analysis of data;
(2) construction and interprettion of experimental variograms;
(3) global estimation of ore reserves; and
(4) local estimation of ore reserves (kriging).

Critical Review of Data

The precise nature of the data used in this study is described in an earlier section. Techniques employed in sampling and assaying are acceptable standard procedures. Furthermore, all sampling was done by qualified geological personnel. We find no reason to question the representativeness of the data. We have ignored data from a central segment of the 7325 level where the vein is extensively oxidized. All other assay data available to us are from essentially unoxidized vein material.

Experimental Variograms

Semi-variograms comprise the fundamental tool for geostatistics. They represent variance between sample pairs as a function of distance (lag) between samples. Experimental variograms are determined for each regionalized variable under consideration, by the formula

$$\bar{\gamma}(h) = \frac{\sum_{i=1}^{n} [Z(x_i) - Z(x_i + h)]^2}{2n}$$

for which Z_{x_i} and Z_{x_i+h} are values at locations x_i and x_i+h, and n is the total number of sample pairs for which the two samples of each pair are separated by distance h. In general, points on an experimental variogram must be calculated by computer because, as the number of samples increases, the number of sample pairs to be considered becomes unmanageable manually. Variograms are commonly prepared for samples aligned in several directions to test for anisotropy, and for various parts of a deposit to test for stationarity. The presence of anisotropy and non-stationarity must be taken into account in obtaining a unifying mathematical model for the variogram that is applicable to an entire deposit, and on which variance estimates will depend.

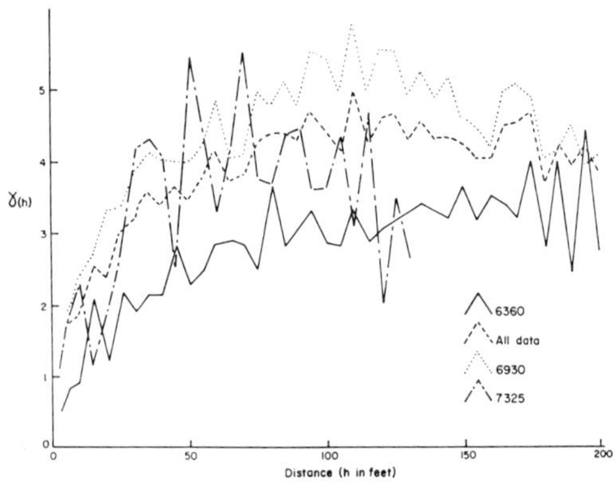

FIGURE 3 — Experimental variograms for vein thickness.

Experimental variograms for the Eagle vein were prepared for vein thickness and accumulation for the 7325 level, 6930 level, 6360 level and total available data (Fig. 3 and 4). Because data are not contained on a perfectly regular grid, it was necessary to group sample pairs into distance classes as follows: 5′±2.5′, 10′±2.5′, 15′±2.5′, etc. The average γ(h) for each class was plotted at the mid-point of the class (e.g. 5, 10, 15, etc.).

All data at our disposal are for three essentially parallel drifts. Consequently, we were not able to check for anisotropy and we *assume* in the remainder of this paper that the deposit is isotropic; i.e. variograms determined along the drifts are comparable to those that would be obtained in other directions. Obviously, this assumption should be checked by the preparation of variograms for different directions of sample alignment, as appropriate data become available.

It should be noted that experimental variograms have been prepared for a planar vein. In actual fact, the vein is somewhat sinuous in plan. We have calculated true distances between paired samples *along the vein* to produce the effect of "unfolding" or "flattening" the vein.

Consider the variograms for vein thickness (Fig. 3). They all show the same general pattern — a rapid rise for short sample spacings, which can be extrapolated through the origin, and a general levelling of values beyond a sample spacing of about 100 feet. A slight break in this trend is apparent in some of the variograms at a lag of about 20 feet. This "break" becomes more apparent on corresponding variograms derived from logarithmic data. Such is commonly the case for data that approximate a lognormal distribution, as is true for the Eagle vein. Such a variogram is characteristic of transition phenomena, in which a well-defined range of influence is exhibited, beyond which samples are independent. In fact, the "break" in the curve at about 20 feet implies the existence of two transition phenomena, characterized by ranges of 20 and 100 feet respectively. In physical terms, these ranges refer to average lengths of structures (lenses) in the vein and indicate the presence of two groups of lenses with average lengths of about 20 feet and 100 feet respectively. An example is shown in Figure 5, where a plan of a vein segment on the 6930 level has been "straightened" by plotting sample centers on a straight line to emphasize the lenticular nature of the vein. Here, a large central lens with a lateral length of about 95 feet is bounded on either end by much smaller lenses with lateral lengths of 20 to 30 feet. Ranges determined from the variogram for vein thickness are thus seen to reflect the mean geometric form of the vein.

It is apparent that all variograms for vein thickness are not identical. The levelling (sill) that exists beyond the range of 100 feet occurs at different values of $\gamma(h)$ from one variogram to another, indicating that the regionalized variable is *not* stationary on a global scale.

A local variation in the sill value of this nature is referred to as a proportional effect and attention will be directed to it again in a later section.

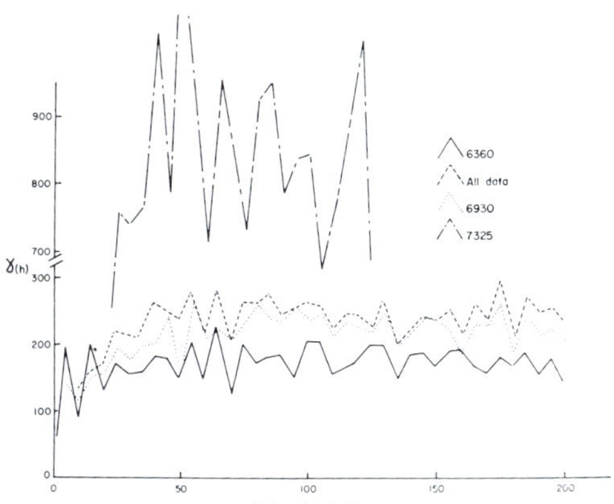

FIGURE 4 — Experimental variograms for accumulation.

In practical terms, ranges estimated from variograms are of the utmost significance. They prove that, on the average, the values of a regionalized variable are *independent* at distances greater than the range (100 feet in the case of vein thickness for the Eagle deposit). In other words, local estimations of ore reserves can be made only within 100 feet of workings for which samples are available. As an example, consider levels 7325 and 6930, separated by 395 vertical feet. Sample data can be extended 100 feet below the 7325 level and 100 feet above the 6930 level, leaving a vertical gap of 195 feet between the levels that can be estimated only by average values representing the entire deposit (global estimates). One aspect of the importance of geostatistics in ore reserve calculations thus becomes apparent.

Variograms for accumulation (Fig. 4) show a pattern similar to those for vein thickness, except that local fluctuations are more pronounced and a discontinuity (nugget effect) exists at the origin. Two intermeshed structures are also shown by these variograms with ranges of 30 and 90 feet. As with vein thickness, the smaller-scale structure is best defined on variograms of logarithms of accumulation values. A proportional effect is also indicated in these variograms by the variation in $\gamma(h)$ values of the various sills. The physical significance of the variograms for accumulation is somewhat more difficult to visualize, compared with those of vein thickness. Accumulations for a "straightened" segment of vein on the 6930 level are shown in Figure 5. The short-range structure is obvious, but the longer-range structure is obscure in contrast to vein thickness plotted for the same segment. Similar limitations to those discussed previously for vein thickness apply to the extension of accumulation data from sample sites.

A brief summary of the results of the study at this early stage, prior to undertaking any mathematical calculations, is in order. Accumulation and vein thickness for the Eagle vein have variograms that exhibit transition phenomena, and the presence of two intermeshed structures. Ranges of the larger-scale structures are about 100 feet in both cases. Beyond this range, the regionalized variables are independent, implying that sample data cannot be extended beyond 100 feet from sample sites in ore reserve estimations. Both variables show the presence of a proportional effect, which means that only local stationarity prevails in the deposit.

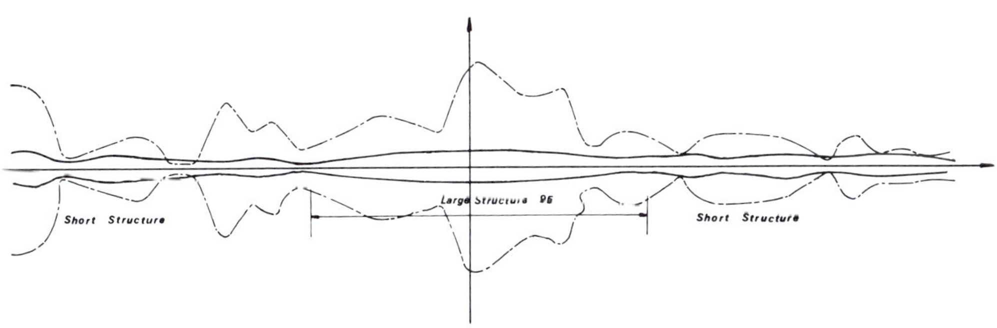

FIGURE 5 — Straightened segment of vein (plan) for a section of the 6930 level. Vein limits (solid lines) and relative accumulation (dash-dot lines) are plotted symetrically about the "straightened" vein axis to emphasize the lenticular structures indicated by experimental variograms.

Variogram Model

For global variance estimates and the more complicated procedure of local ore reserve estimation by kriging it is essential to have the variograms in mathematical forms that are representative of the entire deposit. A variety of mathematical models or schemes have been formulated for this purpose, the most useful being the spherical scheme for describing transition phenomena, such as are indicated in the variograms for the Eagle vein data.

The spherical model is simply a specific mathematical equation that quantifies an experimental variogram. In its simplest form it is defined by an equation containing two parameters, these being the sill (C) and the range (a). Hence,

$$\gamma(h) = C\left[\frac{3}{2}\frac{h}{a} - \frac{1}{2}\frac{h^3}{a^3}\right] \quad \text{for } h < a$$

$$= C \quad \text{for } h > a$$

or

$$\gamma(h) = C_1\,\gamma_1(h) \quad \text{for } h < a$$

Where two intermeshed structures exist, the equation becomes:

$$\gamma(h) = C_1\,\gamma_1(h) + C_2\,\gamma_2(h)$$

Ideal models of the foregoing type were fitted by trial and error to the experimental variograms for vein thickness. Calculated values of the ranges and sills are given in Table 1. An examination of the data shows that the ranges are consistent for all the variograms, but the sills differ considerably (proportional effect). Plots of the sills versus mean values of data used to construct the variograms led to the approximate relationship

$$C_1 = C_2 = 0.15\,m^2$$

as a quantification of the proportional effect. The general model for vein thickness thus becomes

$$\gamma(h) = 0.15\,m^2\left[\left(\frac{3}{2}\frac{h}{a_1} - \frac{1}{2}\frac{h^3}{a_1^3}\right) + \left(\frac{3}{2}\frac{h}{a_2} - \frac{1}{2}\frac{h^3}{a_2^3}\right)\right]$$

$$= 0.15\,m^2\,[\gamma_1(h) + \gamma_2(h)]$$

with a_1 = 20 feet and a_2 = 100 feet.

The significance of this model is more apparent with reference to Figure 6. Here, ideal variograms for small-scale and large-scale structures for total data are shown individually and combined. It is the combined model that describes an experimental variogram in the case of two intermeshed structures.

For accumulation variograms, an added complication of the model is seen as a discontinuity (nugget effect) at the origin. Such a discontinuity is commonly more apparent than real, because it reflects the presence of smaller-scale structures that cannot be observed in detail with the existing sample spacing. In the general equation, however, a third term, C_o, is required to take into account this apparent discontinuity. Thus

$$\gamma(h) = C_o + C_1\,\gamma_1(h) + C_2\,\gamma_2(h)$$

Equations of this type were fitted to all accumulation variograms by trial and error to obtain the estimated parameters in Table 1. Plots of the data in the table led to the relationships $C_o = 2/3\,C_1 = 0.2\,m^2$ and $C_1 = C_2 = 0.3\,m^2$ to take into account the proportional effect noted earlier. The general variogram model for accumulation at the Eagle vein thus becomes

$$\gamma(h) = 0.2\,m^2 + 0.3\,m^2\,[\gamma_1(h) + \gamma_2(h)]$$

$$= 0.3\,m^2\left[\frac{2}{3} + \gamma_1(h) + \gamma_2(h)\right]$$

with a_1 = 30 feet and a_2 = 90 feet.

An example for total data (Fig. 7) shows how the three components of the model combine to describe the experimental variogram.

TABLE 1 — Estimated Parameters for Experimental Variograms, Spherical Scheme

	Parameters*	7325 level	6930 level	6360 level	All levels
Vein Thickness	C_1	2	2.7	1.2	1.9
	C_2	2	2.5	2	2.5
	a_1	20	20	20	20
	a_2	100	100	100	100
	m	3.5	4.7	3.4	4.0
Accumulation	C_o	200	70	60	80
	C_1	350	70	70	90
	C_2	350	90	60	90
	a_1	30	30	30	30
	a_2	90	90	90	90
	m	34.2	18.0	13.9	17.0

*See text for definition of parameters. Subscripts 1 and 2 refer to small- and large-scale structures respectively.

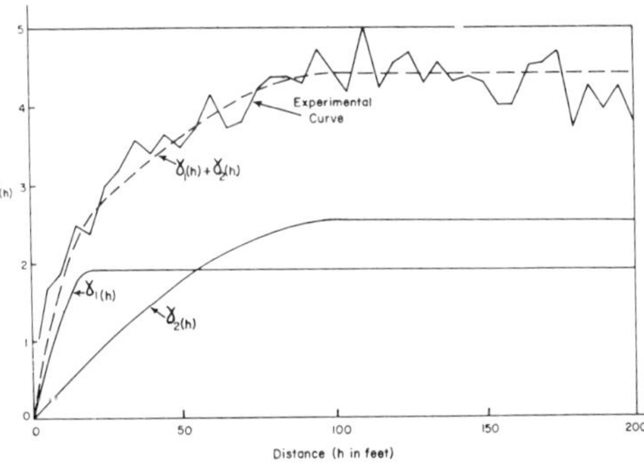

FIGURE 6 — An example for total data for vein thickness illustrating how two intermeshed structures combine to provide a model that describes the experimental curve.

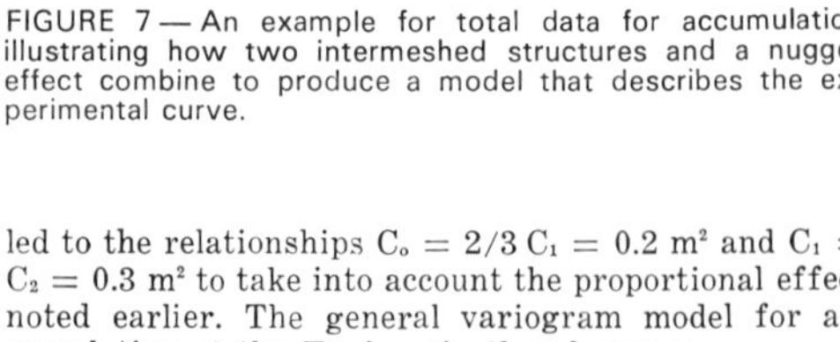

FIGURE 7 — An example for total data for accumulation illustrating how two intermeshed structures and a nugget effect combine to produce a model that describes the experimental curve.

Global Estimators

Global estimators represent an entire orebody or a large part thereof. For regularly located data (approximated in the case of the Eagle vein), an arithmetic average is the best estimator of mean value. If data are strictly randomly dispersed, then under the stationary hypothesis the best estimator still remains the arithmetic mean. However, if data are preferentially and irregularly positioned, individual values must be weighted empirically in proportion to their lengths or areas of influence, as in common practice. It must be made clear, however, that such a zone of influence based only on the experimental distances between adjacent samples is completely arbitrary (i.e. dependent on human bias in locating sample sites). Consequently, there is no reason why such an arbitrary zone of influence should correspond to the physical zone of influence given by the variogram. Mean values obtained for the Eagle vein are 4.02 feet for vein thickness and 17.0 feet per cent for accumulation. These figures give a mean grade of 17.0/4.02 = 4.23 per cent Cu. It should be noted that these values would be slightly different had a weighting procedure been used.

The question now arises as to the precision with which we know these estimates. For estimating global variances a composition procedure is essential. To this end, we have divided the orebody into four rectangular blocks, each with median drifts, as shown in Figure 8. Other patterns could be chosen that approximate the outline of the orebody more closely, but they would involve much more complicated calculations with little changes in the resulting variance estimates. Variance estimates for each block are approximated as a linear combination of a line term (estimation of the drift by the samples) and a slice term (estimation of the panel by the drift, which is assumed to be known). The two terms are shown to be independent (see Matheron, 1971, pp. 74-75). Variances thus estimated for each block are weighted by the squared area of corresponding blocks to obtain the global estimate. The general expression for global variance is therefore

$$\sigma^2 = \underbrace{\frac{C_o}{n}}_{\text{nugget term}} + \underbrace{\frac{\sum_{i=1}^{4} l_i^2 \sigma_i^2(l_i)}{(\Sigma\, l_i)^2}}_{\text{line term}} + \underbrace{\frac{\sum_{i=1}^{4} (D_i h_i)^2 \sigma_{E_i}^2}{[\Sigma\, (D_i h_i)]^2}}_{\text{slice term}}$$

where C_o is nugget effect, n is total number of samples, l_i represents lengths of extension of individual samples in corresponding drift segments of length D_i, $\sigma_i^2(l_i)$ is the mean variance of a single sample in each drift segment l_i, h_i is the vertical dimension of each rectangular block and $\sigma_{E_i}^2$ are extension variances of drifts to corresponding rectangular blocks. Both $\sigma_i^2(l_i)$ and $\sigma_{E_i}^2$ depend on the variogram model, as of course does C_o. Thus, the global estimation variance is also dependent on the variogram model. $\sigma_i^2(l_i)$ and $\sigma_{E_i}^2$ can be determined graphically (see Matheron, 1971), C_o is estimated from the variogram model, and n, D_i and h_i are all known. Calculation of global variance thus becomes a problem in arithmetic.

For vein thickness, $C_o = 0$ and the line term (4×10^{-4}) is found to be negligible. This implies that the sample density is more than adequate for purposes of global estimation. The slice term has a value of 0.038, which is effectively the global estimation variance, and leads to a standard deviation of 0.19 and a relative standard deviation of $(0.19/4.02) \times 100 = 4.8$ per cent.

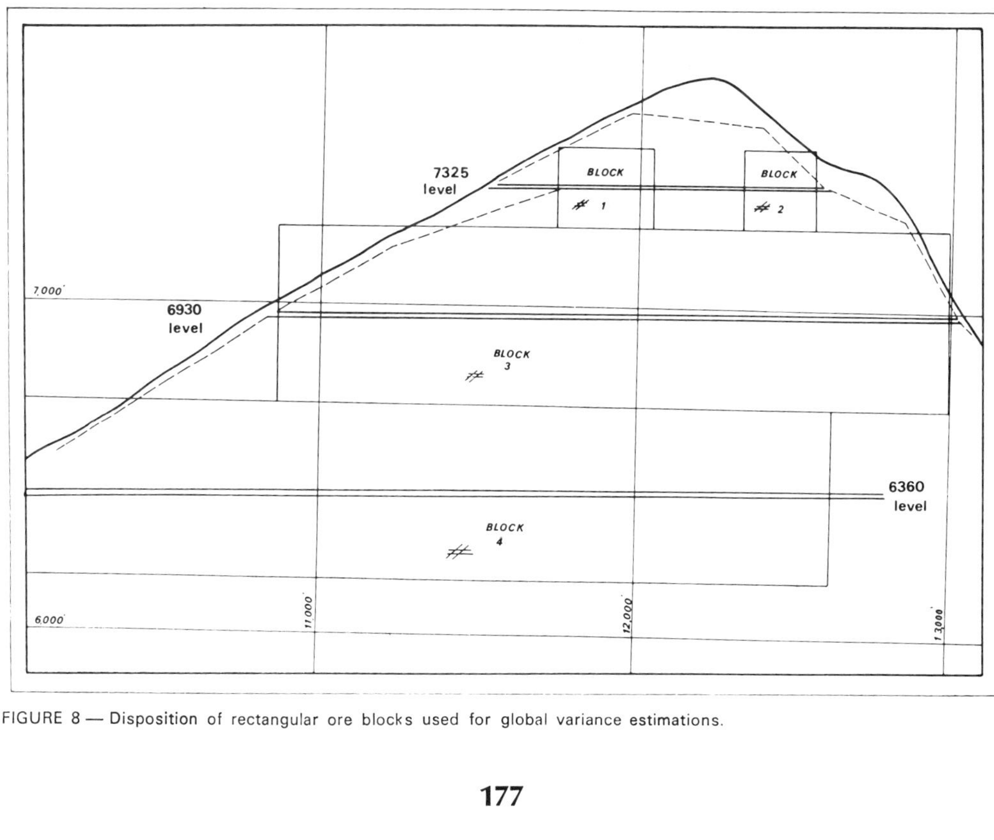

FIGURE 8 — Disposition of rectangular ore blocks used for global variance estimations.

TABLE 2 — Summary of Global Estimators

	Symbol	Value	Symbol	Value	Symbol	Value
Vein Thickness (feet)	$\bar{\rho}$	4.02	σ_{ρ}^2	0.038	$\frac{\sigma_\rho}{\rho} \times 100$	4.8%
Accumulation (feet %)	$\bar{a}$	17.0	σ_a^2	1.33	$\frac{\sigma_a}{a} \times 100$	6.8%
Grade (% Cu)	$\bar{t}$	4.23	σ_t^2	0.072	$\frac{\sigma_t}{t} \times 100$	6.3%

Accumulation variance is determined in the same manner. In this case, the nugget term exists and equals 0.09. The line term is small, but significant, with a value of 0.014 and the slice term is 1.23. The global variance estimate is therefore 1.33, which gives a standard deviation of 1.15 and a relative standard deviation of (1.15/17.0) × 100 = 6.8 per cent. Again, existing sample spacing is found to be adequately close for global estimation purposes. In fact, increasing the present sample spacing to 20 feet would increase the nugget term to 0.18 and the line term to 0.01 to give a variance of 1.42, a standard deviation of 1.19 and a relative standard deviation of 7.0% — very little change indeed!

Average grade as determined from mean values of vein thickness and accumulation is 4.23% Cu. The variance attached to this estimate is calculated from the following equation (limited expansion to the order 1):

$$\frac{\sigma_t^2}{\bar{t}^2} = \frac{\sigma^2}{\bar{a}^2} + \frac{\sigma_\rho^2}{\bar{\rho}^2} - 2\, C_{a\rho} \cdot \frac{\sigma_a}{\bar{a}} \cdot \frac{\sigma_\rho}{\bar{\rho}} + \ldots$$

where $\bar{t}$ is mean tenor with variance σ_t^2, $\bar{a}$ is mean accumulation with variance σ_a^2, $\bar{\rho}$ is mean vein thickness with variance σ_ρ^2 and $C_{a\rho}$ is the simple correlation coefficient for vein thickness and accumulation. $C_{a\rho} = 0.45$ in this case. The global variance of ore grade is therefore estimated as 0.0716, which gives a standard deviation of 0.27 and a relative standard deviation of 6.3%.

Global estimates for vein thickness, accumulation and grade are summarized in Table 2.

Local Reserve Estimation (Kriging)

Kriging, the procedure used to obtain local estimates of ore reserves, is described by Matheron (1967, 1971), and a short description is given by Blais and Carlier (1967). Kriging is one of the most important fundamental methods in geostatistics, with widespread practical applications. In ore reserve calculations, it provides the best local estimators of means and variances for a specific panel size. Means are calculated from the equation

$$Z_K^* = \sum_{i=1}^{s} \lambda_i Z_i$$

where, in this case, Z_i represents the arithmetic means of data within the block to be estimated *and* in certain nearby blocks, and λ_i are corresponding weighting factors. A kriging variance, σ_K^2, is also estimated.

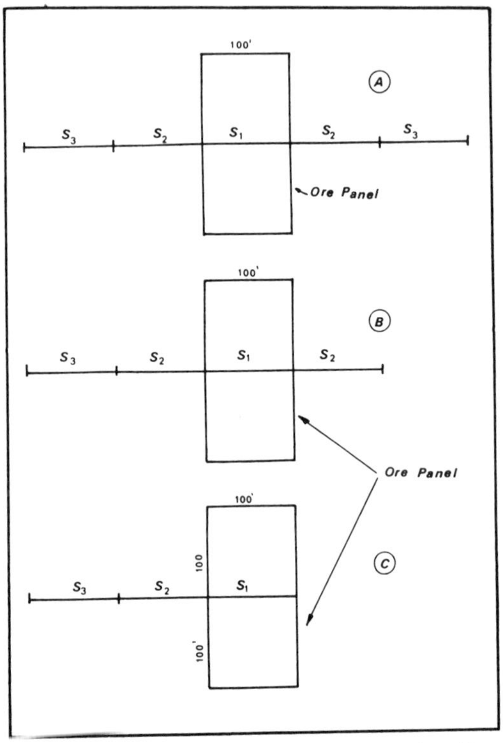

FIGURE 9 — Kriging plans for use with available data Plan 1(A) applies to the bulk of available data, and plans 2 and 3(B and C) to those panels at and near the ends of sampled workings.

Two conditions must be fulfilled so that Z_K^* will be the best possible linear estimator: (1) the kriging variance must be a minimum; and (2) the estimate must not be biased. Given these two conditions, we arrive at a general form of the kriging equations:

$$\sum_{j=1}^{n} \lambda_j \bar{\gamma}(S_i, S_j) + \mu = \bar{\gamma}(S_i, Z) \quad i = 1, 2, \ldots$$

$$\sum_{j=1}^{n} \lambda_j = 1$$

where λ_j are weights, $\bar{\gamma}(S_i, S_j)$ are mean gamma values between data sets S_i and S_j, $\bar{\gamma}(S_i, Z)$ are mean gamma values between Z (the block to be estimated) and data sets S_i, n is the number of data sets and μ is a Lagrange parameter. In order to apply these equations, a kriging plan must be adopted that makes best use of available data. To be more specific, one must define the geometrical relationship of carefully chosen data sets S_i to the block to be estimated. For practical purposes, the number of data sets S_i must be limited so that calculations are manageable.

For the Eagle vein, it is not desirable to krige past a distance of 100 feet from existing samples, because beyond this range samples are independent, and there would be no distinction between local estimators thus obtained. Thus, we have chosen to krige blocks of ore that extend 100 feet along sampled drifts and 100 feet vertically both above and below the drift. The symmetric kriging plan that we have chosen for most such blocks is shown in Figure 9a. In this figure, S_1

refers to sample data along that part of the drift *within* the block to be estimated, S_2 are data in the two adjacent, symmetrically disposed blocks and S_3 are data in the *next* two adjacent, symmetrically disposed blocks. This symmetric plan cannot be applied at and near the ends of drifts and additional kriging plans are required, as also shown in Figure 9.

No effort will be made to show the details of kriging calculations, but some indication of the procedures involved will be illustrated by kriging plan No 1. For this plan, the general kriging equations can be rewritten

$$\lambda_1 \bar{\gamma}(S_1, S_1) + \lambda_2 \bar{\gamma}(S_1, S_2) + \lambda_3 \bar{\gamma}(S_1, S_3) + \mu = \bar{\gamma}(S_1, Z)$$

$$\lambda_1 \bar{\gamma}(S_2, S_1) + \lambda_2 \bar{\gamma}(S_2, S_2) + \lambda_3 \bar{\gamma}(S_2, S_3) + \mu = \bar{\gamma}(S_2, Z)$$

$$\lambda_1 \bar{\gamma}(S_3, S_1) + \lambda_2 \bar{\gamma}(S_3, S_2) + \lambda_3 \bar{\gamma}(S_3, S_3) + \mu = \bar{\gamma}(S_3, Z)$$

$$\lambda_1 + \lambda_2 + \lambda_3 = 1$$

These equations can be most readily solved for λ_i and μ by, for the moment, ignoring the proportional effect. This in no way changes the values of λ_i, but it does affect the μ value. Values for $\bar{\gamma}(S_i, S_j)$ and $\bar{\gamma}(S_i, Z)$ can be determined from geometrical considerations and with the aid of graphs and auxiliary functions (Matheron, 1971), bearing in mind, of course, the general variogram model for the variable under study. The only unknowns then are the λ_i's and μ, which are readily calculated. Results for the three kriging plans are listed in Table 3.

The kriged mean of a block to be estimated can be determined directly from the relationship

$$Z_K^* = \lambda_1 Z_1 + \lambda_2 Z_2 + \lambda_3 Z_3$$

where Z_1, Z_2 and Z_3 are mean values of data sets S_1, S_2 and S_3 respectively.

The kriging variance is calculated from the relationships

$$\sigma_K^2 = \sum_{i=1}^{3} \lambda_i \bar{\gamma}(S_i, Z) + \mu - \bar{\gamma}(Z, Z)$$

In this case, however, the result as calculated from data in Table 3 must be multiplied by a factor km^2 to take into account the sill and the proportional effect. For vein thickness $k = 0.15$ and for accumulation $k = 0.3$. The value of m in each case is chosen as the arithmetic mean of the three data sets, S_1, S_2 and S_3, that form the basis of the kriging plan. The kriging variance thus estimated does not include a line term (estimation of S_1 by the samples), which is found to be negligible in this case. Results listed in Table 3 can be used to make kriged estimates of means and variances of vein thickness and accumulation for any ore panel with dimensions of 100 by 200 feet in the plane of the vein, and for which the disposition of data is comparable to one of the three plans. Such calculations can be done for existing sample drifts and for new workings as adequate sample data become available. Calculations can be done quickly and easily by hand, and more conveniently by using a small desk calculator or computer. Consequently, the method is a practical scheme for use at the mine site. However, a new scheme would have to be worked out for different panel sizes or different dispositions of data relative to the block to be estimated, or both.

TABLE 3 — Summary of Kriging Parameters for Plans 1, 2 and 3

	KRIGING PLAN					
	No. 1		No. 2		No. 3	
	Vein Thickness	Accumulation	Vein Thickness	Accumulation	Vein Thickness	Accumulation
λ_1	.472	.439	.562	.466	.561	.540
λ_2	.245	.275	.289	.357	.197	.223
λ_3	.283	.286	.149	.177	.242	.231
σ_K^2/km^2*	.182	.183	.199	.243	.235	.282
$\frac{\sigma_K}{m} \times 100$	16.5%	23.4%	17.3%	27.0%	18.8%	29.1%

*k = 0.15 for vein thickness and 0.3 for accumulation. In both cases, m is the mean of all data values used in kriging a single panel.

Grade values, of course, cannot be kriged, but must be determined from kriged values of accumulation and vein thickness. Variance of the grade estimate is obtained in an analogous manner to that outlined in the section on global estimations, according to the equation

$$\frac{\sigma_t^2}{\bar{t}^2} = \frac{\sigma_a^2}{\bar{a}^2} + \frac{\sigma_p^2}{\bar{p}^2} - 2 C_{ap} \frac{\sigma_a}{\bar{a}} \frac{\sigma_p}{\bar{p}}$$

where t is the mean tenor with variance σ_t^2 and other variables are defined in the section on global estimation.

Figure 10 shows the distribution of blocks of the Eagle vein for which local estimates can be obtained using data available for this study. Outside these blocks, where the vein is known to be continuous, global estimates apply. It is apparent that with available sample data, local estimates of ore reserves can be made for considerably less than half the known orebody. Furthermore, if local estimates are desired for those zones between levels 7325 and 6360 outside the presently kriged blocks, a minimum of three additional sampled drifts are required, one midway between levels 7325 and 6930 and two spaced approximately equidistant between levels 6360 and 6930.

Although we do not reproduce all results of the kriging study here, several further comments are warranted. At first glance, it would appear odd that some of the weighting factors (λ_3's) are higher than the corresponding λ_2's, indicating that more weight is attached to data S_3 than to data S_2, which is closer to the block being estimated. In fact this is a not uncommon feature of highly continuous regionalized variables and in this case takes into account the lenticular nature of the intermeshed structures.

Kriging is commonly referred to as a "best, linear, unbiased estimator". It is "best" because it produces an estimate of the mean value of a regionalized variable with the minimum possible uncertainty (variance) based on available data. Other standard estimation procedures cannot make this claim and, in fact, standard ore reserve estimation procedures do not normally contain any error estimate, thus providing no measure of the certainty with which mean values are known.

Standard ore reserve calculation methods are generally based on weighted arithmetic means. The use of this procedure tends to overestimate high-grade sections of a deposit and underestimate low-grade sections. This effect is minimized by the use of kriging. An example is shown in Table 4 for six contiguous blocks from the 6930 level. In Table 4, weighted arithmetic means of per cent Cu are compared with average Cu values determined from kriged values of ac-

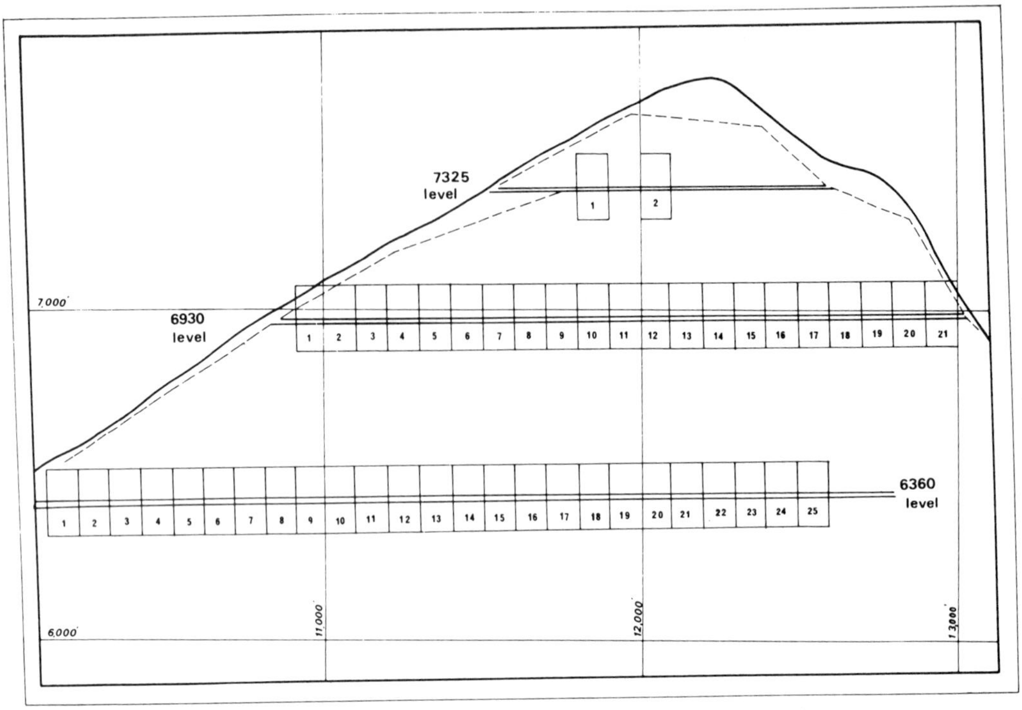

FIGURE 10 — Disposition of 25 ore panels that can be kriged with available data, using kriging plans 1, 2 and 3.

cumulation and vein thickness. It is apparent that the generalization noted above holds for these data.

Furthermore, a measure of uncertainty of a result based on kriging is known. Considering all the kriged blocks, the relative standard deviations, although variable, are about 16.5% for vein thickness, 23.2% for accumulation and 21.9% for per cent Cu. These results do not appear particularly good, but it must be emphasized again that they are the best attainable with available data.

Tonnage Calculations

We do not concern ourselves here with details of tonnage calculations. They can be done in a straightforward manner using the relationship

$$T = \frac{S \times \bar{\rho}}{f}$$

where T is short tons, $\bar{\rho}$ is average vein thickness (in feet) of the block being estimated, S is the area (square feet) of the block in the plane of the vein and f is a factor to convert cubic feet to tons. One should bear in mind that none of these parameters are perfectly known, although such is commonly assumed to be the case. The error in vein thickness has already been discussed on both global and local scales. The error in S is small but significant as a rule, and is generally unknown. Similarly, f has a range of values that depends largely on the variable mineralogical character of the ore. With appropriate data, variances can be determined for the foregoing variables, although in all likelihood the variance of S will have to be assumed.

TABLE 4 — Comparison of Weighted Arithmetic Estimates and Kriged Estimates for Six Contiguous Panels, 6930 Level

Weighted Arithmetic Mean (% Cu)	Average Grade from Kriging % Cu = $\bar{\alpha}/\bar{\rho}$	Variance	Standard Deviation	Relative Standard Deviation
4.01	4.70	1.2	1.10	23%
1.76	3.95	1.22	1.15	29%
3.18	3.82	0.81	.902	24%
3.88	4.24	0.9	.95	22%
9.13	6.08	1.33	1.16	19%
5.70	5.06	1.48	1.22	24%

Given estimates of variances of S, $\bar{\rho}$ and f, a variance for tonnage estimations can be determined as follows:

$$\frac{\sigma_T^2}{T^2} = \frac{\sigma_S^2}{S^2} + \frac{\sigma_{\bar{\rho}}^2}{\bar{\rho}^2} + \frac{\sigma_f^2}{f^2}$$

where T is tonnage with variance σ_T^2, S is area in the plane of the vein with variance σ_S^2, $\bar{\rho}$ is average vein thickness with variance $\sigma_{\bar{\rho}}^2$ and f is a conversion factor with variance σ_f^2. This equation can be applied globally or locally. Note that the ability to determine the variance of a tonnage estimate provides a sound base on which to carry out feasibility studies and to effect tonnage control during mining.

It is important to be aware of the fact that we have dealt exclusively with what can be called "geological reserves", and have made no effort to consider the effects of a particular minimum mining width. This problem can be treated as a simple continuation of our study. The procedures outlined here permit the estimation of geological reserves — grade and volume (or tonnage) — and thus quantity of metal.

The effect of a minimum mining width simply leads to a new, "mining" volume through which the quantity of metal must be distributed in order to calculate the corresponding grade. The "mining" volume is generally larger than the volume of geological reserves, because in some places the vein width is less than the minimum mining width.

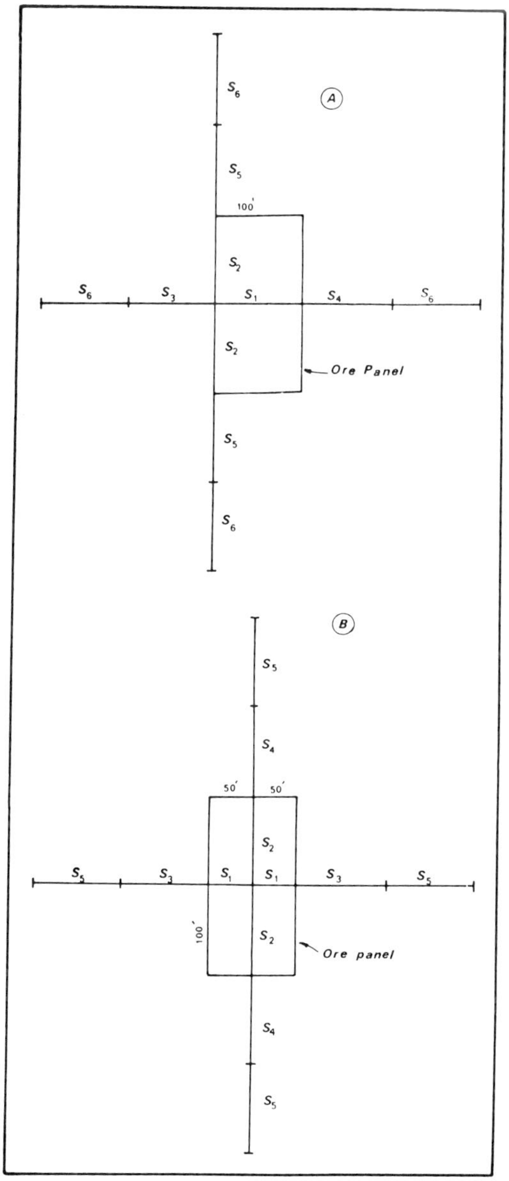

FIGURE 11 — Hypothetical kriging plans, (a) No. 4, (b) No. 5, to illustrate the effect that various additional workings will have on error estimates of vein thickness, accumulation and Cu grade.

Effects of Additional Information on Error Estimates

A potent advantage of geostatistics that can be a useful guide to further development work is the ability to calculate the effects that additional information will have on error estimates. Thus, it is possible to assume the existence of various patterns of new workings for which sample data are assumed available and examine their effect on variance estimates. Consider, for example, the effect that several new, sampled drifts will have on global error estimates. Assume that one new drift is located midway between the 7325 and 6930 levels, and that two new drifts are positioned equidistant between the 6930 and 6360 levels. The new global error estimates, calculated in a manner similar to that described earlier, are shown in Table 5, where the earlier (real) estimates are also shown for purposes of comparison. It is apparent that global estimates based on three additional drifts have been improved by nearly a factor of 2 relative to existing values. Of course, there is no way of knowing what the new mean values will be until the new sample data are actually available. Whatever the mean values, however, they will be significantly better estimates of the entire orebody (nearly twice as good as those existing at present).

By the same token it is possible to calculate local error estimates based on kriging plans that assume the presence of additional sampled workings. Consider the kriging plans for Figure 11, each of which assumes rectangular networks of drifts and raises in the plane of the vein. In both plans, panels of ore 200 feet by 100 feet are used to permit direct comparison with results of kriging plans described previously. Kriging parameters for the two new plans (No. 4 and No. 5) are listed in Table 6. Estimated relative errors for kriging plans 1, 4 and 5 are compared in Table 7 and illustrate the increased precision with which local estimates can be made as the amount of sample data increases. These results have considerable significance. They show, for example, that in order to know local ore grade with a precision of about 13 per cent, a high density of workings is required. In particular, to attain this precision for local estimates *throughout* the orebody, the simplest pattern of sampled workings necessary would be drifts at intervals of 200 feet and raises at intervals of 100 feet (or vice versa). A *decrease* in the density of raises by 50 per cent (i.e. an

TABLE 5 — Comparison of Real and Hypothetical Variances for Global Estimates

	Present Global Estimators	Global Estimators Assuming 3 Additional Drifts
$\sigma_{\bar{\rho}}/\bar{\rho}$	4.8%	2.85%
$\sigma_{\bar{a}}/\bar{a}$	6.8%	4.06%
$\sigma_{\bar{t}}/\bar{t}$	6.3%	3.78%

TABLE 6 — Summary of Kriging Parameters for Plans 1, 4 and 5

	KRIGING PLAN					
	No. 1		No. 4		No. 5	
	Vein Thick-ness	Accumu-lation	Vein Thick-ness	Accumu-lation	Vein Thick-ness	Accumu-lation
λ_1	.472	.493	.327	.295	.123	.105
λ_2	.245	.275	.319	.276	.579	.556
λ_3	.283	.286	.013	.001	.118	.122
λ_4			.060	.056	.081	.075
λ_5			.110	.191	.105	.143
λ_6			.201	.181		
σ_K^2/km^2*	.182	.183	.123	.125	.066	.070
σ_K/m*	16.5%	23.4%	13.6%	19.4%	9.7%	14.5%

* k = 0.15 for vein thickness and 0.30 for accumulation. In both cases, m is the mean of all data values used in kriging a single panel.

TABLE 7 — Comparison of Relative Standard Deviation for Kriging Plans 1, 4 and 5

	KRIGING PLAN		
	1	4	5
$\sigma_\rho/\bar{\rho}$ Vein Thickness	16.5%	13.6%	9.1%
$\sigma_a/\bar{a}$ Accumulation	23.2%	19.4%	14.5%
$\sigma_t/\bar{t}$ Grade	21.9%	18.0%	13.2%

TABLE 8 — Global Error Estimates Based on Two Hypothetical, Dense Networks of Drifts and Raises

Global Plan No.	Separation of Drifts	Separation of Raises	Relative Error % $\frac{\sigma_\rho}{\rho}$	$\frac{\sigma_a}{a}$	$\frac{\sigma_t}{t}$
3	200	200	1.34%	1.94%	1.80%
4	200	100	1.00%	1.68%	1.52%

interval of 200 feet between raises) increases the error estimates significantly. The relative error for grade, for example, increases from about 13% to about 18% (see Table 7).

New global estimates can also be obtained for the hypothetical networks of workings inherent in kriging plans 4 and 5. In the situation where kriging plan 4 can be applied to the entire orebody (requiring a square network of drifts and raises at 200-foot intervals), the global error estimates are considerably reduced from those presented earlier (compare data for global plan 3 in Table 8 with Table 5). By doubling the number of raises, i.e. to an interval of 100 feet, the global error estimates are reduced even further (global plan 4, Table 8).

Summary and Conclusions

Geostatistical methods utilized in this study permit the calculation of mean values of vein thickness and accumulation for the Eagle vein as a whole (global estimates) and locally (kriging). To each of these values can be attached a corresponding variance that is a precise indication of how well the means are known. Utilizing these estimates, it is possible to obtain comparable estimates for grades, both globally and locally. Similarly, tonnage calculations can be made with an indication of their precision.

The study has shown that vein thickness and accumulation are regionalized variables, the variograms of which are the product of intermeshed structures. Two structures are apparent in the experimental variograms for vein thickness with ranges of 20 and 100 feet respectively. These values are *average* lengths of two families of lenses; hence, the variogram gives some quantification to the geometric form of the deposit. Two comparable structures are apparent in the experimental variograms for accumulation, with ranges of 30 and 90 feet respectively. An important practical significance of these ranges is that values of the regionalized variables cannot be extended usefully beyond about 100 feet from sample sites. This conclusion is of obvious importance in designing a sampling plan for the deposit.

The present average sample spacing is about 8.5 feet. Such close-spaced sampling was essential to this study in order that the variogram model for the deposit be well defined. However, for purposes of global estimation, samples could be spaced at 20 feet without significant loss in precision. For local estimation, however, a sample spacing of 10 feet is required to give variance estimates of the quality presented here.

Global estimators presented in the text give the best possible estimates that can be applied to parts of the vein more than 100 feet from existing sample sites. Consequently, they are applicable to large parts of the vein between existing levels, and to vein extensions where there are adequate grounds for believing the vein is continuous. Global means for vein thickness, accumulation and grade are 4.02 feet, 17.0 feet per cent and 4.23 per cent Cu respectively. It must be emphasized that these values apply only to very large blocks of ore and in no way represent local estimators!

Kriging provides the best possible linear estimators for local ore reserve estimations. Three kriging plans are described that utilize data disposed in a manner similar to that on which this study is based. These plans allow the calculation of ore reserves for rectangular panels in the plane of the vein, measuring 100 feet by 200 feet. One plan (No. 1) that applies to the bulk of data in this study gives relative standard deviations of 16.5%, 23.2% and 21.9% for vein thickness, accumulation and grade respectively. Although these errors are large, the estimations are the best that can be obtained from the available data.

Geostatistics can be used to calculate the effects of additional information on error estimates, both global and local. The driving of three additional drifts along the vein, so that successive drifts are about 200 feet apart, reduces global variance estimates by nearly a factor of 2. Comparable calculations can be made, assuming additional data for local estimations, by assuming a variety of reasonable patterns for drifts and raises. Results of various trials show that for the grade error estimates to be reduced to 13 per cent, the simplest development plan required would involve horizontal drifts at 200-foot intervals and vertical raises at 100-foot intervals (or vice versa). Obviously the cost of such workings must be weighed against the necessity of attaining such a precision for local grade estimation. Global error estimates for such a network of drifts and raises are reduced considerably.

All kriging plans are presented in a form appropriate for rapid and simple calculation at the mine site, providing only that sample data are suitably disposed. Additional kriging plans could be prepared easily for a wide variety of geometric patterns of sample data and/or different panel sizes.

This entire study has been based on an assumption of geostatistical isotropy. Obviously, this assumption should be tested by constructing variograms for data aligned in several directions in the plane of the vein. Should the deposit be found to be anisotropic, the conclusions are rendered void and calculations would have to be re-done assuming a more complex variogram model. Nevertheless, the study represents an almost classical case history of the application of geostatistics to a vein deposit and gives insight into the type of information that can be obtained at an early stage of development of such an orebody.

Acknowledgments

This work was carried out while the senior author was on sabbatical leave at the Centre de Morphologie Mathématique, Fontainebleau, France. Financial support was obtained from the Centre de Morphologie Mathématique and the National Research Council of Canada. Discussions and guidance provided by Dr. A. Journel, research director, Dept. of Geostatistics, Centre de Morphologie Mathématique, are appreciated. R. Keays, vice-president of Davis-Keays Mining Co., Ltd., kindly provided access to an earlier feasibility report on the Eagle vein.

References

Blais, R. A., and P. A. Carlier, 1967, Applications of geostatistics in ore evaluation. CIM Spec. Vol. 8, pp. 41-68.

Matheron, G., 1971, The theory of regionalized variables and its applications. Les Cahiers du Centre de Morphologie Mathématique de Fontainebleau, No. 5, 211 p.

Matheron, G., 1967, Kriging or polynominal interpolation procedures. CIM Bulletin, Vol. 60, No. 665, pp. 1041-1045; CIM Trans., Vol. 70, pp. 240-244.

Preto, V. A., 1972, Eagle Mine; *in* Geology, Exploration and Mining in British Columbia, 1972. B.C. Department of Mines and Petroleum Resources, pp. 76-78.

Trimble, R. J., 1972, Mineralogy and computer-oriented study of the Eagle vein complex, Liard Mining District, northeastern B.C. Unpublished thesis, Dept. of Geol., Univ. of British Columbia, 78 p.

Summary Table of Symbols

$\gamma(h)$	=	semivariogram value for sample separation (lag) h.
$Z(x_i)$	=	value of variable Z at location x_i
$Z(x_i + h)$	=	value of variable Z at location $x_i + h$
n	=	number of samples or number of sample pairs, depending on context
h	=	lag or sample separation
a	=	range of variogram, subscripted if more than one model
C_o	=	nugget effect
C_i	=	sill of variogram minus C_o. Subscripted $_1$ and $_2$ if two models are required
m	=	arithmetic mean of a data set
σ^2	=	global variance (i.e. for entire orebody). Subscripts: a = accumulation, p = vein thickness, t = Cu tenor.
$\sigma_i^2(l_i)$	=	mean variance of (linear) drift segment l_i
σ_E^2	=	extension variance of drift to corresponding rectangular blocks
a	=	accumulation
p	=	vein thickness
t	=	tenor
C_{ap}	=	correlation coefficient, accumulation versus vein thickness
Z_k^*	=	estimated mean calculated by kriging of variable Z for a panel of specified size
λ_i	=	weighting coefficients used in kriging
S_i	=	data sets used in kriging and weighted by γ_1's
$(S_i\ S_j)$	=	mean theoretical semivariogram between data sets S_i and S_j
σ_k^2	=	kriging variance
S	=	capital S without subscript refers to surface (or volume in 3-D)
T	=	short tons
f	=	volume per ton (cubic ft/ton)

Part IV

DEVELOPMENT OF RESERVES

Editor's Comments on Papers 22 Through 29

LOCAL ENVIRONMENT OF ORE SHOOTS

Concepts as to the genesis and controls of ore deposits vary widely from one country to another, and even within the same country, dependent upon the character of the prevalent ore deposits in the locality and the breadth of experience of the geologists. Much of the detailed geology of ore deposits was generated in the western portion of North America where ore deposits are closely associated with subvolcanic intrusives in a subaerial environment and where localization of ore shoots was controlled by fracture systems that guided flow of solutions. This interest in establishing the plumbing system for the circulation of ore-depositing fluids dominated thinking in the search for, and extension of, ore shoots. Expansion of thinking to include those deposits formed in a submarine environment grew out of the observations and experience of European, Australian, Japanese, and Canadian geologists.

Although Wisser (Paper 22) refers specifically to vein-type deposits, his generalizations regarding observations and analysis of data are applicable to deposits without regard to genesis. His paper serves as a good introduction to Papers 23 and 24. It goes beyond simply an evaluation of ore to include a careful evaluation of the total environment in which the ore is localized. This is the concept emphasized by Stanton (1972) that ore deposits are often the product of the rock-forming processes and must be understood on that basis.

Structural analysis and pattern analysis as an approach to the discovery of ore shoots within a mine, and in the determination of evidence for the bottoming out of ore shoots, has served as a valuable tool of the mine geologist. Variations of the technique and its applications are outlined by Schmitt (Paper 23) and Conolly (Paper 24). These two papers sparked a great deal of interest, and the method was successfully applied to gold veins in California (Cooke, 1947) and to copper-tin veins in Cornwall (Taylor, 1966; Garnett, 1966). Cooke (1947) applied correlation coefficients to the coincidence of ore values with structural or mineralogical variations that were revealed by the contouring method, taking the first step toward statistical correlations. Computer plots of chemical, mineralogical, structural, and/or geophysical data have greatly simplified utilization of such contouring techniques.

The first author to relate, comprehensively, ore shoots to structural variations was Hulin (Paper 25). He recognized the ore-forming process as a complex one marked by alternating fracturing and mineral deposition with a distinct paragenetic sequence in the depositional history of the ore minerals.

Structures occurring within a bedded sequence may be very subtle, forming during soft sediment deformation or accompanying diagenetic and/or metamorphic processes. The subtle, yet economically important, bedding-plane faults are emphasized by Behre (Paper 26). The effect of these structures is often to localize "blind," or pseudostratiform ore bodies that present a challenge to the mine geologist.

Unraveling the complex history of deposits in deformed rocks remained much of a mystery until the development of structural analysis techniques by Turner and Weiss (1963). The method has been successfully used to correlate deformational history of sulphide ore shoots and the enclosing host rock (see Davis, Paper 10). Perhaps the most successful of its applications has been interpretation of the complex structural history of the Broken Hill ore body in New South Wales, Australia. Excerpts from the paper by Laing, Marjoribanks, and Rutland (Paper 27) illustrate utilization of this technique in the understanding of ore distribution and in the guidance of ore search.

Few ore deposits are without breccias closely associated with the deposit. These breccias, upon close examination, can be identified as resulting from a variety of geological processes often unrelated to faulting. Paper 28 by Locke and Paper 29 by Bryant illustrate two of the many varieties. Bryner (1961) proposed a classification of various types, but continued recognition and study of these structures in mining districts suggests formation by many additional processes.

REFERENCES

Bryner, L., 1961, Breccia and Pebble Columns Associated with Epigenetic Ore Deposits, *Econ. Geology* **56:**488–508.

Cooke, H. R., 1947, Association Indices in Sixteen-to-One Mine, California, *Econ. Geology* **42:**211–250.

Garnett, R. H. T., 1966, Relationship Between Tin Content and Structure of Lodes at Geever Mine, Cornwall, *Inst. Mining and Metallurgy Trans.* **75:**B1–B22.

Stanton, R. L., 1972, *Ore Petrology,* McGraw-Hill, New York, 713p.

Taylor, R. G., 1966, Distribution and Deposition of Cassiterite at S. Crofty Mine, Cornwall, *Inst. Mining and Metallurgy Trans.* **75:**B35–B49.

Turner, F. J., and L. E. Weiss, 1963, *Structural Analysis of Metamorphic Tectonites,* McGraw-Hill, New York, 545p.

22

Reprinted from pages 96, 97–98, 102–104, and 109–110 of *Am. Inst. Mining, Metall. and Petroleum Engineers Trans.* **144:**96–110 (1941)

The Environment of Ore Bodies

BY EDWARD WISSER,* MEMBER A.I.M.E.

(New York Meeting, February, 1938)

The environment of an ore body is taken to mean not only its physical surroundings but every factor, passive or active, that conditioned the ore shoot, saving only the original composition of the solutions that formed the shoot. Collection of data related to the shape and characteristics of veins should be followed by correlation of the data and the formulation of generalizations regarding ore formation. The nature of the criteria used in arriving at these generalizations is discussed. Certain of the criteria, when used with an understanding of their underlying causes, become "ore-guides." In the course of his study the geologist must not confine himself to the immediate surroundings of the veins, but should attempt to work out the geologic history of events culminating in the crucial epoch for any mining district, the period of mineralization. The work may involve study of features far removed from the veins and related to them only indirectly. It commonly means thorough study of the geology of the entire region surrounding the ore deposits.

[*Editor's Note:* Material has been omitted at this point.]

This paper deals with the nature of criteria related to ore occurrence. The question, "Are there any criteria by which one can determine when mining operations have reached the roots of an ore deposit?" concerns the exploration of known ore downward. It aquires a broader interest if this implied situation is reversed. Consider a vein exposed at the bottom of the vertical range of ore body expectancy and as yet unexplored elsewhere. Features indicative of the roots of ore shoots point upward toward ore. If the miner recognizes such features where he has exposed the vein he will raise into ore; otherwise he may sink and will find no ore. Thought of in this light, the question may be expanded: "Are there any criteria by which one can determine whether a vein exposed, thus far, in waste only, is likely to yield ore somewhere, and are there criteria to show, in addition, whether a given exposure in waste lies above, to one side of, or below the place where ore might be expected?"

The mining geologist is inclined to overemphasize the evaluation of ore; he is overly concerned sometimes with the question: How did this ore form and why? His principal function, in practice, is to evaluate waste. As Lindgren[1] says: "When the miner has an abundance of ore he gets along comfortably without the geologist, but when ore resources grow slim, the geologist is called forthwith." When ore is giving out, most exposures are in waste.

It is true that the most successful method of finding ore today is to seek duplicates of settings known to have produced ore. Since such settings may involve changes in the strike of the fracture explored,

---

Manuscript received at the office of the Institute Aug. 15, 1938. Issued as T.P. 1026 in MINING TECHNOLOGY, January 1939.

* Consulting Mining Geologist, San Francisco, Calif. and Manila, P. I.

[1] References are at the end of the paper.

cross-faults, fracture intersections, etc., exploring them may have little to do with the evaluation of waste in terms of ore. Given, perhaps, a fracture about which nothing is known except that it is pre-mineral, and a supposedly favorable but unexplored setting, we head straight for the heart of the setting and often find an ore body. Ore shoots, however, are commonly small targets to hit, and structural or other control of ore shoots is seldom discerned precisely enough so that an exploratory working, say a crosscut, can be so directed as inevitably to cut the heart of an ore shoot. The usual case, even though the exploration is in the end successful, is to cut a "promising" but low-grade vein where we had hoped to cut ore. Most ore shoots lie encased in a much larger body of low-grade vein material. Since the reasons for localization of these large gangue bodies are frequently manifest, it is often not difficult to find these bodies. The problem is to evaluate them. If we can learn how to tell whether a low-grade body of vein matter is likely to carry an ore shoot, and if so, where the shoot probably lies with respect to the exposure, we should be able to find ore much more quickly and much more cheaply than we do at present.

To accomplish this purpose we must study carefully and systematically the environment of ore bodies. We must learn how veins look in ore, at the edges of ore, above and to one side of ore; in short, we must study vein morphology. Through this study we should accumulate data not merely of immediate, "practical" use to the operator, but bearing on broad generalizations as well. There is a present need for reclassification of veins without regard to current theories of genesis, on the sole basis of common features of morphology. While our present classification of hypogene epigenetic deposits into epithermal, mesothermal, etc., rests largely on this method, all generalizations need periodic testing. Lindgren, a pioneer builder of the accepted classification of ore deposits, says:[1]

> Comparative studies are needed of certain classes of veins—for instance, gold-quartz veins—the common features, the relation of depth to structure, the relation of composition of the ores to structure and depth . . . What are the common features of epithermal veins?

Sales, in his discussion of Lindgren's paper, voices a reasonable objection to the study and correlation of vein morphology:

> The economic geologist . . . rarely finds two mining districts with identical geological conditions. Each district has its own problem. Successful search in one district may or may not result in discoveries applicable to others.

Here is the nub of the problem; this conviction discouraged our operator, whose criteria for the roots of his ore shoots failed him. Is not this failure inherent in the use of empirical criteria without some understanding of their underlying causes?

[*Editor's Note:* Material has been omitted at this point.]

Environment of Ore Bodies

It must be plain that in gathering the evidence to form a workable theory of vein formation, we cannot, necessarily, stop at the veins themselves. In the case discussed, the whole district was combed over for evidence bearing on the relative ages of the fractures and veins, and on the nature of the stresses active before, during and after the mineralization. Earlier in this paper the word environment was used in the sense of the immediate surroundings of ore. Evidently in a broader sense it may include the whole district. But the word environment may be used in a sense still broader. In the life sciences, the environment of an organism involves every factor, saving heredity, that affects the form and actions of the organism. There is here a parallel. The analogue to the heredity of an organism is found with ore deposits in the original composition of the solutions that formed them. That composition is usually independent of conditions at the seat of ore deposition. The heredity of an organism may be equally independent of its environment. The environment of an ore deposit comprises not alone its surroundings but all the circumstances, static and dynamic, saving only the composition of the solutions that formed it, that influenced the way in which the ore was formed, and the present appearance of the ore body.

With living organisms, dynamic as well as static factors of environment may be studied more or less directly. With ore deposits, static factors, such as wall rocks, fissures antedating the mineralization—in a word, the passive framework that received the ore—may be studied, perhaps, with little recourse to inference. But dynamic factors of environment, such as variations in the pressure, temperature and concentration of the solutions, and stresses operating during the mineralization, must all be inferred from the traces they may have left behind them. It follows that a study of the factors that make ore or waste, the study of the environment of ore deposits, may never be dissociated from inference.

Method of Portraying Vein Morphology

A great help in the correlation of facts of observation is their convenient presentation. Geologists have failed to utilize to the full the possibilities of a valuable contrivance, the longitudinal vertical or plan projection of a vein. Schmitt[2] uses projections in his method of indicating probable extensions of ore shoots by equal-ore-width or equal-ore-

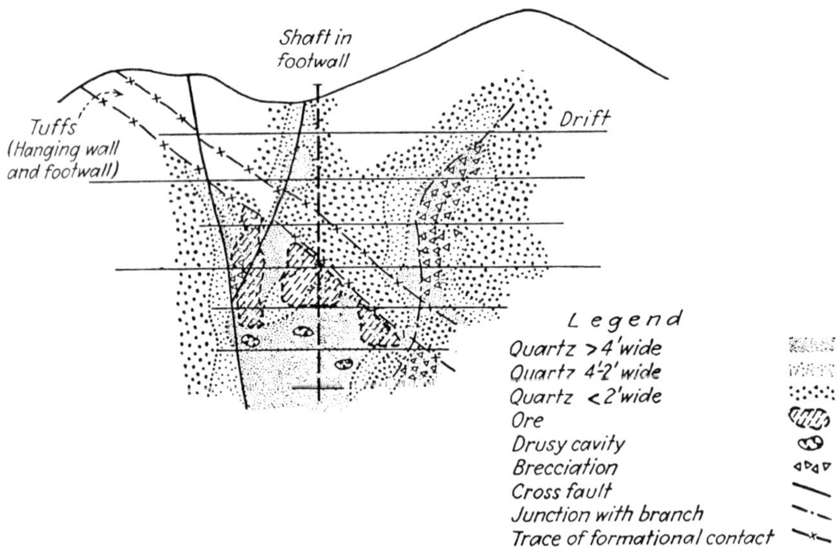

Fig. 1.—Geological vertical projection of a hypothetical vein.

value "contours." Conolly,[3] in an equally valuable paper, shows irregularities in the vein surface, equal-ore-value contours and equal-quartz-width contours upon plan and vertical projections.

Many other features can be shown on such projections. For example, the distribution of ore minerals and that of other metallic minerals such as vein pyrite; the distribution of gangue minerals such as quartz and calcite, and the textural varieties of the same mineral (coarse equigranular quartz, chalcedonic quartz, etc.); variations in the intensity and nature of the alteration halo in the vein walls. Also, the traces upon the vein of cross faults, dikes and veins; lines of junction of vein branches; traces on the vein of rock-formation contacts in the walls; structural peculiarities of the vein, such as zones of brecciation, areas of drusy cavities or of crustiform banding, etc. A few such features are shown on the accompanying vertical projection (Fig. 1). It is rarely convenient to show them all on one projection. They may be grouped into mineralogical, structural features, etc., and shown on tracings to permit superposition. The projections must be used, of course, in conjunction with geologic level maps and vertical cross sections.

These geologic projections bring out conveniently the relations of ore to gangue, and of both to the other features shown. Relations that appear significant become criteria for the exploration of other veins. Such use, however, must be accompanied by critical analysis, as I have shown.

[*Editor's Note:* Material has been omitted at this point.]

Conclusion

Criteria exist in nearly every camp that indicate whether a vein may be productive or not, and point to the location of ore shoots in productive veins. But such diagnostic features may never be used, even between one vein and its nearest neighbor, with perfect safety. Listing criteria of observation with no regard to the causes behind them is fraught with deception. The use of such criteria must go hand in hand with complete study of a district, and, in particular, the exact deciphering of the geologic history of the events that took place at or near the crucial epoch for any mining district, the period of mineralization.

We must continue the systematic collection of data on vein morphology, and make a genuine effort to correlate these data. Above all, we must try to infuse life into the dead factual mass, so that it may act to serve our end, the finding of ore. The vital spark needed lies in the causes back of the empirical facts discovered. So far we have been stubbing our toes on "hard facts"—hard because their meaning is not understood.

Acknowledgments

I am indebted to Prof. C. A. Anderson, of the University of California, for helpful criticism. To those interested in the fundamentals of scientific thinking I can recommend no better guide than the stimulating paper by Douglas Johnson[5] on the role of analysis in scientific investigation. The present paper is largely an attempt to apply to the field of mining geology the principles set forth by Johnson.

References

1. W. Lindgren: Research in Processes of Ore Deposition. *Trans.* A.I.M.E. (1928) **76**, 291.
2. H. Schmitt: Extensions of Oreshoots with Comments on the Art of Ore Finding. *Trans.* A.I.M.E. (1929) **85**, 318.
3. H. J. C. Conolly: A Contour Method of Revealing Some Ore Structures. *Econ. Geol.* (1936) **31**, 259–271.
4. C. D. Hulin: Structural Control of Ore Deposits. *Econ. Geol.* (1929) **24**, 37–38.
5. D. Johnson: The Role of Analysis in Scientific Investigation. *Science* [N.S.] (1933) **77**, 569–576.

23

Reprinted from *Am. Inst. Mining, Metall. and Petroleum Engineers Trans.*
85:318–324 (1929)

Extension of Oreshoots with Comments on the Art of Ore Finding

By Harrison Schmitt, Hanover, N. M.

In the practice of ore finding the geologist is continually confronted with the question of oreshoot extension, so that the general problem seems worthy of systematic investigation. This problem appears when the reserves are estimated and when the advisability of deepening the shaft comes up. Moreover, there is need for lower development cost to keep pace with cost reductions in other departments, or even to offset a rising cost of ore finding or ore following. This can be gained by close geologic control of the development; and much of the geology involved is that bearing on the downward, upward, and horizontal extensions of shoots.

Furthermore, the purchase of a mine or prospect often depends directly on the question of shoot extension; because it is seldom that the ore is fully developed and the purchase price is almost invariably based largely on "possible" and geologically prospective ore. Most current methods for estimating "possible" and the more nebulous classes of ore are crude and indecisive.

In the following pages the extension of oreshoots is the major theme, although the more basic subject, the general art of ore finding, is given some space.

(*A*) Direct Methods of Extension Prediction

(*I*) *Extrapolation and Interpolation of Shoots by Use of Conventional Geologic and Assay Plans, Projections and Cross-sections*

There is little need for stressing the desirability of detailed geologic plans, assay plans, projections, and cross-sections as a basis for the study and extension prediction of shoots. Underground and surface mapping, in most cases, ought to be done in exhaustive detail. All features of obvious bearing should be mapped and it is good practice to record all possible details, some of which may become significant when studied in the ensemble. It should be emphasized that the mapping should be done with an impartial mind, with interpretation reduced to the minimum. The symbols of the map can be made to distinguish observed facts from interpretation.

Impartial, detailed mapping often automatically solves many geologic problems through the completeness of the picture presented. Much less interpretation than is commonly used, is necessary; in fact, in many cases incorrect conclusions are caused by premature judgement that clouds the fact-gathering. A good deal of cold-bloodedness is essential in mapping, for it is surprising how easy it is in geology to find facts to fit almost any opinion.

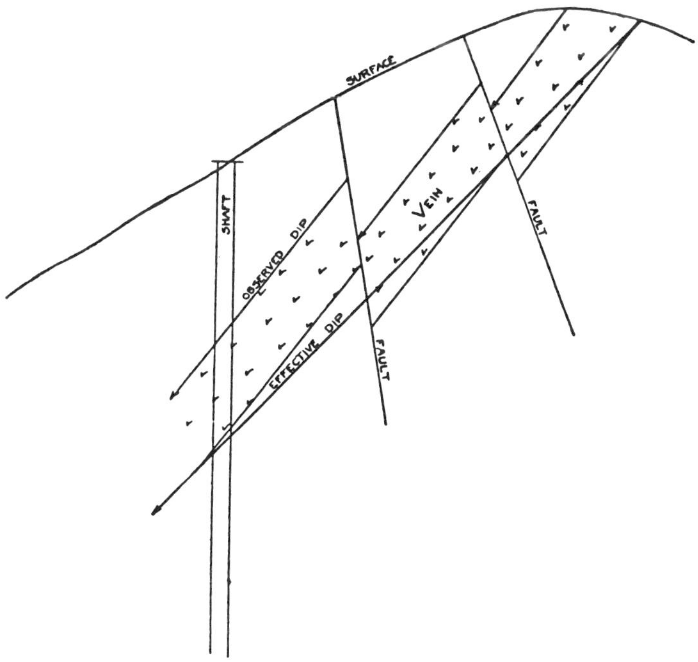

FIG. 1.—IDEALIZED CROSS-SECTION THROUGH ALFAREÑA VEIN AND ORESHOOT AT SANTA BARBARA, CHIHUAHUA; SHOWING MANNER BY WHICH FAULTING CAUSED EFFECTIVE DIP TO DIFFER FROM OBSERVED DIP.

For greatest utility in guiding development by the prediction of extension the geologic detail should be caught up at the headings at short intervals; once a week may not be too frequent during an intensive development campaign.

To get the greatest benefit from the detail much thought and care must be spent in devising a system of maps for recording it. No maps will give an entirely satisfactory three-dimension mental picture, and one can only strive to make them approach this ideal. It is seldom possible, therefore, to have too many cross-sections, projections, and structural contour maps. In many cases it is inadvisable to place the cross-sections at arbitrary intervals. It is good practice to locate them where they will "cut the most information," thus giving the best basis for oreshoot extrapolation.

To get best results on cross-sections, especially where there are low dipping contacts or veins, a definite horizontal plane for mapping on the mine level should be selected and adhered to. This plane may pass through the bottom, top, or middle of the drift. Because it is the most accessible, I prefer to map on the last.

In projecting dips on cross-sections the recorded dip may not be the effective dip of the oreshoot or vein because of distributive faulting or slicing. This was the case at the Alfareña mine at Santa Barbara, Chihuahua (Fig. 1). At Bisbee, Arizona, for the same reason observational dips do not always correspond with the composite dip of the horizon measured.

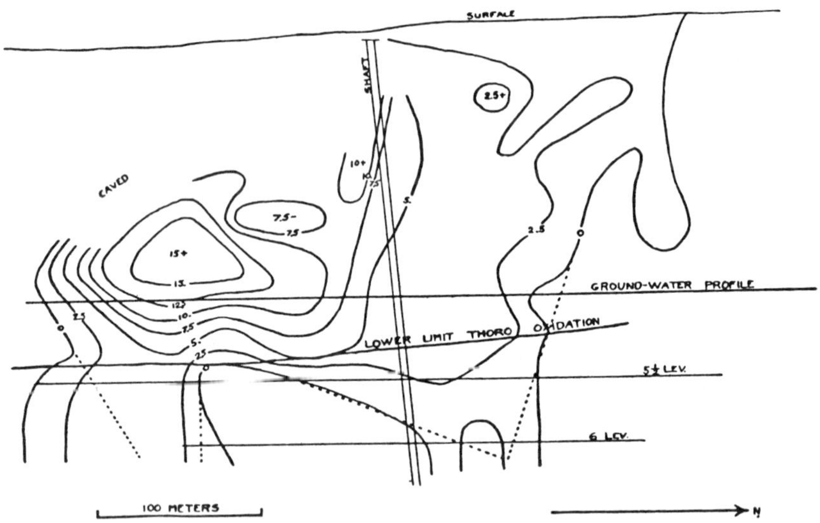

Fig. 2.—Equal-ore-width lines on longitudinal-vertical-projection of an oreshoot on Veta Colorado (Red Vein) near Parral, Chihuahua. The interval is 2.5 meters.

(II) *Extrapolation by the Analysis and Projection of More Detailed Information*

(*a*) *Equal-ore-width Lines.*—Equal-ore-width lines are lines connecting "points" of equal vein width (ore width) on the projection of a shoot. This gives an approach to a three-dimension picture of the shoot. The "iso" lines may be shown on a horizontal, inclined, or vertical projection, depending on the amount of inclination of the vein.

On the Veta Colorado (Red Vein) near Parral, Chihuahua, equal-ore-width lines, projected to a longitudinal vertical plane, were found useful. The known parts of the shoots appeared as a three-dimension picture and their extension downward and horizontally could be predicted with some confidence, because the border or "zero" lines are "reenforced" by the interior lines. Fig. 2 shows a shoot outlined previous to the final development. Below the 5½ level, the predicted extension of the shoot is shown in dotted line and the extension actually developed is shown in solid line.

(*b*) *Equal-ore-value Areas.*—The erratic distribution of the metal content of most shoots makes difficult the direct drawing of equal-metal-value lines. Equal-metal-value areas, however, can be outlined and these give, indirectly, equal-ore-value lines.

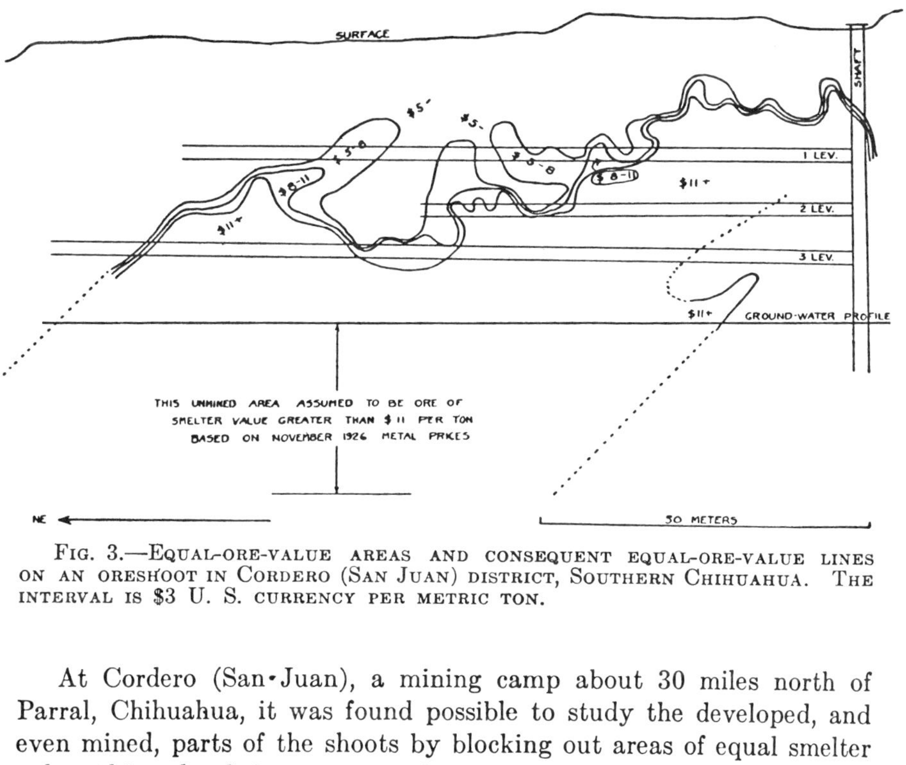

FIG. 3.—EQUAL-ORE-VALUE AREAS AND CONSEQUENT EQUAL-ORE-VALUE LINES ON AN ORESHOOT IN CORDERO (SAN JUAN) DISTRICT, SOUTHERN CHIHUAHUA. THE INTERVAL IS $3 U. S. CURRENCY PER METRIC TON.

At Cordero (San-Juan), a mining camp about 30 miles north of Parral, Chihuahua, it was found possible to study the developed, and even mined, parts of the shoots by blocking out areas of equal smelter value, this value being an approximate measure of the intensity of the mineralization. The analysis of the known parts of the shoots gave a basis for their extrapolation in depth, the interior equal-value-lines "reenforcing" the border or "zero" lines. (Fig. 3.)

This method has as its basis thorough sampling. The areas of equal smelter value are then outlined by reducing the assays of each sample to their smelter value per ton. Lead and silver were the most important metals and bore a remarkably constant ratio, one to the other, so together they made a good basis for this value classification. The proportions of gold, zinc and copper were erratic, but these metals, being of minor importance, could be disregarded.

The Cordero shoots are rather regular in outline, as shown by the projection of one of them (Fig. 3), and so an intelligent guess as to the quantity and quality of the ore in the unexplored extension downward could be made.

(c) *Curves Showing Variation of Metal Content.*—At the Hidalgo mine, Santa Barbara, Chihuahua,[1] exhaustive stope sampling, well recorded, made it possible to construct curves showing the variation of the metal content from near the outcrop to the lowest developed horizon. An analysis of these curves, making allowance for enrichment in the oxidized zone because of leaching of valueless material, indicated a definite trend of values in depth, an increase in this case. The extension of the shoot in depth could be predicted with some assurance.

(B) Indirect Methods of Extension Prediction

(I) *Study of the Relation of Structural Features to the Orebody*

The study of the relation of the structure of the country rock to the oreshoot is one of the first tasks to be undertaken before attempting to predict extension. There is a growing appreciation, not yet fully expressed, of the importance of the association of oreshoots with structural and purely mechanical features of the country rock. In recent years some schools of geology have taught that mineralization tends to avoid strong faults, the theory being that gouge tended to seal them off, preventing circulation and thus deposition. There are a large number of deposits for which this does not hold. For example, in the Parral district, Southern Chihuahua, the strongest veins are in the strongest faults. There in places the mineralization seems to have replaced the fault gouge with silica and ore. Today the important water courses are along the vein-ore-faults, and one can readily imagine that such was also the case during mineralization.

The Veta Colorado in the same district is a fine example of the manner in which orebodies favor structural features. This rich lode is a vein-fault, having a dip slip of 1450 ft. at its richest portion. The best ore is directly associated with the shear plane of the fault zone that has the greatest offset. In further detail, the shoots appear to have originated by the replacement of breccia that filled depressions in the foot-wall of the major shear plane.

Examples of ore in or near strong faults, or structural disturbances of equal importance, are common. Ore recently found at Tintic is associated with conspicuous structural abnormality. At Hanover, New Mexico, the zinc ore is associated with broken "open ground," that is, almost invariably in or near the major faults. There it is felt that if

[1] H. Schmitt: Geologic Notes on the Santa Barbara Area in the Parral District of Chihuahua, Mexico, *Eng. & Min. Jnl.* (1928) **126,** 410.

the location of broken ground of premineral age can be predicted, a favorable chance for ore can be assumed; and ore has been found on this basis.

The prediction of the location of "open ground" usually requires exhaustive preliminary measurements of structure. Very few geologists, and fewer miners, appreciate the possibilities for ore finding where contacts and other details of structure are measured to the foot, and mapped on an appropriate scale. I feel that the time has come when exceptional care in ore geology mapping is in order in many mineral districts. Much new ore can and will be found by such detailed technique, as has been applied at Butte and the Homestake and in a few other districts.

(II) Study of the Regional Habits of Oreshoots

The importance of the study of the regional habit of oreshoots as a basis for discovery or prediction of extension in a given area has been often emphasized. Shoots in a particular district may have characteristic shape; be lenslike, conical, chimneylike, mantalike, etc., or they may bottom at near the same elevation, or they may favor certain beds or intrusives.

For examples I shall again draw on the Parral district. Along the Veta Colorado, and associated veins of the same age, the oreshoots bottom within a few tens of meters of the same elevation. Many of the shoots, furthermore, are V-shaped, with the point of the "V" downward and generally pitching in a common direction.

(III) Study of the Relation of the Accident of Erosion to the Position of Shoots of Lenslike Form

This subject was recently discussed by Blanchard and Boswell.[2] It is especially applicable to lead-silver veins, because their shoots are commonly, if not usually, lenslike. If the form of part of the shoot that has been disclosed by mining be measured, the assumption of lenslike form may aid extension prediction.

(IV) Statistical Studies

Statistical studies of all known, or rather described, oreshoots might make our ideas of extension more quantitative. Thus, shoots of a certain type, such as those of the Tertiary silver veins, bottom at shallow depth. Or do they? What is the average depth, the relation of it to outcrop length, richness, tonnage, or to width of the vein? Hoover[3] has made pertinent suggestions along this line.

---

[2] P. F. Boswell and R. Blanchard: Oxidation Products from Sphalerite and Galena. *Econ. Geol.* (1927) **22**, 421.

[3] H. Hoover: The Principles of Mining, 21–33. McGraw-Hill Book Co., New York, 1909.

By such studies it might be practicable to express possibilities of extension in terms of chance or probability. Thus if a shoot is of a certain type, or size, or richness, for example, what is its chance for a given depth extension? Is it 1 in 10, or 1 in 4?

By going through all available literature on veins in Utah, data on the ratio of outcrop length to extension in depth for eight important shoots or lodes were found. The ratios of outcrop length to depth extension were 6:1, 4:1, 3:1, 5:7, 1:2, 1:2, 1:3, 1:4. The average of these is 15:8, which is near the 2:1 ratio suggested by Hoover as the average for a large number of deposits.

(V) *Study of Vertical and Horizontal Hypogene, and Vertical Supergene Zoning in Shoots*

The extension and zoning of oreshoots of supergene origin has been rather thoroughly studied in the last 20 years. Hypogene zoning has been somewhat neglected. The suggestion has been made that when shoots change their primary mineral character in depth they are liable to play out shortly. This lead and others should be further investigated.

(VI) *Study of the Relation of Oreshoot Extension to the Stage of Erosion of the Source Cupola*

Pioneer work on this subject has been done by B. S. Butler and W. H. Emmons. The theory is that ores are mainly associated with the top of the source cupola and that their chance for extension in depth is dependent on the depth of erosion below the apex of the intrusion. This phase of the subject could well be further studied.

The above are some of the more obvious divisions of the subject of extension. Many others, no doubt, will occur to the reader. For example, many veins split or feather when they are about to play out. In certain districts this is recognized as the end of the ore.

On the Mesabi Range, markers, such as the "algal" horizon, persist through the ore. Where the marker can be followed, it is possible, by measuring the slump, to get a fairly accurate idea of the depth of the ore.

It has been noticed that postmineral faults frequently avoid, and curve around, oreshoots and accompanying walls that have been strongly silicified. It may be useless to drift farther when such a fault is encountered. The fault means the end of the ore, not the offset of it.

Acknowledgments

I am indebted to Augustus Locke of San Francisco, Calif., and A. C. Spencer of the U. S. Geological Survey, Washington, D. C., for criticism of the manuscript of this paper.

24

Reprinted from *Econ. Geology* **31**:259–271 (1936)

A CONTOUR METHOD OF REVEALING SOME ORE STRUCTURES.

H. J. C. CONOLLY.

THE contour method described here is being developed and employed in the examination of a number of old gold mines[1] in Australia. The work is being done for the Western Mining Corporation and the author is indebted to the Corporation for permission to publish the material in this paper. He is also under considerable obligation to Mr. J. D. Campbell, who prepared the accompanying illustrations and advanced constructive criticism of the manuscript. The illustrations and their notes amplify and explain the text, so that a careful study of the figures is necessary in order to understand the contour method and its employment.

As is well known, the concentration of gold is commonly related to the changes in attitude of the host vein, lode, or formation.[2] There are many excellent examples of this relationship in Australia. Commonly, ore is confined to a local flattening of the reef, the change in dip being generally associated with a change in strike. In a few simple cases the ore shoots occupy only those sections of the vein with a particular strike and dip. Essentially, however, the metal concentration is not attached to any particular strike and dip, but is indeed widely related to the whole pattern of changing attitudes, the changes of attitude reflecting the governing structures.

Some connection between the "anatomy" of a vein and its gold content is what might be expected. The vein pattern is determined by the rock environment and structure and by the interplay of shear, tension and compression involved in all faulting and folding movements. The same environment, structure

[1] Most of these old mines are today full of water.

[2] Reference hereafter to vein or reef implies, where necessary, "host vein, lode or formation."

and stresses govern the deposition of gold and open the channels occupied by the ore shoots and followed by the ore solutions. Whatever controls the deposition of gold must as a rule also be reflected in the vein structure.

In the examination of these Australian mines it was found

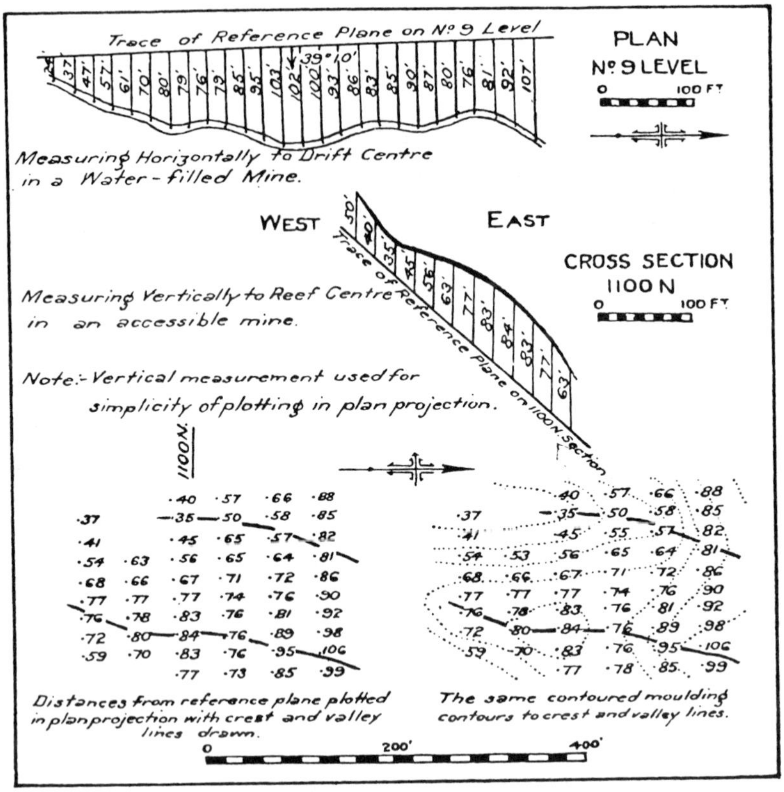

FIG. 1. The vein-contouring procedure.

that the full significance of the relationship between vein structure and gold was quite inadequately portrayed or even not revealed by the ordinary mine plans, sections and models prepared. To learn the structure in these cases and in similar deposits, it is necessary to see its whole pattern, including the slightest changes in strike and dip, exaggerated without distortion, and to see it in three dimensions at a glance, on one plan. This is made possible by the simple expedient of contouring the vein with reference to inclined planes which approximately parallel its average strike and dip. The contours so obtained are then superimposed on plans showing the distribution of gold through the deposit and, taking the study further, also on plans showing the distribution of quartz. Each vein in a vein assembly is treated separately.

The vein-contouring procedure is illustrated in Fig. 1. The trace of the reference plane on the mine-level plan or mine section is drawn. The distance between the reference plane and vein (or drift) center or vein wall is then measured, usually at intervals of twenty feet. Plan or vertical projections are employed for contouring, the distances being written on the projection of the vein position measured. Where the vein is shown on a mine-level plan, the measurements are made horizontally along any convenient direction, usually one of the mine co-ordinates. Where cross sections of the vein are employed, the measurements are made vertically if contouring in plan projection, or horizontally if contouring in vertical projection. Finally, the distances between reference plane and vein (or drift) are contoured in exactly the way in which topography and structure are contoured with reference to a horizontal plane. Crest and valley lines, linking the longest and the shortest distances, are drawn ahead as far as possible to guide contouring, the relatively uniform "gradient" along these lines affording the best locus for the proportional spacing of the contours. These crest and valley lines mark the changes in vein strike and dip and are therefore the most important features of the plan. In water-filled or inaccessible mines, where only old level plans are available, the most careful contouring is necessary to gain any reliable idea of the flatter pitching structures.

The first step in a mine study is to establish whether there is or is not any connection between vein structure and gold (and quartz). This is most safely tested by contouring the vein with reference to an arbitrarily chosen inclined plane, say a plane which parallels the average strike and dip of the most important ore

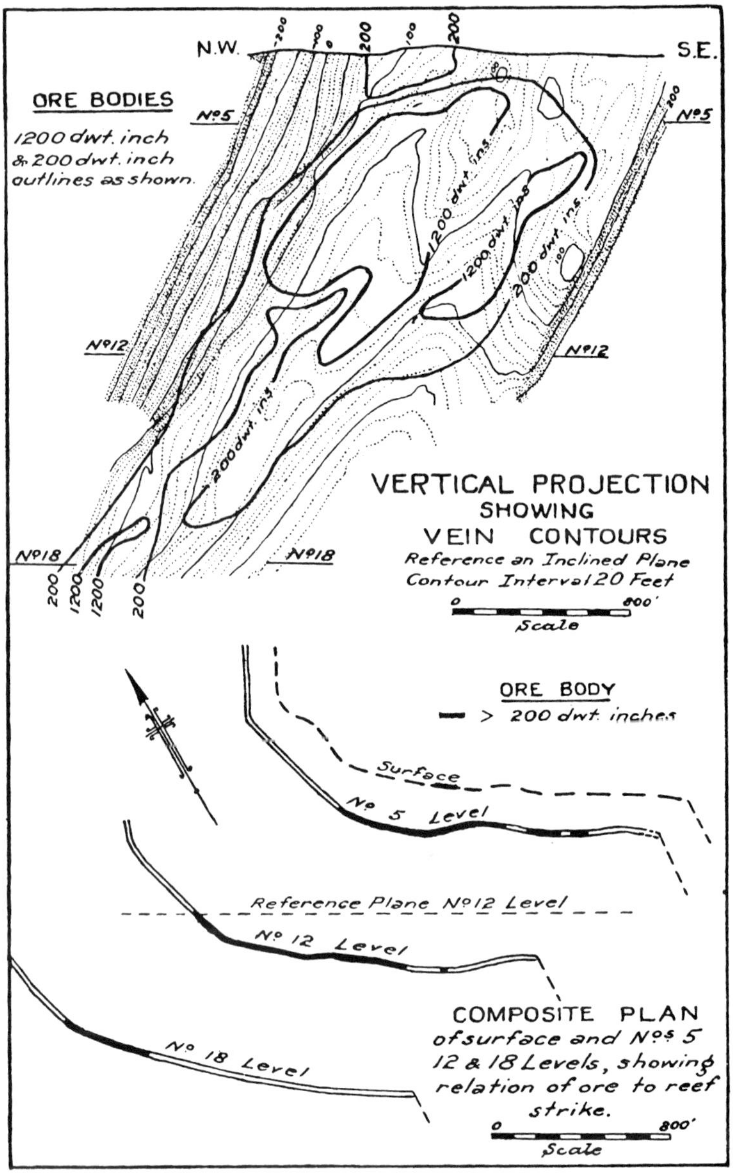
N.W.
S.E.
ORE BODIES
1200 dwt. inch
& 200 dwt. inch
outlines as shown.
Nº5
Nº12
Nº18
200 dwt. ins.
1200 dwt. ins.
VERTICAL PROJECTION
SHOWING
VEIN CONTOURS
Reference an Inclined Plane
Contour Interval 20 Feet
0
800'
Scale
ORE BODY
> 200 dwt. inches
Surface
Nº 5 Level
Reference Plane Nº 12 Level
Nº 12 Level
Nº 18 Level
COMPOSITE PLAN
of surface and Nºs 5
12 & 18 Levels, showing
relation of ore to reef
strike.
0
800'
Scale

FIG. 2.

shoot or of the richest portion of the vein. Figures 2 and 3 illustrate two preliminary tests. As in these instances the coincidence between gold (or quartz) and structure, if any, becomes immediately apparent. With some coincidence proved, the vein contouring is repeated, employing carefully chosen reference planes which will

(*a*) permit of the maximum extrapolation of the vein structure, employing to best advantage the positions given by the remoter reef outcrops and intersections in drill holes and crosscuts. For extrapolation, ridge and valley forms are more serviceable than terraces. In some instances it is necessary, by changing the reference plane, to subdue some of the prominent features which would otherwise mask or conceal the slighter but often equally important structures.

(*b*) resolve the vein structure into a pattern of intersecting loci seen on the contour plans as intersecting ridges and valleys. In most cases there are two or more such loci related to the distribution of gold (and quartz). The theme of intersecting loci in mineral deposits is a familiar one. Ore bodies may form only at their intersections; shoots may follow one locus under certain conditions and another under other conditions, etc.

This selection of reference planes to suit the problem opens a wide field of possibilities, for the plane employed largely determines the spacing of the contours and the actual position and

---

FIG. 2. Preliminary contour test of the Great Fingall Mine, Cue, Western Australia.

In this quartz vein the ore body, pitching southwest, is entirely confined to the middle limb of a prominent S structure. Contouring shows that this middle limb includes three inconspicuous but similar S bends. From the surface to the 5-level the whole ore limb rotates in-plan slightly and relatively anti-clockwise. From the 5- to the 12-level the limb remains in the same relative position of a pronounced S structure, while below the 12 it rotates back slightly and relatively clockwise. The one prominent S structure, with its three inconspicuous but similar structures, and the rotation of the ore limb, are closely related to the distribution of gold. The S bends provide the loci of the whole deposit and the three component ore shoots. The shoots are small and poor where the whole ore limb is rotating anti-clockwise or clockwise, and very large and rich where and only where the ore limb remains in the position of a pronounced S structure between the 5- and 12-levels. In a case like this two drill-hole intersections defining the strike of the reef in unexplored ground would provide information of vital importance in the search for further ore. Note: This mine today is full of water.

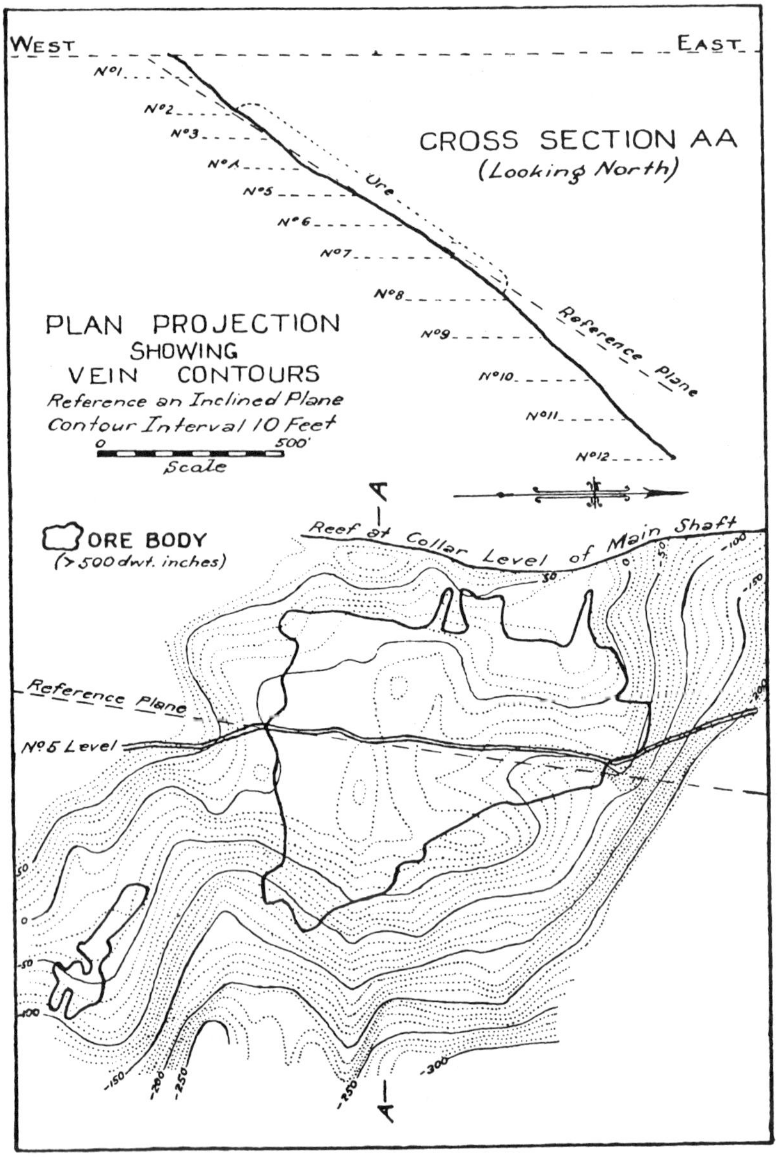
WEST
EAST
CROSS SECTION AA
(Looking North)
Ore
Reference Plane
PLAN PROJECTION
SHOWING
VEIN CONTOURS
Reference an Inclined Plane
Contour Interval 10 Feet
0 500'
Scale
ORE BODY
(> 500 dwt. inches)
Reef at Collar Level of Main Shaft
Reference Plane
No. 5 Level
A—A

FIG. 3.

course of the crest, valley and terrace lines. The moulding of the contours to serve a purpose, while difficult and laborious, is a most essential step, especially where the structure-gold relations will be employed to guide exploration. It necessitates generally a thorough knowledge of the vein structure in plan and section. To date, the selection of reference planes remains very much a process of trial and error. The best planes strike and dip as far as possible between the favorable ore-bearing and the unfavorable barren attitudes. Figures 4 and 5 illustrate the contouring of two quartz veins with reference to planes selected to permit of maximum extrapolation and to resolve the pattern into intersecting ridges and valleys.

With this special attention to vein structure, equal attention must be given to the portrayal of the gold distribution through the deposit. Old stope outlines may be very unsatisfactory and even misleading. The gold contouring procedure is illustrated in Fig. 6. Assay data is employed as far as possible, the gold being measured for contouring in dwt.-inches or dwts. Owing to the erratic distribution of the metal through most deposits, it is probably better and easier to contour the total amount of gold or average gold value in a unit block, measuring say 400 square feet by vein width. In the present work there has not been sufficient information to permit this employment of unit blocks. The amount of gold or average values across the vein at the point sampled has been used (as in Fig. 6) with, on the whole, satisfactory results. In the contouring, two gold contours are employed, the higher, a figure representing about 80 to 90 per cent of the mine average and the lower, a figure approximating the limit of profitable material. The interpolation of the whereabouts of higher grade ore is exceedingly important, for if the contour intervals are well chosen, the higher contour alone may

---

FIG. 3. Preliminary contour test of the Cosmopolitan Mine, Kookynie, Western Australia.

This illustrates a relationship different from Fig. 2 but equally delicate, between vein structure and gold. The quartz reefs here are controlled by an intricate pre-gold fault pattern with a large number of acute fault intersections. The ore shoots are found only on the faults striking N. and NE. and are especially developed where the whole vein system dips more flatly. Note: This mine today is full of water.

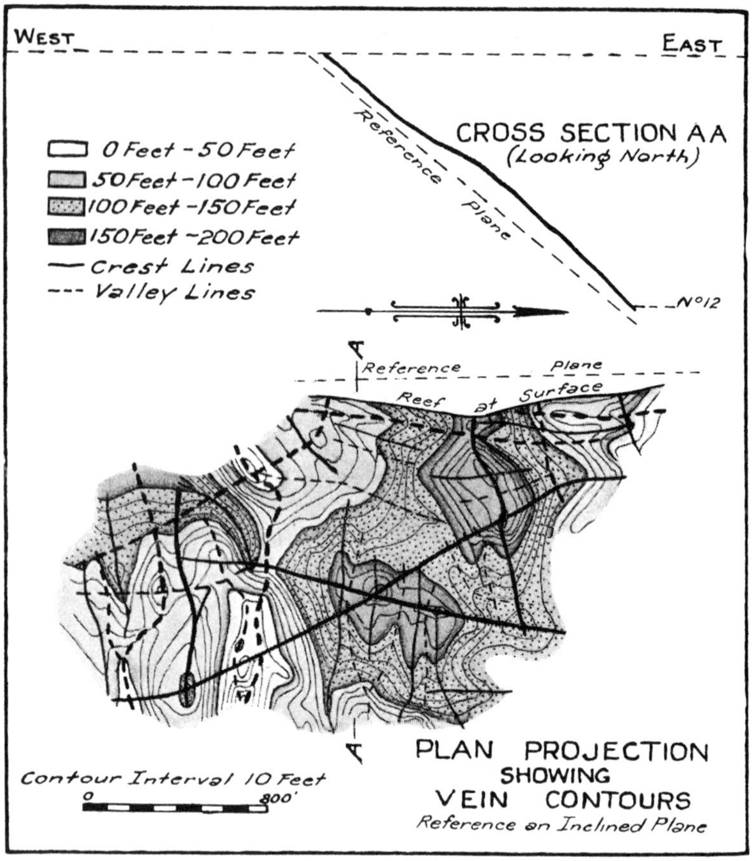

FIG. 4. Final contour test of the Cosmopolitan Mine, Kookynie, Western Australia. This figure shows the final contouring of the same vein illustrated in Fig. 3, to permit of extrapolation and to resolve the pattern into one of intersecting ridges and valleys. Comparison of the two figures reveals the important part played by the reference plane chosen and by careful contouring. Weeks of preliminary study were necessary to produce Fig. 4.

indicate the ore structure and the further extension (or repetition) of the whole deposit. In the gold plan illustrated in Fig. 8, the flat lensing of higher grade material provided the first evidence

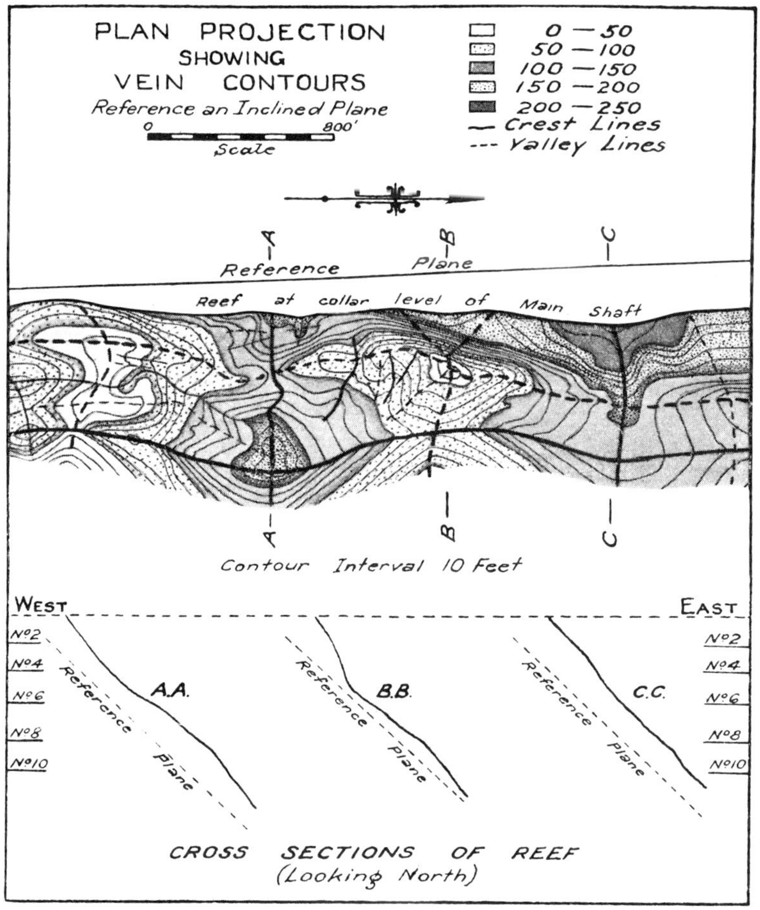

FIG. 5. Final contour test of Mararoa Mine, Norseman, Western Australia.

The plan projection shows broad, evenly spaced ridges and valleys running east and west (down the dip), intersected by a set of narrow spurs and re-entrants trending north and south (with the strike of the reef). The north-south features define, in section, an inconspicuous S-structure whose middle limb pitches flatly north. The same limb includes similar but much smaller structures seen in an alternate flattening, steepening and flattening of the reef. The S bends are relatively pronounced in the east-west valleys, while to a varying degree they are ironed out (but still traceable by contouring) across the east-west ridges.

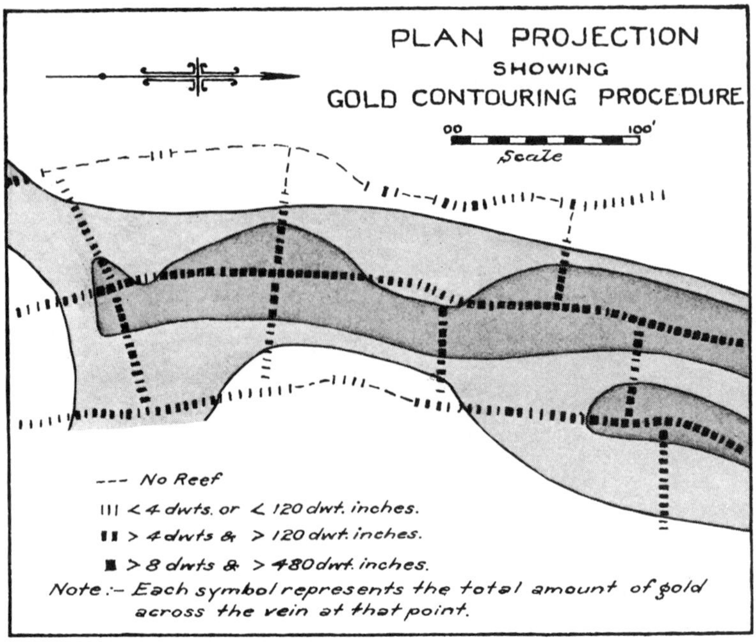

FIG. 6. The gold-contouring procedure.

of the flat pitch of the whole ore zone, an extension quite unexpected because of the steep south pitch of the old stopes.

In the study of auriferous quartz veins, plans are also prepared showing the distribution of quartz through the deposit. Figure 8 shows a quartz plan, the quartz being contoured in exactly the same way as gold.

Finally, the vein contours are superimposed on the gold and quartz plans, the combination showing the relationship between vein structure and gold and between vein structure and quartz. Figure 7 shows the crest and valley lines of vein contours superimposed on the gold plan of that mine. Figure 8 illustrates another mine study with structure lines added to both gold and quartz plans. In both cases the pattern of intersecting ridges and valleys explains to a remarkable and serviceable degree the distribution of gold (and quartz) in the deposit.

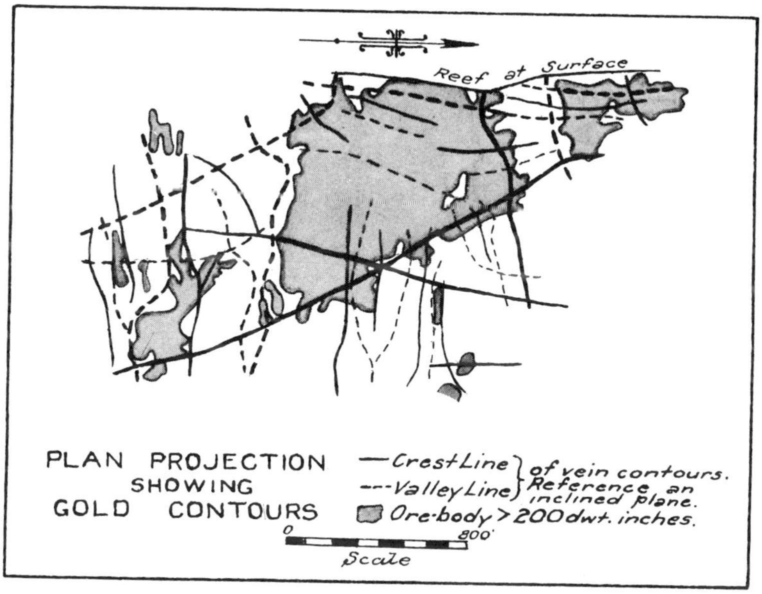

FIG. 7. Final structure-gold plan of the Cosmopolitan Mine, Kookynie, Western Australia.

Figure 4 showed the final contouring of a quartz vein. This figure shows the crest and valley lines of these contours superimposed on a plan showing the distribution of gold through the deposit. There are three ore-governing loci: (*a*) Western slope of a broad ridge trending north-west; (*b*) southern slope of a ridge-valley set running east-west; (*c*) crest and valley lines trending east of north.

Except for small unimportant concentrations of gold with some of the east-west ridges (*b*), the ore is assembled in a south-pitching zone following locus (*a*). Relatively small ore shoots are developed in this horizon where crossed only by the loci (*b*). The larger and richer main ore body fills the western slope of the north-west ridge (*a*) where crossed by a wide southern slope of an east-west ridge (*b*), and where loci (*a*) and (*c*) combine to flatten the dip of the reef. The important lines of loci (*c*) mark in places vein intersections affecting the concentration of gold, with the result that these crest and valley lines do not limit the distribution of ore as sharply as loci (*a*) and (*b*).

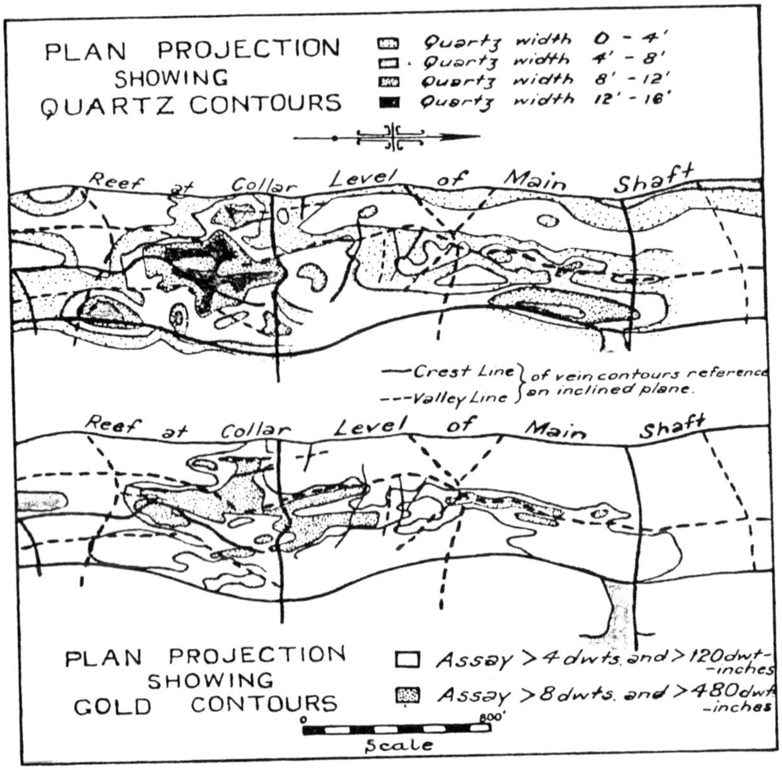

FIG. 8. Final structure-gold and structure-quartz plans of the Mararoa Mine, Norseman, Western Australia.

Figure 5 showed the final contouring of a quartz vein. This figure shows the crest and valley lines of these contours superimposed on the gold and quartz plans of the mine. One ore extension is confined by the north-south spurs and re-entrants which define the middle limb of the S structure (see legend, Fig. 5). Another ore extension follows the east-west ridges. Where the S structures are relatively pronounced, *i.e.* in the east-west valleys, the ore body breaks down into strands following the component smaller bends; where the north-south features are ironed out, *i.e.* on the east-west ridges, the strands coalesce, the ore expanding up and down the vein dip following the loci of the east-west ridges. The north-south features, where they remain at all prominent, dominate the control. The interplay of these north-south and east-west influences and the masking of one by the other are the essential features of an intricate relationship. These structure-gold relations guided the diamond drilling of the reef, the extrapolation of the vein contours being advanced as hole intersections were obtained.

In conclusion, the contour method described reveals in an impressive way the changing attitudes of a vein, lode, or formation, and has shown in the instances studied a surprisingly accurate and delicate control of the distribution of gold and quartz. Extrapolation of the structure-gold relations so observed also provides a practicable means for promoting and guiding exploration for the extension and repetition of known deposits. To date, the work of the Western Mining Corporation along this particular line has advanced sufficiently to launch exploration in the vicinity of two abandoned mines and to complete a diamond-drilling campaign in the vicinity of a third. In each case the warranty to explore and the value of the prospect depended on the gold-structure relations discovered; in each case new ore has been found. This contour technique should also be capable of wide employment in many of the geometrical problems of ore deposits —the interpolation of mine sections, walls of sulphide bodies, faults, rock contacts and rock folds.

25

Reprinted from pages 15–24, 26–27, 30–33, 36–41, and 44–49 of *Econ. Geology* **24:**15–49 (1929)

STRUCTURAL CONTROL OF ORE DEPOSITION.

CARLTON D. HULIN.

"Ore is where it is, not purely by chance, but by the result of definite processes which operated under certain conditions within the crust of the earth." [1]

INTRODUCTION.

A DEFINITE understanding of the factors which have controlled the formation of ore shoots in metalliferous deposits is one of the most important of the objectives of economic geology. Despite this fact, however, but little advance has been made within recent years in our knowledge concerning the origin of ore shoots. And it may be added that our knowledge in this regard is practically nil.

Despite our scant understanding of the causes of ore shoots, however, a voluminous literature descriptive of ore shoots has been developed. A perusal of even a portion of this discloses the fact that ore shoots of the same type are characterized by certain similarities the world over. So strikingly is this true that the writer has long since discarded the old adages that "ore is where it is" and that "each district is a law unto itself in the localization of its ore shoots" and has substituted that quoted at the beginning of this paper.

If it be recognized, as it must be, that ore shoots are not chance phenomena, but result through the action of definite natural causes, it must follow that these causes, subject possibly to some slight local modifications, will hold universally for a given type of ore shoot.

Lindgren [2] has subdivided ore shoots into magmatic segregation shoots; metamorphic shoots; replacement shoots; shoots of ascending circulation; and vadose shoots.

[1] A. M. Bateman, ECON. GEOL., vol. 14, p. 640, 1919.
[2] W. Lindgren, ECON. GEOL., vol. 4, pp. 56–61, 1909.

This paper deals primarily with the "shoots of ascending circulation"; that is, with primary ore shoots of epigenetic deposits which have formed through the action of solutions or emanations. Replacement shoots are not directly considered, although the writer feels that many shoots have been so designated which have in reality formed otherwise.

PRESENT THEORIES OF ORE SHOOT CONTROL.

Various geologists have from time to time advanced theories concerning possible controls of ore shoot localization. While many of the theories that have been put forward appear reasonable when referred to a particular deposit, very few possess the virtue of general applicability. It seems desirable to mention briefly at this time the more important of the current theories of ore shoot control.

The occurrence of ore shoots at vein intersections is so commonplace as to need no further introduction. Though all vein intersections are not loci for ore shoots, many important ore shoots are so located. Such shoots have been explained as resulting from the increased brecciation at the intersection; as due to mixing of solutions at the intersection; or as the result of stagnation of the solutions. No convincing argument has ever been put forward in support of any of these possible causes. More recently Howe has pointed out that the common occurrence of ore shoots at vein intersections suggests a mechanical cause.[3]

The occurrence of ore shoots at the intersection of veins and cross-faults or fissures is closely akin to the occurrence of shoots at vein intersections. Ore shoots may be localized in the vein in the vicinity of a cross-fault, or the shoots may be terminated by the fault. The cross-fault or fissure may be a plane of considerable movement, or may be merely a minor joint plane. Explanations offered to account for the occurrence of ore in the vicinity of cross-faults have been essentially the same as those offered in explanation of the occurrence of ore shoots at vein intersections.

[3] E. Howe, "The Gold Ores of Grass Valley, California," Econ. Geol., vol 19, p. 601, 1924.

Wall rock control has been a popular cause advanced in explanation of ore shoots. Such control may be subdivided into control resulting from the chemical nature of the wall rocks, and control by the physical nature of the wall rocks. Chemical control implies a precipitating or catalytic action by the constituents of the wall rocks and has been suggested as a possible cause of ore deposition by many workers. In various cases basic igneous rocks, ferromagnesian minerals, organic matter and metallic sulphides have been held to be favorable precipitants or catalyzers. Though chemical wall rock control may well be a modifying factor in ore deposition, other and more fundamental causes are held by the writer to be operative, excepting, possibly, the case of replacement ore shoots occurring in relatively soluble rocks. Five different cases have come to the attention of the writer where wall rock control of the ore deposition had been postulated. In each case detailed geologic studies disclosed that other causes were of paramount importance, the possible precipitating effects of the wall rocks being negligible.

The tendency of different wall rocks to behave differently on fracturing has led to the theory of a control dependent on the physical nature of the wall rocks. Practically all known rock types however are somewhere known to be hosts to ore shoots. Insofar as different rocks may fracture in a more favorable or less favorable manner the physical nature of wall rocks may be an important modifying factor, but beyond this it can hardly be considered to be of fundamental importance in controlling the location of ore shoots.

Other factors which have been put forward as possible ore shoot controls include the presence of channels of circulation; the damming action of fault gouge; the precipitating action of jets of gas encountered by the mineralizing solutions; stagnation of the mineralizing solutions; and recurrent mineralization.

The growth of the zonal theory in recent years has offered a new explanation for the deposition of ores; it does not however account for the isolation of ore into definite shoots. According to the zonal theory the deposition of the ore minerals, and indeed of all the materials deposited by the mineralizing solutions, is a direct function of decreasing temperature and pressure resulting from the ascent of the solutions towards the earth's surface.

The inability of current theories to offer a sound explanation for the occurrence of ore shoots must be manifest. Let us picture the common case of a vein containing five or six ore shoots along its length, of which one may occur at a vein intersection, another at an intersection of the vein with a cross-fissure, while the rest are scattered apparently at random along the vein. The shoots have formed at the same time, have received the same mineralization, and shoots and barren portions of the vein are crossed indiscriminately by the same wall rock horizons. Current theories either fail dismally to explain such an assemblage of ore shoots, or force us to assume a different origin for each, which last is manifestly absurd. Ore shoots showing the same mineralization, and occurring in the same vein or region, must manifestly have resulted from the action of the same set of controlling factors.

THE FORMATION OF ORE SHOOTS.

The mode of formation of ore shoots is intimately bound up with the mode of formation of veins. A detailed study of the ore shoots with which the writer is familiar, coupled with a study of those shoots which have been adequately described in the literature, indicates that all shoots of vein deposits, excepting only those formed directly by replacement in easily soluble rocks, have the same fundamental characteristics. An analysis of these characteristics common to ore shoots, which will be considered in detail later, leads to the following conclusions concerning the control of ore deposition and the formation of ore shoots:

The deposition of the minerals of vein deposits, following the zonal theory, is fundamentally a function of concentration, temperature and pressure. As the mineralizing solutions ascend, decreasing temperature and pressure result in supersaturation, with consequent deposition of the various mineral components from time to time.

Ore shoots are developed in these parts of the vein which are open to the passage of solutions at the time of the introduction of the valuable metallic minerals, conditions for deposition being favorable. In general the minerals of the common heavy metals, excepting for the moment the minerals of tin and tungsten, are introduced late in the sequence of mineralization. The early introduced material of the vein usually consists solely of barren quartz. This seals the existing vein fissures, forming barren quartz veins which remain barren unless subsequently reopened. Ore shoots are formed where the early barren quartz veins are brecciated and reopened by intermineralization fault movements at a time immediately preceding the advent of the metallizing solutions. The location of ore shoots is thus directly dependent on favorable local structure of such nature that when affected by intermineralization fault movements brecciation and reopenings of the vein are produced.

In short, *the deposition* of ore is dependent on the physical-chemical conditions of temperature, pressure and concentration of the mineralizing solutions, while the control of *the locus of deposition* is mechanical, being dependent on favorable local structure coupled with fault movements occurring at the proper time during the sequence of mineralization.

FAULTING IN MINERALIZED REGIONS.

The importance of pre-mineral faulting in mineralized regions has been largely underestimated by geologists, and the prevalence of faulting contemporaneous with the mineralization has almost escaped detection. Workers have been prone to class all faults along which vein offsets occur as being post-mineral without further thought.

An offset of a vein by a cross-fault, or the presence in mineralized regions of gouge-filled fissures, themselves unmineralized, cannot be accepted as proof of post-mineral origin for such faults and fissures.

Faults which have successfully developed a filling of gouge are practically immune from the effects of later mineralization.

Fault gouge behaves as a chemically inert substance, and is practically impervious to the passage of solutions. As a result, gouge-filled fissures may retain their pre-mineral characteristics while neighboring fissures free from gouge are being transformed into solid veins. Ransome has recorded this possibility in the literature.[4]

The writer has made detailed studies of some hundreds of vein displacements along cross-faults in which the displacements resulted from pre-mineral dislocation of the vein-fissure, the vein-filling being introduced into the fissure subsequent to the displacement.

The type of evidence which the writer has used in identifying pre-mineral faults includes the following:

(1) Different behavior of the vein on the two sides of the fault. The differences may be in thickness, in value, in the nature (proportion) of the introduced minerals, or any combination of these.

(2) Banding in the vein turning and paralleling the cross-fault at the end of the vein segment. Although this feature is common, it is not prevalent, being frequently destroyed by movements on the cross-fault occurring contemporaneously with the mineralization.

(3) The presence of stringers and veinlets cutting the fault gouge, or partial silicification of the gouge. These features are seldom to be observed more than a few feet from the vein. The veinlets may cut through the gouge, but are more common along the boundary between the gouge and the wall rocks. Such veinlets and stringers are frequently brecciated by slight intermineral or post-mineral fault movements and have unquestionably been commonly classed as "drag ore."

(4) Sealing of the fault line immediately adjacent to the vein by vein-matter. Due to such sealing many faults are practically not discernible where they cut the vein.

(5) Presence of actual veins along the cross-faults. This of

[4] F. L. Ransome, "Relation Between Certain Ore-Bearing Veins and Gouge-Filled Fissures," Econ. Geol., vol. 3, pp. 331–337, 1908.

course is possible only where the fault was more or less open during the mineralization.

Spurr has pointed out that igneous activity, faulting and ore deposition are related phenomena, and that the faulting usually starts before the mineralization, but greatly outlasts the mineralization, reaching a peak long after the mineralization has ceased.[5] A part of the evidence put forward in support of the importance of post-mineral faulting is to the effect that major fault lines are infrequently the hosts to productive veins, in consequence of which they are believed to have developed in post-mineral time.

The preceding argument has brought out, however, that major fault lines may in general be expected to be less favorable sites for vein formation than other minor fissures, due to the amount of movement the major lines have suffered, and the consequent amount of inert and impermeable gouge developed therein.

Some inter-relation between igneous activity, mineralization and faulting appears to be fairly well established. In some mineralized regions there has undoubtedly been a later independent period of faulting superimposed on the faulting related to the mineralization. Considering only the period of faulting which is related to the mineralization, however, it has been the writer's experience to observe that this faulting commences earlier than the mineralization, reaches a maximum by the middle of the period of mineralization, and then gradually declines, commonly however, outlasting the mineralization. This sequence holds for many important mining regions, including the Mother Lode and Randsburg regions of California, the Telluride and Silverton regions of Colorado, and the Pachuca region of Mexico, and is believed to be a general condition elsewhere.

Our knowledge of the behavior of faults gained from the Elastic Rebound Theory[6] is that the total displacements now shown by faults have resulted through numerous small movements, these small movements being spread over a long interval and being separated by periods of fault quiescence.

[5] J. E. Spurr, "The Relation of Ore Deposition to Faulting," ECON. GEOL., vol. 11, pp. 601–622, 1916.

[6] H. F. Reid, "The Elastic Rebound Theory of Earthquakes," Calif. Univ. Pub., Dept. Geol. Bull., vol. 6, pp. 413–444, 1911.

The movements of pre-mineral age result in the development of both faults and vein fissures. The former are produced where excessive movement, or compression normal to the fault, or both, result in a tight fissure filled with inert and impermeable gouge. The latter are formed where small movements, or lack of compression normal to the fissure results in a minimum of gouge with a consequent open and permeable fissure. The distinction between faults and vein fissures is one of degree only, all gradations between the two being found. Veins commonly pass along their strike into unmineralized faults.

The intermineralization movements, those fault movements which continuously recur throughout the period of mineralization, affect both the faults and the vein fissures. Where the gouge-lined faults are affected, the plastic gouge is deformed and squeezed, thereby perpetuating itself. The growing veins, once they have started to form, are fairly immune from the later development of gouge. The early vein matter, on being affected by the recurrent intermineralization fault movements, is brecciated, forming a porous, permeable mass of crushed quartz which is later recemented by the mineralizing solutions. This action may recur innumerable times during the mineralization, and may result either from movements along the vein fissure, or locally from movements along the cross-faults.

Post-mineral continuance of the fault movements is of importance only in that they may cause further displacements of the veins along the cross-faults, shear along the veins, or brecciation of the previously deposited vein matter. Post-mineral movements may to some extent mask the evidence of the intermineralization fault activity.

THE MECHANICS OF VEIN FORMATION.

The extreme importance of pre-mineral and intermineral faulting in connection with mineralization must be manifest. Pre-mineral faulting is important in that it is these movements which develop the vein fissures. Intermineral faulting is of equal importance in that the features of the vein fissures are continuously undergoing modification thereby during the time of the mineralization.

Taber has considered in detail the mechanics of the various processes whereby veins may be formed.[7] Although he postulates six possible modes of origin, a study of his paper reveals that only three mechanical processes of vein formation are involved, each of these three being subdivided into two cases dependent on whether the mineralizing solutions were ascending in the fissure or entered the fissure by lateral secretion. The three possibilities recognized by Taber for vein formation are:

(1) Vein growth by replacement of the adjacent wall rocks.
(2) Deposition in an open fissure.
(3) Deposition in openings produced by the force of crystallization of the growing vein minerals.

For the sake of completeness, two other theories of vein formation which were not considered by Taber must be mentioned. These are:

(4) Deposition in openings produced by the hydrostatic pressure of the mineralizing solutions.
(5) Veindike injection, following Spurr.

Although any or all of these processes of vein formation may be feasible and locally applicable, another mode of vein formation exists which available evidence indicates is universally important, namely:

(6) Vein formation by the repeated filling of small openings produced recurrently by intermineralization fault movements.

For this mode of vein formation the writer has proposed the term *vein formation by accretion.* This process is pictured as the alternate brecciation of previously deposited vein matter followed by recementation and filling of the open spaces. Such brecciation and recementation may be repeated innumerable times during the formation of a vein. By this process veins of any width could

[7] S. Taber, "Mechanics of Vein Formation," *A. I. M. E. Trans.*, vol. 61, pp. 3–41, 1920.

be built up without essential replacement of the adjacent wall rocks having occurred, or without vein fissures open to a width comparable to the width of known veins ever having existed.

The process is as applicable under deep-seated conditions as under shallow conditions, since the openings existing at any one time would be trivial. The only essential condition is numerous fault movements of small magnitude occurring contemporaneously with the mineralization—exactly those movements postulated by the Elastic Rebound Theory.

[*Editor's Note:* Material has been omitted at this point.]

Evidence of the intermineralization brecciation occurring just previous to the introduction of the late metallic minerals is usually the most apparent. Here a change in the nature of the materials being deposited before and after the brecciation lends contrast. Even here, however, replacement of the borders of the early quartz fragments by the later metallic minerals results in rounded and irregular outlines and borders, with a consequent masking of the evidence.

Evidence of the brecciation where only quartz is present can only be observed microscopically in thin section. In some cases the quartz deposited before and after a period of brecciation shows different grain sizes and orientation, so that the effect may be observed under crossed nicols. This effect, however, is only produced where a sufficient time elapses between the deposition of the earlier and the later quartz so that a distinct change in conditions of deposition occurs (*e.g.*, change in temperature and concentration of the solutions).

More commonly the later quartz continues to grow with the same orientation as the earlier quartz of the fragments, with the result that evidence of brecciation cannot be seen under crossed nicols. Frequently in such cases, however, the earlier quartz is more cloudy than the later quartz so that by plane light the cloudy angular fragments of the early quartz can be seen surrounded by clear younger quartz. These patches or veinlets of clear younger quartz are termed " phantom veinlets " and have been previously described by Adams as being common in the gold ores of the Mother Lode of California.[8] The presence of similar veinlets in quartz vein-matter was also noted by Taber.[9]

All degrees of contrast in cloudiness between the older fragments and the younger phantom veinlets are to be observed in ores. Phantom veinlets may frequently be found which can only just be differentiated from the adjacent fragments. By inference, if the deposition of material before and after a period of brecciation is separated by such a short interval that the solutions and physical conditions have not had time to change appreciably, the younger quartz would be identical with the older quartz, and growing onto the older fragments with optical continuity, no evidence whatever of the brecciation would exist. It must be recognized therefore that intermineralization brecciation may be even more abundant than is called for by the known evidence.

[*Editor's Note:* Material has been omitted at this point. Figures 3 and 4 also have been omitted.]

The presence of early intermineralization brecciation of the quartz gangue is usually not apparent in the polished surface by using the ordinary methods with vertically incident illumination, although the effects of the brecciation which just precedes the introduction of the metallic minerals can commonly be seen by using these methods. By using inclined illumination, however, or preferably, using vertical illumination with crossed nicols, the internal structure of the quartz becomes manifest and quite frequently fragments of cloudy quartz can be seen set in a matrix of younger clearer quartz.

The presence of fragments of wall rocks imbedded in veins has attracted considerable attention in recent years, and has been accounted for by replacement; as the result of loose rubble filling the vein fissure previous to the mineralization; as the result of fragments pried from the vein walls by the force of crystallization of the growing vein minerals; and by suspension in a viscous ore magma. The process of vein formation by accretion not only

[8] S. F. Adams, "A Microscopic Study of Vein Quartz," Econ. Geol., vol. 15, pp. 623–664, 1920.

[9] S. Taber, *op. cit.*, p. 17.

offers a more satisfactory explanation for the fragments of wall rocks included in veins, but the presence of such fragments would be expected to be almost essential to this mode of vein formation. During the recurrent reopenings of the vein it would invariably happen that portions of the wall rock which were frozen to the vein would be torn loose from their original positions in the wall and would be surrounded by and incorporated in the vein during the later stages of the mineralization.

The Mother Lode of California, occurring as it does in foliated wall rocks, offers the best example of this process known to the writer. Portions of the Mother Lode contain innumerable thin oriented parallel fragments of the foliated wall rocks. At times these are spaced only a small fraction of an inch apart through vein thicknesses of several feet. This "book structure" could conceivably have resulted only through countless reopenings of the vein fissure contemporaneous with the mineralization, each reopening tearing loose a thin film of the foliated wall rock which had frozen to the vein. The reopenings here commonly tore through the wall rocks rather than through the vein due to the greater strength of the vein matter as compared with the bond holding the lamella of the foliated wall rocks together. Such an effect is commonly manifest in veins occurring in slaty or schistose rocks throughout the world. The same explanation put forward by the writer has been previously advocated by Ransome to account for the origin of the "banded ore" of the Mother Lode.[10]

The writer in field work, underground, has repeatedly been able to trace out the process of fragments being torn loose from the walls and being incorporated in the vein by the later stages of the mineralization.

THE SEQUENCE OF MINERALIZATION.

The fact that the metallic minerals are usually introduced late in the sequence of mineralization, the early-deposited material in general consisting of barren quartz, is becoming more and more clearly recognized by students of ore deposits. In spite of this,

[10] F. L. Ransome, U. S. G. S. Geologic Folio No. 63. Mother Lode District, California, p. 7, 1900.

however, the possible bearing that the late introduction of the ore minerals may have on the formation of ore shoots has so far received no consideration.

In order to give a more detailed picture of the succession of events the following diagrams typical of the sequence of mineralization in three distinct types of ore deposits are presented. These diagrams are designed for qualitative use only. They represent the composite results of studies of numerous thin sections and polished surfaces, combined with field work.

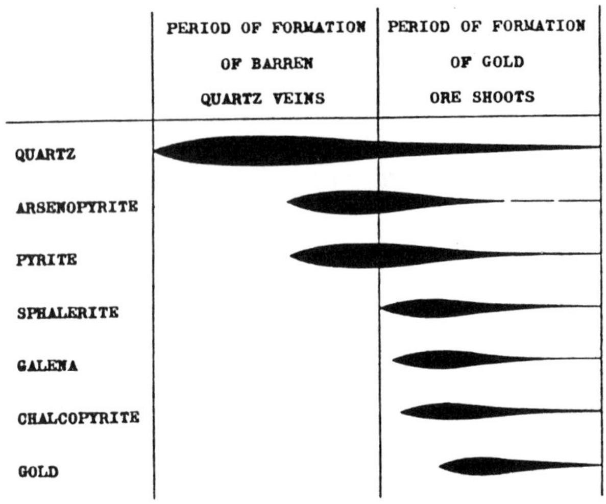

FIG. 4. Sequence of Mineralization—Mother Lode Gold Veins.

The first diagram (Fig. 4) pictures the relative time of introduction of the more common minerals of the mesothermal gold veins as exemplified by the Mother Lode of California. The width of the lines represents in a roughly qualitative manner the relative amounts of each mineral deposited from time to time during the mineralization. Comparison between lines is limited to the time relation only, the widths of the several lines in no sense indicating the relative abundance of the different minerals.

A consideration of the time relations involved in Fig. 4 for the mesothermal gold ores indicates clearly that those veins or portions of veins receiving only the early quartzose mineralization are of necessity barren, while ore shoots can result only where the vein fissure receives deposition of minerals during the later stages of the mineralization.

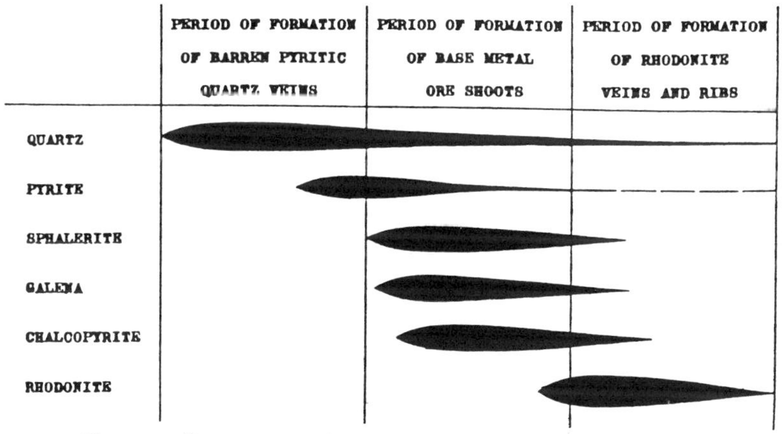

FIG. 5. Sequence of Mineralization—Sunnyside Mine.

The second diagram (Fig 5) is for the epithermal to mesothermal base metal veins, particularly those of lead and zinc, as exemplified by the Sunnyside deposit of the Silverton region of Colorado.

Here again it is manifest that ore shoots must be restricted to those portions of the veins receiving the mineralization of the second stage. Portions of veins receiving mineralization only during the first stage will consist of barren quartz, whereas those portions of the veins formed during the third stage consist essentially of barren rhodonite.

The third diagram (Fig. 6) is typical of the epithermal precious metal deposits, the example here represented being the sequence of mineralization of the California Rand Silver deposit of Randsburg, California. Although the mineralization of the epithermal deposits is usually more complex than is that of deeper-seated types, the same commonplace stages of early barren siliceous gangue and late metallization are characteristic.

Here once more the stage of formation of barren veins is contrasted with the stage of formation of the ore shoots in a clear-cut manner.

Although an early stage of barren quartz coupled with a late stage of metallization is characteristic of the majority of vein deposits, exceptions to this commonplace sequence are not unknown, though they appear to be rare.

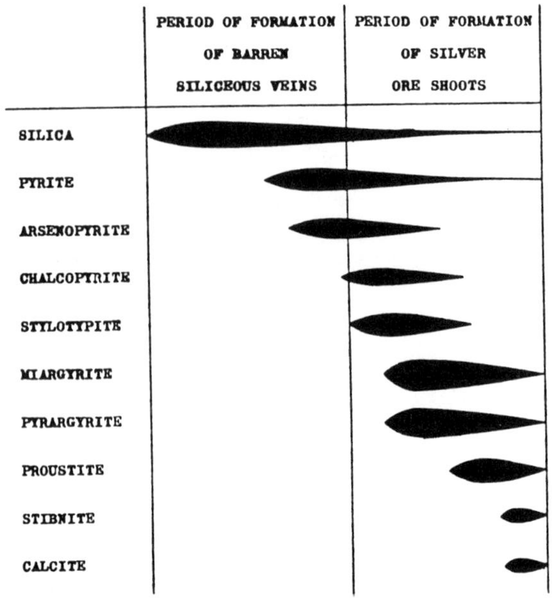

FIG. 6. Sequence of Mineralization—California Rand Silver Mine.

The more common of the known exceptions are represented by one of two conditions, as follows: The ore minerals are introduced contemporaneously with the bulk of the quartz of the deposit; or quartz is largely absent, the bulk of the introduced material consisting of metallic minerals and associated carbonates. In either of these cases, the thesis of this paper must be modified, in general to the effect that the ore, if such be developed, is more or less coextensive with the material introduced. In deposits possessing these latter characteristics, the ore shoots essentially correspond to the "shoots of occurrence" as recognized by Irving.[11]

[11] J. D. Irving, "The Localization of Values or Occurrence of Shoots in Metalliferous Deposits," ECON. GEOL., vol. 3, pp. 143–154, 1908.

[*Editor's Note:* Material has been omitted at this point.]

For most of the metals, however, including gold, silver, copper, lead, zinc and quicksilver, a late time of introduction prevails. For most deposits containing these metals, therefore, definite ore shoots occur which are not co-extensive with, but form only a small part of the total vein filling. The loci of deposition of these ore shoots are due strictly to mechanical control. The shoots form at those structural positions where intermineralization fault movements produce reopenings of the vein fissure at a time immediately preceding and during the metallization.

Arguments in favor of the process of vein formation by accretion have been put forward. The formation of ore shoots follows this same mechanical process of accretion, the locus of ore deposition during the time of metallization being in general repeatedly reopened and brecciated, ore minerals being continuously deposited in the open spaces so produced.

It must be clearly recognized that all favorable structural positions will not be affected by fault movements during the period of metallization and hence will have no contained or related ore shoots. Certain favorable structural positions may suffer reopenings too early or too late to catch the valuable metallization.

The writer has studied many cases of favorable structural positions in veins where repeated reopenings had occurred during the early stage of the mineralization, resulting in large "blow-outs" of barren quartz, but where cessation of the fault movements previous to the advent of the metallizing solutions resulted adversely to the development of ore. In a few such cases slight fissuring during the time of the metallization resulted in the barren quartz of the "blow-outs" being intersected by a few random veinlets rich in sulphides, showing clearly that had the principal reopenings occurred later with respect to the sequence of mineralization a valuable orebody would have resulted.[14]

Occasionally a favorable structural position, even though suffering brecciation at the proper time, will not be so situated that it will be subjected to the action of the metallizing solutions. A few cases have been observed where favorable positions which had been brecciated at the proper time were apparently inaccessible to the metallizing solutions due to isolated positions in the vein and to the presence of impermeable fault gouge in adjacent portions of the vein fissure.

It does not appear feasible to predict in advance of actual exploration whether a given favorable position on a vein has or has not been affected by the requisite fault movements at a time contemporaneous with the metallization.

It is feasible in many cases however to predict positions along a vein where favorable structure may be expected to occur. Manifestly if the position occupied by favorable structure can be predicted in advance of actual exploration, the exploration of that portion of the vein is fully justified. It is of extreme importance therefore to recognize and classify those vein structures which are favorable for the development of ore shoots if acted upon by fault movements at the proper time during the period of metallization.

STRUCTURES FAVORABLE FOR THE DEVELOPMENT OF ORE SHOOTS.

Since the amount of brecciation and reopening required for the development of a base metal ore shoot is considerable, the structures on which base metal ore shoots are dependent are usually quite apparent.

For the precious metal ore shoots, however, the required amount of brecciation and reopening of the vein may be trivial and result from very slight variations in the vein structure. In consequence the structural control of many precious metal ore shoots can be deciphered only with the greatest difficulty, and then many

[14] The writer hopes at another time to describe in detail certain of the illustrative cases studied, which lack of space prohibits here.

times only because of familiarity with ore shoots of base metal deposits controlled by similar but more pronounced structures.

It must be clearly borne in mind that the following structures are effective only when acted upon repeatedly by slight fault movements during the time of metallization.

1. *Variation in Strike of Vein.*—Local variation in strike of a vein fissure, coupled with shear along the fissure, provides an effective locus of deposition for an ore shoot. The mechanics of the control are shown in plan in Fig. 7.

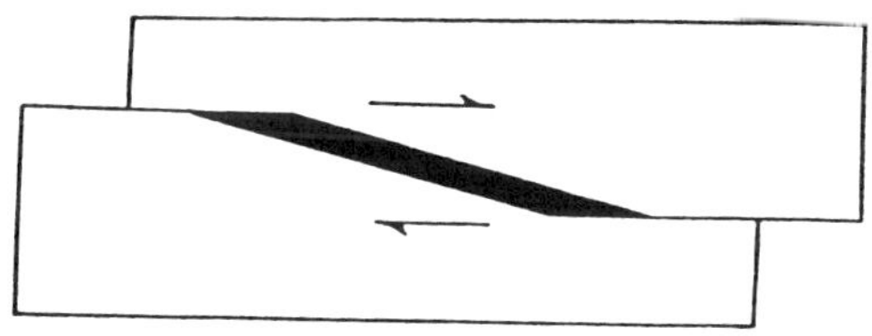

FIG. 7. Mechanics of control of location of ore shoot by variation in strike of the vein fissure.

The magnitude of the reopening produced by this type of structure will be a function of both the variation in strike and the amount of shear.

Where base metal ore shoots are controlled by this type of structure the variation in strike is usually sufficiently pronounced as to be quite noticeable. For the development of precious metal ore shoots however a local variation in strike of the vein fissure of only a degree or so may be a sufficient control. In such cases the variation may be so small and inconspicuous as to practically defy detection.

Several examples of ore shoots controlled by local variation in strike of the vein fissure are known to the writer. One of the clearest cases is the ore shoot of the Arevalo Mine near Pachuca, Mexico, shown in Fig. 8. This ore shoot is the only shoot of importance occurring along the Arevalo fault and it occupies the only position of local change in strike of the fault which is known within a distance of several kilometers in either direction.

At the Sunnyside Mine in the Silverton district of Colorado, two of the most important ore shoots which have been developed are controlled by variation in strike of the vein fissure (Fig. 8, *B*). Minor pre-mineral cross-faults modify these ore shoots to some extent.

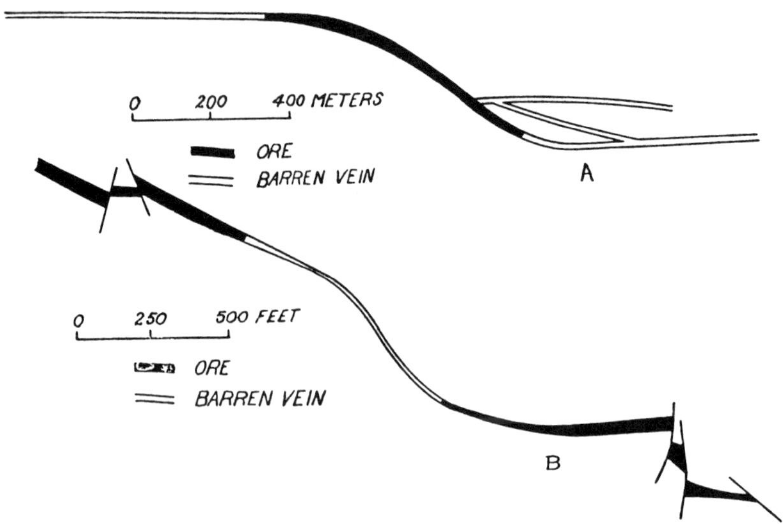

FIG. 8. *A,* Control of location of ore shoot by variation in strike of vein fissure. Arevalo Mine, Pachuca, Mexico. *B,* Ore shoots controlled by variation in strike of the vein fissure coupled with intermineralization cross-faults. Sunnyside Mine, Colo.

The Camp Bird Vein in the same region has been described as containing a series of ore shoots which all occupy positions of local variation in strike of the vein fissure.[15]

2. *Variation in Dip of the Vein.*—Variation in dip of the vein fissure, coupled with a vertical component of shear, may act independently in providing a favorable site for ore deposition; or it may act in conjunction with and modify the effects of variation in strike.

Two cases must be recognized, dependent on whether the vertical component of faulting is normal or reverse. If normal the steeper portions of the fissure are the more favorable, whereas if the movements are of a reverse nature the reopenings are more

[15] J. E. Spurr, *op. cit.*

pronounced in the flatter portions of the vein fissure. These conditions are illustrated in Fig. 9.

The effects of variation in dip on ore deposition are beautifully illustrated in the Washington ore shoot of the Sunnyside Mine (Fig. 10). This shoot, occurring along a normal fault, is con-

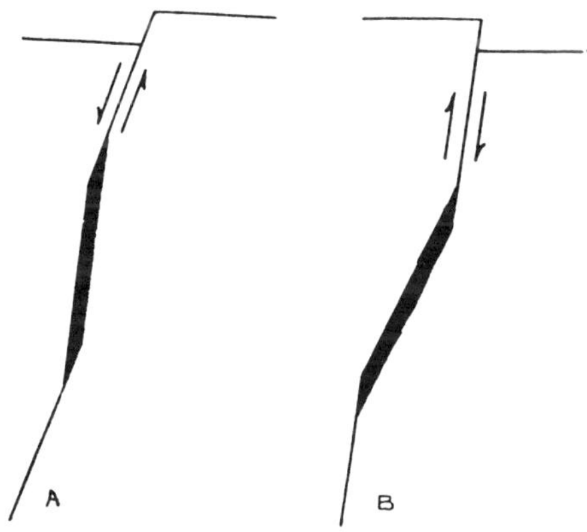

FIG. 9 (left). Mechanics of control of location of ore shoot by variation in dip of the vein fissure. *A*, In a normal fault the steeper portions of the vein are the more favorable. *B*, In a reverse fault the flatter portions of the vein are the more favorable.

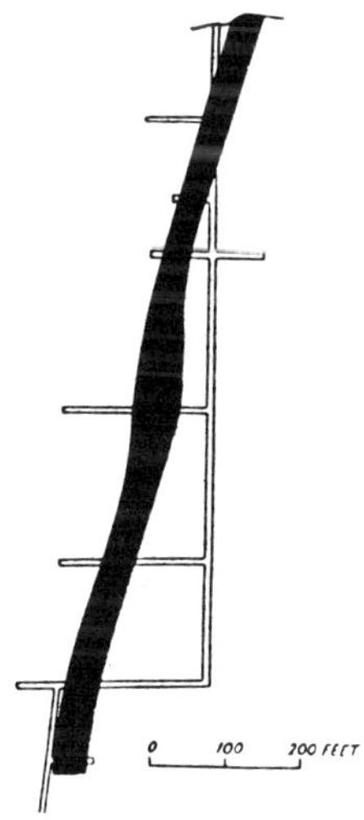

FIG. 10 (right). Modification of ore shoot by variation in dip. The steeper portions of the vein are wider and richer. The vein fissure is known to be a normal fault. Washington Vein, Sunnyside Mine, Colorado.

trolled primarily by variation in strike of the vein fissure. Within the shoot, however, variations in dip occur, the steeper portions of the shoot being both thicker and richer than the flatter portions.

Purington has recorded that in a number of instances at the Virginius and Smuggler Mines in the same region, the veins widen or swell on the steep dips and pinch where the dip is flatter.[16]

[16] C. W. Purington, U. S. G. S. Geologic Folio No. 57, Telluride Quadrangle, Colorado, p. 16, 1899.

3. *Intersection of Veins and Cross-Faults.*—The portions of veins in the vicinity of intersections with cross-faults are commonly favorable positions for ore-shoot formation, but only where the cross-faults have suffered intermineralization movements. Such movements on the cross-faults brecciate the previously deposited material of the vein, providing the necessary reopenings for the passage of the metallizing solutions and for the deposition of the ore minerals.

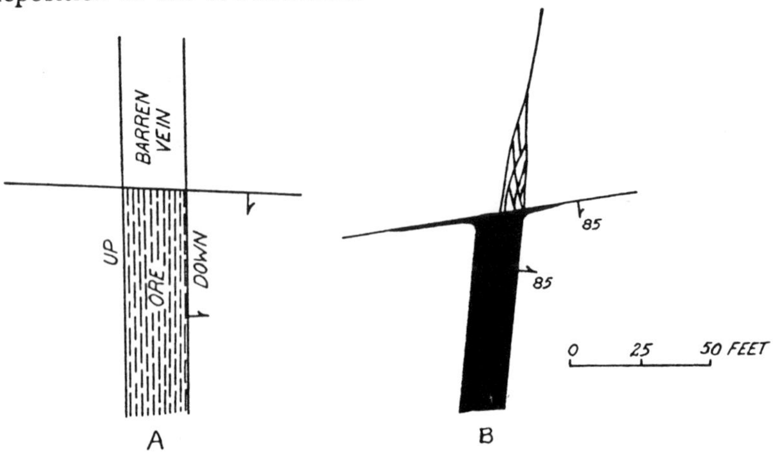

FIG. 11. *A,* Termination of ore shoot by an intermineralization cross-fault. North of the fault the early barren vein remained tight; south of the fault intermineralization movements caused brecciation with the resultant development of an ore shoot. Purisima Vein, Pachuca, Mexico. *B,* Termination of ore shoot by a pre-mineral cross-fault. North of the fault only a small stringer system exists. Santa Inez Vein, Pachuca, Mexico.

The cross-faults responsible for ore-shoot formation may be pronounced lines of movement, or in some cases they may be merely minor joint planes. The amount of movement which occurs thereon may vary within considerable limits. In general base metal ore shoots require considerably more brecciation and reopenings for their formation than do precious metal ore shoots, the latter at times developing where only a very minor amount of brecciation has occurred.

[*Editor's Note:* Material has been omitted at this point.]

5. *Simple Shear along Vein.*—Simple intermineralization shear along a vein may result in reopenings unassisted by any definite structural control. Such shear may produce a thin zone of brecciation following one wall; following the center of the vein; or following any other intermediate position. Occasionally these streaks switch from side to side, occupying first one plane in the vein, then another. Metallization of these brecciated layers gives rise to important ore shoots in precious metal deposits. In such deposits the common tendency for the bulk of the values to occur in one or two thin layers or streaks in the vein, the rest of the vein being low grade or barren, has long been recognized.

In base metal deposits, however, the reopenings produced by simple shear are seldom of sufficient size to permit of the formation of workable ore shoots. The writer has examined numerous base metal veins in the San Juan region of Colorado where the effects of simple shear followed by metallization were quite manifest. The result is a streak of rich sulphides an inch or a few inches wide following the center or wall of an otherwise barren quartz vein. The effect produced, due to the quartz accompanying the sulphides and to replacement by the sulphides of the borders of the adjoining early barren quartz, is not unlike depositional banding, with which it should not be confused.

6. *Torsion Structures.*—Where a fissure is subjected to a torsional stress diagonal to its strike, it tends to gap open, producing a place favorable for ore deposition. It sometimes happens that two or more parallel fissures which are being subjected to horizontal shearing stress are sealed by the early barren quartz stage of the mineralization. Continuance of the shearing stress then

may find the initial fissure lines tightly sealed and stronger than the adjacent wall rocks, in which case the stress may be relieved by the development of new torsion fractures diagonal to the direction of the stress and to the original fissures. These torsion fractures, escaping the early barren stage of the mineralization, and coming into existence about the time of the metallization, are ideal for ore deposition.

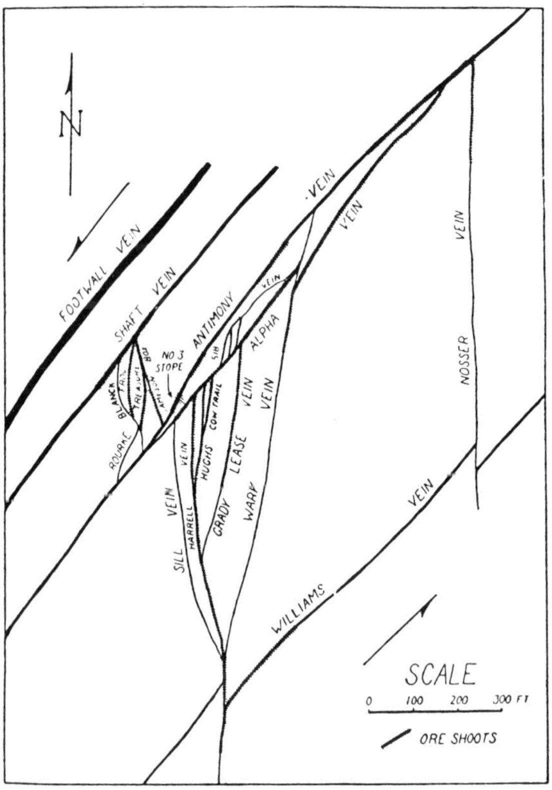

FIG. 14. Ore-shoot control by torsion type of structure. Plan of vein system of the California Rand Silver deposit, Randsburg, California.

The best example of this type of ore-shoot control known to the writer is found in the vein system of the California Rand Silver deposit at Randsburg, California. A plan of this vein system is represented in Fig. 14. The northeastward-striking veins were major lines of shear which were tightly sealed by the early stage of barren quartz and chalcedony. The later stresses were relieved by the development of northward-striking torsion fractures which received the bulk of the metallization. The northeast veins were metallized only where brecciated by the later stresses.

7. *Saddle, Trough and Leg Reefs.*—When a series of sedimentary strata are compressed into alternate anticlines and synclines, openings *tend* to be produced beneath competent beds on the anticlines, above competent beds in the synclines, and shear occurs between competent beds on the limbs of the fold (Fig. 15, *A*). Ordinarily during folding the incipient openings which *tend* to form are filled by the material of the less competent beds

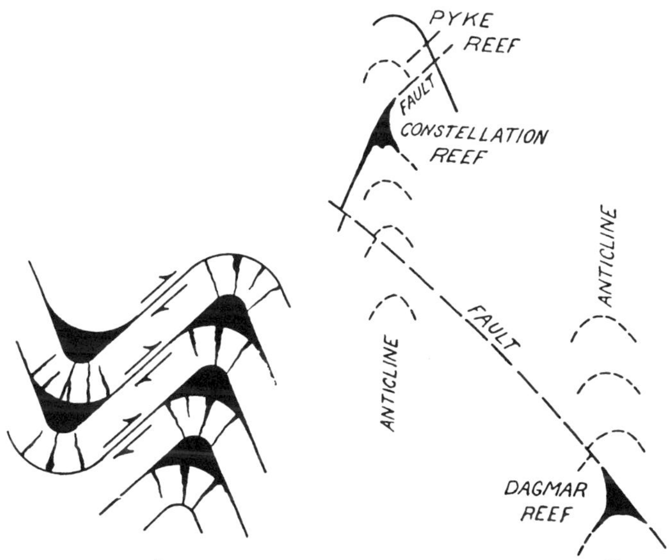

FIG. 15. *A*, Mechanics of formation of saddle reefs, trough reefs, and leg reefs. Leg reefs tend to develop at positions of shear between competent members, while saddle and trough reefs form where incipient reopenings occur as the result of arching of the competent beds. The black indicates localities where reopenings *tend* to occur as the result of folding. *B*, Relation of saddle reefs to faulting. Bendigo, Australia. After Stillwell.

as the result of internal deformation of these beds—thinning on the flanks of the folds and thickening in the arches and troughs.

If the region is undergoing mineralization, however, the incipient openings which *tend* to form may be filled by the products of mineralization. Thus through the repeated production of small openings by the folding and the contemporaneous filling of the openings by the products of mineralization, a saddle reef may conceivably be formed by the process of accretion with little or no replacement being required. That such may be true is attested by the evidence of brecciation present in saddle reefs.[18]

Saddle reefs are most characteristically developed in low plunging folds, but not necessarily so. Ore bodies may occur in the arches of folds which are standing on end, as for example at Jerome, Arizona.[19]

The ore shoots of saddle reefs are characteristically related to intersecting faults (Fig. 15, *B*) or to certain intersecting beds. While chemical control of the ore deposition resultant from the nature of such beds has been the usual explanation put forward to account for the associated ore shoots, it appears probable, and is supported by descriptions, that such beds mark the lines of bedding faults or planes of shear, the shear being consequent upon the folding.

This association with planes of movement is entirely suggestive therefore that the ore shoots of saddle reefs, like those of vein deposits, have formed through the process of accretion, the locus of ore deposition being dependent on the brecciation and reopenings produced by intermineralization faulting—*i.e.*, by the shear resulting from the compression.

8. *Mineral Pipes.*—Locke has recently identified in certain mineral pipes another type of structure which is favorable for the formation of ore shoots.[20] In this type of structure a block of

[18] F. L. Stillwell, "The Factors Influencing Gold Deposition in the Bendigo Gold Field," Commonwealth of Australia, Advisory Council of Sci. and Industry, Bull. 16 (1919).

[19] J. L. Fearing, "Some Notes on the Geology of Jerome," Econ. Geol., vol. 21, pp. 757–773, 1926.

[20] A. Locke, "The Formation of Certain Ore Bodies by Mineralization Stoping," Econ. Geol., vol. 21, pp. 431–453, 1926.

ground subsides as the result of "mineralization stoping," or other cause, producing a cylindrical brecciated zone.[21] The subsidence (known as mineral collapse) and resultant brecciation are recognized as overlapping on the mineralization. In consequence open spaces for the passage of solutions and for the deposition of ore minerals are produced after much of the early barren quartz has been deposited, and these open spaces are still available at the time of the late metallization.

The ore shoots formed in this type of structure are strictly dependent for their location on the open spaces formed by the intermineralization movements and which are existent at the time of the metallization.

9. *Miscellaneous Structures.*—Among the minor types of structure which may be favorable for ore deposition may be mentioned *unfilled vein centers.* Where the early barren quartz stage of the mineralization has not been intensive, the initial fissure may remain incompletely filled at the time of the metallization. In such cases ore shoots may be formed which are coextensive with such openings as still existed at the time of the metallization. Such ore shoots differ from shoots of occurrence insofar as the narrower portions of the fissure may have been completely filled by barren gangue, only the wider portions of the vein still remaining open and receiving ore during the time of the metallization.

While it is felt that the more commonplace types of structure on which ore shoot formation is dependent have been identified, undoubtedly a few types of structure have escaped notice and have yet to be described.[22] Possibly of chief interest among the structures which have not been considered are those of the type repre-

[21] See Locke's Fig. 5.

[22] The deposition of ore at localities where the mineralizing solutions have been retarded or completely dammed is considered by the writer to result from the application of a distinct physical-chemical principle which is not directly related to the subject matter of this paper. The deposition of ore at such localities conforms to the principles of the zonal theory, however, and occurs by replacement or by the process of accretion and under the influence of intermineralization brecciation. The structures involved, in addition to favoring impedance or actual damming of the solutions, conform in their characteristics to the structures considered in this paper.

sented by the disseminated copper porphyry ores of Bingham. Whether, as is held by Locke and as seems probable, the Bingham type of structure represents a modification of the mineral pipe type of structure in which the brecciation affected the whole rock mass instead of merely the periphery of a subsiding rock plug, is not at this time certain. It is certain however that the development of ore in deposits of the Bingham porphyry type was dependent on the openings produced by intermineralization brecciation subsequent to the introduction of much of the early barren quartz, but prior to or contemporaneous with the metallization.

26

Reprinted from *Am. Inst. Mining, Metall. and Petroleum Engineers Trans.* **126**:512-529 (1937)

Bedding-plane Faults and Their Economic Importance

By Charles H. Behre, Jr.,* Member A.I.M.E.

(New York Meeting, February, 1935)

Under the caption "fault," geologists intend to include all mass movements of solid rocks over adjacent rock masses. When these are studied long after their origin, however, circumstances make it possible to recognize as "fault" only formations showing clear evidence of actual movement of one side with respect to the other. Necessarily the chief evidence consists of the offset of a vein, bed, or other key structure. Planes of movement parallel to the bedding, or nearly so, are recognized with difficulty or are overlooked and hence many important faults are wholly unrecorded. Indeed, among standard American references on structural geology, one alone emphasizes the importance of bedding-plane movements and that one only briefly[1].

For some time the writer, cognizant of this difficulty, has paid special attention to such movements, both as observed in the field and as recorded in the literature. It is believed that enough data are now at hand to justify a short discussion. Its purposes are fivefold.

It is desired first to briefly classify and describe at least the more common kinds of bedding-plane faults, and second, to examine the causes and manner of their formation, at least in instances where the subject has not yet been well covered by other investigators. Third, certain by-products of bedding-plane movements are of such conspicuous importance to the economic geologist and mining engineer as to merit mention. Fourth, concrete illustrations should be placed before the reader in order to lend reality and to furnish a basis of comparison with his experience; such illustrations will necessarily be scattered through the text of this paper. They are only in part from the writer's own field observations.

The fifth and prime objective of this paper is to direct attention to the great importance of movements on the bedding planes, in the hope that the recognition of such movements and their significance may simplify mining problems.

---

* Department of Geology, Northwestern University, Evanston, Ill.

[1] References are at the end of the paper.

It is realized at the outset that many movements nearly or quite in the plane of the bedding grade insensibly into internal adjustments such as recrystallization or "flowage." These processes are essentially adjustments by atoms, molecules or crystal units, or at most small aggregates of crystals. This does not constitute true faulting and is not considered here.

Finally, on the basis both of field observation and of deduction, most cases of bedding-plane movement appear to be the result of compressional rather than tensional stresses. Exceptions are conceivable but only a few are known to the writer.

Classes of Bedding-plane Faults

Any plan for natural categories breaks down if due allowance is not made for the gradations between distinguishable facies. In offering the following classification, the presence of such gradations is not denied, but no purpose is served by avoiding systematization merely because of the occasional instances that resist classification.

The following six types of bedding-plane faults appear to be distinguishable, and significantly so; the terms used in designation of the types are not intended for introduction into the literature, but merely to serve as convenient abbreviations for use in this discussion: (1) saddle reef type, (2) steep reverse type, (3) transecting accessory type, (4) flat reverse type, (5) erosion type, (6) zigzag type. Each of these classes will be discussed in some detail.

Most faults are not unwarped planes, of course—indeed, they are anything but simple and planar. In view of their irregularity, it is not surprising that as the fault plane is followed downdip, especially in closely folded strata, occasional coincidence with the bedding planes should be observed; however, often this appears to be purely accidental and such instances are omitted from the classification suggested here.

Saddle Reef Type

The famous saddle reefs of Bendigo, at Victoria, Australia, may serve as type illustrations. Excellent descriptions of these structures were published by Rickard[2,3]. The true saddle reefs are auriferous quartz veins, which fill openings localized near anticlinal crests (Fig. 1). Some doubt may well be expressed as to whether the apical opening actually antedated the deposition of the vein matter, but there can be no question that the present veins represent an opening between the beds, rather than appreciable replacement.

Typically, the limbs of the folds associated with the saddle reef have fairly steep dips; frequently as much as 70°. Of two saddle reefs lying one above the other on the same fold, the lower generally has the wider opening angle, the more gently dipping flanks, and less thickening on the crest. While fractures are found, conspicuous faults are rare and do not generally appear to affect the form of the reef itself.

Origin.—In order to develop openings at the crests of folds, it is obviously necessary for bedding-plane movements to take place. This type of movement is well established in geologic literature. There is some evidence in the same strata of local thickening (at the fold crests) and thinning (on the limbs); but individual laminae bordering the main bedding-plane parting can be traced respectively above and below the vein, which indicates that there was slippage along the surfaces of the country rock now immediately adjacent to the vein. This is clearly a case of bedding-plane movement. The cause for localization of slippage is not clear. It is probably the result in large part of differential elasticity as between different beds, but an analysis of this factor is not attempted here.

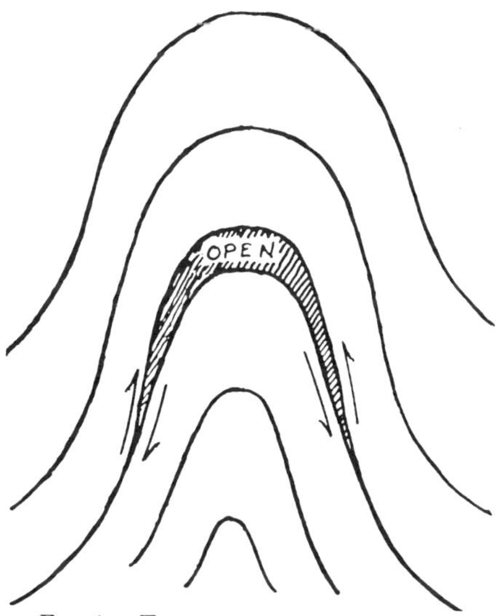

FIG. 1.—TYPICAL SADDLE REEF TYPE OF FOLDING, SHOWING MOVEMENTS ALONG BEDS.

Beds have flatter curves below reef or opening than above. Actual movement is confined to flanks of beds.

The difference in the elevations of the arch crests on the lower and higher beds separated by a given saddle furnishes a means for computing the relative updip movement of the higher beds. In cases available in the literature, this computation is not easily made because of lack of definite data as to the depth of the adjacent trough in a given bed. From one of the scale drawings presented by Rickard, the reef on the 2220-ft. crosscut of the 180 mine indicates an updip movement of 7.25 ft. in a thickness of 276 ft. of strata as measured at the crest of the fold; this computation is based on reasonable assumptions (Fig. 2).

Further light is shed on the progress of the opening to be filled by the saddle reef (or vein) by thickening of the strata at the fold crests, as previously noted. This demonstrates that internal adjustments preceded the actual tearing apart of individual strata and hence also preceded the bedding-plane movements discussed above. In short, a period of relative plasticity was followed by one of greater rigidity (probably the result of anamorphism), such as is necessary for bedding-plane movements. Metamorphism evidently did not destroy the bedding planes, however.

Finally, it is significant that small inverted saddle reefs, occurring in the synclinal troughs, have also been found at Bendigo. Adjustment takes place with a relative movement downward, therefore, as well as upward, but less conspicuously.

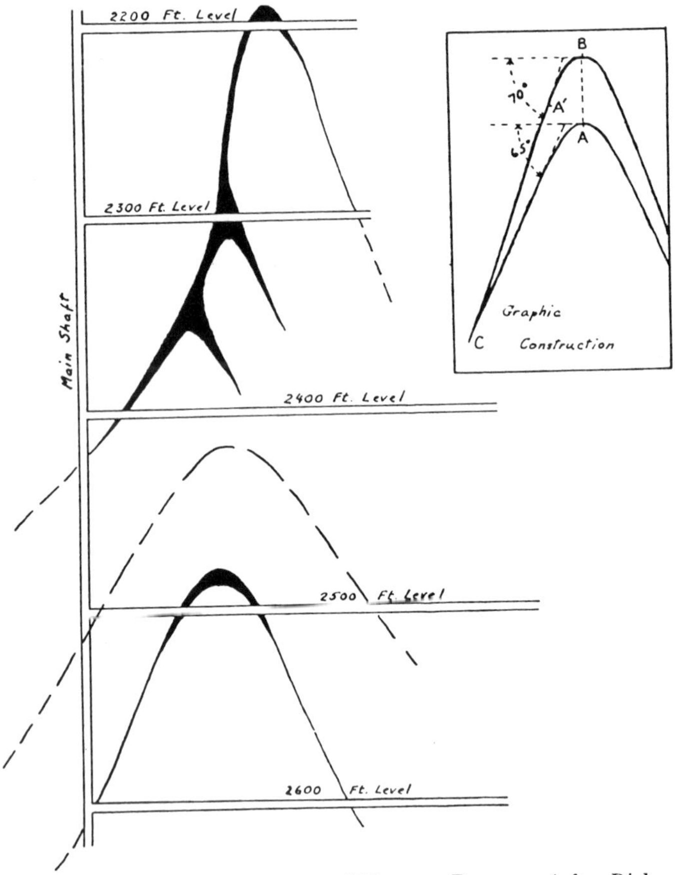

FIG. 2.—VERTICAL SECTION IN THE 180 MINE, BENDIGO (*after Rickard*).
To illustrate saddle reef type of bedding-plane faulting; note that upper folds are tighter than lower ones. Insert shows construction for graphic solution of amount of movement of upper bed *CB* over lower bed *CA*. *A'* is a point in space that, except for separation of upper and lower beds, would rest against *A* on lower bed; *B* is present position of crest point on upper bed; angles of dip of *CB* and *CA* are laid off respectively as observed from mine data, having been generalized with due regard to varying dips of beds as crest of fold is approached. *AB* is known from measurement; the problem is to measure *A'B*, the relative updip movement of bed *BC*.

Examples of Economic Importance.—As mentioned, it is not always possible to classify bedding-plane movements. When, however, in a region of fairly close folding they are neither manifest accessories of transgressive faults nor actually pass into such faults, they may perhaps properly be assigned to the saddle reef type. Examples of economic importance are numerous. The "flat fault" of the Ojuela mine in the

Province of Durango, Mexico, is probably an instance[4]. The ore deposits at Mascot, Tenn., seem to represent another[5] (Fig. 3); they occupy spaces in a breccia discussed below. Both Bateman[6] and Davidson[7] assign much of the localization of sill-like intrusions and also of the copper ores of Northern Rhodesia to bedding faults. Bedding-plane faults have even been found between basalt flows in the Columbia River region[8].

In the Rico district, bedding-plane deposits of uncertain origin have been described; they commonly bear ore and are spoken of as "blankets" or "contacts[9,10]." Some, at least, are certainly of tectonic origin, notably in the Rico-Aspen and Little Maggie mines, though others are regarded as solution effects.

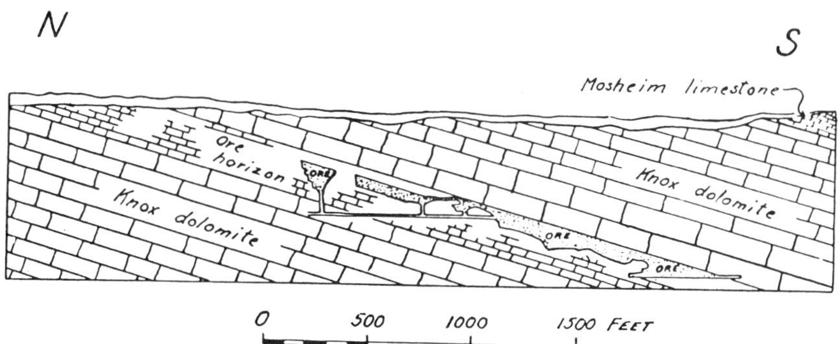

FIG. 3.—VERTICAL NORTH-SOUTH SECTION OF ORE ZONE AT MASCOT, TENNESSEE (*after Newman*).
Note that ore is confined to a single stratigraphic horizon.

Finally, saddle reefs, small but unmistakably showing the typical pattern, have been reported from the Philipsburg district in Montana[11].

By-products of Saddle Reef Type of Bedding-plane Movement.—Aside from the tendency for the beds to open along the separating planes, the saddle reef type of movement is likely to develop accessory fractures, not necessarily in the nature of faults, in the adjacent beds. Such fractures are figured by Rickard in the New Chum Consolidated, 222, and Lazarus mines[2] (Fig. 2).

Especially striking is the potential shattering of the more brittle rocks in one or both walls along the planes of movement. Such shattering is not necessarily confined to the anticlinal crests. A case in point is described by Hewett[12]; the limestone beds above and below the plane of movement are accordant, but they are separated by folded, thickened and shattered interbeds, also of limestone. These secondary features remind us strongly of the drag folds commonly developed under similar conditions in "incompetent" rocks between more rigid layers.

A further illustration of such shattering is to be found in the Mascot mine, already mentioned. The main ore bodies are at a fairly definite horizon, dipping about 20° on the southwest limb of a broad fold of Cambro-Ordovician limestones. The ore occurs in a virtually continuous breccia zone[5,13], not clearly related to any single major fault (Fig. 3). Currier suggests that such breccia horizons are due to shattering of the more brittle beds, but it is clear that he also recognizes the importance of bedding-plane slippage. These two processes are probably closely allied.

At Aspen, bedding-plane breccias in dolomitic limestones resemble those at Mascot. They have recently been restudied[14] and are attributed to movement between beds incidental to close folding; such movements are believed by Vanderwilt to be only measurable in inches, however.

Steep Reverse Type

Reverse faults may be derived from bedding-plane movements and at depth probably pass into such movements. If the picture drawn above for the saddle reef is correct, it may be expected that, as pressure from the sides continues, the fold will finally rupture at some point near the crest. The resulting fracture may constitute a nearly vertical reverse fault near the fold axis but, flattening downward, tends to pass parallel to and into the plane of separation between the beds (Fig. 4). Its vertical trace is concave upward. Downdip it is a true bedding-plane fault and its displacement in that direction may appear to be reduced to zero.

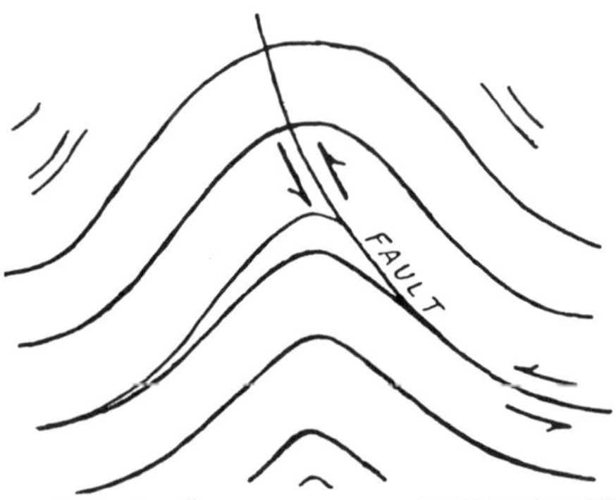

Fig. 4.—Ideal steep reverse type of bedding-plane movement, related to folding.

Mechanics.—In folds having essentially "parallel"* patterns, internal adjustment is negligible and noteworthy tangential compression is likely to result in faulting. In an anticline, if the overburden is great the tendency toward movement near the crest of the fold is largely taken up by internal adjustment; if the overburden is slight, however, as it is near the surface, the bed is more prone to behave as brittle material than at greater depth along the limbs, and hence fractures transverse to the bedding are to be anticipated.

In the troughs of synclines, on the other hand, faulting is less likely, for here the overburden (assuming erosion on the upwarped anticlinal area, concomitantly with folding) is still greater and flowage, as opposed to fracture, more probable. Moreover, lateral compression, in increasing the amplitude of the fold, forces it hard against lower beds, from which

* For use of this term, see Leith[15].

it encounters resistance that also favors internal adjustment and opposes fracture.

Thus, of the various parts of folds, anticlinal crests are least supported by contiguity with underlying beds, are least encumbered and urged into flowage by overlying beds, and therefore are favored points of fracture. Hence a bedding-plane movement tends to pass upward on a fold limb until it intersects the fold axis, whereupon the angle with the bedding increases and a recognizable reverse (or, rarely, normal) fault results. Conversely, a fault that at the surface is a reverse fault passes downward with reduced dip and finally coincides with the bedding plane.

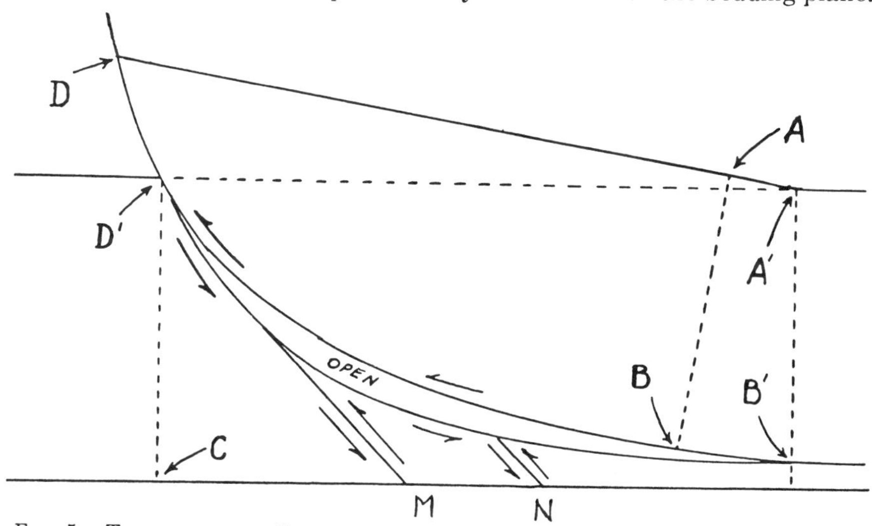

FIG. 5.—TO ILLUSTRATE "CHINKING IN" ALONG A CURVED MAJOR REVERSE FAULT PLANE.

Dotted lines outline an original block ($A'B'D'$) which, after the faulting, occupies the new position ABD. The roughly triangular block below the major fault (line $B'D$) has not changed its position. The consequent opening is likely to be filled up by reverse movement along the two minor and accessory faults at M and N.

Special Considerations and Accompanying Features.—Curved planes steepening upward are likely to result in considerable shattering, for reasons to be discussed next. Assuming vertical movement to be prominent, if the curve of the fault in a vertical trace is a perfect arc of a circle, movement along it does not destroy the matching of opposite sides. But if the curve is not strictly an arc of a circle, a space may develop. Continued pressure from the sides then tends to induce fracturing in one block or the other, and the small accessory faults that result have the effect of "chinking in" the opening. Their movement, if correctly deciphered, probably gives in most cases a clue to the movement on the major fault (Fig. 5), a fact suggested to the writer by others* but not, so far as he is aware, hitherto analyzed.

* T. S. Lovering and G. F. Loughlin, in conversation in 1929 and 1930.

While it is true that the steep reverse fault may be developed as a result of bedding-plane movements near synclinal axes, examples are rare. The writer figured one such case, the Bowden fault, as explored in the Ibex mine at Leadville[16]. Here, however, the opening to be anticipated at the trough of the fold was occupied by an intrusion.

Illustrations of Economic Importance.—In addition to the case just cited, several instances of this type of faulting are known from many mining districts. The pattern is not everywhere ideal, but usually conforms well enough to the description given above. In a north-south section through the Highland Boy, Yampa, and Utah Apex mines, a fold

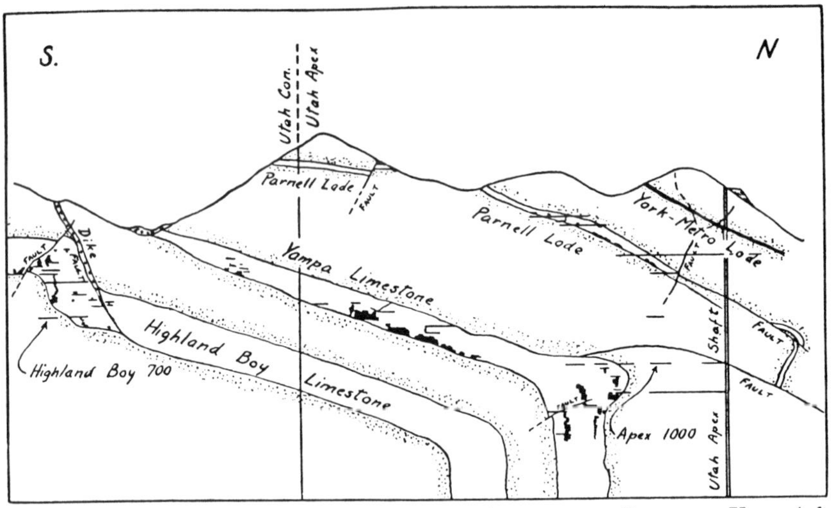

Fig. 6.—North-south section through Utah Apex shaft, Bingham, Utah (*after Hunt*).

To illustrate "fault folds"; one fault dips steeply north near south end of section and drops the limb of the fold; the other is slightly curved and almost horizontal, crossing Utah Apex shaft at about 1000-ft. level.

in the Highland Boy limestone is flanked by a slight fault, followed by a dike[17] (Fig. 6); these relations are unusual in that the fault is normal. Farther north a sudden westward steepening of the Yampa limestone yields a structure that may be regarded as an anticline overturned to the north. It is accompanied by a fault, which follows the upper surface of the limestone nearly to the crest of the fold; thence, it transects the beds. Other related cases are figured by Hunt under the term "fault fold" (ref. 17, Fig. 4 and p. 864). Such features, according to Hunt, "are obviously related in causal stress and in time to late cross faults from which they are structurally inseparable."

The great South Dyer thrust in the Leadville area, below which are localized a few small gold veins, brings pre-Cambrian upon Cambrian rocks (ref. 16, Fig. 1 and p. 46); for much of its extent, it is essentially parallel to the bedding, and at its southeastern extremity it plays out into subsidiary thrusts that dip more steeply and transect the beds up the dip of the fault plane.

A modification of this type of movement is represented by several faults on the western edge of the great Tintic syncline between Eureka and Mammoth, Utah[18]. Here the beds dip very steeply east or stand vertical at the surface, but their dip flattens downward in approaching the synclinal trough, vertical faults, striking with the beds, cross the trough at depth and so offset the beds; the steepening of the bedding up the dip, however, soon brings bedding and faults into coincidence, so that the faults are largely unrecognizable at the outcrop, though they are readily found and their offsets measured underground.

Another special case that probably belongs here is that cited by the author from the Northampton County slate district, Pennsylvania. There, in relatively close folds, the outer, more tightly folded beds are cut off by fault planes parallel to the bedding on the inner, more open part of the fold[19]. As a result, much waste slate has to be quarried in the neighborhood of such faults, to the serious financial loss of the operators.

Transecting Accessory Type

This type of fault was first discussed, as far as this author is aware, by Emmons, Irving, and Loughlin[20]. It is associated with steeply dipping curved faults of larger displacement, and consists of movements on the bedding, the result of which is to permit adjustments along the major fault. An ideal diagram is presented in Fig. 7*A*, which shows both the main fault and the accessory movements along the bedding. Such faults are spoken of as transecting accessory, because they are movements accessory to larger faults that transect the bedding.

Origin.—Where a large fault traverses the beds and is curved in vertical section, the surfaces on either side of the fault plane must adapt themselves to such a curvature. If the rocks involved are so rigid that they do not flow readily, accessory faults may be developed and along these there is movement. If open, the partings along the bedding planes furnish surfaces of readiest movement. The result will be accessory bedding-plane faults like that shown in Figure 7*A*. Alternatives may be a relative movement of the beds in directions opposite to those already figured (see Figure 7*B*) or an actual shattering (Fig. 7*C*). Which of these possibilities occurs depends upon the curvature of the fault; thus, in Fig. 7*B*, a movement toward an arc of larger curvature (that is, one meeting the lower bed at a sharper angle than originally) will tend to cause movement of the upper bed toward the right and a leftward movement of the lower bed; and a bedding-plane fault may result under these conditions also. The same general effect is that shown in Fig. 7*C*; here shattering with drag along the main fault plane has resulted.

It should be clear that bedding-plane movements of this type are favored by marked openings along the bedding planes and by the presence of relatively plastic rock, such as shale, between the more rigid layers.

Illustrations of Economic Importance.—At Leadville, bedding-plane movements of this type are common. Many of them even take place between such relatively rigid rocks as limestone and sills, especially where the limestone has been partly replaced by clayey minerals or the igneous rock of the sill has been altered. (See, for example, ref. 20, p. 65.)

At Weston Pass, not far from Leadville, is what appears to be a transacting accessory bedding-plane fault. It is near and dips toward the

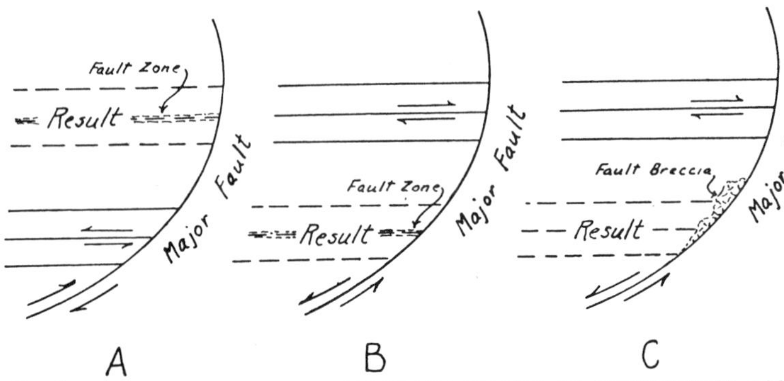

Fig. 7.—Ideal diagrams of bedding-plane faults of transecting accessory type.

Arrows show directions of movement on main fault and directions of movement on accessory bedding-plane faults. Solid lines represent position of beds before movement, dashed lines position of beds after movement. In *A*, the hanging wall moved up; in *B* and *C*, down.

Weston fault, one of the more important thrusts of the Mosquito Range, and is interpretable as accessory to the latter, though the geometrical relations between the two planes of movement are nowhere observable. The bedding-plane movement follows a definite stratigraphic horizon for a distance of about 7200 ft. and is characterized by the presence of a highly silicified breccia[21]. It is a striking fact that a virtually continuous ore horizon occurs about 80 ft. above this plane, suggesting the presence of a still higher zone of shattering similar to that just described[21].

In the Goodsprings, Nevada, district, the Argentena mine shows much ore in open breccia zones roughly parallel to the bedding of the limestones. Very close to the mine is the great Frederickson fault, a thrust dipping under the beds exposed in the workings; this important thrust fault shows movement largely horizontal and parallel to the strike of the fault plane (ref. 12, p. 148 and plate I). The breccia zones that bear the ore may well be accessory to the main movement on the Frederickson thrust. A similar bedding-plane fault with breccia zone is described from the Anchor mine in the same district, but its origin and relationship to larger faults is not clear (ref. 12, 161).

One of the best examples of openings parallel to beds resulting from adjustments near faults is figured by Emmons from Philipsburg, Montana[11]. In this case, however, the faults involved appear to be normal. Here two small silver lodes rigidly follow two bedding-plane fractures between faults (see Fig. 8). Other similar cases are known from the same district (ref. 11, pp. 214, 215).

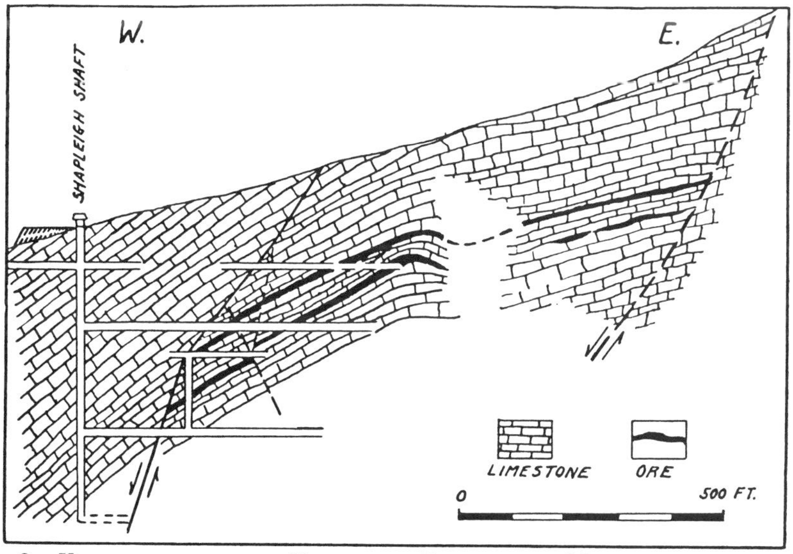

FIG. 8.—VERTICAL SECTION OF HOPE MINE, PHILIPSBURG, MONTANA (*after Emmons and Calkins*).
Silver veins occupying bedding planes that appear to represent planes of movement accessory to faults shown.

FLAT REVERSE TYPE

In the flat reverse type of fault the fault plane dips very gently in the direction from which the thrust advances (underthrusting omitted, as a special case). Such faults are likely to show dips having angles of 15° from the horizontal, or less, and may, as in the folded thrusts observed in the Appalachians and in parts of the Rocky Mountains, even decline to zero, with occasional dip reversals. If the beds also dip gently in such instances, the fault and bedding planes may coincide. Many of the great thrusts belong to this class. The name used here to designate the type refers to the attitude of the fault plane and to the generally reverse nature of the movement.

Origin.—Faults of this type have long been the subject of study by tectonic geologists. An excellent discussion of their origin was presented by Chamberlin and Miller[22]. These authors point out that thrust faults flatten downward, on account of the effect of the overburden and friction, which add a strong vertical component to the pressure; or, to put it differently, the planes flatten downward with increasing rotational stress. If the formation of the fault plane had not been preceded or accompanied by intense folding, the fault might at depth readily become parallel, or nearly so, to the bedding. This is also true where folding produces doming; in such a case, however, the trace of the fault on a vertical plane more nearly approximates that already described for the steep reverse type of bedding-plane fault.

Rich has recently illustrated such faulting in describing the Pine Mountain fault of southeastern Kentucky. He emphasizes its tendency to approximate a bedding-plane fault. Flattening is chiefly attributed by Rich to the presence of nearly horizontal layers of softer, more plastic shales. Oblique rises of the fault plane across the bedding are believed to result only where frictional resistance becomes very great; thereupon the fault plane rises more steeply to a higher level, on which it may meet another shaly series and follow this a way[23].

The two papers referred to discuss fully the problem of low-angle thrusting; they also show that such thrusts may tend to follow the bedding planes. Further consideration here is deemed unnecessary.

Examples.—In addition to the Pine Mountain fault, the planes of numerous large-scale and low-angle faults are, in parts at least, parallel to the beds. Examples include the Lewis thrust in Montana, and subsidiary smaller faults connected with it (ref. 24, Fig. 335 and Plate 53); the Heart Mountain thrust, at least in part as to the floor and very largely as to the hanging wall[25]; some of the flat faults of the Scottish Highlands (see, for example, ref. 26); and the Pulaski fault of the valley coal fields of Virginia[27]. The last named covers in part some of the most valuable coal beds of the state.

Excellent examples on a smaller scale, akin also to the steep reverse type of bedding-plane movement, are given by Reeves for areas bordering the Highwood Mountains. Here well-nigh horizontal beds show movement on the bedding planes; in areas of gentle folding such fault planes change to low-angle reverse faults that cross the fold axes. The movement mentioned is furthered by slippage on shaly beds[28]. It is of economic importance as bearing upon the origin of certain structures that may be petroliferous.

A special case, in which the thrust plane is steep but dips nearly parallel to the beds for an unknown distance, is described from an Arkansas cinnabar district[29,30]; the fault is on one limb of a compressed fold and thus bears some resemblance to the faults at Tintic cited above.

Accessory Details.—For the economic geologist one of the noteworthy features of major thrusts, whether or not they are parallel to the bedding planes, is their tendency to be accompanied by small accessory openings, sometimes in one, and frequently in both walls. In many ore deposits the major faults themselves do not carry the ore; it is borne instead by these accessory fissures. In general, such minor fractures follow two different patterns; which one is followed depends upon whether or not they result from tension or shear. Accessory shear fractures make an acute angle with the major fault, an angle whose apex points in the direction of movement of the opposite side of the fault. Accessory tensional fractures (to which the term "gash vein" is also applied) yield wider crevices so oriented that the apex of the acute angle which they make with the main fault points in the direction of movement by the side in which such crevices are found (Fig. 9). The literature on theories of such features is voluminous. For particulars the reader is referred to any standard text on structural geology.

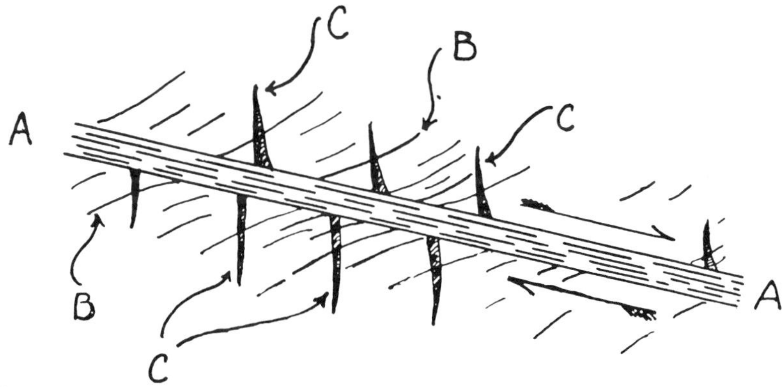

FIG. 9.—IDEALIZED THRUST FAULT (*AA*) SHOWING RELATION BETWEEN ACCESSORY SHEAR FRACTURES (*B*, *B*) AND TENSION FRACTURES OR "GASH VEINS" (*C*, *C*, *C*). NOTE THAT *B*, *B* ARE CLOSED AND *C*, *C*, *C* ARE OPEN.

The formation of gouge is also a conspicuous feature, especially of most larger thrusts but of many minor ones as well. For this reason major thrust faults particularly frequently fail to show the extensive mineralization to be expected in view of their magnitude; they are tight barriers rather than channel ways.

THE EROSION TYPE

The erosion type may be regarded as a modification of the flat reverse type just described. In it the fault plane coincides with a surface developed by erosion. Fortuitously the erosion surface and the bedding also coincide.

Origin and Illustrations.—Such cases seem to have been effected by a fairly constant sequence of events. There is first local bulging as the result of tangential compression. Gradually older, more rigid rocks are exposed on the bulge. Resistance on one flank of such a buttress is reduced by differential erosion, or by overfolding due to further tangential compression from the opposite side, or by both factors operating together.

In the meantime, a stripped plain or peneplain has been developed by erosion in the lee of the bulge. When pressure is renewed, the buttress breaks across the beds and advances over the erosion surface as a relatively flat thrust.

This sequence has been recognized in the southern Appalachians for the Rome and Cartersville faults (especially the latter) by Hayes[31] (Fig. 10); in the northern Rockies for the Lewis overthrust by Willis (ref. 24, pp. 339–340); and for the Heart Mountain thrust by Hewett[25]. The relation between this type of movement on an erosion surface (which perchance virtually coincides with the bedding) and the large-scale bedding-plane movement so commonly observed in the great overthrusts, is

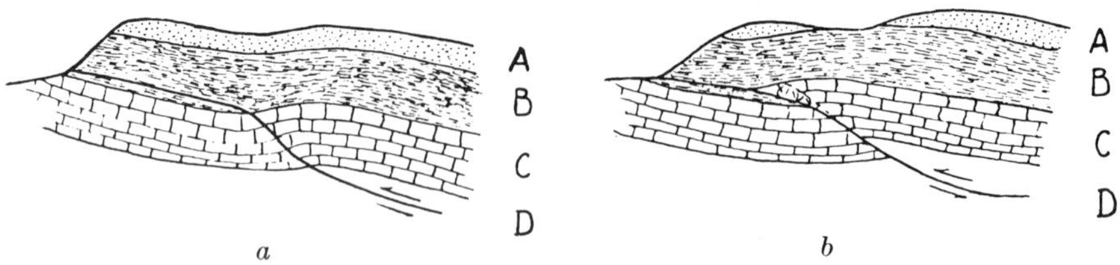

FIG. 10.—HOLLY CREEK SECTION OF CARTERSVILLE THRUST-EROSION FAULT (*after Hayes*).
a, before movement but after erosion; *b*, after movement. A, Cohutta conglomerate; B, Ocoee slate; C, Knox dolomite; D, Conasauga slate.

obvious. Accessory features in the two types are closely similar as well. Faults of this class are not generally important to the economic geologist except by virtue of accessory features.

ZIGZAG TYPE

The vertical trace of a compressional fault sometimes approximates what may be called a zigzag pattern. In such instances the fault plane follows the bedding for some distance, then turns at an abrupt angle and cuts obliquely across the beds, only to follow them again at a higher or lower level. In the larger thrust faults this pattern can seldom be traced, partly because continuous sections along the fault planes are lacking, partly because the irregularities are curvilinear rather than sharply angular in the manner just described. In smaller faults, however, where the vertical displacement is only a few feet, the zigzag pattern is fairly common, as examples will show.

Illustrations.—So far as the writer is aware, Barrell first called attention to such fault patterns, reproduced on a small scale by shear planes and bedding faults, which he concluded were jointly responsible for local mineralization at Marysville, Mont.[32] Another illustration, on a larger scale is cited by Rickard in the Enterprise mine, where a fault transects at a high angle beds that are virtually horizontal up to the level of a certain limestone layer, in which it follows an approximately horizontal course parallel with the bedding for a few feet, and then again cuts across the beds (ref. 33, pp. 942–946).

A case from Philipsburg, Montana, has been well figured by Emmons and Calkins, under the term "roll" (ref. 11, p. 255, Fig. 54).

At Leadville, Colo., there are also numerous illustrations of the zigzag type of bedding-plane fault. One of the best is in the Oro La Plata mine, where, according to Argall, a bedding-plane fault fissure in the "Blue" limestone bears lean ore for a distance of about 500 ft. up the dip; then it turns to rise across the beds at an angle of about 30° and becomes wider, carrying richer ore in this part of its course. In this upper,

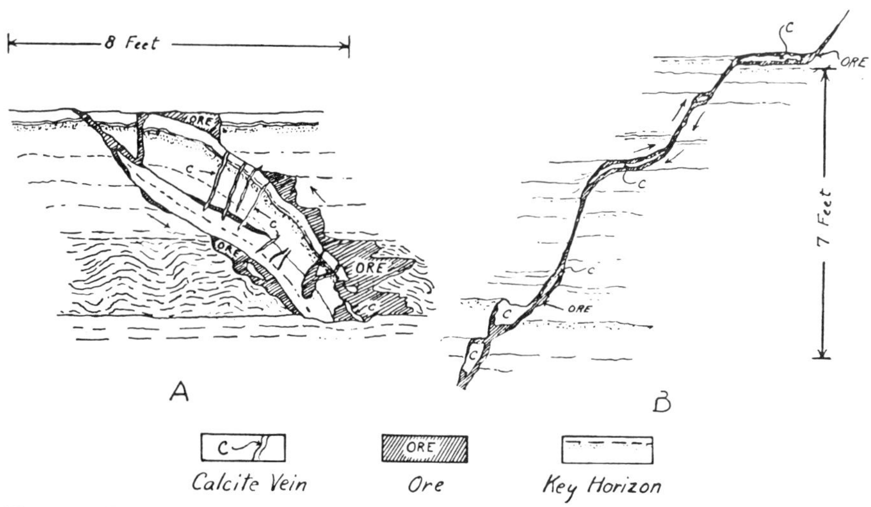

FIG. 11.—DIAGRAM OF TWO VARIETIES OF ZIGZAG BEDDING-PLANE FAULTS AS SEEN IN CRAWFORD MINE, HAZEL GREEN, WISCONSIN.

A, thrust fault in southwest workings (*after Scott*). Fault had course in lower, crumpled, shaly layer, then split upward so that its two branches encircled block, and finally continued to left just beneath topmost bed. Key horizon clearly shows compression, together with rotation of loose block.

B, thrust fault in northwest heading. Key horizons show that left (or hanging) wall has risen, giving bedding-plane partings now filled with ore; hence veins are wider where fault plane is horizontal.

inclined part the displacement is clearly discerned and the dip-slip amounts to about 10 ft. This is a reverse fault[20].

In the lead-zinc district of southwestern Wisconsin, the ore bodies as now worked are chiefly "flats" yielding sphalerite. Closely examined, they show little if any replacement; instead they constitute tabular fissure veins, frequently several hundred square feet in plan though only a few inches or at most a foot or two thick. Careful study shows that at least some of the flats are related to bedding-plane faults of the zigzag type. The plane of movement generally follows bedding partings in the more thin-bedded or shaly lower Galena dolomite, rises obliquely across a higher, more massive layer, and then, instead of passing into the roof along its projection updip, once more turns to the horizontal position and follows the bedding. Scott has described and figured several such cases[34] (Fig. 11).

It is to be expected that in such fractures a movement prevailingly parallel to one of the limbs of the zigzag may serve to widen the fault fissure, thereby separating its sides, while conceivably closing the fissure on the other limb—a principle familiar to mining geologists. (See for example, Emmons[35].) In this district horizontal openings preponderate; they exhibit considerable horizontal extent and generally appear immediately above a prominent shaly zone having the high plasticity that would favor bedding-plane movement. This general association supports the conclusion that horizontal compression, by producing differential movement along zigzag faults, is of far greater significance in the ore deposition of the district than hitherto realized. Indeed, a careful matching up of the vertical displacement with the vertical thickness of the "flats" shows close correspondence between the two.

Origin.—Such features are obviously the result of varying brittleness and plasticity of the materials in a normal sedimentary sequence. Thrust faults thus break across the beds not in an unwarped plane but in a fracture of which the inclination to the horizontal increases with increasing rigidity of the material traversed. If the beds are steeply inclined, faults resembling those at Tintic are more likely to be formed. But where the beds are flat and there is thrusting without a very heavy overburden, fault planes that otherwise would have dips of 45° or so are resolved into two components by the variations in the brittleness of the beds traversed. Since the direction of easiest relief is upward, the hanging wall moves up, rather than forward, and the parts of the fault that appear as more horizontal fissures are more prominent than those more steeply inclined.

Uncertain Classification

In many instances, if not in most, there is uncertainty as to the class to which a particular bedding-plane movement should be referred. Below are listed several examples, which, nevertheless, have been highly important in exploring the respective deposits. In these instances the associated structures, which usually are the main criteria in the classification scheme outlined here, are either not definitely known or not sufficiently described in the literature to permit positive classification.

In the Gogebic Iron Range of Wisconsin and Michigan several bedding faults have been described by Hotchkiss. These faults occur along one or more shear planes, usually in the Yale member, a series of ferruginous cherts and slates. The aggregate movement on this plane or planes is known and the major component is parallel to the strike of the fault. Thus, along the main fault in one locality the displacement amounted to

800 ft. along the strike of the fault plane and 200 ft. up its dip. From the succession of events as worked out by Hotchkiss, it appears that this fault is essentially a flat thrust, antedating folding[36] and thus not assignable to mere bedding plane adjustments of the Bendigo type.

In the Tri-State lead and zinc district, Fowler and Lyden have ascribed to movement on the bedding planes the openings of the "sheet ground" deposits. These are said to be the result of slight but recognizable tangential adjustments that characterize the district as a whole[37].

On a far smaller scale, bedding-plane faulting is shown at Rico, Colo., by the lateral shifting along the bedding of small cross veins (ref. 33, pp. 922–923, 946).

Conclusions

It appears possible to classify the types of bedding-plane faults on the basis of their structural associations and genesis, with due regard to gradations between types. Moreover, there are numerous illustrations available in the literature, especially in regions where evidences of lateral compression are definite or inferred. And finally, many of the cases are of sufficient economic significance to justify the conclusion that bedding-plane faulting should be ranked among the important structural features associated with ore deposition.

References

1. C. N. Nevin: Structural Geology, 108. New York, 1931. John Wiley and Sons.
2. T. A. Rickard: The Bendigo Gold Field. *Trans.* A.I.M.E. (1902) **20,** 475–529.
3. T. A. Rickard: The Saddle Reef. In Types of Ore Deposits (ed. by H. F. Bain), 172–189. San Francisco, 1911. *Min. and Sci. Press.*
4. M. W. Hayward and W. H. Triplett: Occurrence of Lead-zinc Ores in Dolomitic Limestones in Northern Mexico. A.I.M.E. *Tech. Pub.* 442 (1931) 22.
5. M. H. Newman: The Mascot-Jefferson City Zinc District of Tennessee. *Guidebook* 2, XVI Int. Geol. Cong. (1932) 152–161.
6. A. M. Bateman: The Ores of the Northern Rhodesia Copper Belt. *Econ. Geol.* (1930) **27,** 411.
7. D. M. Davidson: The Geology and Ore Deposits of Chambishi, Northern Rhodesia. *Econ. Geol.* (1931) **26,** 139, 151.
8. I. A. Williams: The Columbia River Gorge, Its Geologic History. Oregon Bur. Mines and Geol., *Min. Resources of Oregon* (1916) **2,** No. 3, 112–115.
9. W. Cross and A. C. Spencer: Geology of the Rico Mountains, Colorado. U. S. Geol. Survey 21st *Ann. Rept.* (1920) pt. 2, 107–112.
10. F. L. Ransome: The Ore Deposits of the Rico Mountains, Colorado. U. S. Geol. Survey 22d *Ann. Rept.* (1901) 2, 273–293, 295–297.
11. W. H. Emmons and F. C. Calkins: Geology and Ore Deposits of the Philipsburg Quadrangle, Montana. U. S. Geol. Survey *Prof. Paper* 78 (1913) 183, 184.
12. D. F. Hewett: Geology and Ore Deposits of the Goodsprings Quadrangle, Nevada. U. S. Geol. Survey *Prof. Paper* 162 (1931) Fig. 8.
13. L. W. Currier: Structural Relations of Southern Appalachian Zinc Deposits. *Econ. Geol.* (1935) **30,** 272–276, 282–283.
14. J. W. Vanderwilt: Revision of Structure and Stratigraphy of the Aspen district, Colorado, and Its Bearing on the Ore Deposits. *Econ. Geol.* (1935) **30,** 236–237.

15. C. K. Leith: Structural Geology, 175. New York, 1923. Henry Holt & Co.
16. C. H. Behre, Jr.: Revision of Structure and Stratigraphy in the Mosquito Range and the Leadville District, Colorado. *Proc.* Colo. Sci. Soc. (1929) **12,** 54–56.
17. R. N. Hunt: The Ores in the Limestones at Bingham, Utah. *Trans.* A.I.M.E. (1924) **70,** Fig. 2.
18. W. Lindgren, C. F. Loughlin and V. C. Heikes: Geology and Ore Deposits of the Tintic Mining District, Utah. U. S. Geol. Survey *Prof. Paper* 107 (1919) 83–84, plates IV and V.
19. C. H. Behre, Jr.: Slate in Pennsylvania. Pa. Geol. Survey *Bull.* M-16 (1933) 164–168.
20. S. F. Emmons, J. D. Irving and G. F. Loughlin: Geology and Ore Deposits of the Leadville Mining District, Colorado. U. S. Geol. Survey *Prof. Paper* 148 (1927) 64–65, 189, 190.
21. C. H. Behre, Jr.: The Weston Pass Mining District, Lake and Park Counties, Colorado. *Proc.* Colo. Sci. Soc. (1932) **13,** 63–65, 67.
22. R. T. Chamberlin and W. Z. Miller: Low-angle Faulting. *Jnl. of Geol.* (1918) **26,** 1–44.
23. J. L. Rich: Mechanics of Low-angle Overthrust Faulting as Illustrated by the Cumberland Fault-block, Virginia, Kentucky and Tennessee. *Bull.* Amer. Assn. Petr. Geol. (1934) **18,** 1584–1596.
24. B. Willis: Stratigraphy and Structure of the Lewis and Livingston Ranges Montana. *Bull.* Geol. Soc. Amer. (1902) 331–343.
25. D. F. Hewett: The Heart Mountain Overthrust, Wyoming. *Jnl. of Geol.* (1920) **28,** 554.
26. B. N. Peach, J. Horne, W. Gunn, C. T. Clough, L. Hinxman and H. M. Cadell: Report on the Recent Work of the Geological Survey in the Northwest Highlands of Scotland. *Quart. Jnl.* Geol. Soc., **44,** 426, 427.
27. M. R. Campbell and Others: The Valley Coal Fields of Virginia. Va. Geol. Survey *Bull.* 25 (1925) 76–96.
28. F. Reeves: Thrust Faulting and Oil Possibilities in the Plains Adjacent to the Highwood Mountains, Montana. U. S. Geol. Survey *Bull.* 806 (1928) 167–185, especially Figs. 17–19.
29. J. M. Hansell and J. C. Reed: Quicksilver Deposits Near Little Missouri River, Southwest Arkansas. *Trans.* A.I.M.E. (1935) **115,** 229.
30. N. H. Stearn: The Cinnabar Deposits in Southwestern Arkansas. *Econ. Geol.* (1936) **31,** 6.
31. C. W. Hayes: Overthrust Faults of the Southern Appalachians. *Bull.* Geol. Soc. Amer. (1891) **2,** 141–152, especially 149.
32. J. Barrell: Geology of the Marysville Mining District, Mont. U. S. Geol. Survey *Prof. Paper* 57 (1907) 123, 125 and Plate 12.
33. T. A. Rickard: The Enterprise Mine, Rico, Colorado. *Trans.* A.I.M.E. (1897) **26.**
34. E. R. Scott: Structural Control of Ore Deposition in the Upper Mississippi Valley Lead-zinc District. Master's Thesis (1934), Northwestern Univ. (unpublished) 53–62.
35. W. H. Emmons: Principles of Economic Geology, 185, Fig. 82. New York, 1918. McGraw-Hill Book Co.
36. W. O. Hotchkiss: Geology of the Gogebic Range and Its Relation to Recent Mining Development. *Eng. and Min. Jnl.* (1919) **108,** 537–538, 540.
37. G. M. Fowler and J. P. Lyden: The Ore Deposits of the Tri-State District (Missouri-Kansas-Oklahoma). *Trans.* A.I.M.E. (1932) **102,** 226.

27

Reprinted from pages 1112-1113, 1114, 1116-1118, 1120, 1128-1129, and 1131-1136 of *Econ. Geology* **73:**1112-1136 (1978)

Structure of the Broken Hill Mine Area and its Significance for the Genesis of the Orebodies

W. P. Laing, R. W. Marjoribanks, and R. W. R. Rutland

Abstract

A comprehensive reinvestigation of the Broken Hill mine area shows that the structural evidence is consistent with, and reinforces, interpretation of the orebodies as stratiform premetamorphic deposits.

The fold structure is interpreted in terms of three deformation episodes (F_1, F_2, F_3). F_1 and F_2 were accompanied by high-grade metamorphism, whereas F_3 is associated with retrograde metamorphism and with the initiation of retrograde shear zones.

The principal macroscopic folds of the mine area are shown to be of F_2 age, and the earlier concept of a Broken Hill anticline is reinstated as the Broken Hill antiform of F_2 age. Folds which were thought to disprove the existence of the Broken Hill anticline are now seen as F_3 folds superposed on the F_2 antiform. The orebodies lie in the northwest limb of the antiform, the hinge of which is largely replaced by a synmetamorphic slide, previously recognized as the belt of attenuation.

Evidence from metamorphosed graded bedding confirms the existence of the antiform and also shows that the macroscopic F_2 folds are superimposed on a previously inverted stratigraphic succession. This inversion is the consequence of F_1 deformation.

The three-fold episodes were roughly coaxial, with southwest plunge, over much of the area but have different axial plane orientations. In the North mine area the plunge of the orebody and of the F_2 fold axes is northeasterly, but sillimanite lineations continue to plunge southwest. A new analysis of this region is given and, contrary to recent views, it is shown that the lode horizon, which contains the ore lenses, is conformable with the stratigraphic succession.

These conclusions lead to the establishment of a new stratigraphic succession, in which the orebodies lie close to the interface between a lower, partly volcanogenic sequence and a higher, nonvolcanogenic sequence. The lead lodes lie stratigraphically above the zinc lodes.

It is inferred, therefore, that the ore constituents were originally deposited in a zone of uncertain but probably linear shape, within a limited stratigraphic interval toward the end of a phase of volcanism represented by the Potosi gneisses. There have, however, been substantial mechanical migrations of ore, within the lode horizon, during the deformation history and it is this deformation which controls the details of orebody thickness and shape.

Introduction

During the past twenty-five years there has been an increasing body of opinion in favor of the view that the Broken Hill orebody is stratiform and essentially syngenetic. An exhalative sedimentary process (White, 1968) is usually envisaged, but the relative importance of volcanic and nonvolcanic processes in the generation of the exhalative fluids is debatable (for example, Stanton, 1972; Both, 1973).

Various aspects of the structural relationships have been discussed by many authors and the various hypotheses have been reviewed by Both and Rutland (1976). Most authors have favored structural control of the emplacement of the ore (for example, Andrews, 1922; Kenny, 1932; Gustafson et al., 1950; Lewis et al., 1965; Hobbs, 1966), but King and O'Driscoll (1953, p. 600) first suggested that the "ore deposition preceded rather than followed the development of the structure which the deposit now occupies." This view has been subsequently expanded by Carruthers and Pratten (1961) and Johnson and Klingner (1976), following investigations in the southern mines. The conclusions of Hobbs (1966) and Hobbs et al. (1968) were quite incompatible with this view in inferring not only that the orebody in North mine was probably originally unconformable but also that the folds in the orebody in North mine were unrelated to those in the southern mines.

Studies during the past few years, by the authors and a number of students, go far toward resolving the apparent contradictions between the geochemical and structural evidence relating to the genesis of the orebody. In this paper the basis of a new structural interpretation is presented and the implications for ore genesis are discussed.

During our remapping of the mine area and its environs we have concentrated on the previously neglected metasedimentary rocks and, in particular, on a layered sillimanite gneiss because graded bedding (for facing data), mesoscopic folds, and penetrative foliations and lineations are developed in unusual abundance and clarity in this rock type. Valuable structural data from this rock type have also been gathered for the first time from drill core, using methods described by Laing (1977a). Some earlier studies (Hobbs, 1966; Ransom, 1968; Hodgson, 1974) also investigated the mesoscopic structures in some detail, but they were restricted to relatively small areas. They provided the initial stimulus for the present more extensive work.

The new mesoscopic data have been combined with the previously established information on the shape and distribution of the principal rock units which provides the essential framework for the structural analysis (Gustafson et al., 1950; King and O'Driscoll, 1953; Carruthers and Pratten, 1961; Lewis et al., 1965; Johnson and Klingner, 1976). The structural geometry determined on this basis is closer to that of Gustafson et al. (1950) than to that of subsequent workers. Our interpretation differs from that of Gustafson et al. (1950) in two major respects: we interpret the geometry in terms of the superposition of three principal deformation episodes, the last of which is closely associated with the development of the discrete retrograde schist zones; and we demonstrate that the principal (F_2) macroscopic folds of the mine area are downward facing [1] (Cummins and Shackleton, 1955), i.e., the stratigraphic succession was inverted before the second deformation.

Three major metamorphic and deformational episodes, each resulting in macroscopic folds, can be defined in the mines area. These episodes and the penetrative structures associated with them are labeled F_1, F_2, and F_3. F_1 and F_2 were accompanied by a growth of high-grade metamorphic minerals of sillimanite grade: F_3 was accompanied by a retrograde metamorphic event leading to the growth of biotite, muscovite, and chlorite. It can be shown that the entire mines area lies on the single inverted limb of a regional isoclinal F_1 fold nappe. The hinge of the nappe lies outside the mines area.

Since F_1 folds are isoclinal and recumbent, the attitude of rock units, bedding surfaces, and S_1 schistosity is principally a reflection of the second and third deformational events. Three major F_2 folds are present in the mines area and its immediate vicinity. From northwest to southeast they are the Hanging Wall synform, the Broken Hill antiform, and the Broken Hill synform (Fig. 1). The folds have axial planes dipping at steep to moderate angles to the northwest and, with the exception of a domain between the British shear and Round Hill, plunge to the southwest. In the latter area, F_2 folds plunge to the northeast, the reversal of plunge from the regional plunge direction taking place through the horizontal. The so-called mine sequence occupies the core of the Broken Hill antiform (Fig. 2A), the recognition of which therefore necessitates a complete reinterpretation of the structural and stratigraphic relations of the orebody.

In the following sections the new structural evidence is first outlined before the reinterpretation of orebody relationships is made.

[1] The terms facing and younging have both been widely used to describe the direction of the stratigraphic top of a succession. In such usage the geographic direction of facing or younging changes from limb to limb of folds affecting the succession. Cummins and Shackleton (1955) recognized the usefulness of the additional concept of the facing of folds: "a fold faces in the direction, normal to its axis and in the axial plane, in which the younger beds are met. Beds intersected by a cleavage or schistosity are said to *face* along such a structure, normal to the intersection, towards the younger beds." Thus a set of upright or inclined folds produced during a single deformation episode will be upward facing and the beds will face upward on any axial plane cleavage or schistosity. Folds will be downward facing, if the beds have been inverted during a previous deformation episode (or in certain cases of ductile folding where the axial surfaces formed during a single deformation episode are strongly curved).

To avoid confusion with the geographic usage of facing (or younging), Borradaile (1976) has suggested that facing on axial-plane cleavage or schistosity should be distinguished as structural facing.

[*Editor's Note:* Material has been omitted at this point. Figs. 2 and 6 through 10 have also been omitted.]

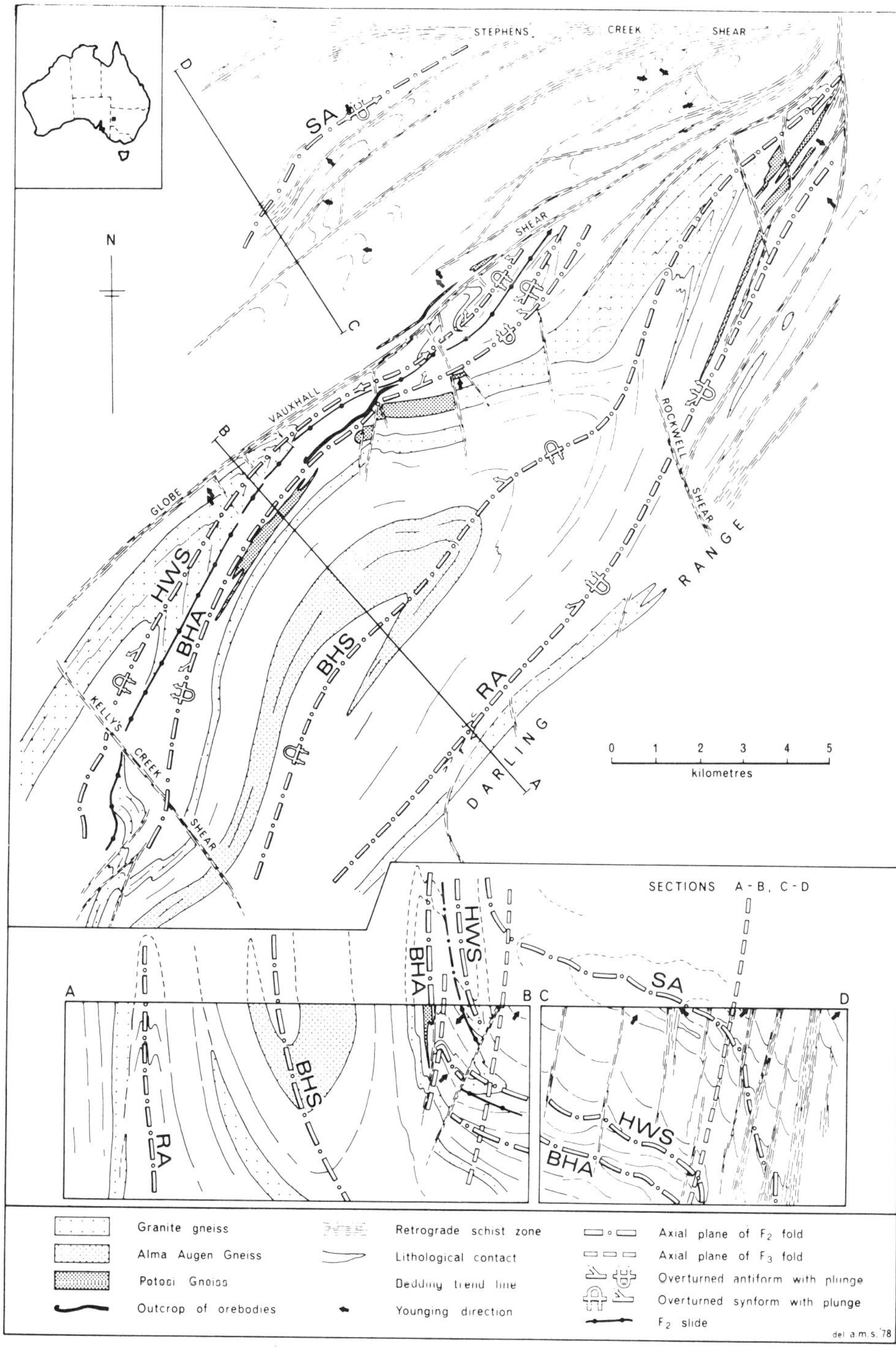

FIG. 1. Geological map and section of the Broken Hill region. For other symbols see Figure 3A.

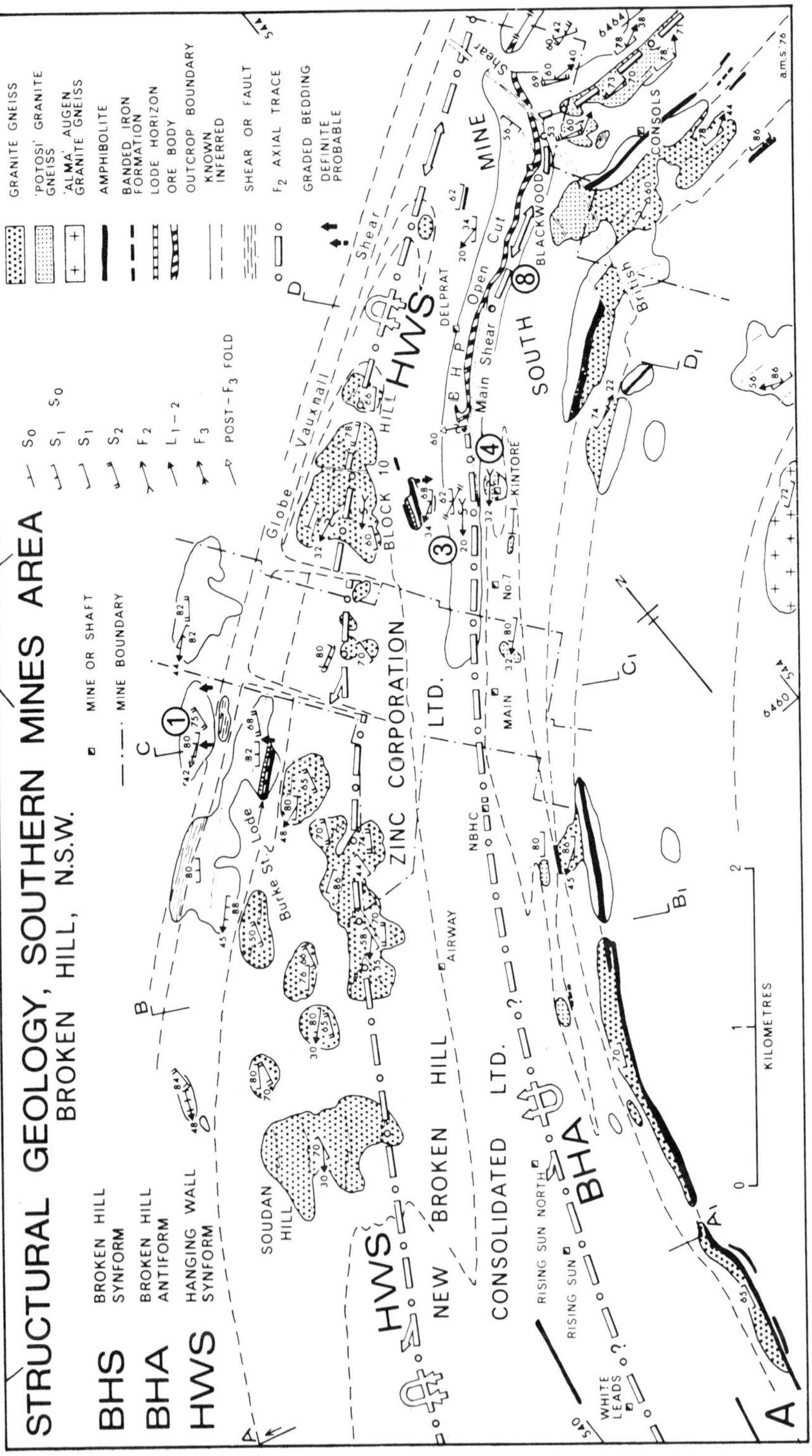

FIG. 3. A. Geological map of the southern mines area. Structural data simplified from detailed mapping. Circled numbers are localities referred to in text.

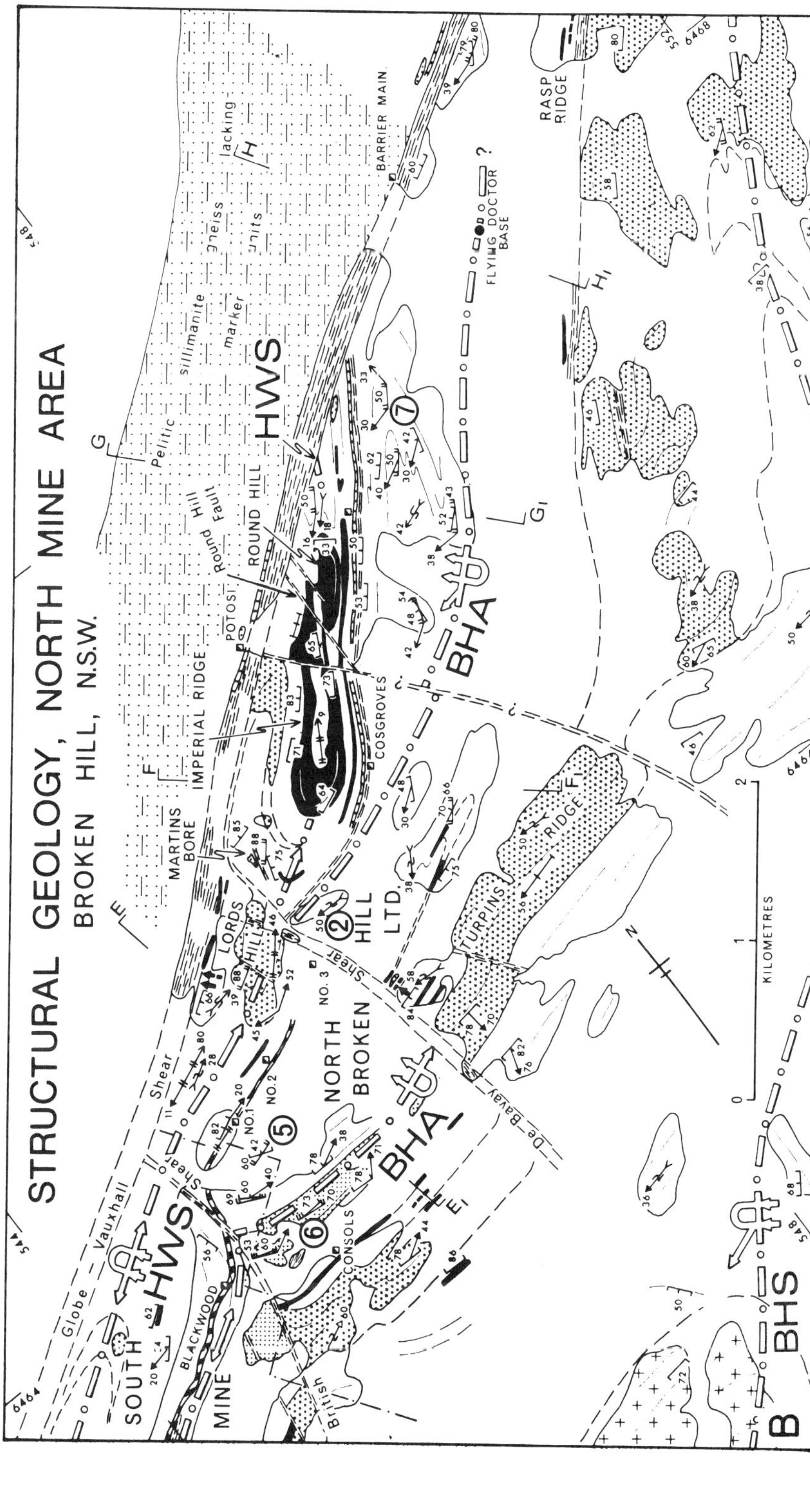

FIG. 3. B. Geological map of the North mine and the northern leases. Structural data simplified from detailed mapping. For legend see Figure 3A.

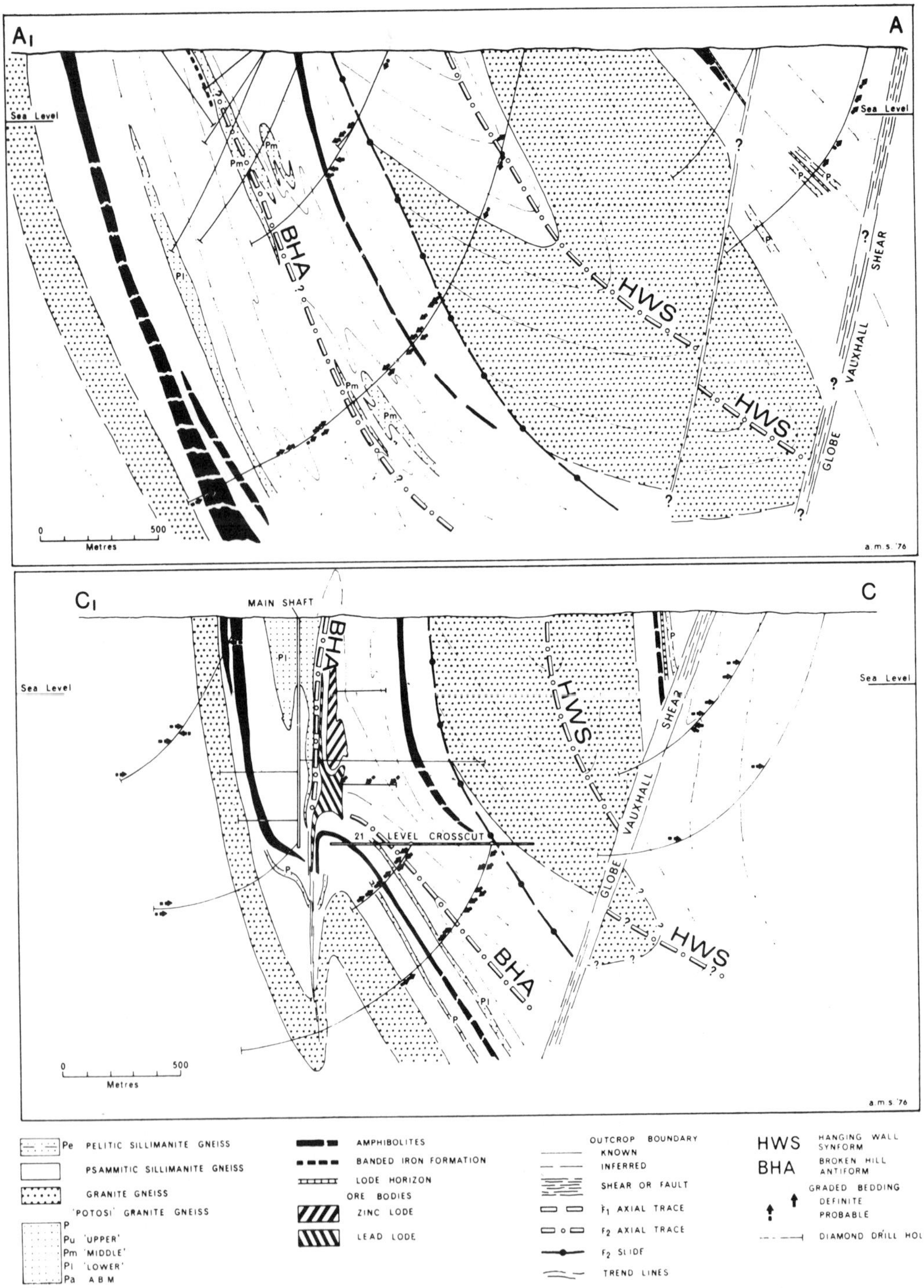

FIG. 4. Geological sections through the southern mines area. Lithological data from CRA Exploration Pty. Ltd; facing data and structural interpretation from this study. For location of sections see Figure 3A.

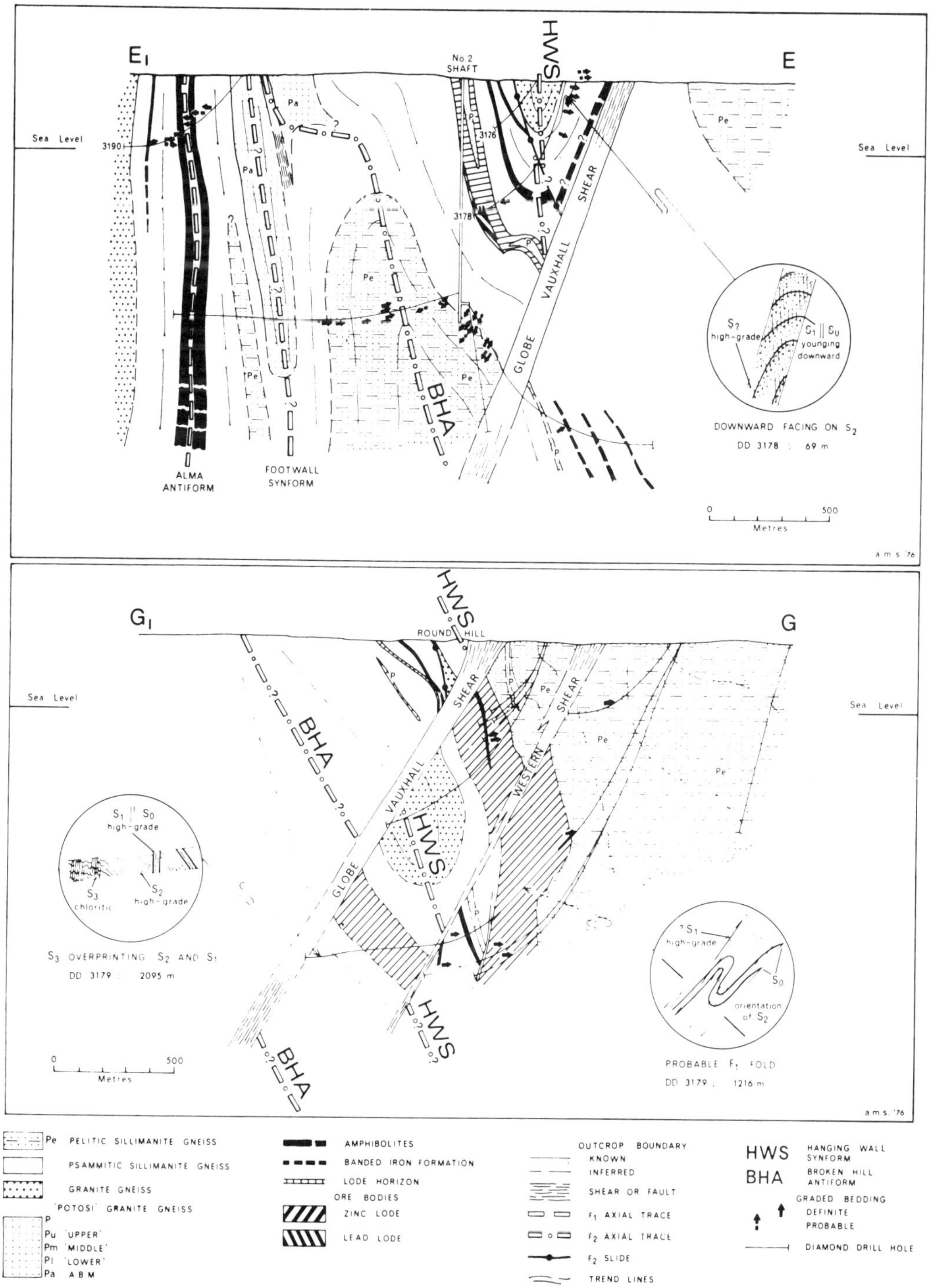

FIG. 5. Geological sections through the North mine and the northern leases. Lithological and structural data from this study, with insets demonstrating the kinds of structural interpretation employed. For location of sections see Figure 3B.

Structural and Stratigraphic Relationships of the Orebody

The macroscopic structure discussed above provides a framework for examination of the stratigraphic and structural relationships of the planar lode horizon and of the elongate orebody that lies within it.

Over much of the mine area the lode horizon is apparently conformable with the major rock units and with sedimentary bedding. It also shows a close association with garnet quartzite, banded iron-formation, and Potosi gneiss. However, in the North mine area it has been suggested that the orebody was unconformable before the first recognizable deformation, and quite probably always unconformable (Hobbs, 1966; Hobbs et al., 1968). Apparent conformability elsewhere is then ascribed to structural transposition (Hobbs, 1966; Ransom, 1968). Contrary to this view, our work indicates that the lode horizon represents a stratigraphically conformable unit. We therefore attempt to establish this for the North mine area, also, before proceeding to further examination of the structural relations (cf. Both and Rutland, 1976, p. 288–290).

Stratigraphic conformability of the orebody

Hobbs (1966) recognized only one phase of development of high-grade schistosity which was axial plane to his group 1 folds. He considered that the northeast plunge of the orebody reflected group 1 fold axes and that group 1 folds in the wall rocks had a quite different southwest plunge. He therefore concluded that the orebody and the wall-rock layering were not parallel prior to the formation of group 1 folds (Fig. 8B) and that the orebody was introduced prior to group 1 folding as a discordant sheet. He qualified this conclusion by recognizing the possibility that if an earlier unrecognized period of deformation had transposed the wall-rock layering but not the orebody into a new orientation, then the orebody could have been originally conformable.

There is now undisputed evidence of a strong deformation (F_1), mainly in the form of a widespread layer-parallel schistosity, which predated the first recognizable deformation of Hobbs (now F_2). However, the recognition of metamorphosed graded bedding in the metasediments has allowed the layering to be definitely identified as deformed bedding which shows consistent facing over large areas. The apparent conformability of the orebody through most of the field is therefore real and not due to transposition.

Hobbs' interpretation of the fold geometry apparently rested on the assumption that the sillimanite lineation which he was able to measure within the axial-plane schistosity also defined the axes of group 1 folds. Our new mapping, especially in the area northeast of the North mine, has shown that this assumption is invalid. Mesoscopic F_2 folds in the layering show very variable plunges within the S_2 plane, although the sillimanite lineation maintains a constant, gentle southwest plunge within the S_2 schistosity (Fig. 8A). Steep northeast fold plunges like those in the orebody are not uncommon (cf. Rutland and Etheridge, 1975).

In the North mine area also, our evidence suggests that both macroscopic and mesoscopic F_2 folds plunge northeast, in spite of the presence of southwest-plunging mineral lineations. The hinge of the Hanging Wall synform in the upper granite gneiss of Lords Hill is well exposed and clearly plunges to the northeast as shown by the mesoscopic folds in layering (Fig. 7, subarea 3). These folds have an axial-plane quartzofeldspathic segregation banding (typical of S_2 in the granite gneiss) which is vertical or steeply northwest dipping. The schistosity in the limited surface exposure in the North mine can also be identified as S_2 rather than S_3, because it has a similar orientation and because it has a typical S_2 high-grade assemblage and distinctive crenulation microfabric.

The North mine orebody occupies, and outlines, a tight asymmetric fold whose plunge and dextral sense of asymmetry are consistent with the adjacent F_2 Hanging Wall synform at Lords Hill. Although there is very little information on whether a layer-parallel schistosity is folded around the hinges of the orebody folds, Hobbs (1966, p. 332), Hobbs and

Vernon (1968, p. 1425), and Hobbs et al. (1968, p. 306) stated that the orebody folds have the same high-grade axial-plane schistosity as the small folds in the wall rocks. It seems clear, therefore, that this schistosity is S_2 and that the orebody folds are F_2 in age.

Thus we conclude that F_2 folds both in the orebody and the host rocks plunge to the northeast and that there are, therefore, no grounds for supposing that the orebody and lode horizon are discordant to the layering in the host rocks. Since the layering has been shown not to be due to transposition, the conclusion can be reasserted that the lode horizon is stratigraphically conformable throughout the field.

The relationships of the F_2 folds to the mineral lineations are, however, of considerable interest for the discussion of the structural control of the orebody and are therefore now briefly described.

Throughout the mines area there is one dominant high-grade mineral lineation defined principally by sillimanite, which plunges moderately to the southwest (Fig. 7). Generally it is parallel to mesoscopic F_2 fold axes but lies within S_1 and is thus labeled L_{1-2}. However, from its occurrence in F_2 folds without an axial-plane schistosity it is believed to have resulted largely from the earlier (F_1) deformation.

Within the domain extending from the British shear to the Imperial Ridge, the dominant high-grade southwest-plunging mineral lineation is oblique to F_2 fold axes. This lineation, however, lies within the axial-plane schistosity S_2 and is therefore labeled L_2. In some critical outcrops (for example, Fig. 10B) the southwest-plunging L_2 is associated with a northeast-plunging high-grade lineation, parallel to F_2 fold axes and lying in S_1 and the layering. The latter is thus inferred to be L_1. Rare F_2 folds contain a high-grade lineation folded around their axes; this lineation is also L_1. Some exposures do not possess an axial-plane S_2 schistosity and contain only a single, northeast-plunging lineation, parallel to the F_2 fold axes. By analogy with the above cases L_2 appears to be absent and the lineation is inferred to be L_1.

To summarize, L_2 has a constant plunge to the southwest. L_1 is generally parallel to L_2, but L_1 plunges to the northeast in a domain of complex lineations between the British shear and the Imperial Ridge. In this domain F_2 folds in the layering are generally coaxial with L_1. Evidence at both extremities of this domain shows that L_1 reverts to the regional southwest plunge through the horizontal. The geometry of the lineations and fold axes in the mines area is depicted schematically in Figure 8A and B (2).

In explanation of these relationships it may be supposed that L_1 and L_2 represent the orientation of the principal elongation strain axes (X) during F_1 and F_2 and that in this complex northern area, the principal strain axes were not colinear from F_1 to F_2. The anisotropy due to the preexisting lineation in the layering was sufficiently strong to control the orientation of the F_2 axes. Where the F_2 fold axes are not parallel to L_2, they presumably represent a direction of strain intermediate between Y (the intermediate axis of the strain ellipsoid) and X.

The relationships are discussed more fully elsewhere (Laing, 1977b), but for the present purpose the important point is that the relationships between northeast and southwest plunges in the North mine area are due to unusual strain effects and do not prejudice the conclusion that the orebody is conformable.

Stratigraphic position of the lode horizon

Thus it now seems beyond dispute that the orebody lies in a restricted stratigraphic interval known as the lode horizon.

Within the lode horizon itself a total of six ore lenses may be present, each with a characteristic mineralogy and generally separated from each other by narrow widths of country rock. Only in the southern mines area are all six lenses developed. In order of structural superposition they are:

B lode A lode Upper No. 1 lens Lower No. 1 lens	Collectively known as the zinc lodes
No. 2 lens No. 3 lens	Collectively known as the lead lodes

The No. 2 and 3 lenses are the only lenses which occur throughout the field and are the only lenses present in significant volume in the central and northern mines. Throughout the field the No. 2 lens always occurs structurally above and to the northwest of No. 3 lens except for local complexities. The individual lenses are highly elongated bodies within the planar lode horizon. The lead lodes are the most extensive and have a strike length of approximately 7 km and a maximum width, measured around the center line of their complex cross-sectional shape, of approximately 1 km.

As has been demonstrated this structural succession of lenses occurs in the common limb of the Broken Hill antiform and Hanging Wall synform, which have been shown to be downward facing. It can therefore be inferred that the structural succession of lenses represents an inverted stratigraphic succession: the zinc lodes are stratigraphically below the lead lodes (see e.g., Figs. 4B and 5A).

The stratigraphic relationship of the lode horizon to the mine sequence as previously defined is clearly greatly altered by the recognition of the Broken Hill antiform. It is now convenient to identify three principal stratigraphic units from top to base:

(3) pelitic and psammopelitic sillimanite gneiss —in the core of the Broken Hill antiform;

(2) psammopelitic sillimanite gneiss containing units of the mine sequence—banded iron-formation, amphibolite, Potosi gneiss, lode horizon; and

(1) granite gneiss—in the core of the Hanging Wall synform.

The stratigraphic relationships within unit 2 cannot be stated with certainty and depend on a number of factors, the most important of these is the interpretation of the precise position of the axial trace of the Broken Hill antiform and of correlations across this trace. Such correlations are rendered difficult by thickness variations, lack of persistence of individual rock units, notably the Potosi gneiss, and by possible facies changes. There are also the possibilities that some amphibolites are sills (which may not be strictly conformable or continuous) rather than flows or tuffs, and that large-scale F_1 and F_2 folds may be present in the southeast limb of the Broken Hill antiform.

However, in the northwest limb of the antiform, the consistent graded-bedding evidence rules out large-scale transposition so that the stratigraphic succession in unit 2 is reasonably clear and consists of the following units separated by psammopelitic gneisses (Fig. 4B): the upper amphibolite, the zinc lodes, and the lead lodes (stratigraphically uppermost).

Within this limb only thin and impersistent bodies of Potosi gneiss are generally present. At the southern end of the field, Hawkins (1968) has shown that the middle Potosi gneiss has a gradational relationship with what is now recognized as the stratigraphic base of the zinc lodes (B lode). Johnson and Klingner (1976, p. 478) suggest that banded iron-formation horizons may occur in the along-strike stratigraphically equivalent position to both zinc lodes and lead lodes.

This evidence therefore suggests that the rocks of possible volcanic origin in unit 2 (Potosi gneiss, granite gneiss, and amphibolites) largely lie stratigraphically below the lode horizon. The principal rocks stratigraphically above the lode horizon (unit 3) are distinctly more pelitic and lack possible metavolcanic rocks. It is therefore possible to envisage a simple model of exhalative formation of the orebodies during the final stages of a phase of volcanism.

This model is apparently reinforced by the well-known spatial correlation, in longitudinal projection, between the orebody and the thickest occurrences of Potosi gneiss—the ABM and lower Potosi gneisses. The latter are essentially linear tablets with plunges corresponding to those of the orebody which, however, lies across the strike. In longitudinal projection near the surface these bodies of gneiss (cf. Fig. 2) would have essentially the same form as the orebody but would project at a somewhat higher level since thick Potosi gneiss is now absent (i.e., projects above ground surface) in the main F_2 axial culmination while the orebody originally cropped out there.

However, the stratigraphic relations of the ABM and lower Potosi gneisses to the lode horizon is problematical, since they are not developed in the northwest limb of the Broken Hill antiform where the succession outlined above has been established. In the southern mines the axial trace of the Broken Hill antiform has been placed in Figures 3A and 4, between the lower Potosi gneiss and the ore lenses, in the belt of attenuation, along which substantial displacements may have occurred. In addition, as has been pointed out above, there is probably considerable structural complexity in the southeast limb of the Broken Hill antiform. These two factors prevent any confident correlation across the antiformal axial trace. Tentatively, the relations can be interpreted in terms of a facies change across the dislocated hinge of the Broken Hill antiform. The lode horizon and associated thin middle and upper Potosi gneisses in the northwest limb may be laterally equivalent to the succession containing the thicker lower and ABM Potosi gneisses in the southeast limb.

If the thickening is primary in origin, then the orebody could be regarded as originally fringing the thickest part of the Potosi gneiss. This concept agrees well with that of Mackenzie (1968) that, on a more local scale, there may be an inverse correlation between orebody and Potosi gneiss development. However, Mackenzie's conclusions would now apply only to the middle and upper Potosi gneisses. Such arguments would support the view that sedimentary

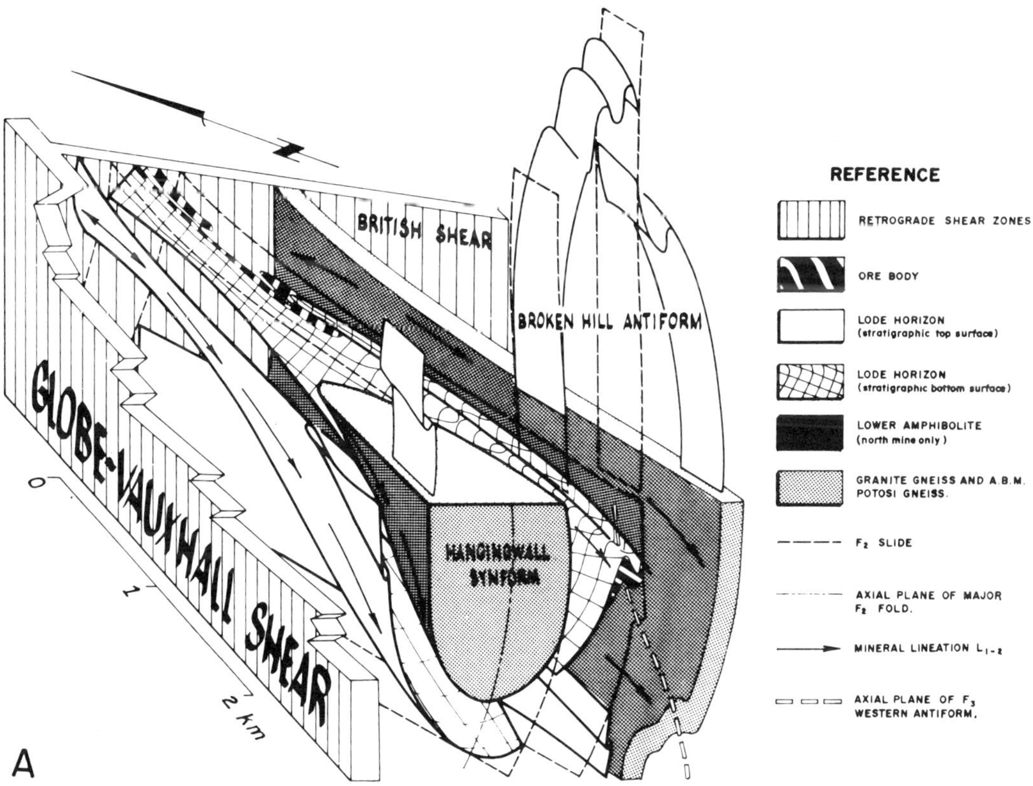

FIG. 11. A. Block diagram of the southern mines area, showing interpreted relationships between the main marker units and the main structures. For simplicity the Globe-Vauxhall shear and British shear are shown dipping vertically and the Broken Hill synform is omitted.

and volcanic facies boundaries were approximately parallel to fold trends, so that the linearity of the orebody may be based on an original sedimentary feature. It can hardly be doubted, however, that the linearity has been strongly accentuated and modified during the subsequent deformation episodes. Indeed it may be that the linearity is entirely a tectonic effect, even though the orebody is restricted to a particular stratigraphic interval and the individual lenses probably correspond to discrete stratigraphic horizons within the composite lode horizon (see e.g., Both and Rutland, 1976).

Structural controls on the orebody

The structural controls on the position of the orebody can be summarized as follows:

1. There is a close relationship of the orebody to a fold culmination of macroscopic F_2 folds. The culmination is best demonstrated by the outcrop pattern of the Hanging Wall synform. Since the No. 2 lens almost always lies northwest of the No. 3 lens in a northwest-dipping succession throughout the field, it is clear that the orebody everywhere lies in the same macroscopic fold limb—the southeast limb of the Hanging Wall synform. Moreover, the plunge variation exhibited by the orebody appears to be similar to that of the Hanging Wall synform. This is especially notable in the anomalous region between the British and De Bavay shears where the present work shows that F_2 fold axes diverge from L_2 lineations. The orebody follows the F_2 fold axes (and the L_1 lineation) rather than the inferred principal F_2 strain direction represented by the L_2 lineation.

2. Within the southeast limb of the Hanging Wall synform (as shown by graded bedding) the orebodies are related to a system of drag folds which are here interpreted as F_2 folds in the northeast and as F_3 folds, coaxial with F_2, in the southwest. The superposed ore lenses are stacked between parallel axial-plane traces of these drag folds and follow their plunge which is everywhere parallel to the dominant

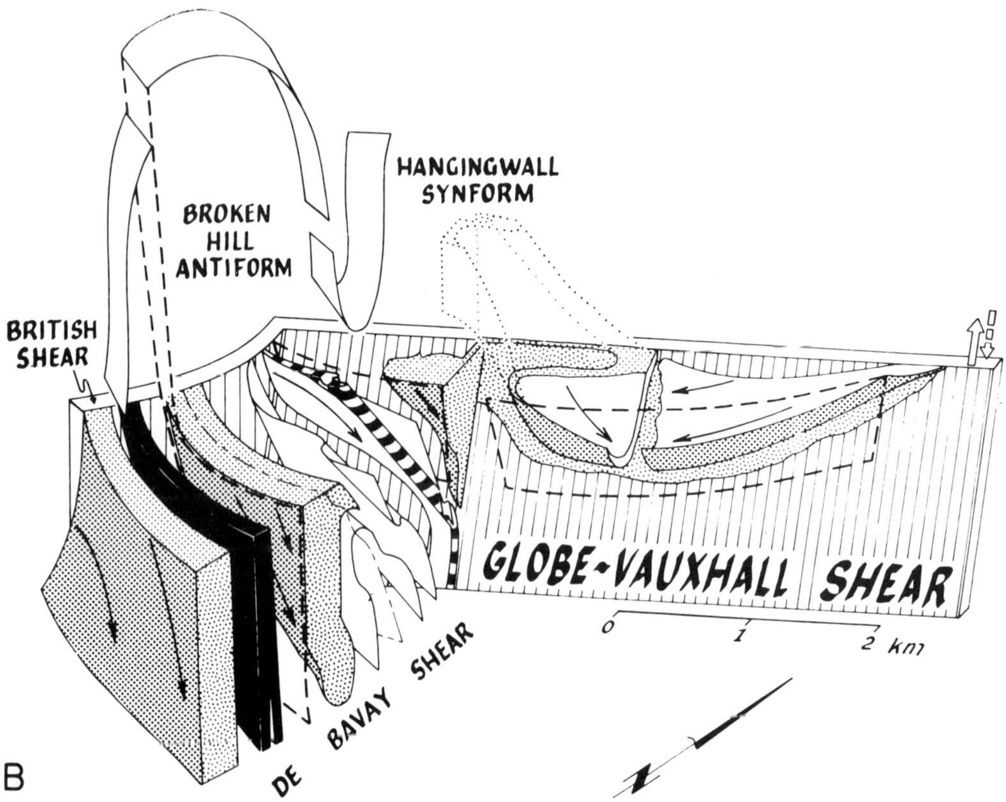

FIG. 11. B. Block diagram of the North mine area, showing interpreted relationships between the main marker units and the main structures. For simplicity the pelitic sillimanite gneiss in the core of the Broken Hill antiform and the Broken Hill synform are omitted.

high-grade mineral lineation within the enclosing rocks. There is a tendency for the ore lenses to be substantially thickened in the fold hinges.

3. One of the more notable features of the structural control of the orebody is the en echelon arrangement of the No. 2 and No. 3 lenses (the lead lodes) and the zinc lodes at the southern end of the field. Early mining development of the orebody was in the No. 3 lens and subsequent development has followed the general F_3 plunge of 20° to 25° to the southwest. However, the reduction of the No. 3 lens and the incoming of the No. 2 lens and of the zinc lodes all tend to be controlled by local plunges in the Western antiform and Eastern synform of about 40° (see O'Driscoll, 1953). The termination of lead lodes in the eastern limb of the Eastern synform follows a similar plunge of about 40°. Mackenzie (1968) remarked that the steep plunge of the overwall of the Western antiform in the lead lodes of the New Broken Hill Consolidated mine is related to the sedimentary pinching out of a body of garnet quartzite within the lode. Johnson and Klingner (1976, p. 477) also noted that the thickly developed portions of both their lower and middle Potosi gneisses plunge southwestward at about 40°. This direction may well correspond to a principal strain axis during F_1 and/or F_2

4. A belt of attenuation is generally present on the southeast side of the orebodies and is attributed to both F_2, superposed F_3, and retrograde schist zone effects which together produced a dislocation in or near the hinge of the Broken Hill antiform at the southwest end of the field. The discontinuous main shear can be correlated with the F_3 event and with development of the retrograde schist zones. The ore shows its greatest vertical extent within the main shear and the ore shoots generally have nearly vertical axes. The late deformation has also controlled the development of steeply pitching rhodonitic zones in the North mine orebodies.

5. The orebody is related to a distinct strike change, which in turn is closely related to the prin-

cipal retrograde schist zones. Thus the plunge culmination in the orebody occurs where the lode horizon becomes parallel to the Globe-Vauxhall shear zone. The northeast-plunging segment of the orebody occurs where the strike has the largest easterly component, between the British and De Bavay shears.

The concept that the linearity of the orebody may be based upon a primary stratigraphic feature is not necessarily incompatible with the fact that the orebody now has a doubly plunging form, approximately within the axial plane of the structures formed subsequent to the major stratigraphic inversion of F_1. An important conclusion of the present work has been that L_1 lineations are essentially coaxial with F_2 fold plunges and with the orebody, not only in the southern mines but also in North mine. The L_2 lineation on the other hand is not parallel to these structures in the anomalous region around North mine, but maintains the more normal southwest plunge.

It follows, therefore, that the anomalous northeast plunge of the orebody in North mine is largely unrelated to strain associated with F_3 or retrograde schist deformations, since L_2 has not been rotated. Rather, the plunge variation apparently reflects an inhomogeneous strain of L_1 during the F_2 deformation. The margins of the zone in which the anomalous plunge developed became the locus of the later British and De Bavay retrograde schist zones.

Thus it is clear that the present linearity of the orebody has existed at least since the F_2 deformation. Some migration of the ore undoubtedly occurred during this phase to produce the observed thickening in F_2 fold hinges. It is possible that a much larger migration of ore took place during this phase so that the linearity of the orebody is essentially an F_2 feature. However, since the ore follows the F_2 fold axes and the L_1 lineation even where these diverge from the L_2 lineation, it seems more likely that the linearity of the orebody was already present during F_1. If this is accepted, it is also possible, but by no means necessary, to suppose that the linearity is based on a primary stratigraphic feature.

Conclusions

The first important result of this work is the reestablishment of an antiformal structure between the Broken Hill and Hanging Wall synforms in broadly the same way as was envisaged by Andrews (1922) and by Gustafson et al. (1950). This has been possible because of the recognition and separation of phases of superposed folding: the folds which were thought to disprove the existence of the Broken Hill antiform (for example Lewis et al., 1965) are now seen as superposed on it.

Even more important, however, is the recognition, for the first time, that the major F_2 folds are downward facing. This leads to a completely new reconstruction of the environment of deposition of the orebody in keeping with the new stratigraphy. The orebody formed toward the end of a period of volcanism and was succeeded by the deposition of a thick unit of nonvolcanogenic sediments.

The evidence that F_1, F_2, and F_3 are generally coaxial largely removes contradictions between the geochemical and structural evidence relating to the origin of the orebody. Furthermore the new interpretation of the anomalous area between the British and de Bavay shear zones reestablishes the stratigraphic conformability of the orebody and the enclosing lode horizon. The way is thus now open for a combined stratigraphic and structural approach to exploration. The general relationships between the major structure and the orebody are shown in the block diagrams of Figure 11.

The stages of evolution of the orebody can be summarized as follows:

1. Deposition of metal sulfide in a zone of uncertain, but probably linear, shape within a stratigraphic interval toward the end of a phase of volcanism.
2. Strong accentuation of any original linearity of the mineralization in a direction parallel to L_1, during F_1, when the stratigraphic succession was inverted.
3. F_2 folding coaxial with L_1, with major dragfolding localized by the orebody. Formation of main plunge culmination and strike variation. Migration of sulfide mineralization into hinge zones.
4. F_3 drag-folding coaxial with F_2 but with different axial-plane dip, modifying main orebody folds at southern end of field. Migration of sulfide mineralization into hinge zones with gentle or moderate plunges.
5. Development of retrograde schist zones probably associated with F_3 and of Main shear ore with steeply pitching ore shoots. Local modifications of trend and plunge of F_2 and F_3 drag folds.
6. Recrystallization of ore in retrograde schist zones.

Implicit in the postulated sequence of events is the general localization of the orebody by primary stratigraphic features. The details of orebody thickness and shape, however, are controlled by the subsequent deformation history. *The present orebody occurs on the northwest limb of the Broken Hill antiform.* The possibility of a similar body occurring in the southeast limb of the Broken Hill antiform due to either stratigraphic or structural controls deserves further consideration in spite of the structural complexity there.

Acknowledgments

We wish to acknowledge many discussions with and whole-hearted cooperation from mines personnel, particularly Ralph Burkett, Duncan Archibald, Wolf Leyh, and Bill Widdop (North Broken Hill Ltd.); David Klingner and Geoff Scott (Conzinc Riotinto Exploration (Australia) Pty. Ltd.), and Trevor Stevenson (Minerals Mining and Metallurgy). These companies gave permission to use and publish mine data, particularly drilling sections. John Anderson and Geoff Scott collected some of the drill core-facing data in the southern mines. We also thank our colleagues in the Department of Geology at the University of Adelaide for helpful discussions, particularly Dick Glen. We are indebted to Milly Swan and Judy Laing for their painstaking drafting and to Mrs. A. Nash for typing the paper and its various revisions.

The research was funded largely by the Broken Hill Mining Managers' Association, to whom we record our thanks. One of us (W.P.L.) was supported by an Australian Postgraduate Research Award.

CENTRE FOR PRECAMBRIAN RESEARCH
MAWSON LABORATORIES
UNIVERSITY OF ADELAIDE
ADELAIDE, S. A. 5001, AUSTRALIA
March 28, April 20, 1978

REFERENCES

Andrews, E. C., 1922, The geology of the Broken Hill district: New South Wales Geol. Survey Mem. 8, 432 p.

Binns, R. A., 1964, Zones of progressive regional metamorphism in the Willyama Complex, Broken Hill district, New South Wales: Geol. Soc. Australia Jour., v. 11, p. 283–330.

Boots, M. K., 1972, The textural, mineralogical and chemical effects of retrograde metamorphism on the main lode horizon, Broken Hill, N.S.W.: Unpub. Ph.D. thesis, Univ. New South Wales.

Borradaile, G. J., 1972, Variably oriented co-planar primary folds: Geol. Mag., v. 109, p. 89–98.

—— 1976, "Structural facing" (Shackleton's rule) and the Palaeozoic rocks of the Malaguide Complex, near Velez Rubio, SE Spain: Nederlandse Akad. van Wetensch. Proc., Ser. B, v. 79, p. 330–336.

Both, R. A., 1973, Minor element geochemistry of sulphide minerals in the Broken Hill lode (N.S.W.) in relation to the origin of the ore: Mineralium Deposita, v. 8, p. 349–369.

Both, R. A., and Rutland, R. W. R., 1976, The problem of identifying and interpreting stratiform ore bodies in high metamorphosed terrains: the Broken Hill example, *in* Wolf, K. H., ed., Handbook of strata-bound and stratiform ore deposits, Vol. 4: New York, Elsevier Sci. Pub. Co., p. 260–325.

Carruthers, D. S., and Pratten, R. D., 1961, The stratigraphic succession and structure in the Zinc Corporation Ltd., and New Broken Hill Consolidated Ltd., Broken Hill, New South Wales: ECON. GEOL., v. 56, p. 1088–1102.

Cummins, W. A., and Shackleton, R. M., 1955, The Ben Lui recumbent syncline (S.W. Highlands): Geol. Mag., v. 92, p. 353–363.

French, D. G., 1976, Structural, petrological and geochemical aspects of the Potosi Gneiss and lower amphibolite at Broken Hill, New South Wales: Unpub. B.Sc. thesis, Univ. Adelaide, 30 p.

Garrety, M. D., 1957, Another lode at Broken Hill . . .?: Chem. Eng. Mining Rev., v. 50, p. 36–41.

Glen, R. A., and Laing, W. P., 1975, The significance of sedimentary structures in the Willyama Complex, New South Wales: Australasian Inst. Mining Metallurgy Proc., no. 256, p. 15–20.

Gustafson, J. K., Burrell, H. C., and Garrety, M. D., 1950, Geology of the Broken Hill ore deposit, Broken Hill, N.S.W., Australia: Geol. Soc. America Bull., v. 61, p. 1369–1438.

Hawkins, B. W., 1978, The zinc lode orebodies of the Broken Hill lode, *in* Radmanovich, M., and Woodcock, J. T., eds., Broken Hill Mines–1968: Australasian Inst. Mining Metallurgy Mon. 3, p. 155–159.

Henderson, Q. J., 1953, North Broken Hill mine: Empire Mining Metall. Cong., 5th, Australia and New Zealand 1953, v. 1, p. 627–649.

—— 1956, Geological exploration at North Broken Hill Ltd.: Australasian Inst. Mining Metallurgy Proc., no. 180, p. 1–37.

Hobbs, B. E., 1966, The structural environment of the northern part of the Broken Hill orebody: Geol. Soc. Australia Jour., v. 13, p. 315–338.

Hobbs, B. E., and Vernon, R. H., 1968, The structural and metamorphic history of the area around the northern half of the Broken Hill lode: Commonwealth Mining Metall. Cong, 8th, Melbourne 1965, Proc., v. 6, p. 1419–1428.

Hobbs, B. E., Ransom, D. M., Vernon, R. H., and Williams, P. F., 1968, The Broken Hill orebody, Australia. A review of recent work: Mineralium Deposita, v. 3, p. 293–316.

Hodgson, C. J., 1974, The geology and geological development of the Broken Hill lode in the New Broken Hill Consolidated mine, Australia. Part I: structural geology: Geol. Soc. Australia Jour., v. 21, p. 413–430.

Johnson, I. R., and Klingner, G. D., 1976, Broken Hill ore deposit and its environment, *in* Knight, C. L., ed., Economic geology of Australia and Papua New Guinea, 1. Metals: Australasian Inst. Mining Metallurgy Mon. 5, p. 476–491.

Kenny, E. J., 1932, The Broken Hill lode—its geological structure: Australasian Inst. Mining Metallurgy Proc., no. 87, p. 217–245.

King, H. F., and O'Driscoll, E. S., 1953, The Broken Hill lode: Empire Mining Metall. Cong., 5th, Australia and New Zealand 1953, v. 1, p. 533–577.

Laing, W. P., 1977a, Structural intrepretation of drill core from folded and cleaved rocks: ECON. GEOL., v. 72, p. 671–685.

Laing, W. P., 1977b, Structural and metamorphic geology of a critical area adjacent to the Broken Hill orebody, Willyama Complex, Australia: Unpub. Ph.D. Thesis, Univ. Adelaide, 232 p.

Lewis, B. R., Forward, P. S., and Roberts, J. B., 1965, Geology of Broken Hill lode, reinterpreted: Commonwealth Mining Metall. Cong., 8th, Melbourne 1965, v. 1, p. 319–38.

Mackenzie, D. H., 1968, Lead lode at New Broken Hill Consolidated Ltd, *in* Radmanovich, M., and Woodcock, J. T., eds., Broken Hill Mines—1968: Australasian Inst. Mining Metallurgy Mon. 3, p. 161–169.

Maiden, K. J., 1976, Piercement structures formed by metamorphic mobilisation of the Broken Hill orebody: Australasian Inst. Mining Metallurgy Proc., no. 257, p. 1–8.

Marjoribanks, R. W., and Laing, W. P., 1975, New observations on the structure of the Broken Hill mine area: Geol. Soc. Australia, Tectonics and Structural Geology Newsletter No. 4, p. 2–4.

O'Driscoll, E. S., 1953, The Zinc Corporation and the New Broken Hill Consolidated Mines: Empire Mining Metall. Cong., 5th, Australia and New Zealand, v. 1, p. 658–673.

—— 1968, Notes on the structure of the Broken Hill lode, and its tectonic setting, *in* Radmanovich, M., and Woodcock, J. T., eds., Broken Hill Mines—1968: Australasian Inst. Mining Metallurgy Mon. 3, p. 87–102.

Offler, R., 1972, A study of the Globe-Vauxhall retrograde schist zone: Broken Hill Mining Managers Assoc. Rept. 1 (unpub.), 30 p.

Pidgeon, R. T., 1967, A rubidium-strontium geochronological study of the Willyama Complex, Broken Hill, Australia: Jour. Petrology, v. 8, p. 283–324.

Ramsay, J. G., and Graham, R. H., 1970, Strain variation in shear belts: Canadian Jour. Earth Sci., v. 7, p. 786–813.

Ransome, D. M., 1968, The relationship of lode shape to wall-rock structure in the southern half of the Broken Hill ore-body: Geol. Soc. Australia Jour., 15, v. p. 57–64.

—— 1969, Structural and metamorphic studies at Broken Hill: Ph.D. thesis, Australian National Univ., 250 p.

Roberts, J. L., and Sanderson, D. J., 1974, Oblique fold axes in the Dalradian rocks of the Southwest highlands: Scottish Jour. Geol., v. 9, p. 281–296.

Rutland, R. W. R., and Etheridge, M. A., 1975, Two high grade schistosities at Broken Hill and their relation to major and minor structures: Geol. Soc. Australia Jour., v. 22, p. 259–274.

Shaw, S. E., 1968, Rb-Sr isotopic studies of the mine sequence rocks at Broken Hill, *in* Radmanovich, M., and Woodcock, J. T., eds., Broken Hill Mines—1968: Australasian Inst. Mining Metallurgy Mon. 3, p. 185–198.

Stanton, R. L., 1972, A preliminary account of chemical relationships between sulfide lode and "banded iron formation" at Broken Hill, New South Wales: Econ. Geol., v. 67, p. 1128–1145.

Stillwell, F. L., 1922, The rocks in the immediate neighbourhood of the Broken Hill lode and their bearing on its origin: New South Wales Geol. Survey Mem. 8, p. 354–396.

White, D. E., 1968, Environments of generation of some base-metal ore deposits: Econ. Geol., v. 63, p. 301–335.

28

Reprinted from *Econ. Geology* **21:**431–453 (1926)

THE FORMATION OF CERTAIN ORE BODIES BY MINERALIZATION STOPING.[1]

Brecciation Due to Removal of Support by the Dissolving Action of Solutions During an Early Stage of Their Activity, Followed by Mineral Deposition in the Brecciated Mass at a Later Stage.

AUGUSTUS LOCKE.

Introduction.

Heretofore, the explanations of certain mineral pipes, of which the famous Pilares is an example, have not been wholly satisfactory. Furthermore, the explanations of a related ore body type, represented by Bingham and Miami, are in a similar condition. The writer has had an opportunity to study the Pilares and several deposits resembling it and has come to a definite conclusion regarding their origin. He has studied also most of the well-known deposits of the other type and feels that their origin is similar to that of the Pilares, though the proof is less clear. Both are attributed, in brief, to removal of rock along trunk channels by rising solutions during an early stage of their activity, collapse and

[1] The writer acknowledges his indebtedness to T. J. Hoover, who permitted the study of specimens in the laboratories of the Stanford Mining School; to Charles Dobbel, H. W. Morse, Nahum Sabsay and M. N. Short, who carried on studies of specimens; to Guy N. Bjorge, who supplied information from Pilares; to H. M. Kingsbury, who supplied a photograph, drawings, and information from Bisbee; to Ira Joralemon, who supplied information from the Copper Creek district near Mammoth, Arizona; to B. S. Butler, L. C. Graton, Adolph Knopf, Waldemar Lindgren, G. F. Loughlin, F. L. Ransome, and A. C. Spencer, with whom the writer has carried on discussions of this subject; to the Phelps-Dodge Corporation, which permitted the publication of information concerning its property in Bisbee and Nacozari; to the Ibex Gold Mining Co., which permitted the publication of information concerning its property in Leadville; and to the United States Geological Survey, which took pictures of specimens and permitted their use here.

brecciation of the rock thus left unsupported, and deposition of ore and gangue minerals in the brecciated mass.

The two types of mineral body mentioned above may both be termed *disseminated mineral deposits;* for both consist of ore and gangue minerals widely scattered through large masses of brecciated rock. Regarding both, the same obvious question must be asked: "Why were the minerals disseminated, rather than aggregated or disposed in linear or in tabular bodies?" Or, to put it one step farther back, "Why was the rock, into which the mineral was to be received, broken into small, angular pieces, with a lavish expenditure of energy, rather than along a few principal planes of movement?"

Because, therefore, the mineral pipe is a special case of disseminated mineral deposit, not formless like the great copper bodies of Bingham, but regular in outline, its well-defined structural characteristics will be used in this paper as the principal evidence on the questions at issue. Answered for the one type, these questions will necessarily, in some degree, be answered for the other.

The Pilares Deposit.

The Pilares mine, Sonora, Mexico, occupies a body of rock which has the familiar fracturing mentioned above. The rock may be described as *crackled:* that is to say, it is broken into small, angular fragments which for the most part lie closely fitted together.

The surrounding surface shows flat-bedded latite tuff, 300 to 500 feet thick, which rests conformably on andesite tuff. The contact between the latite and the andesite is known in detail within the mine and can be used as a marker to indicate any subsidence which may have occurred there. No pre-ore intrusive has been positively identified and much supposed monzonite has proved to be fragmental. The crackled body, a vertical pipe cut in these rocks, has a horizontal section of regular, oval outline, measuring 1000 by 2000 feet; it is developed to the depth of 1800 feet.

Within this pipe, the rock has undergone alteration similar to that of the Bingham deposits, and its crackling is similar, though a little coarser. It carries pyrite, disseminated and on cracks, accompanied by quartz and sericite or chlorite. It carries also chalcopyrite, in separated, thin veinlets, but here and there in

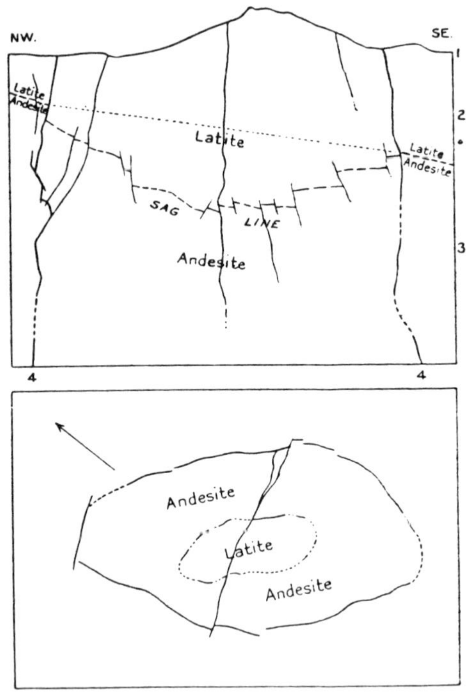

FIG. 1. Plan and section of the Pilares pipe, Sonora, Mexico. 1, surface; 2, latite; 3, andesite; 4, walls of pipe. (Adapted from Bjorge.)

veinlets sufficiently thicker and nearer together to make ore. But most of the ore lies, not in the body of the pipe, but out against its vertical walls; and not in merely crackled rock, but in jumbled breccia or rubble.

This jumbled breccia or rubble forms an envelope surrounding the pipe, some 50 feet thick along the side walls of the oval but thicker in the ends, making great, irregular, steep masses there. It resembles in pattern the broken rock filling a caved stope. Its fragments are similarly unsorted and they similarly range in size from mere slivers to pieces 10 feet in diameter. The space between the fragments, now surviving as quartz-lined vugs, or filled with quartz, pyrite, and chalcopyrite, attains a similar maximum

FIG. 2. Pilares latite ore. ×2/3. (Photo by U. S. Geol. Survey.)

of about 30 per cent. of the volume. The ore has a maximum grade which accords with the filling of 30 per cent. of voids by chalcopyrite. It is a striking fact that the mine, even near the surface where enriched by chalcocite, has yielded little ore of smelting grade. The ore, as broken, is thickly crowded with small rock fragments, like a piece of concrete masonry, and cannot be hand-sorted up to a rich product. It is clear that the mineralization and the movement of the fragments overlapped and that the fragments, though chiefly shaped by fracture, were in some degree shaped by attrition and corrosion. It is clear also that the chalcopyrite arrived, with quartz and pyrite, as the last episode of the mineralization, after the rock had been well altered and the movement had ceased; and that, though partly replacing the rock, it was deposited mainly in the open spaces previously formed by the rotation, attrition, and corrosion of the rock fragments. The statement of the proportions between

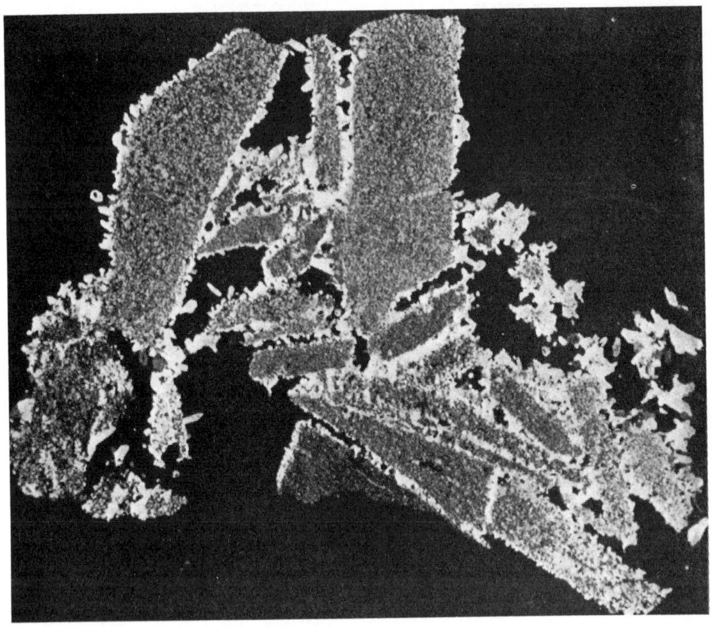

FIG. 3. Pilares latite ore. $\times 6$. The cavities are lined by quartz crystals filled by chalcopyrite. (Photo by U. S. Geol. Survey.)

"replacement" and "open space filling" must depend on the sense in which those terms are used. In the ordinary sense, the quartz, sericite, chlorite, pyrite, and minor chalcopyrite, of which the rock fragments are now made up, would be regarded as "replacements," and the interstitial quartz, sericite, pyrite, and chalcopyrite would be regarded as "open space fillings." The chalcopyrite is conspicuously localized by cracks. It may invade the walls for a short distance or wrap about a rounded-off end of a rock wedge. But the open spaces determine where it is to lie. And no reason has been found, in the very numerous specimens examined, to doubt that the fragments, previous to the advent of the chalcopyrite, were supported against each other; nor to believe that the interstitial chalcopyrite did very much more than fill in the open spaces and pry off small slabs of rock.

FIG. 4. Tourmaline and pyrite cementing fragments of granite which contain pyrite and have outlines more rounded than those shown in Fig. 2. × 3/4. Wickes, Montana. (From the collection of the Department of Geology, Univ. of California.)

The resemblance of the ore breccia to the filling of a caved stope stimulates the search for other evidences of subsidence. Such evidences are yielded clearly. The latite-andesite contact, 300–500 feet deep outside the pipe, sags inside as an inverted cone to the 900-foot level and shows over the entire oval an average subsidence of 200 feet. This subsidence was found by G. N. Bjorge to have occurred on numerous cracks of pre-ore origin. Of these the most continuous are marginal, marking the wall itself:

the individual wall cracks, carrying gouge, persist for distances as great as 500 feet without interruption around the curved, oval outline of the pipe, and when one dies out another is found in a different plane nearby. They make the boundary between the solid wall rock of the surrounding country and the subsided pipe of crackled rock within, and the jumbling of the rock fragments at this boundary may be likened to the drag of ore on the side wall of a bin.

Not only do the jumbled breccia and the cone of subsidence accord with the effects produced in the caving of a stope, but we find that the steep pipe form itself is an equally true analogy. Studies of several months duration, made on the subject of caving in Bisbee by H. M. Kingsbury, E. H. Wisser, and the writer, have shown that when an ore body collapses, whether through mining of the ore or through shrinkage in oxidation, the roof usually fails by caving and domes up until the space below it becomes filled to the maximum thickness of 100 feet with jumbled rock fragments; and that, when the caved material attains a certain critical size, steep cracks, curving in plan and enclosing a circular or oval pipe, spring upward from the margins of the dome. These cracks may ascend even several hundred feet to the surface, undiverted from their course by preëxisting structures. Except in cases of overlapping effects of several ore bodies, they make a curved outline indicating the individual area below in which lies the focus and place of shrinkage.

In short, the pipe form and the steepness of attitude, which are the main peculiarities of the Pilares as contrasted with other disseminated mineral deposits, have been produced in Bisbee by oxidation and mine subsidence. It is true that the Pilares pipe is uncomplicated by overlapping circles of subsidence and that its walls maintain their verticality and their envelope of jumbled fragments with a striking regularity. It is true, further, that the Bisbee subsidence pipes are as a rule less regular and that they come down with little crackling and sagging in the middle. But the general outlines of result in the Pilares and Bisbee cases are similar.

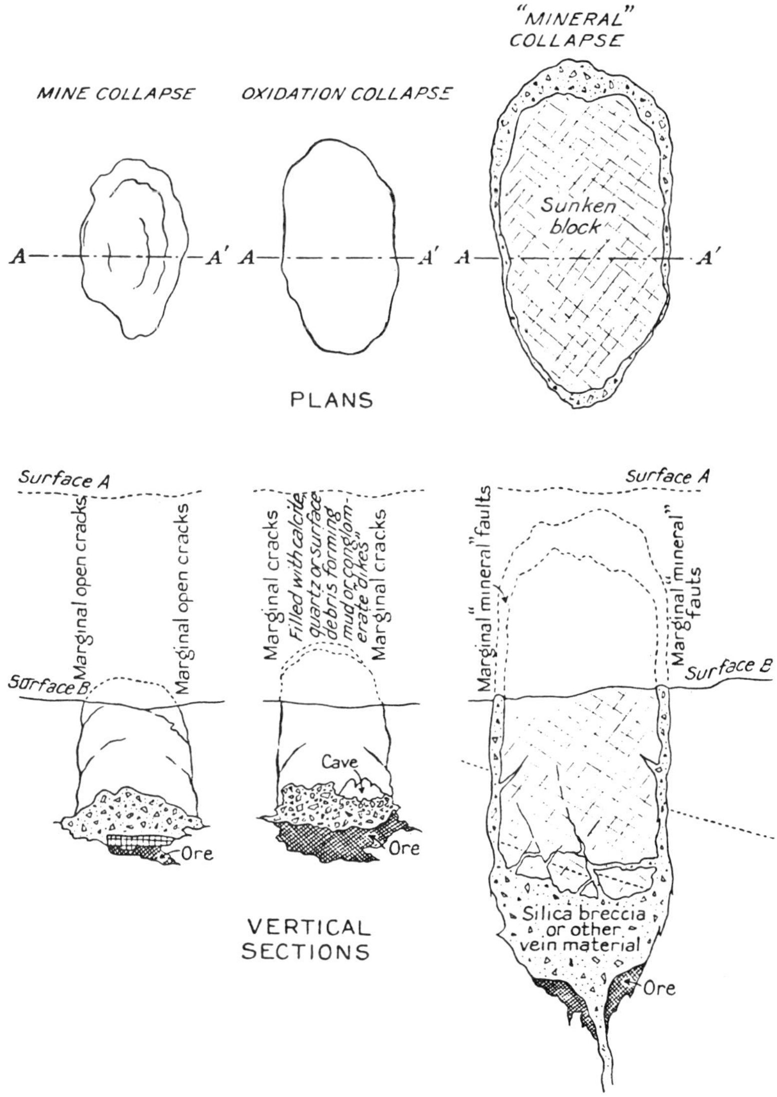

FIG. 5. Sketch plans and sections showing mine, oxidation, and hypogene, or "mineral," collapse. (Drawn by Kingsbury.)

There is indeed reason for astonishment at this similarity, for a whole series of possible results can issue from the same general process. In rock that is plastic, subsidence can make, not a pipe,

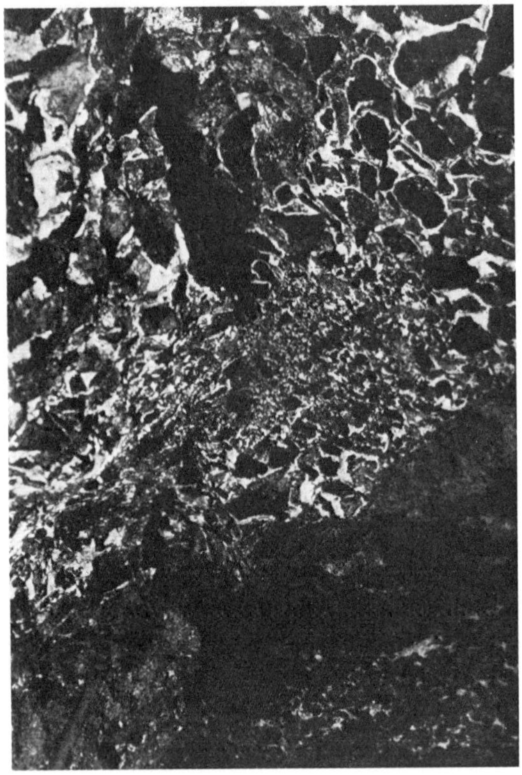

Fig. 6. Jumbled fragments of porphyry, broken by the shrinkage during oxidation of an underlying ore body in limestone, and cemented by calcite and aragonite. (Photo by Kingsbury.)

but a sag-basin without marginal faults. Or, given a pipe, the effects will vary with the speed of the subsidence; for, if rapid, it tends to cause the ground to settle in a solid block; if slow, it tends to cause numerous, successive, small adjustments by which the block is broken in many pieces.[2] Or, the effects will vary with

[2] See Eavenson, Howard, N., "Report of Subcommittee on Coal Mining to Committee on Ground Movement and Subsidence," *Trans. Am. Inst. Min. and Met. Engrs.*, February, 1926.

the size: 100 feet in diameter, the pipe tends to be jumbled throughout; 1,000 feet in diameter, it may drop as a cylinder and may be merely cracked without rotation of its pieces. In any case, the pipe may terminate upward in a dome or arch before it reaches the surface.

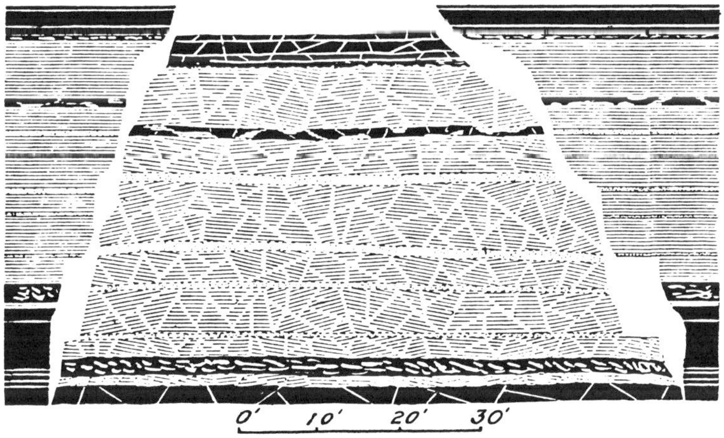

FIG. 7. True section showing subsidence of shale, sandstone and coal strata on removal of pillars in the Pittsburgh seam, Georges Creek region, Maryland. (Adapted from Eavenson.)

If the overlapping subsidence circles be sufficiently complicated, no regular marginal cracks will be established, the pipe form will fail to develop, and the result will be the irregular stockwork or disseminated mineral body which is more common than the pipe itself.

SUBSIDENCE AS A FACTOR IN THE MINERAL PIPES.

The explanations commonly offered for the Pilares pipe are:

1. It was localized on the intersection of two sets of fractures.
2. It is a plug of rock pushed up by the pressure of an intruding fluid and then drawn back like a cork when that pressure declined.

Were the Pilares the only pipe to be accounted for, one or both of these explanations might seem to apply. But there are many

TABLE I.

A Few of the Better Known Mineral Pipes.

	Rock.	Metals.	Atti-tude.	Horizontal Dimensions (Feet).	Depth of Mining (Feet).	Top.	Bottom.
Bassick, Colo.	Volcanic agglomerate	Au, Pb, Zn	85°	30 x 100	1,400	Eroded	Not found
Bull Domingo, Colo.	Schist	Ag, Pb, Zn	80°	50 x 80	550	Eroded	Not found
Catalina, Son.	Tuff	Pb, Zn	Steep	About 300 x 400	About 600	Eroded	At about 500′ rough floor
Cactus, Ut.	Monzonite	Cu	Steep	200 x 800	1,000	Eroded	Diminishes below 600′
Copper Prince, Ariz.	Granite	Cu	Steep	100 x 200	450	Eroded	At 450′ has tapered to vein
Cresson, Colo.	Fragmental volcanic	Au	70°	500 x 800	2,200	Eroded	Not found
Duluth, Son.	Tuff	Cu	80°	300 x 1,000	1,000	Eroded	Not found
Patch, Colo.	Granite gneiss	Ag, Pb, Zn Cu	Steep	400 x 800	1,600	Eroded	Not found
Pilares, Son.	Tuff	Cu	90°	1,000 x 2,000	1,800	Eroded	Not found
South Ibex, Colo.	Quartzite	Au	70°	60 x 120	900	At 450′ ends upward in dome	Not found
Southwest, Ariz.	Limestone	Cu	Steep	1,200 x 2,000	1,000	Eroded	At about 500′ rough floor following bedding.

other pipes and, for a whole group, all having the extraordinary qualities of steepness and of oval or circular outline, they utterly fail. An intersection of fractures would be vertical only here and there. And were the pipes created by pressure below, they would of necessity bend, flare and contract, now going off flat along the contact, and now turning sharply sideways or upward.

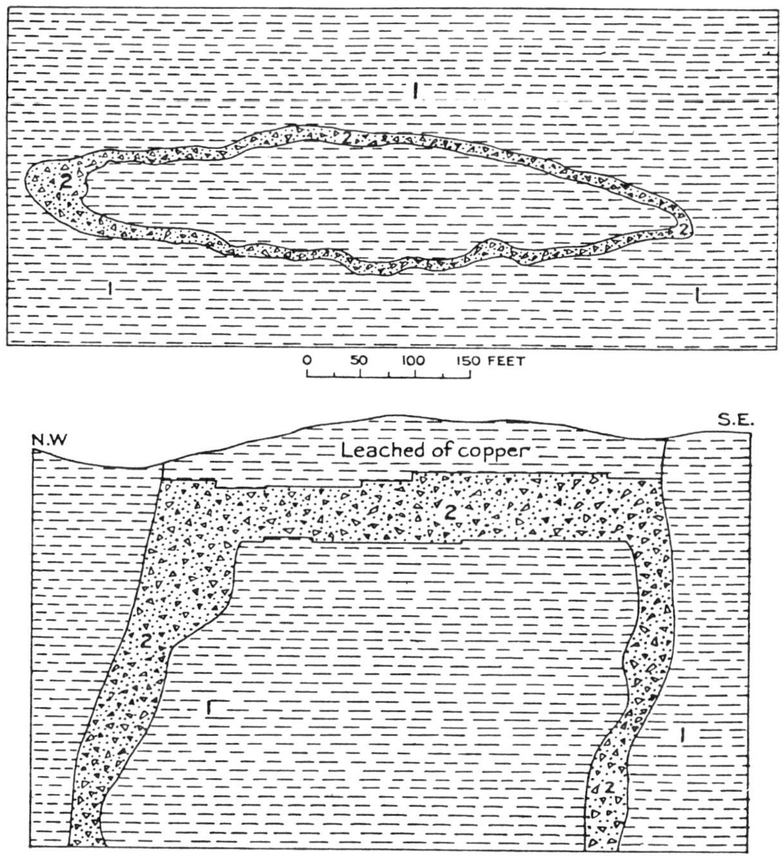

FIG. 8. Plan and vertical section of the Duluth pipe, Sonora, Mexico. 1, tuff; 2, rubble or ore in pipe.

Note: Although this pipe is commonly mentioned as an intrusive plug, H. M. Kingsbury and the writer found the rock within it to be, at least in part, bedded and fragmental.

Their regularity and verticality would not at all accord with expectation, for these qualities could result only from a peculiar succession of accidents, strung together, and many times repeated.

In explanation of the characteristics of steepness and pipe form, the writer has been compelled to seek some other principle and he has found only one that is satisfying. This is the *movement of rock under the impulse of gravity.* In accordance with this principle, the pipes are regarded as having the simple, dominant genetic feature of subsidence, and they are characterized

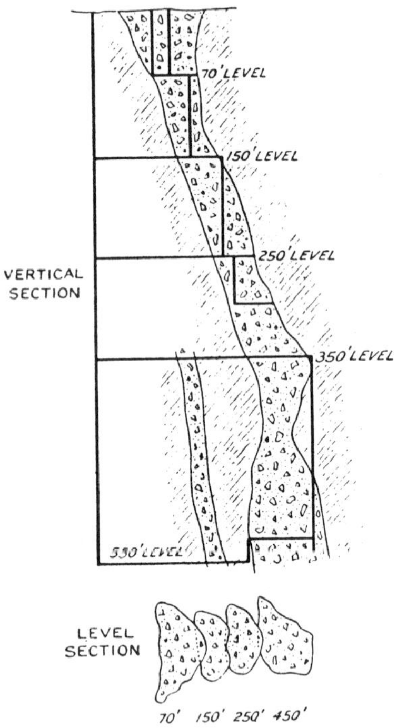

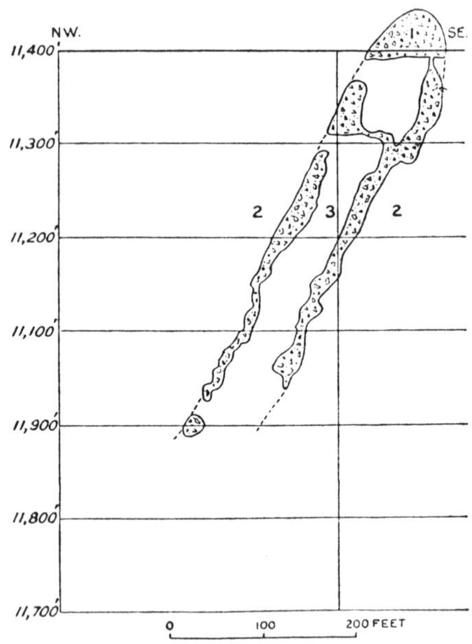

FIG. 9. (Left.) Plan and vertical section of the Bull Domingo pipe, near Lake City, Colorado.[3] The rock is schist.

FIG. 10. (Right.) Vertical section of the South Ibex pipe, Leadville, Colorado. 1, jumbled quartzite ore, cemented by gold-bearing quartz and pyrite; 2, quartzite; 3, quartzite, crackled and mineralized, but without enough gold to make ore.

3 Emmons, S. F., U. S. Geol. Survey, 17th Ann. Rept., Part II., 1897.

as subsidence effects. Thus characterized, the pipes in no case show qualities that fail to fit into the general scheme, and various qualities, found sporadically, and hitherto not understood, become logical parts of the larger picture.

For example, the sagging of the beds at Pilares, where the top of the pipe is eroded, is accordant in idea with the upward termination of the South Ibex pipe in a dome 450 feet below the surface.

DESCRIPTION BY B. S. BUTLER OF THE CACTUS ORE ZONE.[4]

The Cactus ore zone lies entirely within the monzonite. It is a rather irregular stockwork having a . . . nearly vertical dip, so far as shown by present developments, in which the rock has suffered brecciation, alteration and mineralization. This brecciated rock extends . . . a distance of about 2,700 feet. . . .

Within the ore zone (or stockwork) the monzonite has been brecciated or broken into irregular fragments, varying from a fraction of an inch to several feet in greatest dimension. Some of these fragments have been corroded by the ore solution so that they present the appearance of a boulder conglomerate with the spaces between the boulders partly or entirely filled with ore and gangue minerals. Usually, however, the fragments are angular, as in a typical breccia. In general the breccia in the upper levels is coarser than that lower down, the spaces between the individual breccia fragments are larger, and many of the spaces have not been entirely filled with later minerals. In the stopes above the sixth level the breccia was well exposed but is finer than in the upper levels and contains practically no unfilled openings; the ore minerals have also replaced the rock to a greater extent. On the seventh level brecciation is less pronounced and the ore minerals very generally replace the monzonite. The same is true of the eighth and ninth levels so far as is revealed by the small amount of development.

. . . The principal primary metallic minerals are pyrite, chalcopyrite, and hematite. . . The important gangue minerals are sericite, quartz, tourmaline, siderite, anhydrite, and barite. . . .

[4] Butler, B. S., "Geology and Ore Deposits of the San Francisco Region, Utah," U. S. Geol. Survey, Prof. Paper 80, 1913.

THE PROCESS OF SUBSIDENCE.

The easiest process of subsidence to visualize is that seen in stopes. An opening is made which subjects the rock to stresses exceeding its strength; the rock fails; angular blocks fill the opening and by their opposing pressure so reduce the stresses that the caving at last comes to an end.

Subsidence at a depth of 10,000 feet, though similar in being controlled by the free movement of rock under the impulse of gravity, would differ in detail. The elastic deformation of the

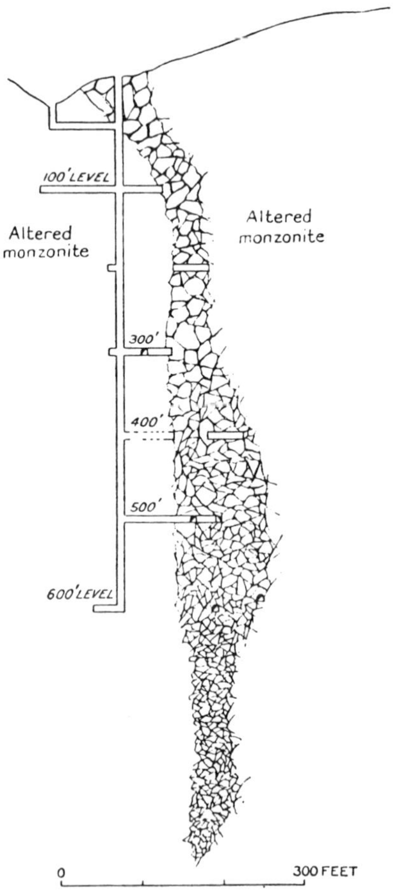

FIG. 11. Plan and vertical section of the Cactus ore zone. (Adapted from Butler.)

rock would be greater and the size of opening that could survive, smaller than that at shallower depth. Were an opening created exceeding that size, as by the solution of a block of limestone or other rock, the unsupported walls would crack and project spalls into the opening, like the walls of deep mine drifts, in which occur "rock bursts"; or, like the walls of the cavities in Bridgman's quartz cylinders, which, when the external pressure was in excess of 4,000 atmospheres, projected into the cavities angular sand grains with such velocity that they cut into and adhered to a copper wire that had been inserted there.[5]

It is of course not supposed that at Pilares, with its cover several thousand feet thick, an opening could have existed equal in area to the present pipe. But, with the supporting pressure, not removed, but merely reduced, widely distributed cracks and minor caves would result and, after a sufficient disturbance of equilibrium, the overlying rock would move downward. The reduction in pressure and the downward movement would, in fact, be simultaneous and would be attended by the successive growth and collapse of caves in large numbers. The now surviving maximum vugs are a foot in diameter and they may be regarded as stable relics and as evidence that all those openings that were larger or that for some reason had weaker walls, were destroyed. It is of interest that the maximum vug in the South Ibex pipe in Leadville is similar in size to that of the Pilares and that the depth of cover there was about 10,000 feet. It is of interest, also, that in the Cresson pipe in Cripple Creek, Colorado, where the depth of cover was only 1,000 or 1,500 feet, the maximum vug was a cave 100 feet in diameter.

In the Pilares, and in most of the others, only the middle section of the pipe is visible and the direct study of that part in which the main shrinkage occurred has not been possible. But in the Catalina in Cananea, and in the Southwest in Bisbee, within a few hundred feet of the present surface, the actual downward terminations are visible; there is a rapid convergence of the walls

[5] Bridgman, Percy, "The Failure of Cavities in Rocks and Minerals under Pressure," *Am. Jour. Sci.* (4), 45, April, 1918.

forming a rude keel below which the silicified breccia descends only in irregular protuberances. No intrusive plug appears in either case and the hypothesis that such a plug has advanced and retreated is inadmissable. The pipes merely end downward, as they do horizontally, against fresher and less fractured country rock. The curved, marginal cracks are proved by mine workings

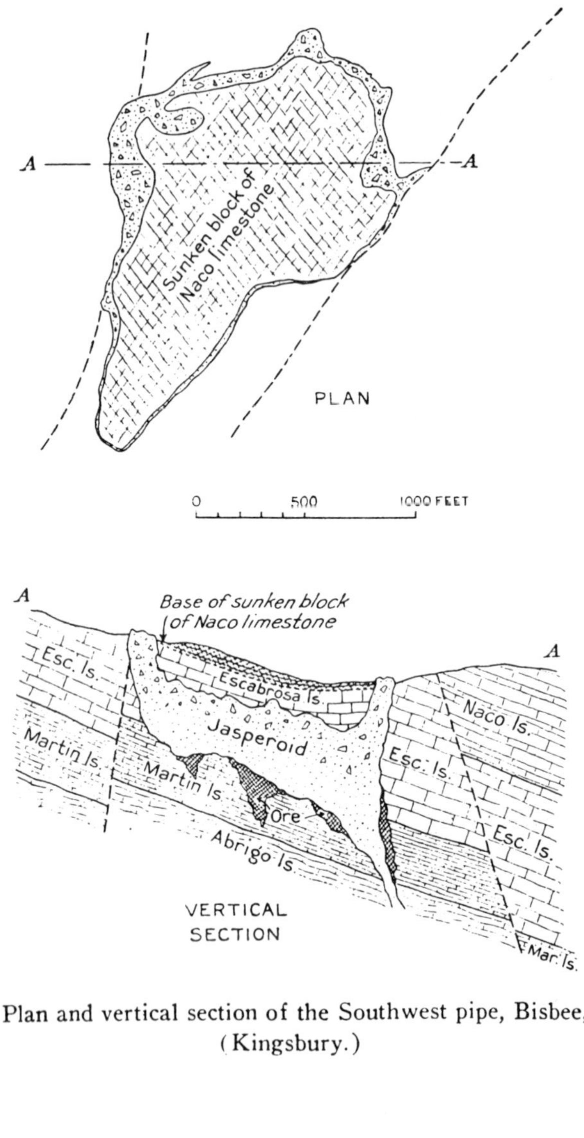

FIG. 12. Plan and vertical section of the Southwest pipe, Bisbee, Arizona. (Kingsbury.)

to terminate with the pipe. The fracturing is not a larger, pre-existing feature along which the pipe was localized, but a property of the pipe itself. The subsidence of these two cylindrical bodies was not something caused by subsidence deeper down, but something due to shrinkage inhering in the pipes and occurring in the region now visible above their rough floors. The facts here indicate the conclusion that is offered as the principal idea of this paper, namely that the mineral bodies in question grew upward through the rock along subsidence cracks: *and that these cracks were created by pressure disturbances attendant on the net removal of rock substance by the mineralization itself.*

FIG. 13. Jasperoid breccia consisting of chert fragments cemented by jasperoid. Southwest area, Bisbee, Arizona. Actual size. (Photo by U. S. Geol. Survey.)

The relations in the Southwest pipe at Bisbee are vital to this discussion; they were determined by H. M. Kingsbury, who has studied them in collaboration with the writer and who will discuss them in detail in a separate paper. Only a general view is needed here. The surface shows a pear-shaped block of flat-lying Naco limestone measuring 1,200 by 2,000 feet, or about the area of the Pilares pipe; surrounding this block are outcrops of jasperoid, a few inches to 100 feet in width, made up of jumbled, angular fragments of jasperoid and chert tightly cemented by jasperoid, and resembling in structure the Pilares ore.

These jasperoid outcrops follow steep, curved faults and these faults are shown by exploration to go down to the margin of a great body of jasperoid that underlies the Naco area. This body is irregular in thickness; where we know it best it measures 250 feet and replaces the larger part of the flat-lying Escabrosa limestone (the member underlying the Naco) which normally measures 600 feet.

These faults, which are analogous to the curved, marginal faults that go up from the outer edges of a caved stope, are contemporaneous with the replacement of the limestone by the jasperoid. To the process by which such faults are in general made the writer has given the name *mineralization-stoping,* and to the faults themselves Kingsbury has given the name *mineral faults.*

The amount of subsidence in the Catalina is not known. The subsidence in the Southwest, though subject to new estimation after further study, promises to lie between 100 and 300 feet. This was a substitution which took place, not volume for volume, but with a shrinkage of over 20 per cent. and with a minimum net removal of 5,000,000 tons of rock.

CONDITIONS FIXING THE LOCATION AND DIMENSIONS OF A MINERAL PIPE.

If a pipe, rather than an irregular mineral body, is to result, there must be a center of subsidence beyond reach of the effects of other centers. Such a center could be initiated with a cavern or with mere cellularity developed, by the solution of rock adjacent to a channel, to the stage of collapse. Further solvent being provided, stoping could proceed, and this would be arched or domed, whether the progress were downward, outward, or upward. Because of the protection of the floor and side walls by the weight of the fallen blocks, the progress would in fact be mainly upward.

To what width the stoping would open out before becoming vertical, would depend on a variety of conditions, such as the initial structures, the volume of the solutions, the channels available for their onward travel, and the condition of stress, the elasticity and the solubility of the rocks. In the Southwest area, in Bisbee, it is evident that the solutions worked their way along the flat bottom contact of the Escabrosa limestone for several hundred feet and continued to dissolve the limestone there, per-

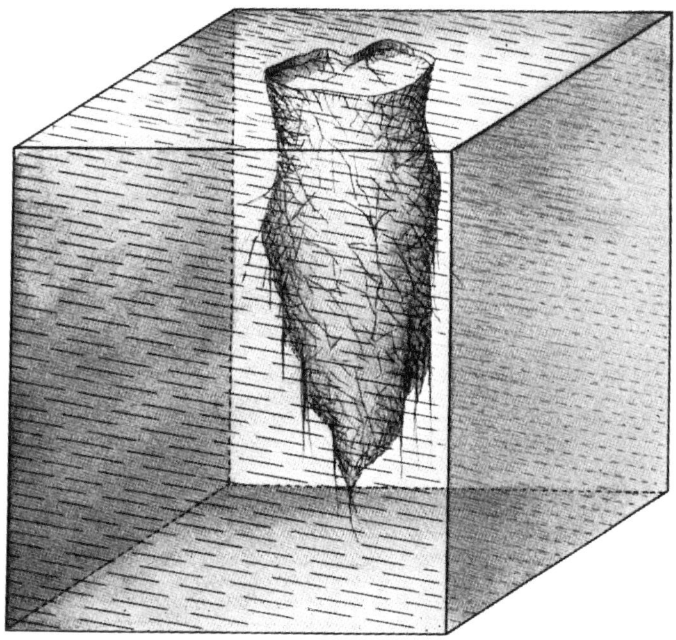

FIG. 14. Diagram of a mineral pipe.

mitting the residual chert layers to fall into the openings, until the whole block sank and steep marginal faults developed. Up to this time, the jasperoid body could have been thin, tabular, and flat. But with the marginal faults available, the volume of solution going upward, and the speed of stoping, could suddenly increase, the jasperoid body could thicken, and the steep marginal "veins" of jasperoid now showing at the surface could be developed.

In all the pipes: the ore came in at the last stage, after the pipe form had been established, and was localized by some feature of the pipe which happened especially to favor it. For example, in the Southwest area in Bisbee, the ore was made under the jasperoid roof at the contact of jasperoid with limestone. In the Pilares, it was made in the vuggy openings of the marginal, jumbled breccia. In the Catalina, of which only the lower end is left by erosion, the whole pipe was jumbled and vuggy and it all became ore-bearing.

The existence of many blind mineral pipes seems certain, but where to find them is a question which has been answered only vaguely. The best rule now available is to look for them where one or more are already known, for they tend to occur in clusters.

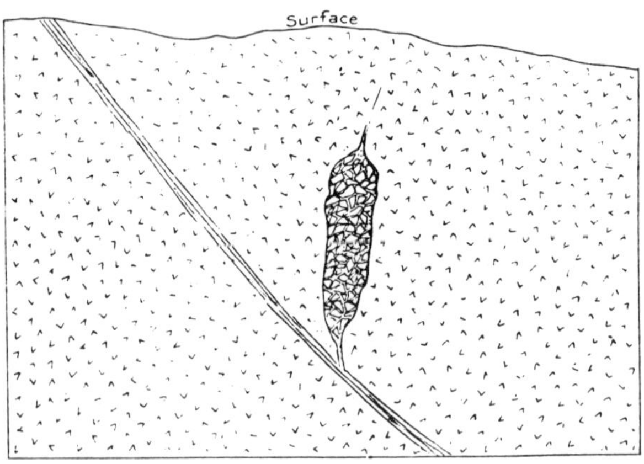

FIG. 15. A mineral pipe stoped upward from a footwall channel and making a structure like that of the Comstock Lode.

All the known pipes lie in brittle rocks and there would seem small reason for expecting them in a soft shale, for example. The pipes occur close to veins and there is evidently some obscure combination of favoring conditions, such as an abundant dissolving fluid so supplied that a single center of solution is established, which determines that a pipe, rather than a vein, or irregular stockwork, is to result. It is hoped that the knowledge of these favoring conditions will improve. As we learn more of the nature of the mineralizing fluid, from such studies as those by Allen, Day, Shepard, and Zies, it will be of interest to ask, for example, how much of the stoping required in the pipes could have been accomplished by a "gas" and how much must have been accomplished by a denser solution. In this way a closer connection might develop between the pipes and the igneous intrusions.

CONCLUSIONS.

For the mineral pipes, subsidence is the only acceptable explanation. As a cause of this subsidence, mineralization-shrinkage and stoping are established for the Southwest and Catalina pipes and are favored for the others.

The irregular, disseminated mineral deposits have in no case had their shrinkage measured or its cause determined, yet their brecciation and that of the pipes are so closely analagous as to indicate that such deposits in general, including those of Bingham, Braden, and Chuquicamata, will be established as cases of the same process, confused by the overlapping of subsidence circles.

The same process gives promise of application also to the tin-bearing and tourmaline-bearing stockworks, and it may even prove in some places to have been a preparatory stage for igneous intrusion.

The process, as visualized in the pipes, is not incompatible with the movement of rock fragments by pressure developed by growing crystals; but such pressure would be exerted within a body already opened up by subsidence, and would be of minor effect in the final result. Nor is the process incompatible with volume-for-volume replacement, in a particular sense; for, occurring between two flat markers, the stoping can start above the lower and ascend to a dome below the upper and leave both undisturbed.

With volume-for-volume replacement in its strict sense, however, the facts of mineralization-stoping as described in this paper fail to accord. They indicate caving consequent on the net removal of rock substance, and the final mineral body shows a large increase of porosity, though its outlines may be coextensive with those of the rock body for which it has been substituted.

The facts as described fail to accord also with the idea of mineralization by a fluid that makes the place for new mineral by pushing the rock walls apart. In view of these facts, neither this idea nor that of strict volume-for-volume replacement are to be seriously considered except in those cases where shrinkage is proved not to have occurred.

29

Reprinted from pages 1-2 and 4-12 of *Econ. Geology* **63:**1-12 (1968)

Intrusive Breccias Associated with Ore, Warren (Bisbee) Mining District, Arizona

DONALD G. BRYANT

Abstract

The rocks of the Warren (Bisbee) mining district consist of Precambrian quartz-sericite schists overlain by 6,000 feet of predominantly calcareous Paleozoic sediments, and 3,000 feet of Lower Cretaceous clastic beds. Intrusion of igneous rocks, intrusive breccias and associated hydrothermal activity occurred in the Late Jurassic. Only superficial effects of the "Laramide" compressional deformation prevalent in Southeastern Arizona are present.

An intrusive breccia is a heterogeneous mixture of angular to rounded, even spherical, fragments in a clastic matrix which has been transported to its present position along preexisting structures. The intrusive breccia is in the form of dikes, sills, and irregular masses varying in size from thin films to pipes several hundred feet in diameter. The contacts between the intrusive breccias and the intruded rocks are sharp and smooth. The fragments, derived from all the pre-Cretaceous rock formations except the massive replacement ores, range in size from minute particles to blocks over a hundred feet in diameter. The fragments have been transported upward for distances ranging from a few inches to several thousand feet. The matrix is composed of microscopic comminuted rock flour produced by attrition. Alteration effected during the mobilization of the breccias was weak to non-existent.

The intrusive breccias originated during the upward passage of hydrothermal fluids through fault or shatter zones. The fragments and fine-grained material of the fault zones were incorporated into the flowing stream, and transported upward by fluidization, saltation and suspension. During passage of the fragments, the channels were enlarged and smoothed as more and more debris was incorporated into the moving fluid. Deposition of the intrusive breccia occurred when the fluid pressure differential was insufficient to transport the fragments, or the channels became choked. The ore solutions continued along or adjacent to the same channels.

The intrusive breccias are used as guides to ore. The usefulness of intrusive breccias as ore guides results from the continuous permeability of fault zones throughout the magmatic history of the Warren mining district.

Introduction

In the Warren (Bisbee) mining district, ore is associated in time and space with breccia masses composed of heterogeneous rock fragments. The fragments, derived from underlying formations, have been transported upward, in many cases for several thousand feet to their present location along pre-existing zones of structural weakness. The matrix is comminuted rock flour derived from the fragments by attrition during transport. The name "intrusive breccia" has been applied to these breccias to indicate the method of emplacement. Recent mining activity, both underground and open pit, has exposed many intrusive breccias for detailed mapping. This study was undertaken to determine, if possible, the origin of intrusive breccias and their value as guides to ore.

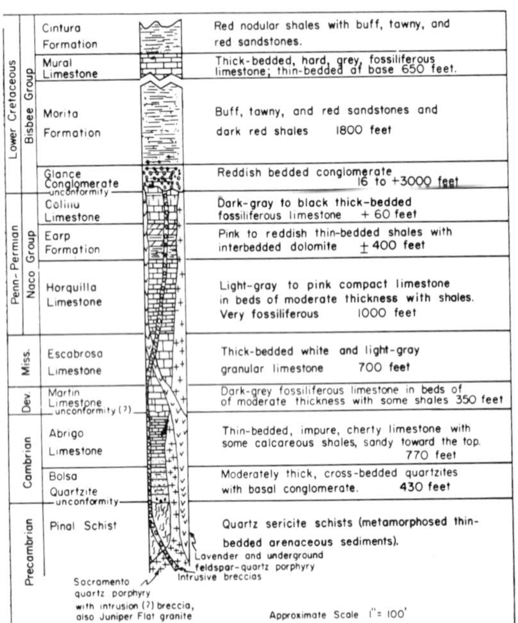

Figure 3

COLUMNAR SECTION OF THE ROCKS OF THE WARREN MINING DISTRICT

The present investigation has yielded considerable information not only on the distribution and nature of intrusive breccias but also on the igneous and hydrothermal history of the Warren mining district. The criteria presented concerning the relationship of ore to intrusive breccias may be of value in other districts having similar rocks.

Field work in the underground mines at Bisbee was done from May, 1956 to September, 1962. The

[*Editor's Note:* Material has been omitted at this point.]

Intrusive Breccias

Definition

The term Intrusive Breccia, as used in this paper, is defined as a heterogeneous mixture of angular to rounded fragments in a matrix of clastic material that has been mobilized and intruded into its present position along pre-existing structures. Although intrusive breccias encompass "pebble dikes" as described by Farmin (8), the new term is proposed because of the variation in the form of the intrusives and the shape of the fragments. The intrusive breccia is differentiated from intrusion breccias and protoclastic breccias because of the absence of igneous matrix and the lack of control by igneous intrusive contacts.

Description

Configuration and Size.—Dikes are the characteristic form of the intrusive breccias; although sills, pipes, and irregular bodies are not uncommon. Commonly, intricate networks have been formed by the coalescence of different types of intrusive breccias as shown in Figure 4. The pipes and irregular masses generally are in areas of intense shattering produced by fault intersections or drag folding of brittle rocks. Pipe-like breccias also fill the openings created by movement along curved faults.

In some places, intrusive breccia dikes have shown considerable continuity; a few have been traced for several hundred feet along strike and dip. Along strike and dip of the controlling fractures, the intrusive breccias pinch and swell with the structures, but continuity is retained. Similar changes are recognized in the sills and irregular bodies.

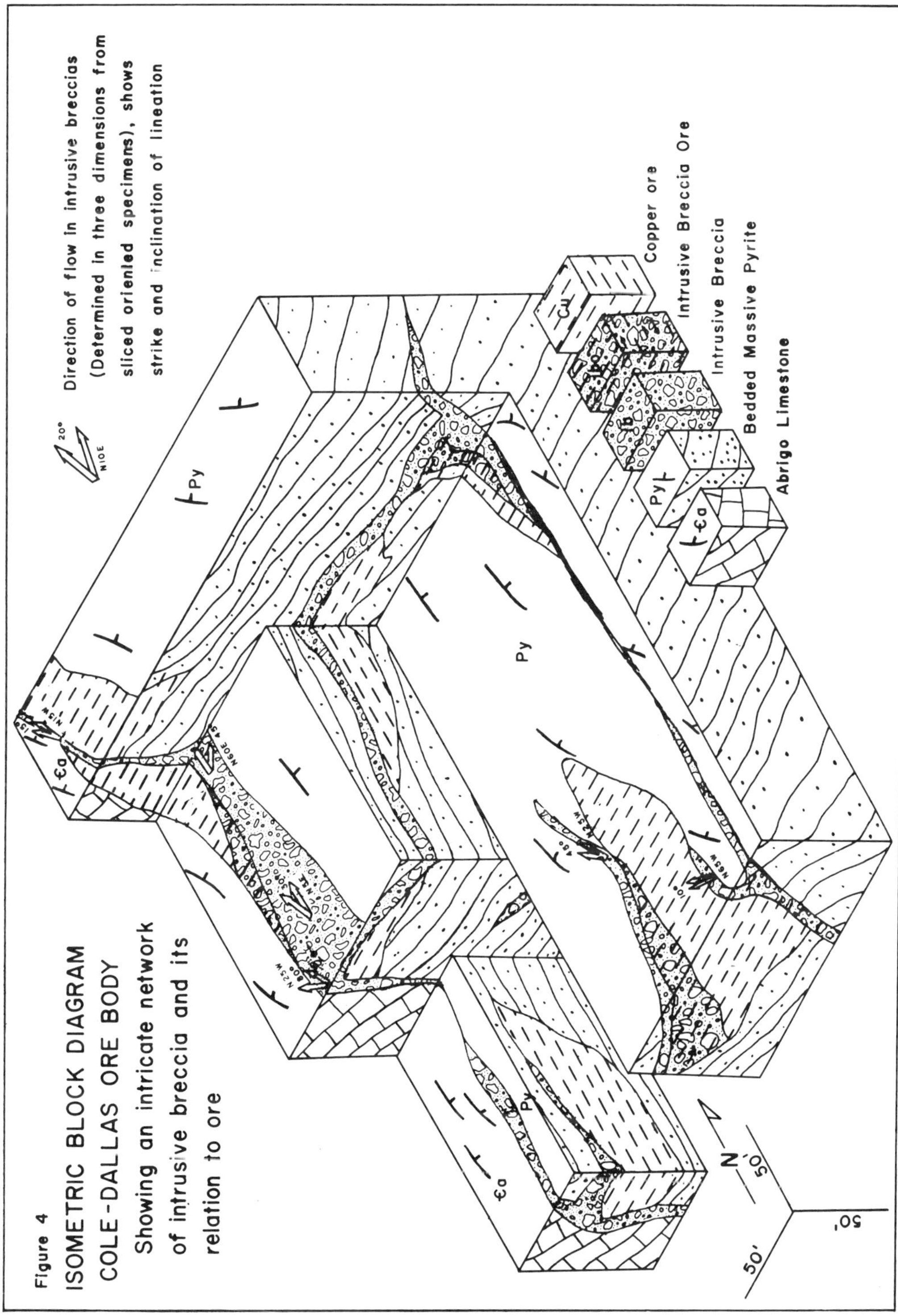

Figure 4
ISOMETRIC BLOCK DIAGRAM
COLE-DALLAS ORE BODY
Showing an intricate network of intrusive breccia and its relation to ore

The intrusive breccias vary in form from thin-filmed dikes and sills to large circular pipe-like masses. The largest of these bodies is about 500 feet in diameter; it is exposed in the center of the Lavender pit. Underground, dikes may be 1 to 10 feet wide and sills are commonly less than a foot thick, except in areas of intense deformation. Irregular intrusive breccia masses have dimensions of several tens of feet and vary in attitude from vertical to horizontal depending on the controlling structure. The contact between intrusive breccia and wall rock is invariably sharp and commonly smooth.

Internal Composition and Structures.—Fragments of all Precretaceous rock types exposed in the Bisbee area, except massive ores, are observed in the intrusive breccias. Rock types included are Pinal Schist, all of the Paleozoic formations, altered Jurassic igneous rocks, massive siliceous pyrite and Contact (?) Intrusion breccia. In some exposures, rounded fragments of ore are observed, but microscopic investigation indicates that these fragments resulted from the preferential replacement of rounded limestone pieces after, not before, consolidation of the breccia.

The fragments, subangular to well-rounded or spherical, are variable in size, and range from microscopic to blocks at least 100 feet in diameter. The subangular fragments are transported only short distances and, in some cases, have not been rotated. The lack of rotation is indicated by parallelism of the bedding in the wall rocks and the fragments. The variation in size and shape of the fragments depends on the original competency of the rock types, the original shape, and the amount of turbulence and distance transported.

Pinal Schist and Bolsa Quartzite fragments are common in intrusive breccias cutting upper Paleozoic limestones, indicating a minimum stratigraphic displacement of 3,000 feet. These fragments are often noted in the same exposure with fragments that show little or no transport.

Many of the large porphyry blocks in the Lavender pit intrusive probably were not transported great distances but may have floated in the breccia. The actual displacement or movement of the large boulders cannot be determined but their near-parallelism —like beads of a necklace—suggests original continuity. The mobilized intrusive breccia moved upward past the large porphyry blocks and caused some rounding by abrasion. These large boulders are truly detached and suspended in the intrusive breccia. This relationship as shown by the arrows on Figure 2 was observed many times in three dimensions during detailed mapping of the Lavender pit.

The largest rounded fragment observed during this study is an elliptical boulder measuring 11 feet along the major axis and 7 feet along the minor axis. This boulder, which is approximately 1,500 feet from its closest known possible source, weighs 80 tons. Although it is exceptional, rounded fragments 2 to 3 feet in diameter are common in the intrusive breccias in the Sacramento Hill Complex.

The spherical boulders, generally quartzite or silicified porphyry, are not easily explained by simple turbulent attrition. The presence of the onion-skin structure and small residual ridges on these spherical boulders suggest hypogene exfoliation as noted at Tintic by Farmin (9).

The matrix is comminuted rock flour consisting predominantly of angular quartz fragments approximately 0.005 mm in diameter surrounded by submicroscopic material. Particles distinguishable as limestone, quartzite, schist, porphyry, and pyrite are present as grains ranging in size from about 0.005 mm to megascopic. As shown in Figure 5, many of these microscopic particles are rounded. The heterogeneous composition of the matrix is maintained even in intrusive breccia exposures where the larger fragments are derived from the immediate wall rocks. Photomicrograph 5a was taken of a thin section of the matrix of an intrusive breccia with unrotated limestone fragments. No indication that the matrix was originally igneous or fused was observed in any of the thin sections studied.

The internal structure of the intrusive breccias is characterized by flow structure or lineation adjacent to the walls, and by eddy current patterns peripheral to the fragments. The flow structures are delineated by parallel orientation of the fragments and linear particles of the matrix. The direction of transport as indicated by fabric studies varies from horizontal to vertical, even within a single intrusive breccia body. The lineation, shown by arrows on Figures 4 and 6, were determined on oriented specimens cut on three mutually perpendicular axes. In dikes or sills less than 1 or 2 feet thick, the lineation persists from wall to wall. In the central part of the thicker units, however, a persistent lineation was not observed.

Syngenetic Alteration and Mineralization.—Recognition of alteration associated with the mobilization and emplacement of intrusive breccias is difficult. The intrusive breccias are generally associated with the ore bodies and any syngenetic alteration may be obscured by the later mineralization. The extent of alteration in the fragments before mobilization is unknown. The products of alteration are commonly soft and would be removed by abrasion during transportation. These same factors would also prevent distinguishing alteration in the wall rock.

In the few exposures of breccia that appear to be unrelated to massive replacement ore, alteration is weak to non-existent. Limestone fragments are pe-

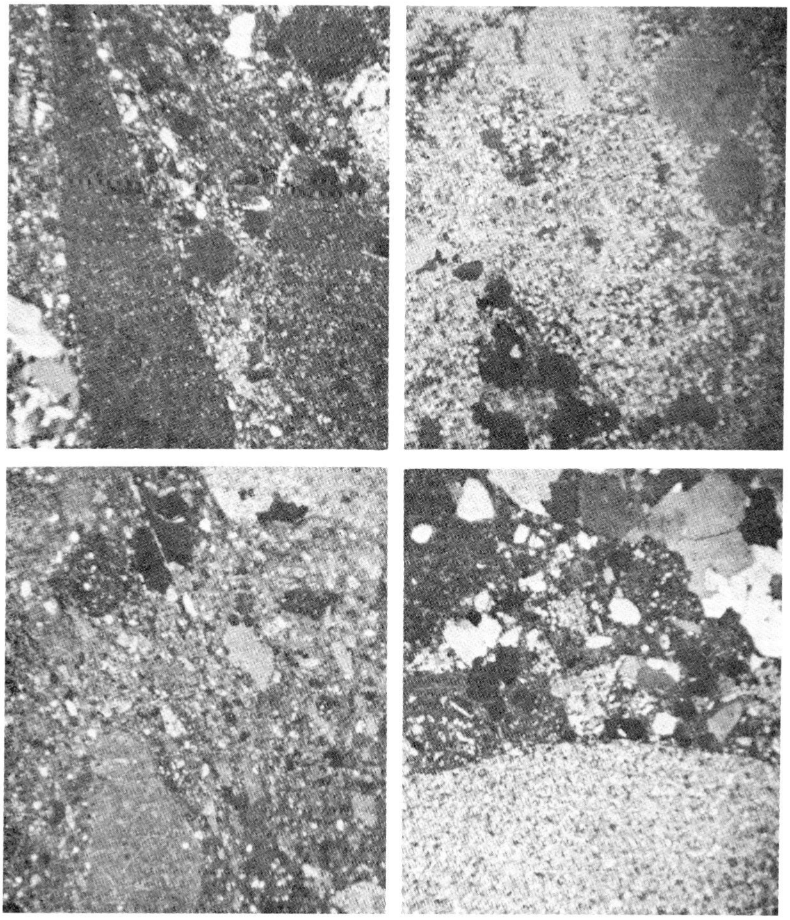

FIG. 5. Photomicrographs of intrusive breccia matrix. 5a (upper left). Intrusive breccia matrix with predominant limestone fragments. Note the variety of rock types. ×10 x-nicols. 5b. (upper right). Intrusive breccia matrix with strong lineation of elongated fragments indicating flow. ×10 x-nicols. 5c (lower left). Intrusive breccia matrix with spherical fragment diverse lithologies. ×10 x-nicols. 5d (lower right). Intrusion (?) breccia with rounded fragments. Note the apparent assimilation of matrix by silica and chlorite. ×10 x-nicols.

ripherally recrystallized and porphyry pebbles are slightly bleached. A thin skin of tolcose or chloritic alteration has been observed on pyritic silicified porphyry fragments.

In all the intrusive breccia polished sections studied, microscopic colloidal (?) blebs of sulfides were observed in the matrix. Chalcopyrite, bornite, pyrite and galena have been identified. No alteration effects have been observed that could be attributed to the bleb mineralization. The ubiquitous distribution of the mineralization and lack of alteration may be indicative of a syngenetic origin of this low-grade mineralization.

Detailed Descriptions.—The following examples have been selected from more than a hundred studied at Bisbee to illustrate some of the diagnostic features of intrusive breccias. In all of the illustrations, the close association of ore and intrusive breccia is evident.

Many intrusive breccias are associated with ore in the Cole-Dallas orebody. Along the western periphery, an intricate network of dikes and sills laces the mineralized limestone. Evidence that intrusive breccias changed from dike to sill to dike, as shown in Figure 4, was actually observed in several localities and inferred in others. The distribution of ore in

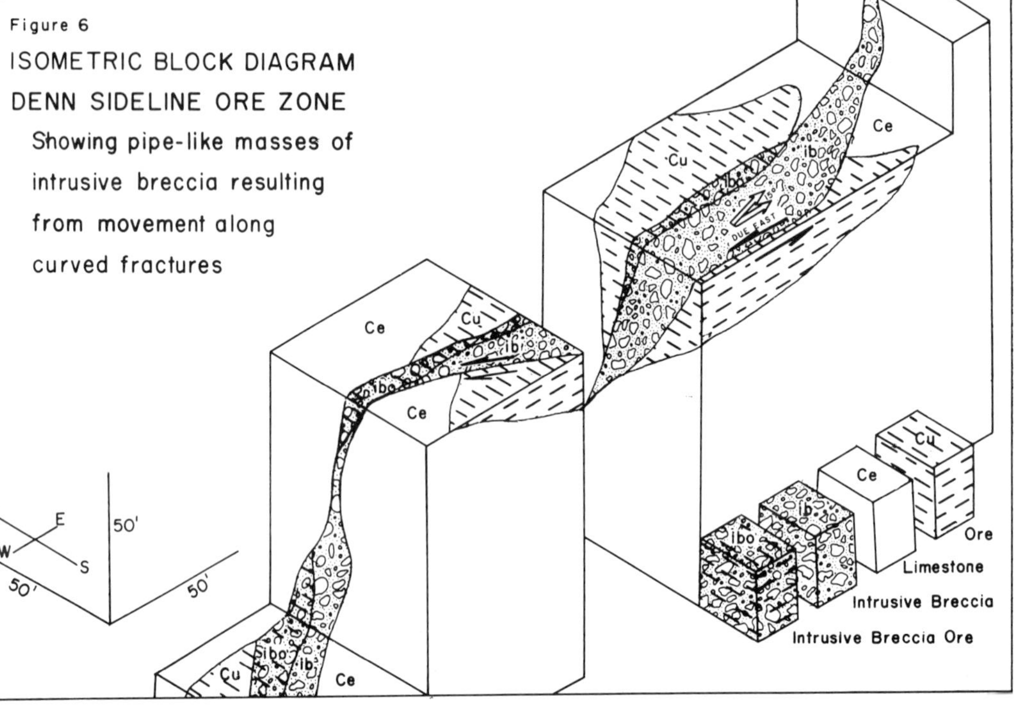

Figure 6
ISOMETRIC BLOCK DIAGRAM
DENN SIDELINE ORE ZONE
Showing pipe-like masses of intrusive breccia resulting from movement along curved fractures

relation to the intrusive breccia indicates that the relative permeability of the breccia is an important factor in localization of ore. In areas where the breccia is permeable, massive replacement ore is found in the breccia in addition to the adjacent limestone. But individual ore bodies also appear to be the result of impounded ore solutions below impermeable intrusive breccias. The observed intimate association of ore and breccia leaves little doubt that common channelways were used. As indicated on Figure 4 by the lineation arrows, the flow of material was not always upward but varied from horizontal to vertical. The variability of flow direction, the common changes of channelways, and the extent and thinness of the sills is indicative of an extremely mobile fluid rather than a mushy viscous mixture.

The Crown King ore zone consists of many different ore bodies localized along a fault zone trending northeast, which is 125 to 250 feet wide. Intrusive breccias along the Crown King fault zone persist for hundreds of feet along the strike and dip. In the multitude of fractures within the fault zone, the intrusive breccias are along the more persistent of the ore-bearing fractures. The intrusive breccias pinch and swell and range in thickness from a mere film to six feet. The ore bodies are localized near the thicker sections of the intrusive breccias. The variation in thickness of the intrusive breccias is an indication of the width of the fractures and therefore of the accessibility to hydrothermal solutions. Although apparent terminations of intrusive breccias have been observed, careful study of three dimensional exposures indicates that the fine-grained matrix extends beyond the exposure. When unable to transport the larger fragments, the fluid was still able to transport the fine-grained clastic material.

The shape and distribution of many intrusive breccias are controlled by the type and direction of movement on the fractures, especially if curved. In the Denn Sideline area, for example, several breccias were localized in restricted channels along curved fault planes. The channels are the open or excessively fractured areas produced by movement along the sinuous faults. The intrusive breccias emplaced along these channels are mostly pipe-like in form. The attitude of the pipes depends on the distribution of the curves along the fault plane and the direction of movement. The example illustrated in Figure 6 is an inclined pipe and results from strike-slip movement. Intrusive breccia pipes with attitudes ranging from horizontal to vertical have been mapped.

An early stage in the development of intrusive

breccias is distinctly recorded in exposures on the 2566 Level, Junction mine. The breccia is in broken to shattered Escabrosa limestone along the Mountain Maid ore zone. Initially, the Escabrosa was a mosaic or crackle breccia with blocks of limestone separated by gouge and small fragments. With the emplacement of the intrusive breccia, the gouge and small fragments were incorporated into upward streaming fluid and replaced by fragments from below. The resulting rock is a limestone mosaic breccia, cemented by intrusive breccia, in which the bedding of the limestone blocks shows continuity with the wallrocks. Within the same zone, where the original fracturing was more intense, the limestone blocks have been rotated and transported upward or even completely displaced beyond the limits of the exposure.

Intrusive breccia also intrudes discordantly all the rock units of the Sacramento Hill complex. In addition to the predominant or typical type (ib) containing fragments of many different rock types, a mappable variation or facies (pib) is shown on Figure 2. This variation contains only fragments of altered pyritic Sacramento quartz porphyry in the usual detrital matrix. In all likelihood, this breccia, like the breccia in Escabrosa limestone discussed above, was formed in moderately fractured ground. The contact of this facies with typical intrusive breccia is gradational but it is sharp with all other units.

Origin of Intrusive Breccias

Many theories have been advanced for the origin of breccia masses similar to the intrusive breccias of the Warren mining district. Investigation of the literature indicates that the origin of breccias with rounded fragments associated with hydrothermal mineralization has been and is the subject of considerable controversy. Some of the controversy stems from the confusion between the origin of the fragments and the origin of the agency causing transportation of the fragments. The existing theories that have been or could be considered applicable are listed below:

- I Explosive action of expanding fluids—Emmons, W. H. (7) and Walker, R. T. (23)
- II Tectonic activity—Waters, A. C. and Campbell, C. D. (24)
- III Igneous intrusion—Waters, A. C. and Krauskopf, K. B. (25)
- IV Solution and replacement—Butler, B. S. (2) and Kuhn, T. H. (11)
- V Subsidence—Perry, V. D. (14), Speers, E. C. (20) and Locke, Angustus (12)
- VI Fluid intrusion or fluidization—Reynolds, D. L. (17) and Farmin, Rollin (8).

With the exception of the fluid intrusion or fluidization theory, the proposed hypotheses are subject to serious objection at Bisbee. The theories and objections are discussed below.

(I) Explosive action results from the violent and sudden release of fluids, probably gases, which have collected under high pressure. The common result is a vertical circular opening or a diatreme. The force of the expansion envisaged is so violent that the opening is necessarily vertical and circular to provide immediate relief of pressure. The rock formerly occupying the diatreme is broken and redistributed by the expanding fluid. The diatreme is generally filled with fragments, some rounded, of igneous material, country rock or both. The Warren intrusive breccias do not belong in this category for at least three reasons—brecciation occurred before mobilization; pressure relief occurred along preexisting structures, not through new openings; and the linear flow structures indicate that the intrusive breccias traveled in different directions, from horizontal to vertical, not directly upward as expected in an explosion.

(II) Breccias are commonly formed by and during tectonic activity such as faulting and folding. In general, the fragments, which may be rounded, are found at the place of origin or are dragged a determinable amount according to the fault movement. In the Bisbee intrusive breccias, however, fragments have been transported hundreds of feet not only along faults with little or no offset of the walls, but also along bedding planes.

(III) Intrusion or protoclastic breccias, also referred to as contact breccias, formed by the push or drag of viscous intruding magma are generally restricted to the igneous contact. The matrix is commonly igneous and the fragments may be broken and sheared wall rock or chilled border zone. If brecciation is intense, the matrix may be clastic. Bisbee intrusive breccias, however, are not restricted to igneous contacts and do not have an igneous matrix. The Contact intrusion (?) breccia in the Sacramento Hill intrusive complex has been so intensely altered that its origin is indeterminable.

(IV) Solution and replacement breccias are developed by the modification of broken rocks by ascending solutions along the fractures. These breccias are composed of angular to rounded boulders surrounded by vuggy ore and/or gangue material rather than finely comminuted rock. When evidence of orientation or position is available, the boulders show little or no displacement. According to this theory, the boulders are rounded in situ by the corrosion or replacement of the corners and edges by hydrothermal solutions. Objections to the application of this theory to Bisbee intrusive breccias are that the fragments are definitely transported and abraded and evidence of corrosive or extremely active hydrothermal fluid is lacking.

(V) Subsidence breccias are caused by mineralization stoping, shrinkage of crystallizing magmas and possibly the up and down pumping action of some igneous intrusions. The evidence of predominant upward movement in the Bisbee intrusive breccias eliminates consideration of subsidence as a major contribution to the origin of the breccias.

(VI) Of all the theories proposed, only fluid intrusion or fluidization explains adequately the observed characteristics of Bisbee intrusive breccias. The concept proposes the suspension and transportation of solid particles in an upward flowing fluid that has a lower density than that of the particles. The fluid—gas, liquid or both—is derived from a magmatic source but is not actually magma. The hypothesis proposed below for the origin of the Bisbee intrusive breccias is a modification of the fluidization concept and is not intended to explain other occurrences except by analogy.

The intrusive breccias of the Warren mining district originated during the final stages of the Jurassic magmatic activity and were probably a hydrothermal phenomenon. The fluids were generated at an unknown depth from an assumed magmatic source and ascended toward the surface of the earth. The fluids, under the influence of a natural pressure gradient, rose along fracture zones. At a depth of possibly 6,000 feet, the rising fluids began to suspend and transport the fragmental material produced during earlier faulting. The change from a passive to an active transporting phase may have resulted from either an increase in fluid density or a change from tight shears to open fractures. The non-commital term "fluid" is used because the actual state of the transporting media is not known. Initially the fluid moved upward along joint and fracture surfaces picking up the finely fractured pieces and gouge. Then as the fluid removed more and more of the interstitial broken material, it suspended larger fragments that were rotated, mixed together, and transported in the upward moving stream. Attrition during mixing and transport produced the rounded boulders and the finely pulverized matrix material. The intrusive breccias were deposited when the flow pressure lessened or the channels narrowed and became choked. The fluid medium, however, continued to flow and carry the fine-grained matrix material. Fluid transport stopped before the matrix material was completely removed from the interstices of the impounded boulders. The hydrothermal solutions responsible for the massive copper and lead-zinc sulfide orebodies were introduced after the intrusive breccias were sufficiently consolidated to permit fracturing of fragments. Depending on the permeability of the intrusive breccia, ore replaced the adjacent limestone, the intrusive breccia or both.

The ability of a fluid to transport boulders has been amply demonstrated by dredging and commodity transport pipelines (6, 10). The theoretical fluid velocity required to affect transportation is slightly greater than the settling velocities. Richards (18) determined the average settling velocity of galena and quartz in water to be about 75 inches per second for particles about one inch in diameter. The actual velocities used in dredging and industrial operations, however, varied depending on the type of transport, viscosity of the fluid media, and frictional losses. Assuming heterogeneous horizontal flow, the empirically determined velocity of water necessary to transport particles one inch in diameter is 25 feet per second. In the Florida phosphate industry, boulders up to six by nine inches have been transported by saltation and heterogeneous suspension at velocities of 18 feet per second. Although experimentation by Zenz with vertical systems is not complete, the results indicate that "the velocity required to convey materials horizontally is several times as great as that required to convey vertically" (26, p. 319). Calculations relating to the velocity of the fluid are available only for systems with fragments up to one inch in diameter. Field evidence indicates that the intrusive breccias contained 50 to 80 percent fine-grained material. The presence of this material probably decreased the viscosity and increased the specific gravity of the transporting medium, thus enhancing its ability to carry large fragments. The decrease or, at least no increase of viscosity due to the addition of fine-grained particles in excess of 35 percent has been experimentally verified by C. H. Shack, and others (19). Including the effects of friction by fragments on each other and on the walls, a reasonable assumption of minimum fluid velocity for intrusive breccias should be 25 to 50 feet per second. These velocities are not unreasonable when compared with known hot springs activity.

The amount and source of the fluid needed is probably the most common objection to the fluid intrusion hypothesis for the origin of intrusive breccias. A very rough approximation of the total volume of intrusive breccia in the Warren district is 600 million cubic feet in the Sacramento complex and 200 million cubic feet in the underground workings. To cause this mass of breccia to behave like a liquid, the voids resulting from an initial state of open packing must be filled with fluid. Mead (13) states that if grains are spherical and uniform in size, 25.9 percent of the volume is open space. In order to cause movement within a mass of unconsolidated material, without deformation of the fragments, the volume of the voids must be increased. In the Bisbee intrusive breccias the percentage of initial void space is less than 25.9 percent because the fragments are irregular in shape and size. Considering the high percentage of very fine-grained matrix material, 10 percent void space is a reasonable assumption. Thus if the total volume of intrusive breccia was conveyed without friction or fluid losses and assuming water to be the fluid medium, the amount of water needed is 2.3 cubic kilometers. This amount of water is insignificant when compared to the tremendous volume of dilute hydrothermal solutions needed to deposit three and a half million tons of copper and several hundred million tons of pyrite!

Although the preceding discussion assumes that water is the transporting medium and that the ores are deposited from dilute hydrothermal solutions, the possibility that a vapor could be equally effective should not be overlooked. Recent work by Krauskopf (unpublished manuscript) indicates that sufficient copper sulfides to make ore can be transported in the volatile state. The magmatic water under this condition would probably be vaporized. The presence of water as a vapor, however, does not invalidate the hypothesis of fluid intrusion. The process of transporting and mixing discrete particles by streams of rapidly moving gases has become increasingly important in the chemical industry during the last decade (8). Although particles equivalent in size to the fragments of the intrusive breccias have not been studied, extrapolation of the available data indicates that the transportation of large particles is not unreasonable.

Summary and Conclusions

The intrusive breccias of the Warren mining district were intruded during the Jurassic igneous activity and copper mineralization. The rock fragments were transported, rotated, and abraded into rounded boulders. The matrix is comminuted rock of the same composition as the enclosed fragments. The intrusive breccias are observed in a wide variety of shapes, including dikes, sills, pipes, and irregular masses. The contacts with the intruded rocks are sharp and smooth. Internal flow structures consisting of aligned, elongated fragments are common and indicate flow directions that vary from horizontal to vertical.

The intrusive breccias are a phase of the complex igneous history of the Warren mining district. The sequence at Bisbee was first quartz porphyry intrusion, hydrothermal activity, then repeated intrusion of porphyries and more hydrothermal activity. The first hydrothermal stage was the introduction of enormous quantities of pyrite and silica. The next hydrothermal event following intrusion of the later porphyry was the emplacement of the intrusive breccias with a low-grade copper mineralization. Following the intrusive breccias was the later mineralization, which deposited the massive sulfide deposits of copper, lead, and zinc.

The sequence of igneous events was punctuated by lapses of time during which the igneous rocks and breccias consolidated. Deposition of the late massive sulfides, however, appears to have been from one solution.

Structural activity seems to have originated and ended with the initial intrusion. Thus, the channelways responsible for the deposition of the massive siliceous pyrite, intrusive breccias, and late feldspar-quartz porphyry were still open at the time of the deposition of massive copper, lead, and zinc sulfides.

Intrusive breccias are a hydrothermal phenomenon. The mobilizing fluids are derived from the same source as the igneous rocks and ore metals. The hydrothermal fluid ascends from the deep magmatic mass toward the surface along zones of structural weakness. The cause of the ascension is a natural pressure gradient resulting from differences in confining pressures and vapor pressures developed by the crystallizing magma. As the fluids passed through broken zones, the rock fragments were suspended by the fluid and transported toward the surface. Abrasion between the moving, rotating fragments produced the characteristic rounded boulders. The boulders and enclosing matrix were deposited when the pressure gradient became too weak to convey the rock material or the channel became too small to allow passage of the fragments. The ubiquitous syngenetic low-grade copper mineralization in the intrusive breccias was probably carried by the mobilizing fluid of the intrusive breccias. Whether the transporting fluid was gas or liquid is unknown.

The importance of intrusive breccias as ore guides in the Warren mining district results from the invariable association of intrusive breccias with ore. In many mining districts breccias similar to the Bisbee intrusive breccias are used as guides to hydrothermal activity but not necessarily to ore. A perplexing problem in many of these districts is why some breccia masses contain ore or are associated with ore and others are not. The answer probably depends upon the structural conditions prevalent during ore deposition. Certainly, at Bisbee, circulation was maintained along or through the intrusive breccias during ore deposition. Why the conditions in the Warren mining district differed from the other districts is not known.

Molybdenum Corp. of America,
Questa, New Mexico,
November 29, 1966
Present address:
P.O. Box 607,
Louviers, Colorado

REFERENCES

1. Bonillas, Y. S., et al., 1916, Geology of the Warren mining district: Am. Inst. Mining Metall. Engineers Trans., v. 55, p. 284–355.
2. Butler, B. S., 1913, Geology and ore deposits of the San Francisco region, Utah: U. S. Geol. Survey Prof. Paper 80, p. 172–178.
3. Bryant, D. G., 1964, Intrusive breccias and associated ore of Warren (Bisbee) mining district, Cochise County, Arizona, Stanford Univ. unpub. doctoral thesis, 149 p.
4. Bryant, D. G., and Metz, H. E., 1966, Geology and ore deposits of the Warren mining district: *in* Geology of the Porphyry Copper Deposits, Southwestern North America: Univ. of Arizona Press, Tucson, Arizona, p. 189–204.
5. Creasey, S. C., and Kistler, R. W., 1962, Age of some copper-bearing porphyries and other igneous rocks in southeastern Arizona: U. S. Geol. Survey Prof. Paper 450D, p. 1-5.
6. Durand, R., and Condolios, P., 1952, The hydraulic transport of coal and solid material in pipes: Colloquium on hydraulic transport proc., London, 70 p.
7. Emmons, W. H., 1938, Diatremes and certain ore-bearing pipes: Am. Inst. Mining Metall. Engineers Tech. Pub. No. 891, p. 170–180.
8. Farmin, Rollin, 1934, "Pebble dikes" at Tintic, Utah: Econ. Geol., v. 29, p. 356–370.
9. Farmin, Rollin, 1937, Hypogene exfoliation in rock masses: Jour. Geology, v. 45, p. 625–635.
10. Koch, L. W., 1961, Solids-carrying pipelines—what to consider in their preliminary design: Eng. Mining Jour., v. 163, p. 74–78.
11. Kuhn, T. H., 1941, Pipe deposits of the Copper Creek area, Arizona: Econ. Geol., v. 36, p. 512–538.
12. Locke, Augustus, 1926, Formation of certain ore bodies by mineralization stoping: Econ. Geol., v. 21, p. 431–453.
13. Mead, W. J., 1925, The geologic role of dilatency: Jour. Geology, v. 33, p. 505–523.
14. Perry, V. D., 1961, The significance of mineralized breccia pipes: Am. Inst. Mining Metall. Engineers Preprint 61178, 35 p.
15. Ransome, F. L., 1904, The geology and ore deposits of the Bisbee Quadrangle, Arizona: U. S. Geol. Survey Prof. Paper 21, 168 p.
16. ——, 1914, Description of the Bisbee Quadrangle, Arizona (1904 revised): U. S. Geol. Survey Folio 112, 19 p.
17. Reynolds, D. L., 1954, Fluidization as a geological process, and its bearing on the problem of intrusive granites: Am. Jour. Sci., v. 252, p. 577–613.
18. Richards, R. H., 1907, Velocity of galena and quartz falling in water: Am. Inst. Mining Metall. Engineers Trans., v. 38, p. 210–235.
19. Schack, C. H., et al., 1957, Measurements and nature of the viscosity of water suspensions of some common minerals: U. S. Bur. Mines Rept. Invest. 5334, 16 p.
20. Speers, E. C., 1957, The age relation and origin of common Sudbury breccia: Jour. Geol., v. 65, p. 497–514.
21. Tenney, J. B., 1932, The Bisbee mining district, *in* Ore deposits of the Southwest: Internat. Geol. Cong., 16th Washington 1932, guidebook 14, p. 40–67.
22. Triscka, Carl, 1938, Bisbee district: Arizona Bur. Mines Bull. 145, p. 32–41.
23. Walker, R. T., 1928, Mineralized volcanic explosion pipes: Eng. Min. Jour., v. 126, p. 895–898, 939–942, 976–984.
24. Waters, A. C., and Campbell, C. D., 1935, Mylonites from the San Andreas fault zone: Am. Jour. Sci., v. 229, p. 473–503.
25. Waters, A. C., and Krauskopf, K. B., 1941, Protoclastic border of the Colville Batholith: Geol. Soc. America Bull., v. 52, p. 1355–1418.
26. Zenz, F. A. and Othmer, D. F., 1960, Fluidization and Fluid-Particle Systems: New York, Reinhold Publishing Corp., 513 p.

Part V

MINING APPLICATIONS

Editor's Comments on Papers 30 Through 38

ROCK MECHANICS

Physical characteristics of the ore and host rock are of great importance in the determination of mining methods and the location of mine workings and surface installations. Surface and underground geological mapping and logging of core must supply the necessary information for the interpretation and prediction of rock behavior.

Analysis of rock/soil conditions as the basis for design of stable slopes or underground opening entails evaluation of rock stresses and the ability of the material to withstand the stresses (strength). The basic difference between analyses for soils and those for rocks is the identification of failure surfaces in rock materials.

Clark (Paper 30) cites a number of examples in which expensive mistakes were made because adequate determination of rock characteristics was not made in advance of the determination of mining methods. The traditional conflict between mining engineer and geologist is gradually becoming a thing of the past as the geologist directs more attention and makes substantial contributions to operational problems.

Proper recording of various geological data relevant to operational decisions needs to begin during the exploration phase of drilling and sampling. Pillar and Drummond (Paper 31) outline the types of data to be included: a complete mineral inventory, geomechanical characteristics of the ore and host rock, and hydrological conditions. Details for the analysis are suggested in subsequent papers in this section.

Hoek (Paper 32) presents a simple summary of the basic concepts of rock failure, support, and the measurement of in situ stress conditions. These result from geological parameters and become the responsibility of the geologist to record, analyze, and predict. An extensive list of references in Hoek's paper serves as a guide to the geologist in obtaining more detailed information on both techniques and theory.

Logging of drill core is normally oriented toward geological rather than geomechanical aspects; both are essential. Paper 33, compiled by a committee of the Geological Society of London Engineering Group, of which J. L. Knill was chairman, is a valuable guide to the acquisition of data for engineering design purposes.

Swaisgood and Versaw (Paper 34); Swaisgood, McMahon, and West (Paper 35); and McMahon (Paper 36) demonstrate how various mining activities require different types of geological guidance and how this guidance can assist design and reduce cost. Additional details can be gained from the *SME Mining Engineering Handbook* (Obert, 1973; Loofbourow, 1973).

Application of theory and techniques to a real-life situation is the best test of validity. The testing program carried out by the Kennecott Corporation at the Kimbley Pit, described by Halstead, Call, and Rippere (Paper 37), enabled testing and evaluation of examination techniques.

Rock mechanics and ground stabilization problems in stratified rocks present problems related to the strongly anisotropic nature of the rock material. Denkhaus (Paper 38) relates stratification, fracturing, and rock strength and brittleness to the design of mining systems, and develops models that assist analysis.

REFERENCES

Loofbourow, R. L., 1973, Ground Water and Ground-water Control, in *SME Mining Engineering Handbook,* vol. 2, I. A. Given, ed., American Institute of Mining, Metallurgical and Petroleum Engineers, New York, pp. 1-55.

Obert, L., ed., 1973, Elements of Soil and Rock Mechanics, in *SME Mining Engineering Handbook,* vol. 1, I. A. Given, ed., American Institute of Mining, Metallurgical and Petroleum Engineers, New York, pp. 1-52.

30

Reprinted from *Eng. and Mining Jour.* **149:**94-96 (August 1948)

Geological Studies Can Avoid Costly Mistakes

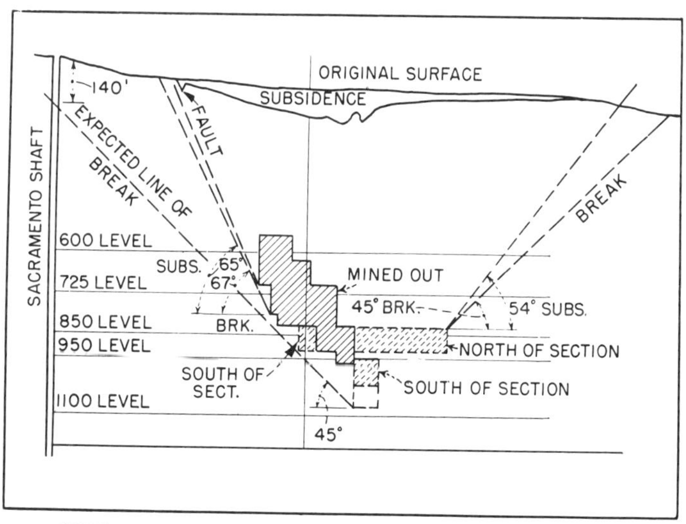

SUBSIDENCE at Copper Queen mine, Phelps Dodge Corp. (Trans., A.I.M.E.)

GEORGE B. CLARK
Assistant Prof. of Mining Engineering
University of Illinois

Laboratory testing appears a promising field for research. Data on physical properties of rocks will help with problems

EXPERIENCE HAS DEMONSTRATED that it is wise to make detailed geological studies before undertaking mine development and to maintain geological investigation during development and mining operations. Failure to do this has caused some properties substantial losses.

The following are important operational factors (other than ex-

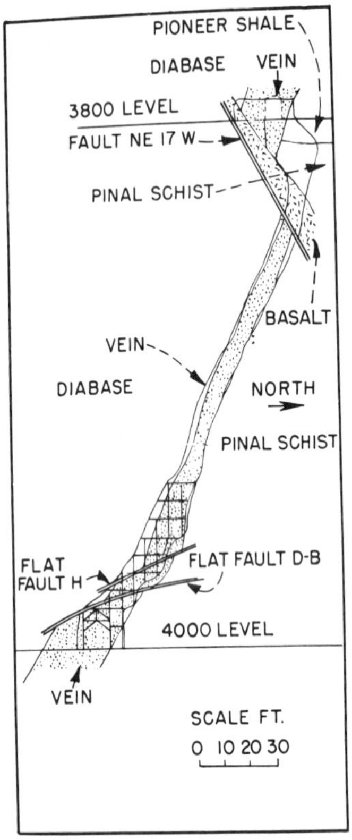

MAGMA VEIN in cross section showing geological structure which affects mining operations (After Michell)

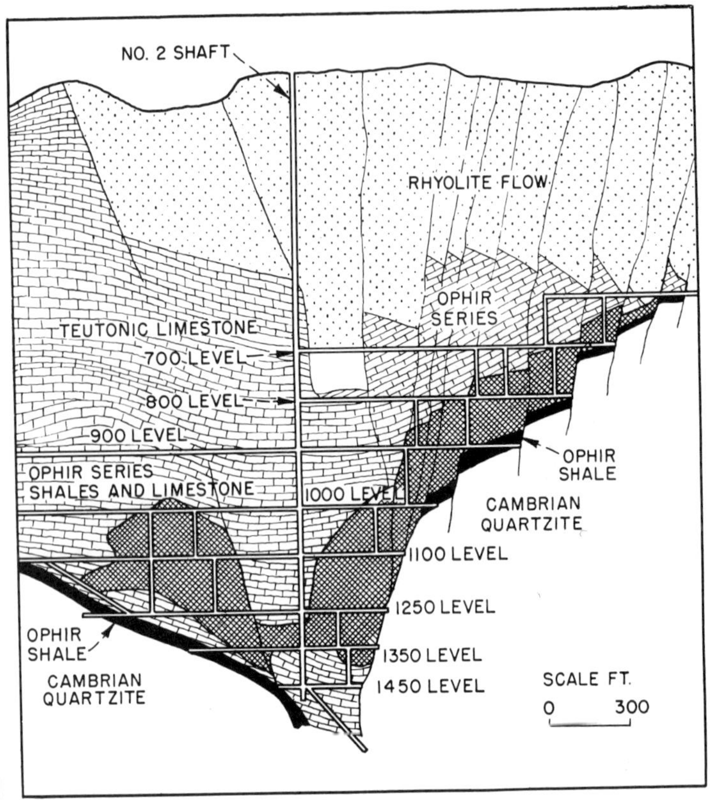

TINTIC STANDARD MINE, an ideal section on N 45 deg. E line through No. 2 shaft (U.S.B.M., Inf. Cir.). When operations began different ideas were held about the geology from those later developed.

ploration) which are closely related to geology: 1. Placing preliminary openings to best advantage. 2. Placing permanent openings in solid, firm rock. 3. Stope planning. 4. Maintenance of mine openings. 5. Location of openings to avoid water. 6. Furnishing information for mine valuation. 7. Safety of mining operations. 8. Choice of mining method.

In many of our mining schools, mining and geology are taught as separate courses by separate departments, and it is left up to the individual student in most cases to see the correlation between the two subjects. Unfortunately, most students have not had enough experience or educational background to bring the principles of the two together in order to form an accurate concept of their relative and joint importance. Consequently, very few graduates leave school with the realization that certain phases of mining engineering are dependent upon an understanding of geology.

A recent statement from an official[1] of a large mining company declared: "We are very much interested in this subject as I know of several instances where the selection of the wrong method of mining cost a great deal of money. Closer cooperation between mining and geological departments would have disclosed the existence of unfavorable structural conditions, and a more suitable mining method would have been adopted."

The following factors affecting a choice of mining method are important: 1. Size and shape of orebody and amount of overburden. 2. Disposition, location, strike and dip of deposit. 3. Strength and physical character of ore. 4. Strength and physical character of surrounding rock. 5. Grade and type of ore. 6. Drainage. 7. Economic factors such as cost of supplies, safety, equipment, labor, type of ore treatment, etc.

Of these factors only the last is relatively unaffected by geological conditions. The others are directly related. Any of these conditions may vary locally throughout the deposit so that different methods may be required to mine different sections of the ore. On the other hand, the existence of uniform geological conditions throughout the deposit may make the employment of only one method most economical for a mining property.

When new deposits are found, the most effective methods of mining should be employed as early as possible to keep the over-all cost per unit of metal produced at a minimum and to permit maximum extraction of ore. In production operations, as well as in prospecting and exploration, a knowledge of local geology will play a vital part in the successful solution of the problem involved in producing ore at the lowest possible cost per ton. Regardless of the structural simplicity or the complexity of any ore deposit, the method of mining which can be employed to extract the ore most economically will be

1. Personal communication, R. J. Ennis, Vice Pres. and Genl. Mgr., McIntyre Porcupine Mines, Ltd., Schumacher, Ontario, Can.

determined by geological factors.

For example, Sales[2] says of mines in the Butte district, "The structural features affecting choice of mining methods are joint planes, faulting, including structure faults, alteration of the granite within and along the veins, and numerous small slips and movement planes associated with the orebodies. . . . It is the alteration and fault movement which cause the Butte veins to require timber."

Changing Methods Is Costly

The mines, large and small, which have been forced to change their method of mining at high cost after a short period of operation are numerous. A few examples will illustrate the points outlined above.

The Tintic Standard mine in the East Tintic district of central Utah has been one of the important silver and lead producing mines in this country. The orebodies consist of massive deposits in limestones and shales overlain by altered limestone and porphyry. The whole structure is cut by a series of step faults (post mineral). The complete geology was not known or considered of vital importance when operations were begun, so it was assumed erroneously that the replacement orebodies were similar to those in the West Tintic district, which required relatively little support in stoping and mining operations. The deposit was mined by simply following the rich high-grade ore, which developed stopes as irregularly shaped cavities in larger masses of low-grade material. The system of mining employed both open stopes and light square-sets, the latter being used mainly as scaffolding. No effort was made to mine the ore systematically or to provide support for the hanging wall.

Later, improved treatment facilities made it possible to mine lower grade ores, so in 1921 filling of stoped areas was commenced to attempt to preserve the unmined ore. However, subsidence had already begun, resulting in severe caving within and adjacent to the mined ground. Both shrinkage stoping and the Michell system of mining were tried without success. Square-set stoping with back filling was finally tested and it proved successful. But continued subsidence remained a problem in shaft maintenance and maintenance of underground openings.

The lesson learned from the experience at this mine is obvious: If the geology of the orebody had been studied and understood as it became exposed by underground openings and before mining operations had caused a large amount of subsidence and caving, over-all costs per ton of mining ore could have been reduced in terms of a considerable saving of time, ore and capital during the entire subsequent history of the mine.

The objectives of geological work at the Magma mine in Arizona illustrate recent progressive trends in applied geology at some mines, and are summarized as follows:[3] 1. Giving direct assistance to the mining department by predicting possible size, shape, grade and location of orebodies. 2. Providing intelligent guidance for exploration programs, especially diamond drilling in undeveloped sections of the mine. 3. Contributing to the understanding of the general geology of the deposit, particularly the ore genesis, structure and mineral relationships. 4. Providing assistance in stoping operations. Small post-ore faults which have displaced the vein about 15 to 20 ft., present obstacles to efficient mining operations. Raises and stopes must be planned to permit extraction of the ore on both sides of a fault without crossing through the fault itself, as crossing the fault zone would require stoping or raising in fractured, heavy ore which would be difficult to support.

Structure Studies Needed

Present day knowledge of roof-control in coal mines is very limited. As an example, fatalities from roof-falls in coal mines in Illinois represent a large percentage of the total fatalities in the coal mining industry.[4] General types of roof instability which result from geological influence point emphatically toward the importance of rock structure in roof control.

Numerous mines have been forced to abandon operating shafts because of subsidence of ground over stoped-out areas. Other shafts have been saved from destruction by the presence of a fault between the shaft and the subsiding blocks. The Sacramento shaft at the Copper Queen mine[5] continued in operation because of a steeply dipping fault which interrupted subsidence before it reached the expected line of break.

In addition to these examples, it could be shown quite readily that mining operation problems in every mine are either simplified or complicated by the character of local rock structure. In most standard works on mining this relationship is taught or exemplified by vague implication when its importance should dictate a considerable amount of emphasis. There is a tendency to present recitation of details of mining practices over the world with little or no complete discussion as to the physical reasons for their use, except perhaps a brief statement that rock or ore is "heavy" or firm, fractured or solid.

Physical Data Important

The importance of geological studies to mining operations would be better appreciated if textbooks and teaching methods stressed the quantitative side of geology more effectively, particularly the applications of physics and chemistry to geological problems. Admittedly, the fact that ore deposits are seldom, if ever, in homogeneous, unfractured material, and thus a large part of laboratory data cannot be directly applied to underground conditions. Nevertheless, laboratory testing, such as is now being conducted by the U. S. Bureau of Mines and by Harvard and Columbia universities, represents one of the most promising fields for research. Data on strength of rocks and other physical properties such as creep, compressibility, density, porosity and strength when subjected to stresses under ordinary and high confining pressures, will assist materially in throwing more light on rock-burst problems, support problems, roof-falls and many others. An appraisal of stresses in rock around mine openings on a more quantitative basis appears as a possibility which may not be too remote. This type of valuable information has a definite place in both mining engineering education and mining engineering practice. Students of today, who are the men of science of tomorrow, should acquire a background in this type of *fundamentals* and their application if progress is to be made.

2. Personal communication, Reno H. Sales, Anaconda Copper Mining Co., Butte, Mont.

3. Michell, Wilson D., Applied Geology at the Magma Mine, Superior, Ariz. AIME TP 2214, Sept., 1947.

4. Roley, Rolf W., Certain Geological Conditions That Cause Coal Mining Fatalities. Illinois Engineer, pp. 7-9, Jan., 1947.

5. Kantner, W H., Surface Subsidence Over the Porphyry Caving Blocks, Phelps Dodge Corp., Copper Queen Branch. AIME Trans. 109: 181, 1934.

31

Reprinted from *Canadian Inst. Mining and Metallurgy Bull.* **68:**114–116 (1975)

Importance of Geological Data in Planning Underground Ore Extraction

C. L. Pillar, Vice-President, Operations,
Placer Development Ltd.,
Vancouver, B.C.
A. D. Drummond, Assistant Exploration Manager,
Canex Placer Ltd.,
Vancouver, B.C.

Abstract

Brief comment is made on the importance of accurate and factual geological data to the engineering team planning the development of a mining enterprise, particularly when underground extraction is involved. The assembling of the geological information and the manner in which it can be used to best advantage is discussed.

Introduction

IN MANY CASES, substantial time may elapse between the exploration stage and a serious detailed planning stage which has the objective of putting a mineral property into production. This is always true when properties change ownership and new partners are brought in for joint-venture situations. Such conditions make it imperative that the recording of complete and accurate geological data for the property is done during the active part of the exploration phase and accomplished by competent geologists and engineers. This work should not be left to helpers and technicians. Furthermore, it should not have to be redone years after the actual drilling was accomplished, when either those involved have parted company or those present have lost the touch and feel of their part of the exploration program and must rely on a memory that is, at best sketchy in detail. All of this emphasizes the importance of thorough geological evaluation and reporting during the exploration program.

To evaluate and comment on the application of geological data by a mine planning department, we must assume (a) that sufficient diamond drill holes have outlined the existence of a potentially economic ore deposit and (b) that the next step is to obtain the required information for a detailed feasibility study. This paper will comment on:

(1) what types of geological information are required;

(2) how it can be obtained; and

(3) the application of this information by the mine planners.

Basic Information Required

There are three major areas in which the mine planning team requires accurate and reliable information from the exploration group: (1) a detailed mineral inventory; (2) the strength and competency characteristics of the ore and host rocks; and (3) the amount and quality of subsurface waters. Each of these three areas will be examined in detail.

Mineral Inventory

Generally, the over-all tonnage and grade is defined into a geological reserve which is classified by the distance to the nearest drill hole. However, the geological reserve can be very much larger than the actually extractable or mining reserve, which is usually determined by the mine planning team. It is obvious that the mine planner must have more information than merely a geological concept which has produced a three-dimensional model of the mineralized body. What are the other items of information which should be supplied in addition to total tons, average grade and shape?

A complete inventory of all economic mineralization should be calculated during the exploration program. For example, in the case of a copper body, the tonnage, grade and spatial distribution of that mineral zone which is characterized by oxide, sulphide and the overlap of sulphide plus oxide should be given. Oxide zone here is defined as the zone in which secondary copper minerals occur. Similarly, primary sulphide copper-molybdenum mineralization should be defined in terms of the overlap or non-overlap of these two metals.

The content of secondary metals which contribute to the value of the ore should be recorded and the minerals present identified. This is particularly important for precious metals and supplementary metals, such as molybdenum, nickel, cobalt, cadmium, bismuth and mercury. All of this information can be readily obtained by geologically logging and assaying all drill core within the mineral deposits, and the data can readily be shown on plans and sections.

This information is necessary to calculate ore on a gross value for the recoverable marketable minerals present. A full and detailed economic assessment of the ore-grade distribution throughout the deposit allows the planning team, where physically practical, to schedule early production from the most valuable sections of the ore deposit. Such procedures favour rapid return of capital investment.

Associated with mineral inventory is a need for geological definition when considering bench-scale metallurgical testing, because there may be several different mineralogical types of ore which may occur at different positions throughout the deposit. In the definition of mineralogical types for the metallurgist, the geologist, during the exploration stage, must be observant and recognize the differences in the mineralized zone which may cause changes in the milling process. A detailed mineralogical study of mill products as well as head material must not be academic in nature, but must be keyed to the needs of the engineers who design the concentrating plant.

This data, coordinated with metallurgical recovery, can be a valuable aid in scheduling future mine production.

Strength Characteristics of Ore and Host Rock

The strength or degree of competency of the ore in the deposit and that of the adjacent rock formations where development openings will be driven are of extreme importance to the mine planner. This is the major factor in deciding the most favourable mining plan to be used in extracting the ore deposit. The degree of rock competency will determine the structural effect of ore extraction on the mineral deposit and its adjacent rock formations.

To evaluate the degree of competency of the ground involved, one must include compressive, tensile and shear strength, as well as the frequency of jointing and parting planes in the rock. An evaluation of the strength characteristics of the rock should be made and amply described from the drill cores by qualified geologists during the exploration program. This itself can aid in the answer to a host of the planning team's questions where multi-million-dollar decisions must be made, based solely on rock competency evaluation and experienced mining judgment.

As rock strength is a complex factor with many variables, the mineralized zone should be subdivided geologically to outline the spatial distribution and inter-relationship of: (a) rock type, (b) hydrothermal alteration, (c) sulphide and silicate zoning, (d) veining intensity, and (e) distribution of fracture, fault and breccia zones.

All of this information is available through detailed logging of the drill core by an experienced geologist during the drilling stage of the project. While logging, the geologist may set aside specimens to determine the physical strength of a rock unit and, when in ore, he will designate appropriate sections of core to determine the work index of that particular rock assemblage; this will contribute not only to mine planning but to metallurgical design as well.

To utilize the interrelated variables which affect rock strength, the geologist must consider some of the following questions:

1. Is the rock cavable with satisfactory fragmentation for underground extraction?
2. Is the overburden cavable?
3. How large a span will stand unsupported without subsidence in both ore and overburden?
4. How much ground area can be opened at one time and stand unsupported as a vertical wall and as a horizontal back?
5. What will the transportation and handling characteristics of broken ore and waste be — in transfer raises, pockets, chutes, cars, conveyor belts, crushers and screens?

The definition of rock type, for example, will denote what range of strength that rock might have and, consequently, the size of opening that will stand unsupported. If the rock is quartz monzonite and is fresh, the strength characteristics for quartz monzonite can be used, but these same characteristics can not be used if the rock has been hydrothermally altered by feldspathic decomposition to clays or if the rock has been intensely veined or fractured. The best information on the strength characteristics is obtained when the geologist logging the core can see the inter-relationship of the various physical features and can make the appropriate notations.

Let us take this example one step farther. A hydrothermally altered, veined, quartz monzonite may be cut by a number of strong, continuous and, hence, definable zones of weakness; perhaps a 10-foot gouge zone in one locality or a 50-foot zone of parallel shear surfaces in another. The geologist's interpretation of a specific structure or, more important, of where several structures meet within the mineralized zone can be of importance to the mine layout, because it may be less than desirable to have major development openings in the only well-sheared rock on that particular level.

Alternately, it is just as important to recognize those portions of the mineralized zone which would have a low work index; i.e., relatively weak rock. If, for example, the best of the sulphide body is composed of many different rock types, each of which exhibits differing chemical responses to alteration and to fracture density, the mine planner must be aware of this geological condition. A work index which gives an indication of the crushability of the various rocks is a necessary parameter in mine scheduling and hence is important to the mine planner, but its importance at an early stage must be recognized by the exploration geologist, because this type of information is of use not only to the mine planning team but also to the plant designers. Normally, the average work index of the ore will set the design capacity of the concentrator. Ore with a work index 10% higher than the design average means 10% less concentrate production by virtue of reduced mill throughput; inversely, the lower the work index, the higher the mill throughput. Grade, metallurgical recovery and work index of the ore to be extracted are major criteria for the planning team in scheduling mineral concentrate production.

It is acknowledged that the best assessment of rock competency from diamond drill cores can only be approximate. However, it can be a very useful guide in directing the development of initial concepts for the mining plan before actual mine development allows direct ground evaluation. Data on the competency of the rocks that will be encountered, with their structural and shear zones of weakness, are highly important to the planning team and if reasonably accurate will be of major help in arriving at correct decisions for the mining method and development plan to be used.

Subsurface Waters

The planning team must have a fairly accurate appraisal of the amount of ground water that will be encountered in driving major primary development openings and extracting the ore deposit. Accurate knowledge in this area will aid the planning team in estimating the cost and time required for primary development.

This type of information can be assessed by the exploration team. During exploration drilling, all of those occasions when the hole suddenly starts to make water or, alternatively, lose water can be determined from the drill contractor's daily log and noted on the sections given to the mine planner. In those holes which make water, the flow can be measured by noting the time required to fill a gallon container. Analysis of the water can give a good indication of acidity (pH), as well as metallic and soluble salt content. The temperature of the water should also be recorded. A thermometer, potentiometer and clean bottles are all that is required to get this type of information, which must then be correlated with the geological sections and plans to determine which shear zones, faults, rock formations or fractured zones will contain appreciable quantities of water.

The best possible interpretation of the ground-water conditions will aid the mine planner in assessing the following questions:

1. Can the deposit be successfully drained; if so, what equipment will be needed and how long will drainage be required prior to initial mining?

2. What effect will the underground water have on ground stability?

3. How much additional ground sealing or support is required to combat the estimated water flows which may be encountered during the primary development stage?

4. What is the amount and quality of the ground water in the area that can be used as a potential source for treating plants and as potable water?

5. What is the temperature of the underground water and what will its effect be on the development cost and time schedule?

Summary

A mine planning team requires definite and accurate geological information in order to properly plan the development and extraction of any ore deposit. The deeper-seated deposit requires information that is more thorough in extent and more accurate in content, due to the fact that design concepts must be developed and early planning decisions made based solely on the geological information supplied by the exploration group. This source of information must be obtained from a complete and thorough geological evaluation of the surface area, the drilling operation and the drill cores. At this stage, there has been no opportunity to examine the underlying rock formations or obtain bulk samples from mine development workings.

The data required from the exploration group can be divided into three major areas and should be as complete and reliable as conditions permit.

1. An inventory of all the economic mineralization in the deposit as to tonnage, grade and location, which is commonly known as the geological reserves. This reporting should be backed by proper storage and indexing of all diamond-drill cores made readily accessible for further reference as may be required.

2. The strength and competency of the ore in the deposit and in the adjacent rock formations where development openings will be driven. The competency of the ground involved is the major factor in deciding the most favourable mining plan and determining the size and location of the mine's primary development.

3. The amount and quality of ground water that is in and around the ore deposits, and the underground flows that will be encountered in driving development headings and extracting the ore. This is a major factor in the time and cost required to develop and extract the ore deposit.

It is highly important that the exploration team be acutely aware of the needs of the mine planner in order that the necessary care and attention will be used to obtain and report the required data. A competent mining engineer, preferably a senior member of the future mine planning team, should make frequent inspections and become thoroughly familiar with the project during the exploration drilling program.

32A

Reprinted from *Mining Mag.* **114**:236–237, 239, 241, 243, 245, 247, 249, 251, 253, 255 (1966)

Rock Mechanics—An Introduction for the Practical Engineer

Part I—Theoretical Considerations

E. HOEK, Ph.D., M.Sc. (Eng.), B.Sc. (Eng.).*

THE purpose of this series of articles is to present some of the principles of rock mechanics to the reader who is interested in knowing something about the subject without having to become too involved in technical details. The contents of the articles is based, almost entirely, upon the author's own experience and, hence, does not represent a complete review of the whole subject of rock mechanics. Consequently, the formulation of an overall rock mechanics philosophy has been assiduously avoided and the reader is left to draw his own conclusions from the facts as they are presented.

Definition and Scope

In 1963 the Rock Mechanics Committee of the American National Academy of Science adopted the following definition [1]†:

Rock mechanics is the theoretical and applied science of the mechanical behaviour of rock, it is that branch of mechanics concerned with the response of rock to the force fields of its physical environment.

It is convenient to subdivide rock mechanics into the following branches:

(a) *Structural rock mechanics*, which is concerned with the stability of engineering structures in which the material is predominantly rock.

(b) *Comminution*, which is concerned with the reduction of rock to small fragments by the application of external forces as in drilling, blasting, cutting and grinding.

Both these branches of rock mechanics involve the control of rock deformation and fracture processes. In the first case, excessive rock failure** must be avoided in order to preserve the stability of the structure and, in the second case, rock fracture must be induced with the minimum input of external energy. Major disasters, such as the Malpasset and Vajoint dam failures, and the Coalbrook mine disaster, serve to illustrate the importance of rock fracture in practical engineering terms.

Rock Fracture—Griffith Theory

Rock mechanics research in South Africa was initiated some 15 years ago in an effort to provide an understanding of the rockburst hazard which occurs in many deep-level gold mines.

The effects of a typical rockburst are illustrated in Fig. 1. It has been defined[2] as damage to underground workings caused by the uncontrolled disruption of rock associated with a violent release of energy additional to that derived from falling rock fragments. The main causes of rockbursts are associated with the energy changes induced by mining in the rocks surrounding large excavations and these causes have been reviewed elsewhere[2].

From the rock mechanics point of view, the main characteristic of a rockburst is the fact that it occurs in hard, brittle, highly competent rocks. Consequently, in studying the fracture behaviour of these rocks, it was considered justifiable to study the behaviour of the rock material itself, treating it as a homogeneous, isotropic solid and ignoring the effect of major geological discontinuities. The deficiency of this approach, when applied to the fractured and geologically discontinuous rocks which occur on or near the earth's surface will be immediately obvious to the reader. Never-the-less, it is believed that an understanding of the basic mechanism of the fracture of rock material can be of assistance in formulating a rational behaviour pattern for rock masses.

Griffith's theory of brittle fracture[3], modified by McClintock and Walsh[4] to allow for the predominantly compressive stresses in rock mechanics, has been found to provide a

---

* Senior Chief Research Officer, Rock Mechanics Division, National Mechanical Engineering Research Institute, South African Council for Scientific and Industrial Research, Pretoria, Republic of South Africa.

† Superscripts refer to the list of references given at the end of this paper.

** In this context, failure is taken to mean either excessive deformation or fracture.

reliable theoretical basis for the prediction of rock fracture phenomena[5]. This theory is based upon the assumption that fracture initiates at inherent cracks and discontinuities within the material and that propagation of these cracks occurs as a result of the tensile stress which is induced at the crack tip under load. Brace[6] has shown that fracture in hard rock usually initiates in grain boundaries which can be regarded as the inherent discontinuities required by the Griffith theory.

Griffith's original theory was concerned with brittle fracture under conditions of applied tensile stress and he based his calculations upon the assumption that the inherent crack, from which fracture initiates, could be treated as an elliptical opening. When applied to rock mechanics, in which the applied stresses are predominantly compressive, this simplifying assumption is no longer valid and the theory has to be modified to account for the frictional forces which occur when the crack faces are forced into contact. This modification was carried out by McClintock and Walsh[4] who made further simplifying assumptions concerning the mechanism of crack closure. These simplifying assumptions have recently been theoretically validated by Berg[7].

The extent to which the modified Griffith theory defines the fracture behaviour of rock is illustrated in Fig. 2. Published triaxial strength test data for the fifty rock and concrete types listed in Table 1 are included in this graph. In order to render the test results comparable and to minimise differences caused by different testing techniques, specimen sizes and environmental conditions, the values are plotted on dimensionless scales which are obtained by dividing each test result by the uniaxial compressive strength of that particular material.

A further illustration of the usefulness of the Griffith theory in defining the fracture behaviour of hard rock is given in Fig. 3. In this figure a theoretical Mohr envelope is fitted to Mohr fracture circles obtained from triaxial tests on specimens of a typical South African quartzite.

In spite of the encouraging agreement between theoretical and experimental results, illustrated in Figs. 2 and 3, it would be incorrect to suggest that the Griffith's theory provides a complete description of the mechanism of rock fracture. It must be emphasised that its derivation is such that it is only strictly correct when applied to fracture initiation under static stress conditions[5]. It is largely fortuitous that it can be so successfully applied to the prediction of the fracture of rock specimens since, once fracture has initiated, propagation of this fracture and ultimate disintegration of the specimen is a relatively complex process[8 9 10]. Fortunately, it appears that the forces involved in fracture propagation are closely analogous to the friction forces assumed in the modified Griffith theory and hence the general form of the equations which define fracture propagation is very similar to that of the equations which define fracture initiation.

The original and modified Griffith theories, when expressed in terms of the stresses at fracture[5], contribute little to the understanding of rock fracture under dynamic stress conditions, the energy changes associated with fracture or the deformation process of rock. However, since the theoretical concept of fracture initiation from inherent cracks has proved so useful in describing the observed fracture behaviour of rock, this concept is being extended to the theoretical study of energy changes and deformation processes in rock[11 12].

The author is particularly fascinated by the belief that the processes which govern the failure of large, fissured and discontinuous rock masses are very similar to those which operate during the failure of a small rock specimen[10]. It is hoped that a rational theoretical description of the movement of interlocking blocks of rock in a large rock mass will eventually be built up.

Fig. 1. Effects of a rockburst in a deep-level South African gold mine.

Additional Factors Governing Fracture

The Griffith theory was derived on the assumption that the material contains a random distribution of uniform cracks and that the inherent physical properties of the material remain constant. It is interesting to consider to what extent the theoretical concepts of the Griffith theory can be modified to cover cases in which the above assumptions do not apply.

Fracture of Anisotropic Rock

An extreme example of a rock in which inherent cracks are not randomly distributed is slate. If it is assumed that slate contains two crack systems, one preferentially oriented system of large bedding plane cracks and one randomly oriented set of small grain boundary cracks, it becomes possible to calculate the stress levels at which fracture would initiate under various conditions[13 14].

Fig. 4 illustrates the remarkable agreement between the predicted and observed fracture behaviour of slate specimens subjected to uniaxial compression. It will be noted that the highest strength of slate can be as much as four times its lowest strength, depending upon the orientation of the bedding planes to the direction of applied load.

Table 1

Summary of Triaxial Test Results on Rock and Concrete

Graph Point	*Material*	* *lb./sq. in.*	*Tested by*
1	Marble	13 700	Ros and Eichinger
2	Marble	18 000	Ros and Eichinger
3	Marble	20 000	Von Karman
4	Carthage Marble	10 000	Bredthauer
5	Carthage Marble	7 500	Bredthauer
6	Wombeyan Marble	10 000	Jaeger
7	Concrete	2 380	McHenry and Karni
8	Concrete	3 200	Akroyd
9	Concrete	6 000	Jaeger
10	Concrete	5 700	Fumagalli
11	Concrete (28 day)	3 510	Balmer
12	Concrete (90 day)	4 000	Balmer
13	Granite Gneiss	25 500	Jaeger
14	Barre Granite	24 200	Robertson
15	Granite (slightly alt)	10 000	Wreuker
16	Westerly Granite	33 800	Brace
17	Iwaki Sandstone	1 780	Horibe and Kobayashi
18	Rush Springs Sandstone	26 000	Bredthauer
19	Pennant Sandstone	22 500	Price
20	Darley Dale Sandstone	5 780	Price
21	Sandstone	9 000	Jaeger
22	Oil Creek Sandstone	**	Handin
23	Dolomite	24 000	Bredthauer
24	White Dolomite	12 000	Bredthauer
25	Clear Fork Dolomite	**	Handin
26	Blair Dolomite	**	Handin
27	Blair Dolomite	75 000	Brace
28	Webtuck Dolomite	22 000	Brace
29	Chico Limestone	10 000	Bredthauer
30	Virginia Limestone	48 000	Bredthauer
31	Limestone	20 000	Jaeger
32	Anhydrite	6 000	Bredthauer
33	Knippa Basalt	38 000	Bredthauer
34	Sandy shale	8 000	Bredthauer
35	Shale	15 000	Bredthauer
36	Porphy	40 000	Jaeger
37	Siouxyr Quartzite	**	Handin
38	Frederick Diabase	71 000	Brace
39	Cheshire Quartzite	68 000	Brace
40	Chert dyke material	83 000	Hoek
41	Quartzitic shale (Dry)	30 900	Colback and Wiid
42	Quartzitic shale (Wet)	17 100	Colback and Wiid
43	Quartzitic sandstone (dry)	9 070	Colback and Wiid
44	Quartzitic sandstone (wet)	4 970	Colback and Wiid
45	Slate (primary cracks)	4 300	Hoek
46	Slate (secondary cracks)	15 900	Hoek
47	Dolerite	37 000	CSIR
48	Quartzite (ERPM Footwall)	31 000	CSIR
49	Quartzite (ERPM Hangingwall)	43 200	CSIR
50	Glass	91 000	CSIR

*Uniaxial compressive strength
** Presented in dimensionless form by McClintock and Walsh.

An important practical conclusion which can be drawn from Fig. 4 is that a comparison of the results obtained from *compression* tests on core drilled normal to and parallel to the bedding planes does not necessarily determine whether the material is anisotropic—a procedure sometimes advocated by those concerned with "practical" tests. In the case of slate, the compressive strength of specimens drilled normal to and parallel to the bedding planes is almost the same, and if the strength of the specimen in which the bedding planes are oriented at 30 deg. to the direction of applied load is not taken into account, one may be tempted to conclude that slate is isotropic. This extreme example is included to demonstrate the dangers involved in drawing conclusions from inadequate test data.

Slate is a relatively simple example of an anisotropic material, having only one set of dominant planes of weakness. The same theoretical process as was applied in the case of slate could be extended to describe the behaviour of a material such as coal which may have two or more cleat systems in addition to its bedding planes.

Influence of Environment upon Rock Strength

It is frequently assumed that the strength of rock is not significantly influenced by the temperature or humidity of its surroundings. It has, however, been demonstrated that this assumption, particularly in the case of the influence of humidity, can be seriously in error[15].

The influence of temperature upon the strength of rock is probably not significant within the normal range of temperatures encountered by the civil or mining engineer. However, at great depths where the temperatures approach the melting point of some of the rock constituents, the reduction in strength may be significant and could be of importance to those concerned with the overall behaviour of the earth's crust and with the origin of deep-level earthquakes.

The influence of moisture upon the strength of rock is so important that the author advocates that tests on coal and soft rocks should be carried out on site. In order to minimise changes in the moisture content of the specimen, the tests should be carried out as soon as the specimens have been removed from the parent rock. The practical details of this type of test will be discussed later.

The influence of moisture content upon the strength of quartzitic shale is illustrated in Figs. 5 and 6 which are reproduced from a paper by Colback and Wiid[15]. It will be seen that a saturated specimen of this quartzitic shale is only half as strong as a dry specimen. Colback and Wiid have postulated that this reduction in strength is due to a reduction in the molecular cohesive strength of the rock material when moisture is present.

The practical importance of the influence of moisture upon the strength of rock in structural rock mechanics is the danger of a normally stable structure becoming unstable in wet conditions. In comminution, the strength reduction obtained under wet conditions results in more efficient cutting or drilling.

Influence of Fluid Pressure

In addition to the strength reduction associated with a high moisture content, a further threat to the stability of a rock structure occurs when water is present under pressure. This fluid pressure reduces the compressive stress acting across a fissure or fracture plane and hence the frictional resistance which causes interlocking of blocks of rocks can be reduced. In an extreme case, one block can be literally floated off another by the buoyance effect of water pressure. The role of fluid pressure in determining the strength of a rock mass is fairly well understood[16] and its influence can be allowed for in strength calculations.

Time-dependent Failure of Rock

One of the least understood aspects of the mechanical behaviour of rock is the influence of time upon its deformation

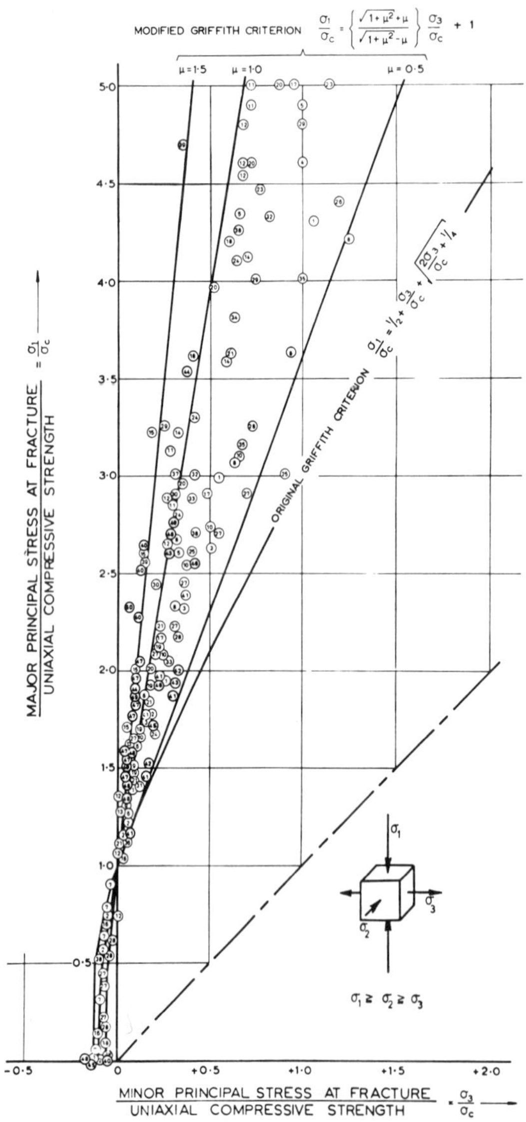

Fig. 2. Triaxial fracture data for the fifty rock and concrete materials listed in Table 1.

and fracture. It is convenient to consider this subject under two headings:

(a) Weathering which is the gradual deterioration of an exposed rock surface. This deterioration may take place in the absence of applied stress and is due mainly to physical and chemical processes which are governed by the environment to which the rock is exposed;

(b) time-dependent mechanical behaviour which involves the deformation or fracture of rock under conditions of constant applied stress*.

In most practical cases, both of these phenomena play a part and it is usually impossible to establish which of the two is the main cause of failure.

Experience in South Africa shows that coal pillars usually fail a number of years after they have been mined. Examination of the pillars and measurements of time-dependent deformation suggests that both weathering and time-dependent failure are important. Consequently, one must conclude that a full understanding of the time-dependent behaviour of rock must involve a knowledge of both weathering and time-dependent failure. To the best of the author's knowledge, no complete and systematic study of the process of rock weathering has ever been undertaken and this deficiency presents an interesting challenge to rock mechanics research workers.

Time-dependent mechanical behaviour of rock has been fairly intensively investigated, both theoretically [17] and experimentally[18], but a great deal more remains to be done before this knowledge can be effectively applied to practical problems. For the engineer faced with the problem of designing a rock structure in which weathering or time-dependent failure may be important, the most realistic approach appears to be to use the results of short time laboratory tests with a liberal allowance for the possible strength reduction with time. The extent to which the strength will reduce with time depends upon so many unknown factors that no definite rules could be suggested but the author believes that under severe conditions, a reduction of 50 per cent over a ten year period is possible.

Influence of Specimen Size

It is accepted that the strength of a brittle material is dependent upon the size of the test specimen and yet very little reliable quantitative data on this effect is available. In the author's opinion, the most rational approach to this problem is that adopted by Protodiakonov[19] and his most important conclusions are presented graphically in Fig. 7. The parameter 'm', as defined in Fig. 7, depends upon the nature of the material and also upon the state of stress to which the specimen is subjected. From the few experimental data which are available the author has made the following estimates of the value of 'm':

Coal and soft rocks subjected to compression	$5<m<10$
Coal and soft rocks subjected to tension	$10<m<50$
Hard rock subjected to compression	$2<m<5$
Hard rock subjected to tension	$5<m<10$

While these estimates must be treated with extreme caution, they do enable one to arrive at an order of magnitude for the acceptable specimen size.

Hence for a material such as coal subjected to compression, the specimen size should be approximately 50 to 100 times

* Frequently referred to as "creep", a term which the author avoids because of the possible confusion with the process of creep in metals which need not be the same as in rock.

the spacing between discontinuities if the strength of the specimen is to be within 10 per cent of the strength of the rock mass. Since the spacing between discontinuities may, as a first approximation, be equated to the cleat spacing which can be of the order of 2 in., the specimen diameter required is estimated at between 100 and 200 in.

This size of specimen would daunt even the most courageous exponent of large-scale testing and yet, if Protodiakonov's deductions and the author's estimates are reasonably accurate, it may have to be accepted that most tests on coal are carried out on specimens which are far too small to give meaningful results.

In a series of large-scale tests planned by the Coal Mining Research Controlling Council of South Africa in conjunction with the Council for Scientific and Industrial Research it is proposed to test coal specimens to 8 feet cube (the technique to be used will be discussed later). The results of these tests should contribute towards a more reliable evaluation of the problem of the size effect in strength testing.

In the case of hard rocks such as the Witwatersrand quartzites, the value of 'm' for compression is believed to lie between 2 and 5 and the spacing between discontinuities, in this case assumed to be the grain boundaries, may be of the order of one tenth of an inch. Hence, from Fig. 7, the diameter of specimen required to give a strength value within 10 per cent of that for the massive rock (excluding the effect of major discontinuities) is between 1 and 2 in.

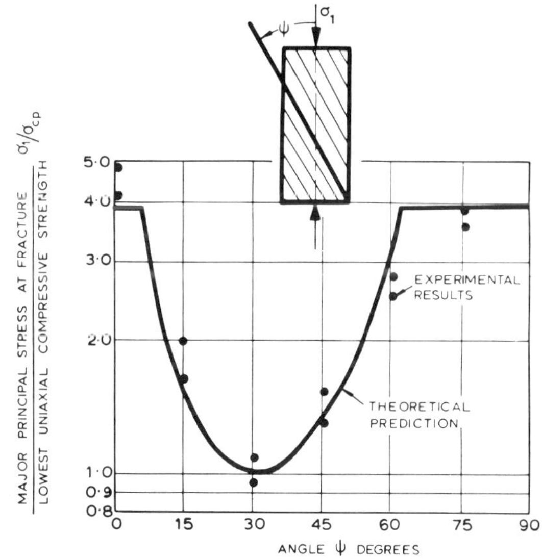

Fig. 4. Relationship between bedding plane orientation and strength of slate.

Stability of Rock Structures

In designing a rock structure such as a dam foundation or underground excavation, the most important consideration is the stability of the entire structure. Local rock failure at points of high stress or in zones of exceptionally low strength may only be significant if this failure forms part of a sequence of events which leads to collapse of the structure.

The obvious question which the practical engineer will ask is—what are the main factors which govern the stability of a rock structure and what information on these factors and their interaction does the science of rock mechanics provide?

Two Main Problem Types

In discussing this question it is necessary to distinguish between two main types of structural rock mechanics problems:

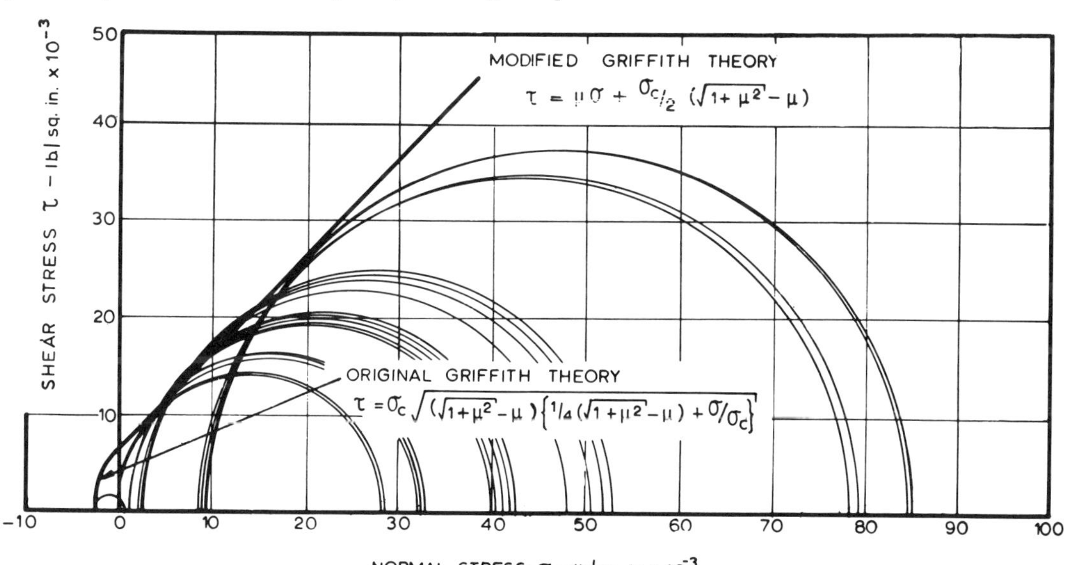

Fig. 3. Mohr fracture diagram for typical Witwatersrand quartzite with a uniaxial compressive strength $\sigma_c = 30{,}000$ *p.s.i. and a coefficient of internal friction* $\mu = 1\cdot00$.

(a) Underground excavations in solid homogeneous rock, such as the massive quartzites which occur in the deep-level gold mines in South Africa, in which the stability of the rock surrounding the excavation is primarily dependent upon progressive failure of the *rock material*.

(b) Rock structures such as dam foundations and surface excavations in which the material is so faulted and fissured that the stability of the structure depends upon the movement of interlocking blocks within the *rock mass* rather than upon failure of the rock material.

Progressive Failure of the Rock Surrounding an Excavation

When a rock specimen is loaded in a hydraulically actuated testing machine, the behaviour of the specimen at the point of fracture is obscured by the behaviour of the testing machine. The release of the energy stored in the hydraulic system usually results in violent disintegration of the rock specimen with the consequent loss of all post-fracture data. On the other hand, an element of rock which forms part of a rock mass and which is subjected to an identical state of stress to that applied by the testing machine will behave in an entirely different manner once the point of failure has been reached. In this case, disintegration of the rock element may be arrested by the transfer of load onto an adjacent element which previously carried a lower load. In this way, local fracture induces a redistribution of stress which may or may not result in further fracture.

In an attempt to understand this process of progressive failure, the author has studied the initiation and propagation of fracture in rock under non-uniform stress conditions such as those which occur around underground excavations[20].

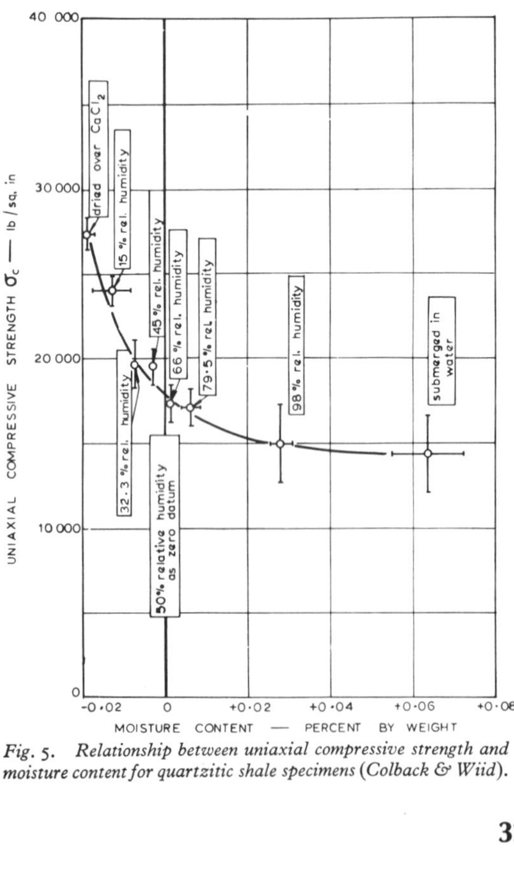

Fig. 5. Relationship between uniaxial compressive strength and moisture content for quartzitic shale specimens (Colback & Wiid).

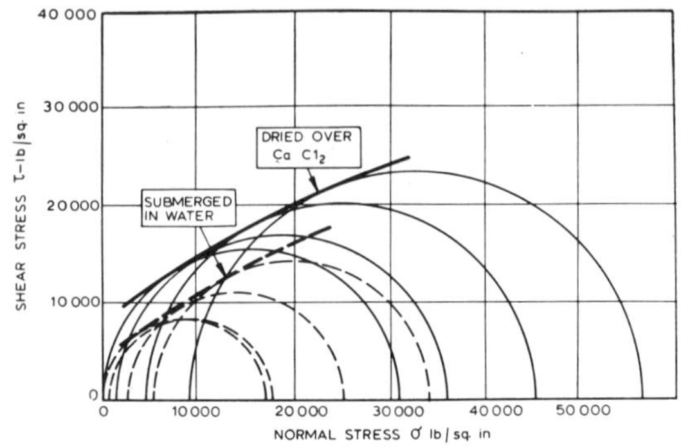

Fig. 6. Mohr fracture envelopes showing the effect of moisture on the compressive strength of quartzitic shale (from Colback & Wiid).

An example of this type of analysis is presented in Fig. 8 in which the theoretical fracture contours in the rock surrounding square and elliptical tunnels are compared. In deriving these contours it was assumed that fracture initiation in the material, assumed to be a homogeneous quartzite, is defined by the Mohr envelope illustrated in Fig. 3. This fracture criterion is combined with the stress distribution in the material surrounding the excavations[21] to give the fracture contours and the critical crack trajectories[20].

The fracture contours are plotted in terms of the vertical pressure P which would have to be applied to the material surrounding the excavation in order to cause initiation of fracture at any point. Hence, if the fracture contour has a value P=10,000 p.s.i., all the material enclosed by this contour, *i.e.* the material in which the fracture contours have lower values, would be prone to failure if the vertical applied pressure P was 10,000 p.s.i.

The vertical applied pressure P is due primarily to the weight of the overburden, increasing at a rate of approximately 1 p.s.i. per ft. of depth. The local pressure P which acts upon a particular tunnel may, however, be intensified by mining activities in the vicinity of this tunnel. Hence, a tunnel situated in a pillar may be subjected to a vertical applied pressure P of 15,000 p.s.i. while the pillar itself may only be subjected to a nominal pressure, due to its depth below surface, of 7,000 p.s.i.

The critical crack trajectories define the most dangerous crack orientation at any point in the material surrounding the tunnel. An inherent flaw or discontinuity oriented in this direction would fail at a lower stress level than that required to initiate failure from any other flaw, of similar size, but at a different orientation, in the vicinity.

The significance of the fracture contours and critical crack trajectories, illustrated in Fig. 8, is interpreted in more practical terms in Fig. 9. In both cases, the lowest fracture contour occurs in the roof (and floor) of the excavation hence ailure can be expected to initiate at these points. Since the critical crack trajectories in this region lie parallel to the vertical axis of the excavation, a vertical crack will occur as shown in Fig. 9a.

Experimental studies of the formation of vertical roof and floor cracks[22] have shown that, apart from a redistribution of the stress in the proximity of the crack, the stability of the excavation is not endangered by the formation of these cracks. It must also be pointed out that the tendency for these cracks to form is markedly reduced by the presence of lateral applied stress due to the restraint of the surrounding rock[23].

The initiation of sidewall fracture, illustrated in Fig. 9b is dependent upon the geometrical shape of the excavation. In the case of the square tunnel, the high stress concentration in the sharp corners results in fracture initiation at a vertical applied pressure P > 7 500 p.s.i. The critical crack trajectories suggest that this fracture would propagate in such a way that "slabbing" of the sidewall would occur.

The occurrence of this sidewall slabbing is illustrated in Fig. 10 which shows a haulage in quartzite at a depth of 9,500 ft. below surface in a South African gold mine. Fig. 10a shows the haulage in an area remote from mining activity and Fig. 10b shows the sidewall fracture associated with an increase in stress level due to extraction of the reef immediately above the haulage.

In the case of the elliptical tunnel, sidewall failure can be anticipated at P > 13,000 p.s.i. and this failure would probably take the form of sidewall "scaling" as shown in Fig. 9b. The fact that this scaling occurs at such a high stress level is important. Practical experience in deep-level mines which use both square and elliptical tunnels confirms that the elliptical tunnel, at the same depth, has a lower tendency to sidewall failure and requires less maintenance and support in the form of rock bolting. Evidence from model studies[20] suggests that the next stage in failure may be associated with the stress redistribution in the roof and floor and may follow the pattern illustrated in Fig. 9c, resulting in the final fracture configuration illustrated in Fig. 9d.

It must be emphasised that this analysis has been carried out on the assumption that the excavations are situated in solid, homogeneous quartzite which is subjected to vertical pressure only. The presence of a fault or fissure or of marked anisotropic strength behaviour would have a significant

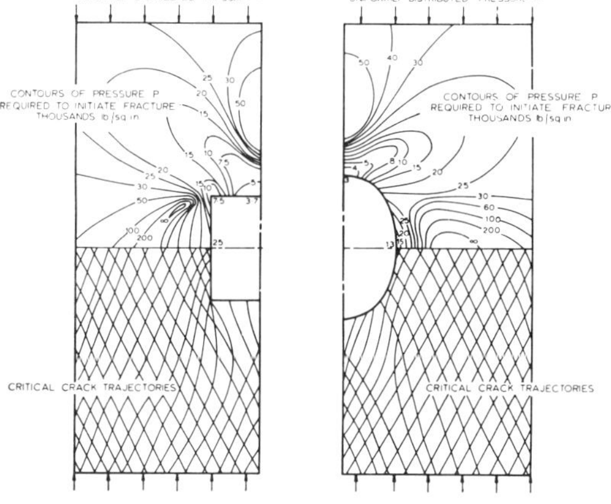

Fig. 8. Fracture contours and critical crack trajectories in quartzite surrounding square and elliptical tunnels.

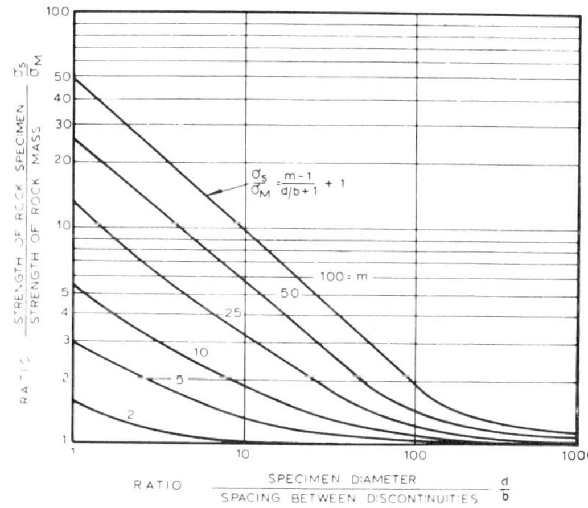

Fig. 7. Relationship between specimen size and strength according to Protodiakonov.

influence upon the analysis and could invalidate the conclusions reached.

A further difficulty, which has emerged from fracture studies on models [20], is that the stress distribution in the rock around an excavation is drastically altered by failure of the rock. In order to overcome this difficulty it is necessary to re-analyse the stress distribution for each new boundary condition created by propagation of the fracture. While this form of analysis has been done[20] it is, at present, far too complex and tedious to be of practical value. This example has been quoted to illustrate an approach which the author believes may lead to a better understanding of the problem of the stability of rock structures.

Another approach, also still in its early stages, has been adopted by Deist[24] and Cook[25]. This approach is based upon the assumption that, beyond a certain stress level (failure), the load carrying capacity of an element of rock decreases linearly with increasing strain. This assumption has been experimentally verified by Cook using a specially designed "stiff" testing machine[26]. Deist has carried out theoretical studies into the stability of the fractured-solid rock boundary around excavations and, while, as yet only simple cases have been analysed, promising results have been obtained.

A recent paper by Fairhurst and Cook[27] suggests that a factor which may play an important part in determining the stability of the rock around a mining excavation is the buckling of sidewall slabs such as those illustrated in Fig. 10b.

It will have become obvious to the reader that, as yet, the engineer who is faced with a problem of designing a stable rock structure has very little practical information at his disposal. At the same time, it will be appreciated that a start has been made in determining the most important factors governing the stability of rock structures and that the problem, having been defined, will be subjected to concentrated study until an acceptable practical solution has been achieved.

Stability of a Rock Mass

To the best of the author's knowledge there are, as yet, no general rules for analysing the stability of a rock mass.

In any case, rock structures in which the rock is fissured and faulted can take such a wide variety of forms that the only logical approach is to treat each individual case on its merits.

A practical example of such treatment, derived from factual experience, is presented in Fig. 11 which illustrates a rock slope in which several geological discontinuities are known to exist. A cursory examination of this figure shows that only one of these discontinuities (marked as a heavy line) represents a threat to the stability of the system. The forces acting on this potential failure plane are as follows:

F_W—The total weight of the rock wedge
F_N—The normal force acting across the plane
F_S—The shear force acting along the plane
F_P—The total force due to fluid pressure (if present) within the rock mass.

The condition for unstable equilibrium is

$$F_S = \mu (F_N - F_P)$$

where μ is the coefficient of friction which is effective along the potential failure plane.

The requirement for stability of the slope can be expressed as follows:

$$\mu > \frac{\cos\theta}{\sin\theta - F_P/F_W}$$

From the limited amount of experimental data available, it is believed that the coefficient of friction μ between two dry rock surfaces is in the region of 0·7. This value could be considerably lower if the potential failure plane contains soft filling material or if the rock surfaces have deteriorated.

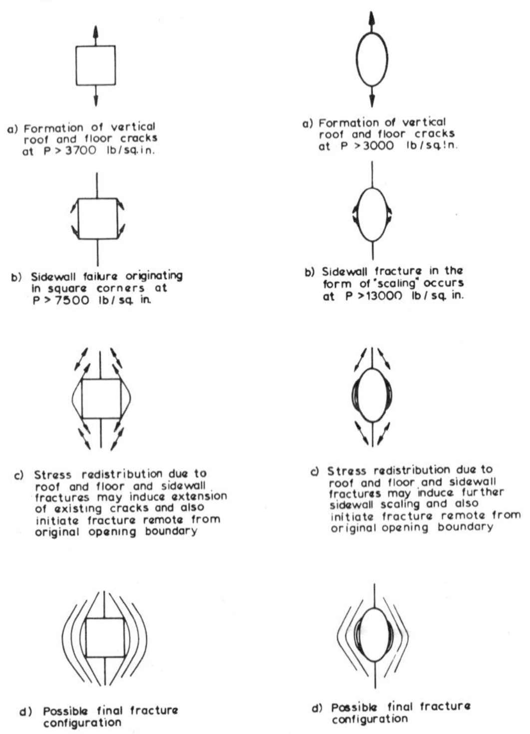

Fig. 9. Possible fracture sequence for square and elliptical tunnels in hard quartzite subjected to vertical pressure P only.

Control of Fracture in a Rock Structure

Rock has several physical characteristics which distinguishes it from other materials and, of these, the most important structural characteristics are:

(a) The large difference between the tensile and compressive strengths.

(b) The rapid increase in strength with increasing lateral or confining pressure.

(c) The tendency for the structural properties of some rocks to deteriorate with time due to weathering.

Bearing these characteristics in mind, the engineer has the following techniques at his disposal for making optimum use of his materials:

(i) The design of the structure to avoid, if possible, concentrations of tensile stress.

(ii) The use of support methods which increase the strength of the structure by increasing lateral pressure or restraint in zones of high compressive stress.

(iii) The protection of surfaces which are liable to weather.

Design of Rock Structures for Optimum Stress Distribution

In mining, the geometry of the orebody being mined dictates the layout of the major excavations within the mine. The only practical control which the engineer can exercise in this case is in support methods and in the sequence of mining[2]. However, in the case of secondary excavations such as service haulages, airways, pump chambers, etc., the mine designer usually has a reasonable amount of freedom which he can use to position these excavations for the most favourable stress distribution.

The calculation of the stress distribution around excavations is a complex process[21] which it would certainly not pay the average engineer to attempt to learn. Fortunately, there are several relatively simple model techniques which can be used, either by the engineer himself or by a university or research organisation which could undertake this work on a contract basis, to give solutions which are sufficiently accurate for practical design purposes. These techniques are discussed in Part 2.

In the case of civil engineering structures such as dam foundations or open-cast excavations, the engineer is usually in a position to design the structure for optimum stress conditions (in relation to the structural material at his disposal).

In this case, the low tensile strength of rock, particularly if it contains unfavourably oriented flaws and fissures, must be kept in mind and the structure designed to minimise tensile stresses as far as possible. Since this subject falls outside the scope of the author's experience, the interested reader is referred to the work of Müller[28] and Serafim[29] for further details on the application of rock mechanics to the design of civil engineering structures.

Use of Support in Rock Structures

In this context, support denotes the use of materials other than rock to improve the local properties of the rock and thereby improve the stability of the structure as a whole. Hence, for example, lining a shaft with concrete is frequently used to inhibit local deterioration and failure of the rock surface.

Generally, structural support is used (a) to support broken rock and to prevent it from breaking further and (b) to improve the strength of a rock mass by increasing the lateral stress in selected areas.

In the first case, the use of timber, steel or concrete structures or of rockbolts is familiar to any engineer who has been concerned with mining or tunnelling. The main purpose of this support is to minimise the danger of falling rock and to prevent loose pieces of rock from choking a passage-way. The use of this type of support depends so much upon local conditions that it would be meaningless to attempt to formulate general rules. On the other hand, the use of support to increase the strength of rock is a less well known concept and deserves close examination.

One of the principal features which distinguishes rock from other structural materials is the rapid increase, with increasing confining pressure, of the stress required to cause failure. In the case of quartzite, the stress at fracture increases by 6 units for every unit increase in confining pressure. Hence, for a quartzite with an unconfined compressive strength of 30,000 p.s.i., the stress required to fracture a specimen, subjected to a confining pressure of 1,000 p.s.i., would be 36,000 p.s.i.

If therefore, it is required to increase the strength of rock in a particular area, this can best be achieved by artificially increasing the triaxial compressive stresses acting on this area. This increase can be achieved by the use of such devices as rockbolts or steel arches.

High-tensile steel rockbolts have been used successfully in South Africa to inhibit sidewall failure of excavations. Rockbolting can be particularly effective when joints are present in the rock since bolts placed normal to the joint plane increase the resistance of the joint to shear movement by increasing the friction forces and also by the pinning action of the bolts themselves. The use of steel arches in tunnels provides effective support for broken roof rock and also increases the strength of the sidewall. The volume increase associated with early sidewall fracture results in a packing up of broken rock between the steel arch and the solid rock surface. The restraint provided by these rock fragments is usually adequate to prevent further sidewall failure.

Protection of Rock Surfaces from Weathering

Many rock types deteriorate rapidly when exposed to the atmosphere—a phenomenon commonly known as weathering. Hence a rock which appears to have desirable structural properties when freshly exposed may be unacceptable after having been allowed to weather for a few months. In some extreme cases in the author's experience, soft rock specimens left exposed to the atmosphere between preparation and testing have disintegrated into heaps of rubble within a few days.

In most cases, fairly simple remedial measures will prevent or at least effectively inhibit weathering. Frequently, painting or spraying the exposed surface with a thin mixture of cement and water will provide sufficient protection. Rocks

[*Editor's Note:* Figure 10a has been omitted.]

which are particularly prone to weathering may require a fairly thick covering of cement or concrete or some similar sealant. A commercially available unit which sprays foam plastic on to exposed rock surfaces has been found effective in some South African coal mines.

Summary and Practical Conclusions

The following characteristics distinguish rock from other commonly used structural materials, apart from concrete which behaves in a similar manner.

(a) Rock failure is generally of a brittle type, *i.e.* it occurs with little prior warning and is not accompanied by large non-elastic (plastic) deformations.

(b) The tensile strength of rock is usually of the order of one tenth of its unconfined compressive strength.

(c) The strength of rock increases rapidly with increasing confining pressure.

(d) Some rocks exhibit marked anisotropic strength behaviour due to the existence of preferentially oriented weakness planes such as the bedding planes in a sedimentary rock.

(e) Most rocks exhibit some degree of time-dependent deformation or fracture behaviour.

(f) Some rocks suffer from a serious strength reduction when wet.

(g) Fresh rock surfaces exposed to the atmosphere are liable to deteriorate in time (weather).

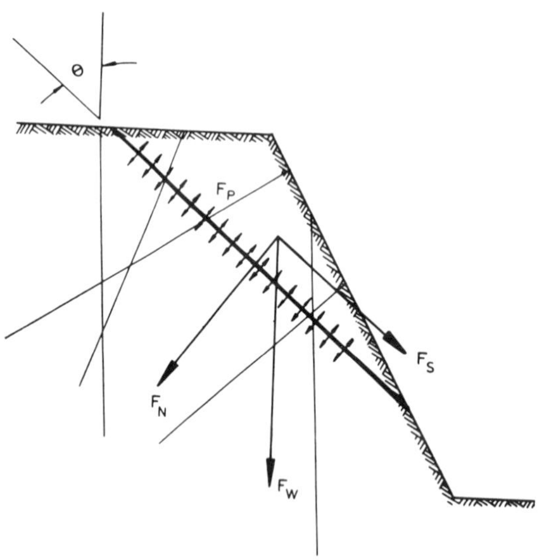

Fig. 11. *Forces acting on a rock slope containing a geological discontinuity.*

Fig. 10*b*. *Sidewall slabbing in a square tunnel in quartzite which has been subjected to a pressure increase due to over-mining. Depth below surface* 9,500 *ft.*

The most effective use of the structural characteristics of rock can be achieved if these physical properties are kept in mind and if necessary precautions are taken in designing a rock structure. Some of these precautions are:

(1) The structure should be designed to avoid or, at least, minimise zones of high tensile stress.

(2) Support techniques such as the use of rockbolts can be used to improve the stability of a rock structure by increasing the confining pressure and hence the compressive strength of the material.

(3) When a rock structure contains geological weaknesses in the form of bedding planes, fissures, faults, joints or dykes, due allowance must be made for the directional strength variation associated with these features.

(4) In calculating the strength of a rock structure, allowance should be made for the possible reduction in strength with time.

(5) Exposed surfaces of rocks which are particularly liable to weathering should be protected, as far as possible, from exposure to the atmosphere.

Current activity in rock mechanics, throughout the world, is such that significant advances in knowledge can be anticipated during the next decade. The interested reader is therefore advised to keep in contact with rock mechanics literature and with the proceedings of the conferences which take place from time to time.

List of References

(1) JUDD, W. R. Rock stress, rock mechanics and research. *Proc. Intnl. Conf. on the State of Stress in the Earth's Crust.* Santa Monica 1963, ed. W. R. Judd. Elsevier Publishing Co., New York, 1964, pp. 5-53.

(2) COOK, N.G.W. et al. Rock mechanics applied to rockbursts—a synthesis of the results of rockburst research in South Africa up to 1965. *Journal South African Institution of Mining and Metallurgy*, Vol. 66, No. 10, 1966, pp 435-528.

(3) GRIFFITH, A. A. Theory of rupture. *First Intnl. Cong. Appl. Mech.* Delft, 1924. Technische Boekhandel en Drukkerij, ed. J. Waltman Jr., 1925, pp. 55-63.

(4) MCCLINTOCK, F. A. and WALSH, J. B. Friction on Griffith cracks in rock under pressure. *Intnl. Cong. Appl. Mech.* Berkeley, 1962. American Soc. Mech. Engrs., New York, 1963, pp. 1015-1021.

(5) HOEK, E. Rock fracture under static stress conditions. *Ph.D. Thesis*, University of Cape Town, 1965, also South African Council for Scientific and Industrial Research Report MEG 383, 1965.

(6) BRACE, W. F. Dependence of fracture strength of rocks on grain size. *Penn. State Univ. Mineral Expt. Sta. Bull.*, No. 176, 1961, pp. 99-103.

(7) BERG, C. A. Deformation of fine cracks under high pressure and shear. *J. Geophysical Res.*, Vol. 70, No. 14, 1965, pp. 3447-3452.

(8) BRACE, W. F. and BOMBOLAKIS, E. G. A note on brittle crack growth in compression. *J. Geophysical Res.*, Vol. 68, No. 12, 1963 pp. 3709-3713.

(9) HOEK, E. and BIENIAWSKI, Z. T. Brittle fracture propagation in rock under compression. *Intnl. J. Fracture Mechanics*, Vol. 1, No. 3, 1965, pp. 137-155.

(10) HOEK, E. and BIENIAWSKI, Z. T. Fracture propagation mechanism in hard rock. *Proc. First Intnl. Congress on Rock Mechanics*, Lisbon, 1966.

(11) WALSH, J. B. The effect of cracks on the uniaxial compression of rocks. *J. Geophysical Res.*, Vol. 70, No. 2, 1965, pp. 399-411.

(12) WALSH, J. B. The effects of cracks on the compressibility of rocks. *J. Geophysical Res.*, Vol. 70, No. 2, 1965, pp. 381-389.

(13) HOEK, E. Fracture of anisotropic rock. *J. S. Afr. Inst. Min. Metall.*, Vol. 64, No. 10, 1964, pp. 510-518.

(14) WALSH, J. B. and BRACE, W. F. A fracture criterion for anisotropic rock. *J. Geophysical Res.*, Vol. 69, No. 16, 1964, pp. 3449-3456.

(15) COLBACK, P. S. B. and WIID, B. L. The influence of moisture content on the compressive strength of rock. *Proc. Third Canadian Symposium on Rock Mechanics*, Toronto, 1965.

(16) NICOLYASEN, O. The role of fluid pressure in the failure of rock and other porous media. Unpublished report *Bernard Price Institute of Geophysical Research*, July, 1965.

(17) ROBERTSON, E. C. Viscoelasticity of rocks. *Proc. Intnl. Conf. State of Stress in Earth's Crust.* Santa Monica, 1963, ed. W. R. Judd. Elsevier Publishing Co., New York, 1964, pp. 181-224.

(18) HARDY, H. R. (Jnr.). Study of the inelastic behaviour of geologic materials. *Ph.D. Thesis.* Virginia Polytechnic Institute, June, 1965.

(19) PROTODIAKONOV, M. M. Methods of evaluation of cracks and strength of rock systems at depth. *Proc. 4th Intnl. Conf. Strata Control and Rock Mechanics.* New York, 1964, (Presented in Russian). South African Council for Scientific and Industrial Research Translation No. 449.

(20) HOEK, E. Rock fracture around mining excavations. *Proc. 4th Intnl. Conf. on Strata Control and Rock Mechanics.*, New York, 1964.

(21) SAVIN, G. N. *Stress concentration around holes.* Translated from Russian in Intnl. Series of monographs in aeronautics and astronautics, Pergamon Press, New York, 1961.

(22) HOEK, E. Experimental study of rock stress problems in deep level mining. *Experimental Mechanics*, ed. Rossi, Pergamon Press, London 1963, pp. 177-193.

(23) DENKHAUS, H. G. The application of the mathematical theory of elasticity to problems of stress in hard rock at great depth. *Ass. Mine Mngrs. S. Africa*, Vol. 1958/59, 1948, pp. 271-310.

(24) DEIST, F. H. A non-linear continuum approach to the problem of fracture zones and rockbursts. *S. Afr. Inst. Min. Metall.*, Vol. 65, No. 10, May, 1965.

(25) COOK, N. G. W. A note on rockbursts considered as a problem of stability. *S. Afr. Inst. Min. Metall.*, Vol. 65, No. 10, May, 1965.

(26) COOK, N. G. W. Failure of rock, *Intnl. J. Rock Mech. and Min. Sci.*, Vol. 2, No. 4, 1965.

(27) FAIRHURST, C. and COOK, N. G. W. The phenomenon of rock splitting parallel to a surface under a compressive stress. *Proc. 1st Intnl. Congress on Rock Mechanics*, Lisbon 1966.

(28) MULLER, L. Application of rock mechanics in the design of rock slopes. *Proc. Intnl. Conf. on State of Stress in the Earth's crust*, Santa Monica 1963. Ed. W. R. Judd, Elsevier publishers, New York, pp. 575-598.

(29) SERAFIM, J. L. Rock mechanics considerations in the design of concrete dams. *Proc. Intnl. Conf. State of Stress in the Earth's crust.* Santa Monica 1963. Ed. W. R. Judd, Elsevier publishers, New York, pp. 611-645.

32B

Reprinted from *Mining Mag.* **114:**409, 411 (1966)

ROCK MECHANICS—AN INTRODUCTION FOR THE PRACTICAL ENGINEER

Part II—Laboratory Techniques in Rock Mechanics

E. Hoek

[*Editor's Note:* In the original, material precedes this excerpt.]

Summary and Conclusions

Because of the practical difficulties associated with carrying out a complete rock mechanics investigation in the field, laboratory studies on simplified physical or theoretical models play an important part in the solution of such problems. Provided that sufficient care is taken in planning, executing and interpreting the laboratory investigations and provided that the practical requirements are kept in mind at all times, the results of these investigations can be very useful in designing rock structures.

Numerous techniques for the determination of the stress distribution in rock structures exist. When the problem is too complex to be solved mathematically, the photoelastic technique can be used to give results, the quality of which is directly proportional to the effort expended on manufacturing and loading the model and analysing the resulting fringe patterns. In studying the mechanical properties of rock in the laboratory, attention to detail in specimen preparation and testing techniques is important if meaningful results are to be obtained.

One of the most important problems in rock mechanics is the stability of structures in which some fracturing of the rock has taken place. Hence, an underground excavation may be stable in spite of the fact that the periphery of the excavation is extensively fractured. The interaction of the solid and fractured rock and the changes in stress distribution which are associated with fracture propagation are most conveniently studied by means of laboratory models.

In carrying out these model studies, it is essential that the laws of similitude be adequately understood and rigidly adhered to. Failure to comply with these requirements reduces a physical model to the status of a demonstration model and the results of such a study are only of qualitative significance.

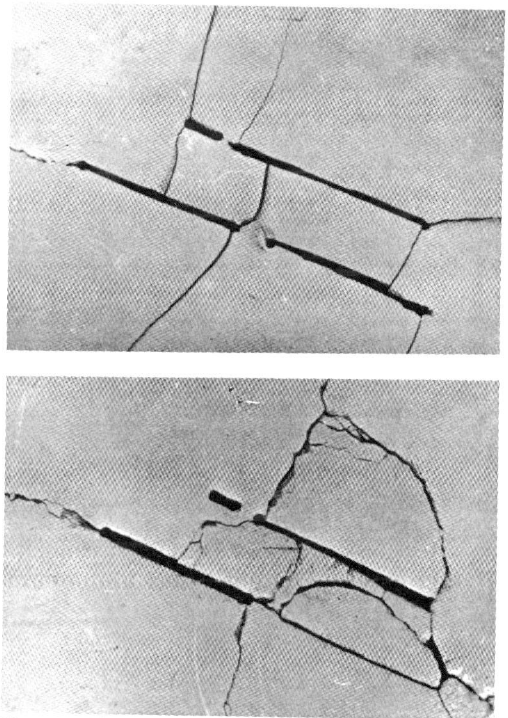

Fig. 16. Comparison between the propagation of fracture in plaster of paris models subjected to uniformly distributed edge loading (upper photograph) and centrifugal acceleration (lower photograph).

List of References

(1) COOK, N.G.W. et al. Rock Mechanics applied to Rockbursts—a synthesis of the results of rockburst research in South Africa up to 1965. *J. S. African Inst. Min. Metall.* Vol. 66, No. 10, 1966, pp. 435-528.

(2) DENKHAUS, H. G. The application of the mathematical theory of elasticity to problems of stress in hard rock at great depth. *Ass. Min. Mngrs. S. Africa.* Vol. 1958/59, 1958, pp. 271-310.

(3) SAVIN, G. N. *Stress concentration around holes.* Translated from Russian in International Series of monographs in aeronautics. and astronautics. Pergamon Press, New York 1961.

(4) SALAMON, M. D. G. Elastic analysis of displacements and stresses induced by mining of seam or reef deposits. *J. S. African Inst. Min. Metall.* Vol. 64, Nos. 4, 6 and 10, 1964.

(5) FROCHT, M. M. *Photoelasticity.* Vols. I and II. John Wiley and Sons Inc. New York, 1941.

(6) LEVEN, M. M. Epoxy resin for photoelastic use. *Photoelasticity* edited by M. M. Frocht, Pergamon Press, New York, 1963, pp. 145-165.

(7) SANFORD, R. J. The validity of three-dimensional photoelastic analysis of non-homogeneous elastic field problems. *Brit. J. Appl. Phys.* Vol. 17, 1966, pp. 99-108.

(8) DIXON, J. R. Problems of materials for composite models in photoelastic frozen stress work. *Brit. J. Appl. Phys.* Vol. 13, 1962, pp. 64-67.

(9) KARPLUS, W. J. *Analog Simulation,* solution of field problems. McGraw-Hill Book Co. Inc., New York, 1958.

(10) ALLISON, I. M. Stress separation by block integration. *Brit. J. Appl. Phys.* Vol. 16, 1965, pp. 1573-1581.

(11) HOEK, E. Experimental study of rock stress problems in deep level mining. *Experimental Mechanics* ed. Rossi, Pergamon Press, London, 1963, pp. 177-193.

(12) HOEK, E. A photoelastic technique for the determination of potential fracture zones in rock structures. *Proc. 8th Symposium on Rock Mechanics.* Minneapolis, September 1966.

(13) HOEK, E. The design of a centrifuge for the simulation of gravitational force fields in mine models. *J. S. African Inst. Min. and Metall.* Vol. 65, No. 9, 1965, pp. 455-487.

(14) PROTODIAKONOV, M. N. Methods of evaluation of cracks and strength of rock systems at depth. *Proc. 4th Intnl. Conf. Strata Control and Rock Mech. New York.* May 1964 (in Russian). S.A.C.S.I.R. Translation No. 449, January 1965, 36 pp.

(15) COOK, N. G. W. A note on rockbursts considered as a problem of stability. *S.A. Inst. Min. and Metall.* Vol. 65, No. 8, 1965.

(16) PAULDING, B. W. Techniques used in studying the fracture mechanics of rock. Fifth Pacific Area Nat. Meeting ASTM, Washington, 1965, in press.

(17) DEIST, F. H. A non-linear continuum approach to the problem of fracture zones and rockbursts. *S.A. Inst. Min. and Metall.* Vol. 65, No. 10, May 1965.

(19) COLBACK, P. S. B. An analysis of brittle fracture initiation and propagation in the Brazilian test. *Proc. First Intnl. Cong. on Rock Mech.* Lisbon, September 1966.

(20) HOBBS, D. W. The tensile strength of rocks. *Intnl. J. Rock Mech. Min. Sci.* Vol. 1, 1964, pp. 385-396.

(21) PAULMANN, H. G. Messung der Festigkeits-Anisotropie tektonischen Upsprungs en Gesteinsproben. *Proc. First Intnl. Cong. on Rock Mech.,* Lisbon, September 1966.

(22) HOEK, E. and BIENIAWSKI, Z. T. Brittle fracture propagation in rock under compression. *Intnl. J. Fracture Mech.,* Vol. 1, 1965, pp. 135-145.

(23) HOEK, E. and BIENIAWSKI, Z. T. Fracture propagation mechanism hard rock. *Proc. First Intnl. Cong. on Rock Mech.* in Lisbon, September 1966.

(24) ROSENBERG, L. D., KAZANTSEV, V. F., MAKAROV, L. O. and YAKHIMOVICH, D. F. *Ultrasonic Cutting.* Translated from Russian by J. E. S. Bradley and published by Consultants Bureau Enterprises Inc., New York, 1964, 142 pp.

(25) ZANDMAN, F. Analyse des Constraintes par Vernis Photoélastiques. *Groupement pour l'Advancement des Méthodes d' Analyse des Constraintes.* Vol. 2, No. 6, 1956, pp. 3-14.

(26) HOEK, E. and BIENIAWSKI, Z. T. A large field reflection polariscope. *S. African Mech. Engr.* Vol. 12, No. 8, 1963, pp. 222-226.

(27) THEOCARIS, P. S. Moiré Fringes, a powerful measuring device. *Appl. Mech. Rev.* Vol. 15, No. 5, 1962, pp. 333-339.

(28) HOEK, E. Rock fracture around mining excavations. *Proc. 4th Intnl. Conf. on Strata Control and Rock Mech.,* New York 1964. Columbia Univ. Press., New York, 1965, pp. 334-348.

Reprinted from *Mining Mag.* **115:**4–5, 7, 9, 11, 13, 15, 17 (1966)

Rock Mechanics—An Introduction for the Practical Engineer

Part III—Rock Mechanics in the Field

E. HOEK, Ph.D., M.Sc. (Eng.), B.Sc. (Eng.).

THE successful solution of a rock mechanics problem usually depends upon contributions made by both the research worker and the practical engineer. In order that the research worker should obtain results of practical value, and, having obtained such results, be able to communicate them to the engineer in intelligible terms, it is essential that he should remain in constant contact with the field problem. On the other hand, effective co-operation can only be achieved if the practical engineer is familiar with the principal features, the potential and the limitations of the techniques which are available to the rock mechanics research worker.

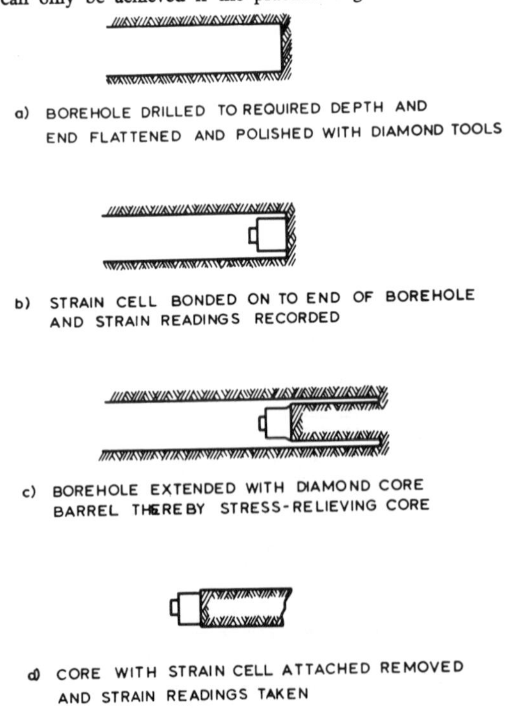

Fig. 1. Sequence of operations in Leeman's method for measuring rock stress.

This part of the paper describes some of the techniques which are available for the investigation of rock mechanics problems in the field.

Analysis of Field Records

An obvious, although frequently neglected, means of investigating rock mechanics problems is the analysis of field records. Analysis of site records will often reveal the presence of common factors which warrant further investigation.

The statistical analysis of records of depth below surface, type of excavation, proximity to geological discontinuities, etc., collected from the sites of rockbursts in South African gold mines, has greatly assisted research workers in achieving an understanding of the rockburst phenomenon[1]*.

Similarly, the analysis of mine records has revealed factors which may be important in controlling the collapse of coal pillars in bord and pillar workings.

Obviously, the value of any such analysis depends upon the quality of the records kept and also upon the purpose for which the analysis is intended. In general, field records should not be overlooked as a potential source of information.

Geological Observations

In a field problem involving the behaviour of a large rock mass, the strength and deformation characteristics of the rock material may be of secondary importance as compared to the presence of geological discontinuities such as faults or dykes. Hence, a rock slope, in which the material may have excellent mechanical properties, may be unstable due to the presence of a single critically oriented fault. Consequently, any rock mechanics investigation in the field should include a geological examination of the site.

* Superscripts refer to the list of references at the end of this paper.

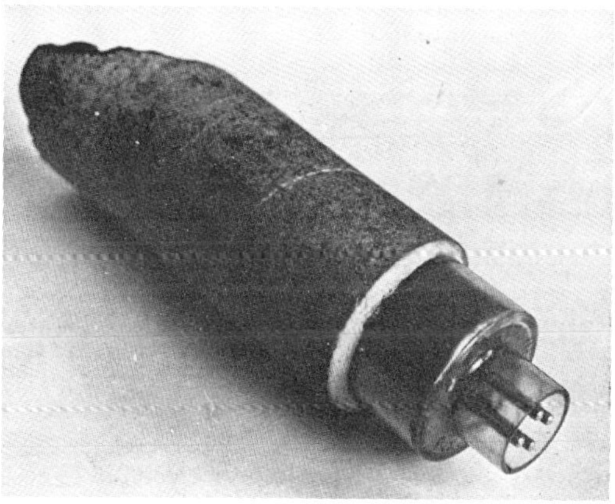

Fig. 2. A CSIR strain cell bonded on to a rock core recovered from a borehole.

Fig. 3. The measurement of stress in a deep level South African gold mine.

The type of observations required of a geologist depend upon the nature of the problem being studied and may include the classification of rock types, measurements of the frequency and orientation of faults, fissures, cleats or bedding planes and the determination of the extent of weathering.

In order that these geological observations should be of practical value to the engineer, they should be quantitative rather than qualitative and, where possible, a sufficiently large number of measurements should be made to permit statistical evaluation of the results.

Stress, Strain and Deformation Measurements

The application of any theory to the prediction of rock behaviour can only be justified if field observations are in agreement with the theoretical predictions. Since stresses, strains and displacements play an important part in practically all rock behaviour theories, the measurement of these quantities is an important task in field rock mechanics.

Precise Levelling

One of the most successful techniques to be used in South Africa for the study of rock deformation is the precise levelling of survey bench-marks, installed either on surface or in underground excavations. Using high quality levelling instruments, the careful observer can detect displacements of a few hundredths of an inch[2]. Since displacements of several inches frequently occur in large rock structures, the resolution of these precise levelling techniques is adequate for most practical purposes.

An important practical consideration in installing bench-marks for precise levelling is that they should be anchored sufficiently deep in the rock to avoid the influence of local surface conditions. Where long-term measurements are required, care should be taken that the bench-marks are manufactured from corrosion resistant material and that any exposed portion of the bench-mark is protected from accidental damage due to the passage of equipment and even heavily booted personnel.

In addition to the measurement of vertical displacements by levelling, extensometers can be used to measure horizontal movements between adjacent bench-marks.

An important advantage of precise levelling and extensometer measurements is that the movement of the bench-mark is influenced by a comparatively large volume of rock surrounding the point at which the bench-mark is anchored. Consequently, local small scale discontinuities in the rock do not have a significant influence upon the behaviour of the bench-mark.

A disadvantage associated with the use of bench-marks is that they can only be installed in tunnels or on surface where there is sufficient room for a man to gain access.

Extensometer measurements in boreholes

When access for the installation of bench-marks cannot be provided, displacement measurements of comparable accuracy can be obtained from wires clamped in boreholes. Spring-loaded clamps which apply a high clamping force against the walls of the borehole can be used to hold the end of a tensioned measuring wire in a predetermined position and any displacement of the clamp will result in a measurable movement of the wire. If provision is made for the wire to pass through the body of the clamp, a number of clamps along the length of the borehole can be used to measure differential movements in the rock mass.

This type of installation has been extensively used in South Africa for the measurement of subsidence in coal mines[3]. Clamps have been successfully installed in vertical boreholes at depths in excess of 600 feet below surface measuring stations.

Strain gauge measurements

The electrical resistance strain gauge is used extensively for the measurement of local strains in engineering components and it is not surprising that numerous attempts have been made to apply it to the measurement of strains in rock structures.

One of the most successful techniques for this type of measurement has been devised by E. R. Leeman of the South African Council for Scientific and Industrial Research[4,5]. This technique involves bonding a strain gauge rosette on to the flat end of a borehole. The gauge is protected by a rubber filled plastic moulding which also secures the electrical connections which link the strain cell with its installing tool. The sequence of operations involved in obtaining a set of strain measurements from a point in a rock body is illustrated diagrammatically in Fig. 1.

The advantages of this technique are its simplicity and the relatively low cost of the instrumentation and strain cells. These cells have been successfully installed at depths in excess of 160 feet in rock bodies. The large amount of practical experience gained in South African gold and coal mines has contributed to the perfection of the instrumentation and the application of the technique.

A disadvantage associated with this technique is that the gauge measures the strain in a small volume of rock and the results can be significantly influenced by local small scale discontinuities in the rock. Since calculation of the stress state in the rock from the three measured strains involves a knowledge of the elastic properties of the material, local variations in these properties can give rise to serious errors in the calculated stresses.

In order to overcome this problem, work is currently being carried out on apparatus which is designed to load the core on to which the strain cell is bonded in such a way that the stresses, which existed before the cell was overcored, are restored. In this way, the strain cell is used only as an indicator of the stresses in the rock and, since these stresses are measured directly on hydraulic pressure gauges, the elastic properties of the rock do not enter into the calculation of results. If this technique proves successful, it is intended that it will be used on site to load the cores as soon as they are removed from the boreholes and thereby provide the user with an immediate measure of the stress in the rock.

A further disadvantage of the above technique is that the three strain measurements are all in the plane of the flat borehole end. These three strain measurements are not adequate for the determination of the complete three-dimensional state of stress in a rock body unless the borehole is drilled parallel to one of the principal stresses. In many practical problems, the principal stress directions can be assumed with sufficient accuracy to permit use of the technique but, where such assumptions cannot be made, it is necessary to use three boreholes to determine the complete state of stress. These three boreholes must be drilled from different directions towards the point at which the stress is to be determined and the drilling of such a configuration of boreholes presents many practical problems.

A recent development by Leeman[6] is an instrument which is designed to measure the complete state of three-dimensional stress in a rock body from a single borehole. This technique involves bonding three rosette strain gauges on to the cylindrical surface of the inside of a borehole. The rock to which this cell is bonded is then overcored with a larger diameter core barrel, resulting in a stress-relieved cylinder.

Miscellaneous measuring techniques

In addition to the measuring techniques already discussed, there are several less refined field techniques which warrant brief consideration.

It has frequently been observed, when drilling in highly stressed rock, that the core breaks into discs such as those illustrated in Fig. 4. Some investigations into the reasons for this discing have been carried out[7,8] and it has been suggested that the thickness of the discs could be used as an indirect indication of the stresses in the rock mass. To the best of the author's knowledge, no serious attempt has been made to apply this technique in the field but it does represent an interesting possibility.

Flat pancake jacks, consisting of two thin steel sheets welded together along their edges, have been used for various field applications ranging from the measurement of stress[9,10] to the measurement of load in the timber packs used as support in mine excavations. Apart from the difficulties associated with successfully welding the periphery of the plates together, these jacks are simple and effective in operation.

Numerous dynamometers for measuring the load in rock bolts are available. One of the most effective, in the author's opinion, is a dynamometer which makes use of a photoelastic disc transducer which has been developed by Sheffield

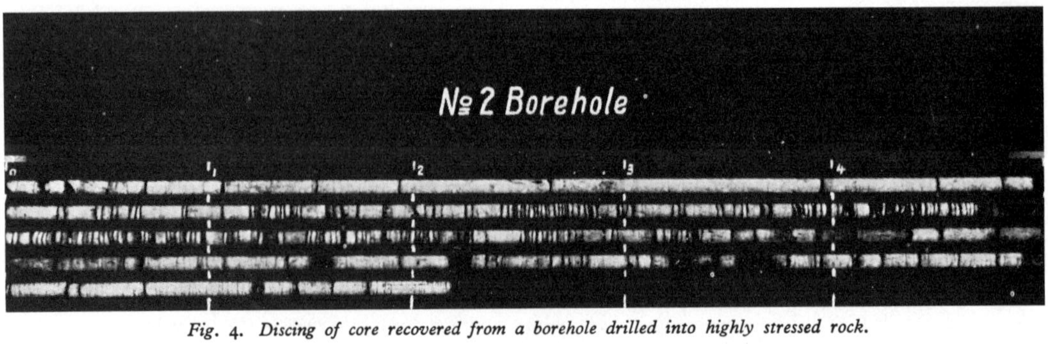

Fig. 4. Discing of core recovered from a borehole drilled into highly stressed rock.

University. The principal requirements of a rock-bolt dynamometer are that it should be inexpensive and simple to install and operate. Such an instrument can be used in large numbers to obtain the load distribution in a pattern of bolts—an important consideration in mining operations where large numbers of bolts are used for support purposes.

Closure and ride, the vertical and horizontal components of the displacement of the roof and floor of a mining excavation, play an important part in the calculation of the overall behaviour of the rock surrounding the excavation[1]. An instrument for measuring closure and ride is illustrated in Fig. 5. This particular instrument is no longer in use, having been succeeded by devices which permit the remote measurement and recording of the required quantities. It has, however, been included in this discussion because it represents the simple and yet effective type of instrument which is considered ideal for many field applications. In general, such instrumentation should be rugged, simple to install and operate and should make use of the simplest possible mechanical components which will do the job required.

As stated in the introduction to part I of this paper, the discussions are limited to the author's own experience and are not intended to be a complete review of the entire subject of rock mechanics. Consequently, in this discussion on field instrumentation, only a few of the many instruments which are available have been mentioned. The interested reader will find a wealth of publications on similar instrumentation and is advised to choose the instrumentation which is best suited to his particular needs.

Fig. 5. An instrument used by the South African CSIR for the measurement of closure and ride in a stope in a deep-level gold mine.

The Detection of Discontinuities in Rock

As already mentioned in this paper, the presence of a geological discontinuity such as a fault or a dyke can have a significant influence upon the behaviour of a rock mass. Similarly, fracture planes induced by stress changes in the structure can play an important part in determining the stability of the structure. The detection of these discontinuities, whether inherent or induced in the rock, is a task of major importance in field rock mechanics.

An obvious approach is the systematic mapping of the visible traces of these discontinuities on the exposed surface of an excavation. While this method can give useful information, it suffers from severe limitations in that major discontinuities can exist in the close proximity of the excavation and, because of their orientation, cannot be detected by visual examination.

An improvement on the visual examination of exposed surfaces can be achieved by examining the inside surfaces of boreholes drilled into the rock. For this purpose, several types of borehole camera, utilising both photographic and television principles, have been placed on the market. One of the most effective instruments of which the author is aware is a simple inclined mirror surrounded by a cluster of lights. This device is pushed into the borehole to the required depth and the image of the borehole surface is observed in the mirror by means of a telescope.

Any visual examination, whether of the surface of the excavation or of the interior of a borehole, suffers from the disadvantage that, if sufficient measurements are to be made to be of practical value, the method becomes excessively tedious and time-consuming.

Sonic and seismic techniques

In an effort to overcome the limitations of the visual inspection techniques discussed above, several devices which detect the passage of elastic waves through the rock have been evolved.

One of these devices, designed by the Canadian Department of Mines and Technical Surveys[11], measures the time taken for a pressure wave to travel from its source to a detector. The time taken for such an elastic wave to travel a given distance depends upon the nature of the material and, the more discontinuities which are present in the rock, the longer will be this travel time.

This instrument consists of two almost identical probes which are placed in boreholes in the rock. Hydraulically actuated pistons grip the probe in the borehole and force the transducer into intimate contact with the rock surface. The elastic wave is generated by knocking the rods holding one of the probes in the borehole with a hammer. A timing signal is generated when this pulse leaves the probe and enters the rock and is terminated when it leaves the rock and enters the detector probe.

In operation, these probes are placed at various distances in several boreholes drilled outwards from the excavation. Travel times for the various probe positions are recorded and a network of travel times is thereby established. Variation in travel times in various portions of the rock mass indicate the presence of discontinuities.

An instrument developed by the South African Council for Scientific and Industrial Research utilises a reflected rather than a transmitted pulse[12]. In this instrument, a single probe is placed in a borehole and this probe contains both transmitter and detector units. Ultrasonic pulses generated by the transmitter are reflected back from discontinuities in the rock and their presence can be established by observing the signals received by the detector unit on an oscilloscope screen.

In research into the problem of rockbursts which occur in deep-level South African gold mines, extensive use has been made of seismic techniques similar to those used in the detection of earthquakes[13, 14]. A network of seismometers installed in boreholes in various locations in a mine is continuously monitored by a multi-channel tape-recorder. Any seismic event such as a rockburst radiates elastic waves and, by recording the arrival times of these waves at the various seismometers, the focus of the event can be determined. Use of this seismic technique in South Africa has contributed greatly to the understanding of the rockburst phenomenon[1].

Similar techniques, in which the passage of elastic waves generated by explosions are recorded, have been used for prospecting and for the study of the elastic properties of rock masses[15]. Since the author has no personal experience of these techniques, they are merely mentioned for the information of the reader.

Determination of the In Situ Properties of Rock

A knowledge of the deformation and strength properties of rock is fundamental to any design of rock structures. As already discussed in previous parts of this paper, small scale tests in the laboratory do not always give results which are representative of the properties of the rock mass. Consequently, the determination of the properties of the rock material and the rock mass *in situ* is an important aspect of field rock mechanics.

Field tests on small scale specimens

Although most tests on small scale specimens can be carried out in the laboratory, there are cases in which it is preferable to carry out these tests on site. An example of such a case involves tests on rocks which deteriorate so rapidly when removed from the parent rock body that specimens cannot be successful transported to the laboratory.

In research carried out by the South African CSIR on behalf of the Coal Mining Research Controlling Council, this problem arose in carrying out properties tests on small scale coal specimens. At one stage, five tons of coal were transported a distance of about seventy miles to the CSIR laboratories. No core specimens were recovered from this material in spite of very elaborate precautions taken by the laboratory technicians responsible for the drilling.

As a result of this experience, it was decided that all future tests on materials such as coal, mudstone, shale and similar rock materials would be carried out on site, as soon as possible after removal of the core from the borehole.

The mobile drilling rig, illustrated in Fig. 6, is used by the South African CSIR for both stress measurement drilling and the recovery and preparation of small scale test specimens. This drill consists of an identical unit to that illustrated in Fig. 7 of part II of this paper. The drilling machine is mounted on a trailer for highway towing and is fitted with a specimen preparation unit—visible above the motor at the front end of the machine in Fig. 6. This unit, almost identical to the specimen end grinding machine illustrated in Fig. 8 of part II of this paper, is driven by a Vee belt from the main motor of the drilling machine. It can be fitted with either a diamond saw for cutting the core into suitable lengths or a diamond cup-wheel for grinding the specimens ends.

Specimens can be prepared with this equipment within a very short time of the removal of the core from the hole and, once prepared the specimens can be tested immediately in equipment such as that illustrated in Fig. 10 of part II of this paper.

In addition to the use of this equipment in special circumstances such as those discussed above, the author believes

Fig. 6. Mobile equipment used by the South African CSIR for drilling and specimen preparation in coal mines.

that there is considerable justification for carrying out all properties tests on site. The reason for this belief is that the liaison between the site and the laboratory is usually imperfect. This frequently results in the field and laboratory personnel being at cross-purposes and in the practical engineer being presented with a mass of tables and graphs which are of little value to him in the solution of his problems.

A more realistic approach, admittedly not always possible, is that the person responsible for the solution of a particular problem should himself carry out the necessary properties tests with equipment such as that described above. In this way, he can evaluate the results as he goes along and he need only carry out those tests which are directly relevant to his problem. In cases where it is not justified to provide such services on the mine or on the civil engineering site, the services of a research organisation, university or a consulting concern can be used.

Fig. 7. Drilling coal specimens for properties testing.

Large scale rock properties tests

When a rock mass contains a large number of fissures, faults, cleats or bedding planes, it is usually difficult to obtain a reliable estimate of the strength and deformation characteristics of this rock mass from tests on small scale specimens of the rock material. Under these conditions, serious consideration must be given to conducting large scale tests *in situ.*

Unfortunately, large scale tests are almost always extremely expensive and fraught with practical difficulties. This explains why so few tests of this type have been carried out and why it is necessary to reserve such tests for special circumstances.

In a series of large scale tests currently being conducted by the South African CSIR on behalf of the Coal Mining Research Controlling Council of South Africa, it is planned to test coal specimens of up to eight feet cube *in situ.*

The specimens for these tests are cut from the corner of a coal pillar by means of a universal coal cutting machine as illustrated in Fig. 8. A typical specimen is illustrated in Fig. 9 and it will be noted that the specimen remains attached to the floor, having been cut free on five sides. Hydraulically actuated jacks are inserted between the top and bottom faces of the horizontal cut at the top of the specimen and the specimen is loaded against the roof and floor.

By preparing the vertical faces of the specimen first and then installing strain measuring instrumentation before the horizontal cut is taken, a measure of the elastic recovery of the stress-relieved specimen can be obtained. This provides an estimate of the load carried by the specimen before commencement of the test. The jacks are then inserted and the specimen loaded to fracture, axial and lateral strain measurements being taken at intervals during the loading.

The test illustrated in Fig. 9 was part of a preliminary series designed to establish the feasibility of this testing technique and to study the major problems associated with the preparation and testing of coal specimens *in situ.* As a result of the experience gained during these tests, it is

Fig. 8. A universal coal cutter being used to prepare large scale coal specimens.

planned to construct further jacks which will eventually be used to load specimens with a loaded surface measuring eight feet by eight feet.

Similar tests, carried out with a different aim, have recently been reported by Lama[16] and it appears that, for coal, this technique is highly effective.

The problem of conducting large scale tests in surface excavations such as open pits and quarries is further complicated by the fact that the specimens can only be loaded horizontally if the rock itself is to be used to support the reactions of the loading jacks. The magnitude of the loads involved in large scale testing is such that the design of the "testing machine" in the surrounding rock presents more of a problem that the preparation of the specimen itself.

From the brief discussion presented above, the reader will appreciate that large scale tests can only be carried out on a very limited scale. Consequently, an important objective in rock mechanics is to establish a reliable relationship between large scale and small scale specimens in order that the results of small scale tests may be extrapolated to predict large scale behaviour. Such a relationship has been suggested by Protodiakonov (see Fig. 7 of part I of this paper) but the experimental determination of such a relationship is a task which will require the efforts of a dedicated team of research workers with a very large budget to work with.

Indirect properties testing in the field

The practical engineer does not always require an accurate value for the strength or deformation characteristics of the rocks in which he is working. In many cases, a large number of approximate values obtained at various points on the site is preferable.

One of the most convenient and effective instruments for determining an approximate value for the uniaxial compressive strength of a rock is the Schmidt hammer—originally designed for use on concrete. The relationship between the Schmidt Type N Impact Hammer reading and the uniaxial compressive strength of rock and coal, according to Skutta[17], is presented in Fig. 10. It is estimated that, if the operational procedure is in accordance with the German Standard Specification DIN 4240[18], the uniaxial compressive strength can be estimated to an accuracy of ±2,000 lb./sq.in. (±150 Kp./sq.cm.).

Fig. 9. A large scale coal specimen under test.

Since the drillability of rock is reasonably closely related to its uniaxial compressive strength[19], it is also possible to use the Schmidt hammer to give an estimate of the rate at which a particular rock could be drilled[20].

This part of the paper has been concerned with instrumentation and techniques for determining stresses in rock structures, the extent of fracture in rock, and the determination of deformation and strength characteristics of

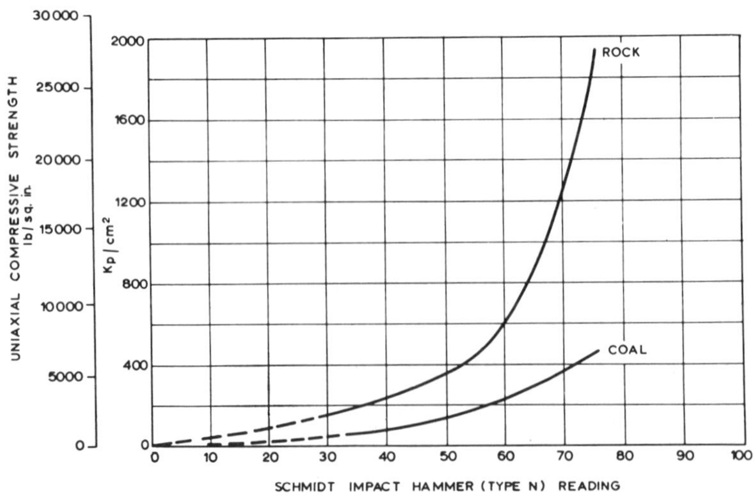

Fig. 10. Relationship between the uniaxial compressive strength of rock and the Schmidt impact hammer reading (according to Skutta).

rock *in situ*. This paper is not intended to be a complete catalogue of all available instruments and techniques and the examples which are given are intended to illustrate the underlying principles of rock mechanics testing in the field.

It will be appreciated that field measurements alone will seldom suffice for the solution of a practical rock mechanics problem and that, ideally, the results of such measurements should always be correlated with theoretical and laboratory results. If this is done, a rational pattern of rock behaviour can gradually be built up and the experience gained in solving each problem can contribute to future solutions.

List of References

(1) COOK, N.G.W. et al. Rock mechanics applied to rockbursts—a synthesis of the results of rockburst research in South Africa up to 1965. *J. S. African Inst. Min. Metall.* Vol. 66, No. 10, 1966, pp. 435-528.

(2) ORTLEPP, W. D. and COOK, N. G. W. The measurement and analysis of the deformation around deep, hard-rock excavations. *Proc. 4th Intnl. Conf. on Rock Mech. and Strata Control.* New York 1964. Columbia Univ. Press, New York, 1965, pp. 140-152

(3) SALAMON, M. D. G. and OROWECZ, K. I. Mathematical analysis of the stress and displacement distributions around bord and pillar workings. *Proc. 1st Intnl. Cong. on Rock Mechanics*, Lisbon, 1966.

(4) LEEMAN, E. R. Remote measurement of rock stress under development in South Africa. *Eng. and Mining J.*, vol 165, No. 9, 1964, pp. 104-107.

(5) LEEMAN, E. R. The measurement of stress in rock. Parts I to III. *J. S. African Inst. Min. and Metall.*, vol 65, No. 2, 1964, pp. 48-114 and vol. 65, No. 4, 1964, pp. 254-284.

(6) LEEMAN, E. R. and HAYES, D. J. A technique for determining the complete state of stress in rock using a single borehole. *Proc. 1st Intnl. Cong. on Rock Mechanics.* Lisbon, 1966.

(7) JAEGER, J. C. and COOK, N. G. W. Pinching-off and discing in rocks. *J. Geophysical Research*, vol. 68, No. 6, 1963, pp. 1759-1765.

(8) OBERT, L. Effects of stress relief and other changes on physical properties of rock. U.S. Bur. Mines. R.I. 6053, 1962.

(9) JAEGER, J. C. and COOK, N. G. W. Theory and application of curved jacks for measurement of stress. *Proc. Intnl. Conf. on State of Stress in Earth's Crust*, Santa Monica, 1963. Ed. W. R. Judd. Amer. Elsevier Publishing Co. Inc., 1964, pp. 381-395.

(10) MERRILL, R. H. *In situ* determination of stress by relief techniques. *Proc. Intnl. Conf. State of Stress in Earth's Crust*, Santa Monica, 1963. Ed. W. R. Judd, Amer. Elsevier Publishing Co. Inc. 1964, pp. 343-378.

(11) LAROCQUE, G. E. A sonic unit for the determination of *in situ* dynamic properties and for outlining fracture zones. *Proc. 6th Symposium on Rock Mechanics*, Rolla, 1964, pp. 358-380.

(12) LUTSCH, A et al. Ultrasonic reflectometer for detecting fracture zones in gold mines. *Trans. S. African Inst. Electrical Engrs.*, March, 1962, 5 p.

(13) COOK, N. G. W. The application of seismic techniques to problems in rock mechanics. *Int. J. Rock Mechanics and Mining Sciences*, vol. 1, No. 2, 1964, pp. 169-179.

(14) COOK, N. G. W. *A study of the failure of rock surrounding underground excavations.* Ph.D. Thesis Univ. Witwatersrand, 1962, 129 p.

(15) WANTLAND, D. Geophysical measurement of rock properties *in situ*. *Proc. Intnl. Conf. State of Stress in Earth's Crust*, Santa Monica, 1963. Ed. W. R. Judd, Amer. Elsevier Publishing Co. Inc., 1964, pp. 409-450.

(16) LAMA, R. D. Rockbursts—a comparison of the *in situ* mechanical properties of coal seams. *Colliery Guardian*, Jan. 1966, pp. 20-25.

(17) SKUTTA, E. Einfache gesteinmechanische Untersuchungen als Grundlage der Ausbauplanung. Glückauf, Jh. 98, Heft 25, S, 1962, pp. 1455-1461.

(18) DEUTSCHE NORMEN. *DIN* 4240. Kugelschlagprüfung von Beton mit dichem Gefüge. April 1962, 5 p.

(19) WHITE, C. G. The development of a rock drillability index. *VII Symposium on Rock Mechanics.* Penn State Univ. 1965, vol. 1. pp. 182-204 also Ph.D. Thesis, Colorado School of Mines, 1965, 155 p.

(20) FURBY, J. Tests on rock drillability. *Mine and Quarry Eng.*, July 1964, pp. 292-298.

33

Reprinted from pages 1, 2-3, and 3-24 of *Quart. Jour. Eng. Geology* **3:**1-24 (1970)

The logging of rock cores for engineering purposes

Geological Society Engineering Group Working Party

[*Editor's Note:* The table of contents has been omitted.]

1. Introduction

1.1. GENERAL

Many engineering geologists devote a significant proportion of their time to the logging of boreholes and, specifically, to the interpretation of rock cores in engineering terms. A variety of techniques have been devised for logging such cores but in the United Kingdom the systems which have been adopted are often inadequate in relation to modern requirements. The Code of Practice for Site Investigations (C.P. 2001 1957, 94) provides a minimal statement on the methods of preparing logs from borings; no specific advice is given on rock core description. It is apparent, also, that the maximum potential yield of data from many boreholes is not attained and, in consequence, the benefit from drilling not fully realized.

1.2. PURPOSE

This Working Party of the Engineering Group of the Geological Society of London was set up with the objective of recommending methods which can be used in the logging of rotary cored boreholes sunk primarily for civil engineering purposes. It was agreed that the report should deal with the description of cores, and methods of handling and storage; only aspects of the drilling technique and programme relevant to the appreciation of the ground conditions would be included. Detailed consideration of drilling, *in-situ* testing and borehole inspection methods would be omitted.

[*Editor's Note:* Material has been omitted at this point.]

2. General requirements of a borehole log

2.1. DEFINITION

A borehole log should provide an accurate and comprehensive record of the geological conditions encountered together with any other relevant information obtained during drilling. The Working Party examined a wide variety of logs which have been, or are, used in this country and overseas.

2.2. TYPES OF LOG

The simplest log considered consisted of a statement of the rocks encountered with information on the levels at which changes in strata occurred. Although such a log is totally inadequate as an engineering record it is surprising how many logs do not extend far beyond these basic requirements. In contrast, one of the most detailed methods of recording borehole data is that devised for the special requirements of the Snowy Mountains Authority (Moye 1967) which satisfies most requirements. This log contains separate columns which are provided for recording rock description, core loss, casing, core size, elevation and depth, pictorial log, fracture log, rock structures, water-level and any changes in water-level during drilling, and permeability based on *in-situ* tests. The Working Party considered that the complexity of this log made it inappropriate for general use. Reference was also made to the system used by the U.S. Bureau of Reclamation at the Folsom Dam and described by Kiersch & Treasher (1955) and the methods of description recently presented by Deere (1964, 1967).

2.3. INFORMATION TO BE RECORDED

The information to be recorded can be summarized under the following headings:

(i) Basic information including project name, location, timing, Employer, Engineer and Contractor;

(ii) Drilling method and progress;

(iii) Description of type and condition of material encountered;

(iv) Key and miscellaneous comments.

2.3.1. *Basic information*

The basic information provided must be adequate to locate the borehole in time and space, to define the authorities responsible for all aspects of the boring and to assist data storage and retrieval. The project name and locality together with the name of the drilling contractor or sub-contractor should be provided.

2.3.2. *Drilling*

Adequate information is not generally provided at present on the drilling method and borehole progress. The machine, core barrel and bit should all be described. Detailed records of core sizes and changes, use of casing, standing water levels, losses of circulating fluid and penetration rate can also provide useful supplementary information on the ground conditions.

2.3.3. *Description*

The main part of the log consists of the description of the geological features of the core relative to a depth scale. The condition of the sample recovered should be defined in terms of both the proportion actually recovered and the state of fragmentation. The descriptive geology should include definition of the rock type, its alteration state and relative strength, together with information on the natural discontinuities and rock structures. Later correlation of boreholes is assisted by incorporation of a simplified pictorial (symbolic) log (C.P. 2001).

2.3.4. *Symbolism*

In order to ensure economical use of the available space in the log it is advisable to make use of symbols which need only be identified on the first sheet of any series related to one borehole (C.P. 2001). Additional information covering, for example, special events occurring during drilling, or instrumentation installed, can be provided on the end sheet of a log. It was concluded that at least one spare column should be provided on a log to ensure that *in-situ* test results, locations of samples extracted or other information can be recorded.

2.4. FACTORS DETERMINING LOGGING METHODS

It was recognized, in part from the variety of logs which are in current use, that techniques of logging are often determined by such factors as the method of drilling, type of project, cost or the client's requirements.

2.4.1. *Overburden*

Many boreholes in rock may be drilled through a cover of overburden by percussion or other methods and, in consequence, the log should be designed to incorporate the results of both soft- and hard-ground drilling. In view of this decision, it was not considered appropriate to distinguish between the logs prepared for rotary-drilled and percussion-drilled holes.

2.4.2. *Log format*

The type of information which should be given by a particular set of cores is determined to a significant degree by project requirements. Nevertheless certain basic information is required on all logs and a similar format to all log sheets, irrespective of project, would be a useful advance.

2.4.3. *Cost*

The cost of logging cores according to the proposals presented in this report is up to about 5 per cent of the cost of the drilling to the client. In special cases the cost of logging may be greater and reach up to 10 per cent or more. It is not considered, therefore, that cost is a significant factor in determining logging procedure. There could be a general improvement in logging methods if this work were treated and charged separately and not as part of the drilling or reporting costs. It should also be recognized that good drilling requires good supervision. This is best provided by frequent, or possibly constant, attendance of the geologist at the drill rig.

2.4.4. *Cost benefit*

The cores, once recovered, should be retained for an adequate period before being destroyed. It is common practice to hold the cores at least until construction of the project has begun. Once the cores are disposed of, the borehole log remains the sole tangible product of an expensive operation; this fact should not be forgotten.

2.4.5. *Simplified logs*

Simplified borehole logs are prepared for a variety of purposes. The Working Party was opposed to the use of a simplified log as the sole and final record of drilling unless an additional comprehensive log of the same borehole was also available. The simplified log if used should be clearly identifiable and contain an indication of that information which has been omitted.

2.4.6. *Daily record*

The drillers' daily record sheet is an important source of technical and contractual information; reference is made to the technical requirements to be taken into account in the design of such sheets in 4.2.1. It cannot be over-emphasized that the success of drilling, and consequently the degree of success in correct logging, is largely based on the quality of the driller, his skill at drilling, the care he gives to recording basic data and the attention given to explaining special requirements to him.

2.5. INFORMATION STORAGE

There is an increasing need for storage of borehole information so that rapid retrieval of data is possible. It is suggested that the minimum information which needs to be stored in a computer index includes the grid reference, stratigraphical formation encountered, depth, and the location of the full record. Secondary information to be stored might include the project name and a record of the graded quality of borehole.

2.6. STYLE

The style of log proposed has been designed for A4 paper and is presented in SI terms; it is not appropriate to continue the use of Imperial Units at this time.

3. The handling, labelling and preservation of rock cores

3.1. CORE RECOVERY

Good sample recovery may be obtained by skilled operators using modern well maintained drilling equipment and supervised by experienced engineering staff. The value, however, of such expertise may be lost if subsequently the cores are poorly labelled or mishandled before geological examination and laboratory testing. Cores of suitable size for testing are expensive to obtain and should be treated with care at least equal to that presently accorded to soil samples which are generally recovered more cheaply.

3.2. CORE EXTRACTION

After being brought to the surface the core should be removed from the barrel by methods designed to produce minimal additional disturbance. Extraction by hammering the barrel or by explosive extrusion under high air or water pressures are clearly undesirable and may damage not only the core but also the barrel; rotary core drilling specifications should deprecate such practices. Extraction under steady, carefully applied pressure or the use of split inner barrels is to be preferred, particularly the latter which allows examination of the core in the barrel. The use of plastic and rubber liners for inner barrels aids removal of the core under gravity and can improve sample recovery.

3.3. CORE BOXES

The core box into which the core is transferred must be of sound robust construction to withstand the weight of the core and any full boxes that may subsequently be placed upon it and also sufficiently watertight to protect the core from rain on site and in transit. The boxes should be designed, in relation to the requirements of the job, to be handled by either one or two men. It should be noted that, in many cases, a geologist is required to log cores without assistance and heavy core boxes are both awkward and dangerous to handle by one man. A box designed to be handled by one man should hold about 4 m of core in SF (113 mm diameter) size and no more than about 15 m of AX (31 mm diameter) size core; a convenient size for handling and examination is 1 m long internally by about 0·5 m to 0·7 m wide. Boxes to be handled by two men could contain up to about twice this length of core. Boxes should be made to hold the particular size of core tightly in place in rows separated by wooden slats. A strong metal hinged lid fitted with a padlock, hasp and staple for closing is necessary and end rope handles should be attached for lifting. The top and bottom of the box should be reinforced by cross strips which aid both stacking and retrieval. Core boxes are commonly constructed of wood; marine plywood, being durable and not liable to splinter, is to be preferred. Solid wood should have a minimum thickness of 10 mm. Hardboard is unsuitable for use in permanent boxes. Rehousing core samples in lighter duty boxes, for storage, is acceptable providing that the core samples are transferred by experienced personnel and that subsequently the boxes are treated with greater care. Core boxes may be painted to give protection to the wood. In certain climatic conditions wood is liable to rapid deterioration and appropriate precautions would then be required. Plastic guttering and corrugated sheeting have been successfully adapted for storage of cores. Improvements in core box design may be achieved by experiment with moulded plastic material.

3.4. CORE BOX LABELLING

As the core is extruded it should be laid in the box reading in book fashion, that is with the shallower core to the left deepening to the right; the highest row of core should be placed next to the lid hinge. The box should be identified by the site name, borehole number, core box number, depth of bottom and top of the core included and the drilling contractor's name on both the outside and inside of the box. This information is provided in the most permanent fashion by painting on the box or fixing stamped metal labels. The box will be more readily identifiable when stacked if the opening longside and ends are also marked. Labels made from durable materials and permanently marked may be used but have a more limited life. The depth of the top and bottom of the total core contained, and the separate core runs, should be recorded within the box by painting on wooden blocks appropriately dimensioned to fit between the dividing slats. Alternatively, metal or other durable labels may be used. In order that zones of core loss may be readily identified, a wooden dowel, cut to the appropriate length and suitably identified, could be inserted either in the section where loss is known to have occurred or at the base of the particular core run.

3.5. TRANSPORTATION

Almost all cores are removed from the drilling site to a place for examination and storage. During transportation cores may easily be damaged by careless handling; for this reason it is preferable that the initial logging be carried out at the drill site. Before being moved from the borehole site, boxes should be securely fastened, preferably by wire straps rather than nails; packing materials may be added. Damage most commonly results from careless loading and unloading, particularly if cores are moved as part-loads in general transport lorries by personnel unaware of the value of the cores; it is preferable to use special transport manned by informed personnel.

3.6. STORAGE

Cores in long-term storage should be kept indoors to minimize deterioration due to loss of moisture; in some countries, core sheds may require special construction. Boxes should be stacked to allow ready identification and easy removal. During the course of an investigation cores may be examined several times and not always by the same person or with the knowledge of all concerned. In consequence, the advantage of an 'examination record' for each borehole or box is recommended. This would consist of a card pinned inside the core box recording the date and purpose of each examination and the name of the examiner, together with a note of any samples removed or of particular sections broken. Repeated inspections of the cores permit the opportunity to assess any long-term deterioration of the rock. Colour photographs provide a useful and convenient method of recording the appearance of cores and are of considerable value as a permanent record and means of rapid reference.

3.7. SAMPLE PRESERVATION

It is the weakest and most deformable strata that are generally of greatest interest in engineering practice. These materials are usually the most liable to deterioration after extraction from the core barrel and before logging. Deterioration of such strata should be reduced by sealing the cores in microcrystalline wax, aluminium foil or plastic. The selection and preservation of such cores has to be left to the driller who should be able from training, experience and specific instructions to identify and select samples for preservation. It is desirable for a geologist to be in attendance during drilling if an important section is to be recovered so that he may inspect the core as it is removed from the core barrel. The core run should never be completely sealed so that inspection of some of the material recovered is possible. If it is necessary to cut the sealing, the samples should be re-sealed after identification and description. When a continuous sequence of weak strata is encountered, samples should be preserved at specified intervals. These should include some of a size suitable for later testing.

4. Information to be recorded on the borehole log

4.1. GENERAL INFORMATION

The essential information which needs to be recorded on the log is as follows:

4.1.1. *Borehole number*

This should be used only once on a site and kept as simple as possible without extraneous ciphers.

4.1.2. *Location*

(i) Site, including project name, town, country or state name where necessary.

(ii) Grid Reference of the borehole should preferably be stated to 1 m and at least 10 m accuracy. In the United Kingdom the reference should be related to the National Grid; overseas, or at sea, appropriate local, temporary site or world co-ordinate systems should be applied.

(iii) Elevation relative to O.D. for the ground or sea-bed level or underground datum at the borehole site to an accuracy of 0·1 m.

(iv) Orientation of the borehole given as an angle to the horizontal (+ve upwards, −ve downwards) and azimuth (0° to 360° clockwise relative to Grid North).

4.1.3. *Drilling technique*

The following should be stated:

(i) The method of penetration and the flush system.
(ii) The make of machine with the model number and feed type if other than hydraulic.
(iii) The type of core barrel and bit.

4.1.4. *Contract details*

The following should be noted:

(i) Name of site investigation contractor.
(ii) Name of client or authority.
(iii) Job reference number.
(iv) Name or cypher of logger.

4.1.5. *Miscellaneous*

There should be an opportunity for relevant miscellaneous information to be included in the log.

4.2. DRILLING PROGRESS

4.2.1. *Daily record information*

Apart from the data listed above, the following technical information should be recorded by the driller on his daily record sheet:

(i) Depth of hole at start and end of working day or shift as relevant.
(ii) Depth of start and finish of each core run.
(iii) Depth and size of casing at start and end of each core run.
(iv) Core diameter (B.S. 4019, 1966 or metric dimension) and changes in core size.
(v) State of bit.
(vi) Time to drill each core run.
(vii) Gain or loss of water, mud or air flush; type of cuttings.
(viii) Standing water-level at start and end of each working period.
(ix) Simplified description of strata.
(x) Total core recovery with information as to possible location of core losses.
(xi) Sample locations.
(xii) Details of delays and breakdowns.
(xiii) Details of *in-situ* tests and instrumentation installed.
(xiv) Backfilling and grouting.

This list excludes items which may be required for contractual record purposes.

4.2.2. *Training*

Accurate core logging requires care and vigilance by the driller in the recording of data, particularly that information which only he is able to collect. This may necessitate increased liaison between driller, engineer and geologist and tactful education of drilling staff. The driller must always be given precise written instructions regarding the particular problems and requirements of any investigation.

4.2.3. *Non-recoverable information*

All the items listed under 4.2 may be relevant to the final borehole log but items (iii), (v), (vi), (vii), (viii), (xii), (xiii) and (xiv) are particularly important in that they cannot be deduced from later inspection of the cores.

4.2.4. *Water*

Particular mention should be made of the value in the detailed study of flush returns and standing water-levels during drilling. Loss of water during drilling is a useful measure of permeable conditions as are variations in the 'make' of water during air-flush drilling. Standing water-levels should be carefully related to the ground conditions and location of the casing. It is important to ensure that such levels have become adequately stabilized and to record whether the levels are rising, falling or static.

4.3. STATE OF RECOVERY OF CORE

4.3.1. *Core fracturing*

The state of the rock cores recovered is a valuable indication of the *in-situ* condition and probable engineering behaviour of a rock mass. This must be assessed against the drilling methods used and, on individual projects, it is desirable that standardization of drilling techniques should be adopted to permit comparisons. In any core recovered, fractures may be of natural or artificial origin. Even in the soundest rock, particularly at the ends of core runs, some artificial fragmentation is to be expected and should be recognized. Therefore, some degree of artificial fracturing is acceptable until experience has determined the optimum association of machinery and techniques on a specific site or in particular materials. Records of artificially induced fractures may assist in the assessment of rock excavation.

4.3.2. *Core recovery*

During the drilling process, the bit cuttings are removed in the flush system. The sample which passes up into the core barrel may be divided into five parts:

(i) solid core greater than 0·1 m in length (see discussion of RQD below),
(ii) solid core less than 0·1 m in length,
(iii) fragmental material not recovered as core,
(iv) additional material which may have been lost from the previous core run. This may be the core stump left when the barrel was pulled, material dropped from the core barrel during its withdrawal from the hole or cuttings which have settled when circulation of drilling fluid was stopped.

In addition, core may have been lost by:

(v) erosion of soft or friable material, resulting in a reduction in diameter or length of the core or both. This eroded material may be entirely removed by the flushing system.

The material which is placed in the core box consists of items (i), (ii), (iii) and (v) above and (omitting (v) from the subsequent discussion) is strictly defined as the *total core recovery*. If no material falls into class (v), then the total core recovery is 100 per cent in that there is no loss of sample. The material which is recovered as solid core pieces at full diameter ((i) and (ii) above) is strictly defined as the *solid core recovery*. It must be stressed that total and solid core recoveries are only equivalent when no fragmental material is recovered. This arises either when the rock is solid or loss of sample is represented wholly by material carried away in the flushing system. Too often in current practice the core recovery recorded is a driller's assessment of the total core recovery which may provide a misleading impression of the state of the core in engineering terms. It should be noted that, for convenience, recoveries are measured in terms of length and expressed relative to core run length. Any significant diminution of core diameter should be noted.

4.3.3. *Fracture state*

Various criteria may be used for quantitative description of the fracture state (s.l.) of the cores; these are the solid core recovery, fracture log, RQD and Stability Index. The simplest measure is the solid core recovery, particularly when contrasted to the total core recovery. A fracture log is a count of the number of natural fractures present over an arbitrary length; it is conventional to omit obvious artificial fractures and discontinuities resulting from deterioration of the core.

4.3.4. *R.Q.D. and Stability Index*

Deere *et al.* (1967) have presented an argument for separating the solid core recovered into fractions greater and less than 4 in. in length. Some support is provided for a correlation between the percentage solid core recovered greater than 4 in. in length (defined by Deere *et al.* as the Rock Quality Designation = RQD) and various physical parameters of the rock mass. Although RQD is slightly more time consuming to determine than either total or solid core recovery, or the fracture log, it is a useful additional measure and the Working Party considered that a modified, metric RQD based on a 0·1 m size could be adopted (Fig. 1). Ege (1968) has proposed a more sophisticated method for assessing rock quality by developing an equation summing a number of different factors to give the Stability Index as follows:

> Index number = 0·1 × core loss (length drilled less total core recovery) + number of fractures per foot + 0·1 × broken core (core less than 7·5 cm in length) + weathering (graded 1 to 4 from unweathered to severely weathered) + hardness (graded 1 to 4 from very hard to incompetent).

The index number was then related to a rock-grade classification from 10 (good rock, index <8) to 1 (incompetent rock, index >18); faulted rock was described as ϕ. It has been demonstrated by Deere *et al.* (1969) that RQD and the Stability Index can be closely related for specific conditions. It should be noted that none of these methods takes account of changes in diameter of core. Fractures in the core should be recorded by one, several or a variant of these four methods but they must be applied with discretion taking care to avoid their use as infallible guides in the place of proper consideration.

4.4. DESCRIPTIVE GEOLOGY

4.4.1. *Geological information*

The following factors have to be incorporated in a log for adequate engineering geological description:

(i) systematic description,
(ii) alteration and weathering state,
(iii) structure and discontinuities,
(iv) assessment of rock material strength,
(v) other features, including stratigraphy.

4.4.2. *Systematic rock description*

The following standard sequence of systematic description is proposed:
i—weathered state; ii—structure; iii—colour; iv—grain-size (C.P. 2001); (iv(a)—subordinate particle size; iv(b)—texture; iv(c)—alteration state; iv(d)—cementation state as relevant); v—rock material strength; (vi(a)—mineral type as relevant); vi—ROCK NAME.

It is considered that the qualifications are more important in core descriptions than the actual rock name and, for this reason, the name was placed last. Such a system is appropriate to an engineering description where classification by mechanical properties is more significant than classification by mineralogy and texture. The following examples are provided for illustrative purposes:

Fresh	foliated	dark grey	coarse	very strong	hornblende	GNEISS
i	ii	iii	iv	v	vi(a)	vi

Moderately weathered	thickly bedded	cream	medium-grained	strong
i	ii	iii	iv	v

dolomitic	LIMESTONE
vi(a)	vi

Completely weathered	thinly flow-banded	mid-grey	very coarse
i	ii	iii	iv

porphyritic	kaolinized	weak	tourmaline	GRANITE
iv(b)	iv(c)	v	vi(a)	vi

4.4.3. *Weathering classification*

The following system of weathering classification is proposed:

IA—Fresh (F): no visible sign of weathering;

IB—Faintly weathered (FW): weathering limited to the surface of major discontinuities;

II—Slightly weathered (SW): penetrative weathering developed on open discontinuity surfaces but only slight weathering of rock material;

III—Moderately weathered (MW): weathering extends throughout the rock mass but the rock material is not friable;

IV—Highly weathered (HW): weathering extends throughout rock mass and the rock material is partly friable;

V—Completely weathered (CW): rock is wholly decomposed and in a friable condition but the rock texture and structure are preserved;

VI—Residual soil (RS): a soil material with the original texture, structure and mineralogy of the rock completely destroyed.

It will be noted that this scheme is broadly based on that developed for the Snowy Mountains Authority (Moye 1955), but has been adapted for a general range of rock types rather than simply for granitic rocks for which it was originally devised. In the case of weathered rocks with a significant clay content, the material may exhibit plasticity rather than friability; in consequence, some care may be required in assessing the weathering state of such rocks.

4.4.4. *Discontinuities*

The fracture log and RQD do not provide information on orientation, type or relative importance of any discontinuities which may be present. Orientation is preferably treated graphically or symbolically (Fig. 3) and the log could include a notation for relative dominance. Comments on discontinuities and veins in relation to their type, orientation, frequency, infilling, degree of open-ness and surface features should be incorporated following the systematic description.

4.4.5. *Discontinuity spacing*

Considerations of discontinuity spacing should lead to an appraisal of rock mass structure; this may be assisted by field observation. In the case of sedimentary rocks, where bedding may be the dominant discontinuity, it is possible to recognize and define a bedding spacing from borehole cores. A number of spacing classifications were considered and one developed by Binnie & Partners appeared to satisfy most requirements. This system has the advantage that the scale is related to that used in the mechanical analysis of soils (C.P. 2001, 1957). The proposed classification is as follows:

Description	Bedding plane spacing	Soil grading
Very thickly bedded	>2 m	Boulders
Thickly bedded	0·6 to 2 m	Boulders
Medium bedded	0·2 to 0·6 m	Boulders
Thinly bedded	60 mm to 0·2 m	Cobbles
Very thinly bedded	20 mm to 60 mm	Coarse gravel
Laminated	6 mm to 20 mm	Medium gravel
Thinly laminated	<6 mm	Sand and fine gravel

In the case of the general fracturing (discontinuities such as joints, fissures, etc.) of the rock mass it is not normally possible to assess the discontinuity spacing from a single set of borehole cores. If inclined holes are specially drilled it may then be practicable to determine and define the three-dimensional fracture state. A descriptive classification of fracture spacing is therefore not applicable to cores and such information should be recorded in the form of the fracture log. It should be noted, however, that the scale of general fracturing characteristic of rocks is usually about one order of magnitude greater than that for bedding; the system used for bedding, referred to above, should not, therefore, be applied directly to fracture spacing.

For igneous and metamorphic rocks the separation of the integral rock discontinuities (such as foliation, flow-banding, etc.) may be described by adaptation of the bedding plane spacing scale given above, e.g. medium foliated gneiss. It is suggested that 'close' and 'very close' are applied to that part of the scale where a sedimentary rock would be described as 'laminated' or 'thinly laminated'.

Systems for description of the gross homogeneity of the rock mass and the continuity of the rock substance, normally only applicable to exposed rock masses, have been given by Coates (1964) and Burton (1965). Terms such as blocky, intact, uniform, etc., may be used providing these terms are defined in the preamble to the borehole records and are based on *in-situ* inspection of the rock mass or by deduction from several boreholes.

4.4.6. *Strength*

A scale of strength, based on uniaxial compressive tests, is recommended as follows:

Strength MN/m² (1 MN/m² = 146 lb/in²)	lb/in²	Term
<1·25	<182·5	very weak
1·25 to 5	182·5 to 730	weak
5 to 12·5	730 to 1825	moderately weak
12·5 to 50	1825 to 7300	moderately strong
50 to 100	7300 to 14 600	strong
100 to 200	14 600 to 29 200	very strong
>200	>29 200	extremely strong

Any rock with a strength significantly less than 1·25 MN/m² should be described with reference to soil mechanics practice (C.P. 2001, 1957).

The scale proposed above is based largely on that developed by Deere & Miller (1966). It should be noted that field assessments of rock strength can be aided by simple tests, such as the use of a hammer or penknife, and readily supplemented by the Schmidt hammer or ultrasonic velocity determinations. It was considered that a scale based on deformation characteristics of rock materials could not be obtained without extensive rock testing.

4.4.7. *Ancillary geological information*

The systematic description of the rock cores may need to be supported by ancillary geological information aiding correlation between boreholes. This might include, for example, dip, stratigraphical location, sedimentary structures, identification of fossil bands and coal seams or qualifications relating to genesis.

4.5. ROCK GRADE CLASSIFICATION

4.5.1. *Objective*

The purpose of a rock grade scale is to provide an engineering classification of a rock mass for a specific purpose or project. Such a grading system is a subjective scale and, if incorporated in a log, is the only part of the log which is a statement of opinion.

4.5.2. *Types of grade scale*

Various grade scales have been devised for different purposes (Knill & Jones 1965; Ward, Burland & Gallois 1968; Chandler 1969) and each of these has been based on an appraisal of the strength and texture of the rock material, the weathering or alteration state and the discontinuity system; it should be noted that a grade scale is quite distinct from the weathering scale described earlier. The scale may be devised incorporating the results of any *in-situ* or laboratory tests. There may be some superficial conflict between the rock material strength description and the assessment of rock mass behaviour on the grade scale. Thus a rock such as chalk, which is likely to be defined as 'weak' on the proposed rock material strength scale, may be described as having superior rock mass properties on a grade scale set up for foundation design purposes for a site in southern England. Conversely a fresh granite which would be high on the strength scale could, by virtue of its discontinuities, be low on a grade scale for assessing tunnel overbreak.

4.5.3. *Limitations*

The subjective nature of such scales becomes most apparent when they are devised for different purposes, although a scale for foundation assessment tends to provide the most balanced appraisal of the general engineering characteristics of the rock mass. Provided that their limitations are appreciated, grade scales of this type can be a valuable adjunct to logging methods.

5. Proposed borehole log

5.1. ARRANGEMENT

The general layout of the proposed log sheet follows the conventional pattern. Information concerning the location and method of drilling of the borehole is presented at the top of the sheet and other relevant data are at the bottom of the sheet. As the log itself is plotted in a series of columns with a binding margin on the left, it is sensible to group the more important information on the right of the sheet. Such information would include the borehole number and location, relevant figure or reference number and the pictorial log. To the left of the symbolic log appear the geological description and depth scale, followed by information on the state of recovery of the core, information obtained from drilling and *in-situ* test results. It is stressed that the basic layout of the proposed log may be adapted to suit particular requirements and three varieties of the log sheet are illustrated by Figs 1, 3, 4, Pl. 1. An element of choice must be left because of the varying requirements of specific logs.

5.2. DATA PLOTTING

Percentage recoveries may be presented numerically or graphically. In the former case the values given will be exact and would allow detailed comparison to be made if the accuracy of the information permits; this precision cannot be equalled by graphical representation on the small scale of the borehole sheet log. Graphical representation, however, permits a more ready comparison of variations between the recoveries (water, core, RQD, etc.) recorded (Fig. 1) and thus will aid assessment of engineering properties. Should recoveries vary within small percentage limits in a borehole (say between 75 and 100 per cent) the lateral percentage scale may be expanded to fit these limits of variation within the available column space, thus permitting greater accuracy of presentation. In this case, however, attention should be drawn to the changed scale. Reference should also be made to a common graphical system of plotting core recovery in which each core run is represented by a narrow column; the proportion of core which is recovered is represented by blacking-out the equivalent length of the column. As this system is plotted on the vertical scale it tends to give a more precise record of recovery than the laterally measured graphical scale. The method is, however, less satisfactory when more than one type of recovery is being plotted and also has the disadvantage that the position of core loss is given an implied location which could give rise to erroneous conclusions.

5.3. LAYOUT OF LOG

5.3.1. *General*

The basic layout of the log can now be described with reference to section 4.

2.7cm binding margin

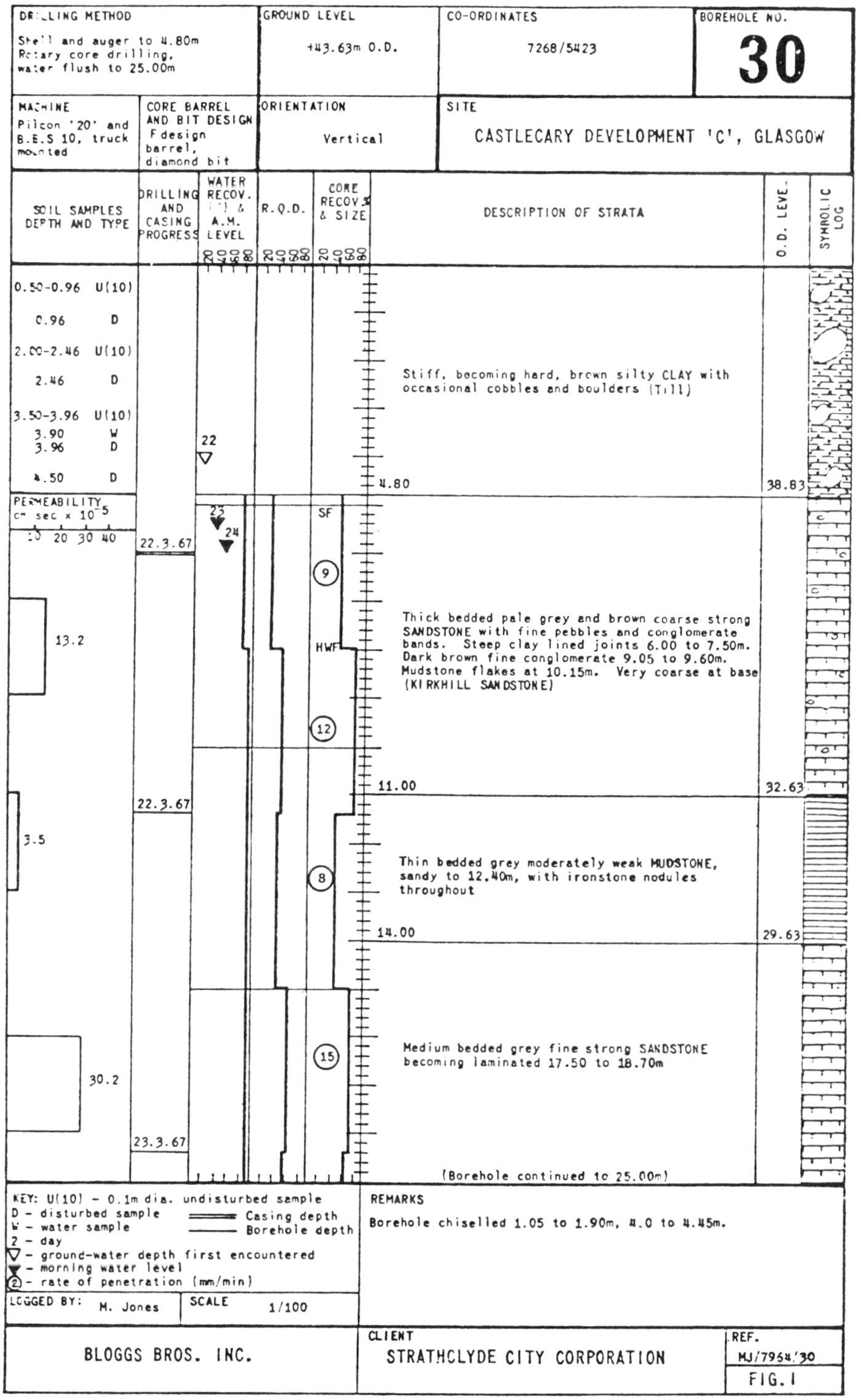

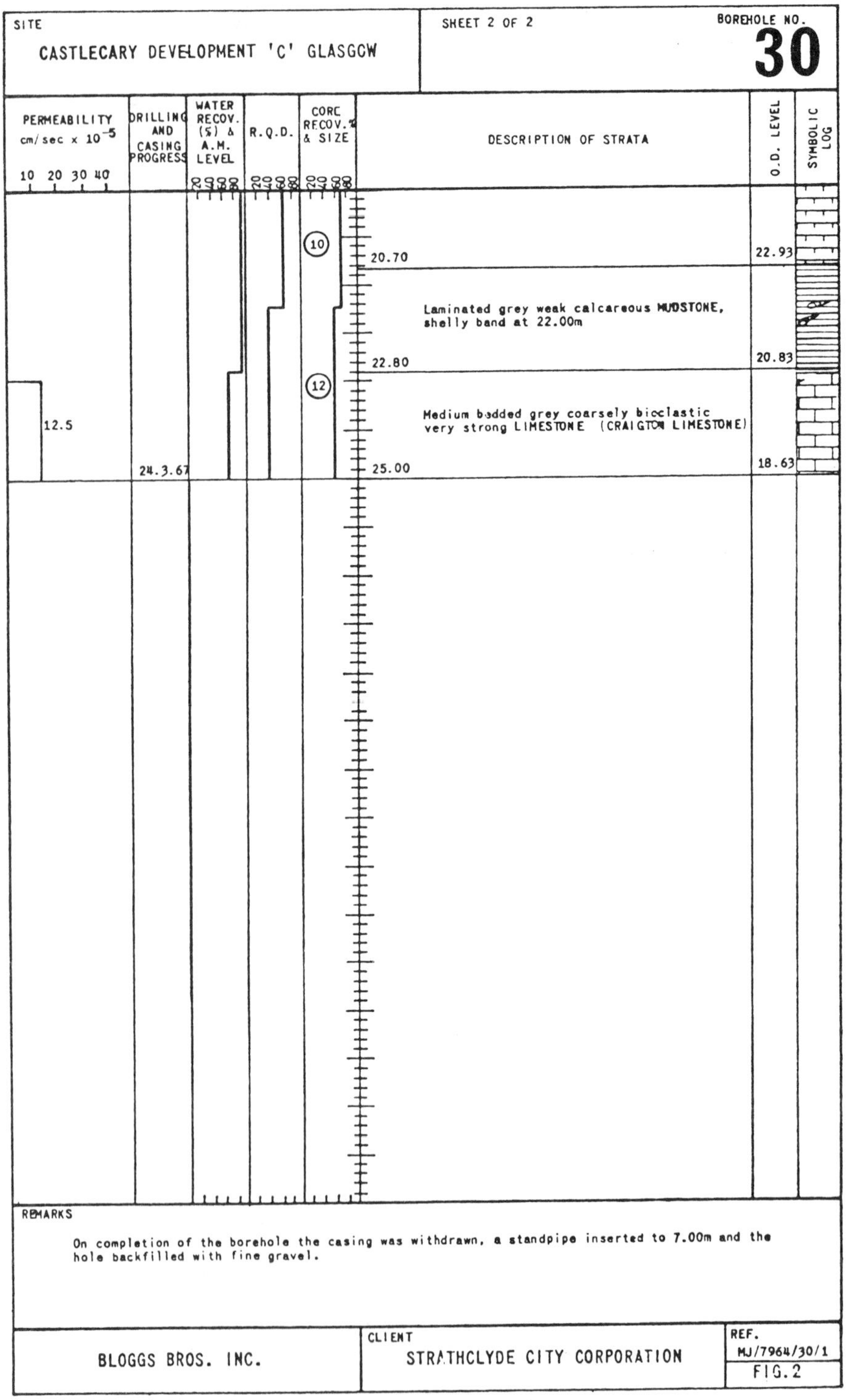

2.7cm binding margin

1 cm binding margin

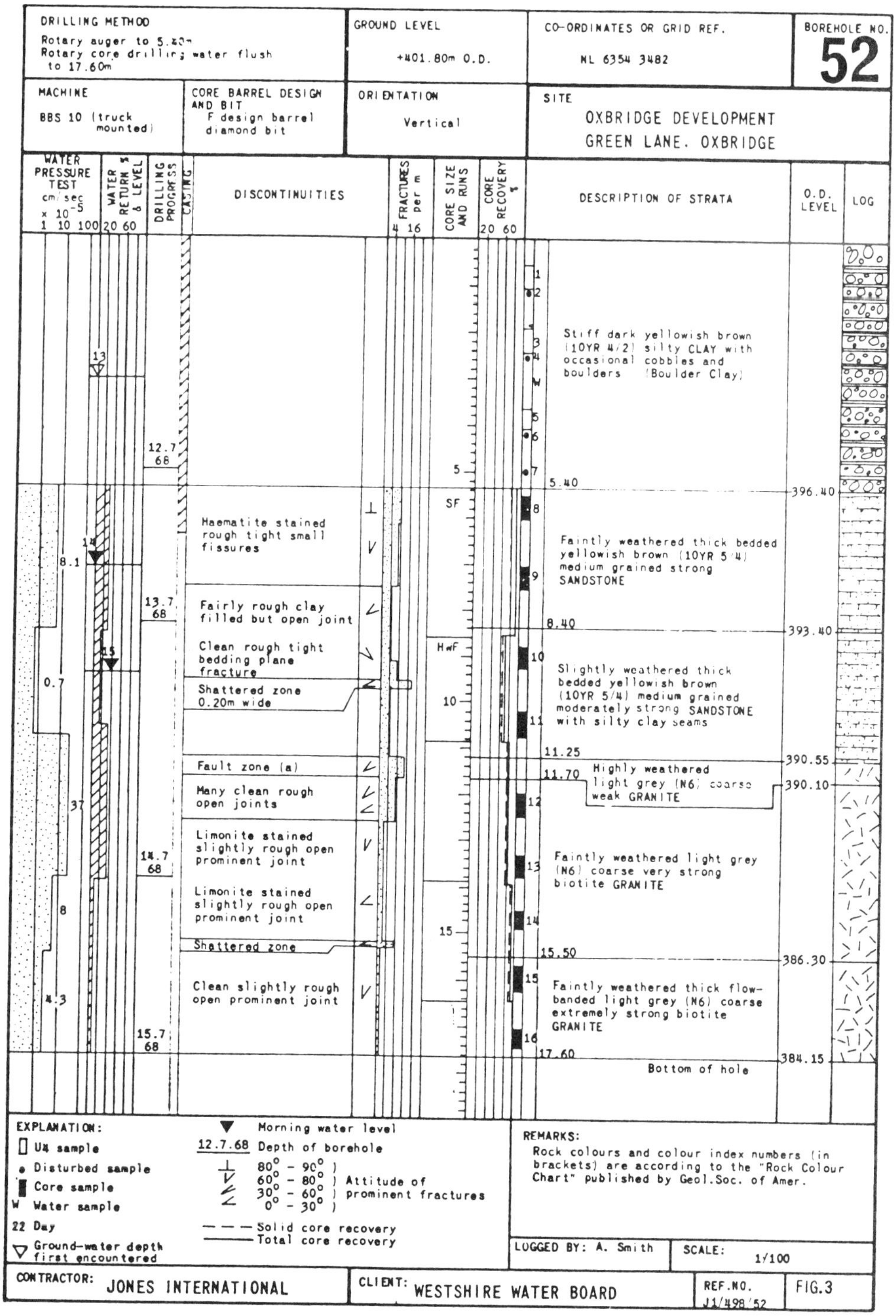

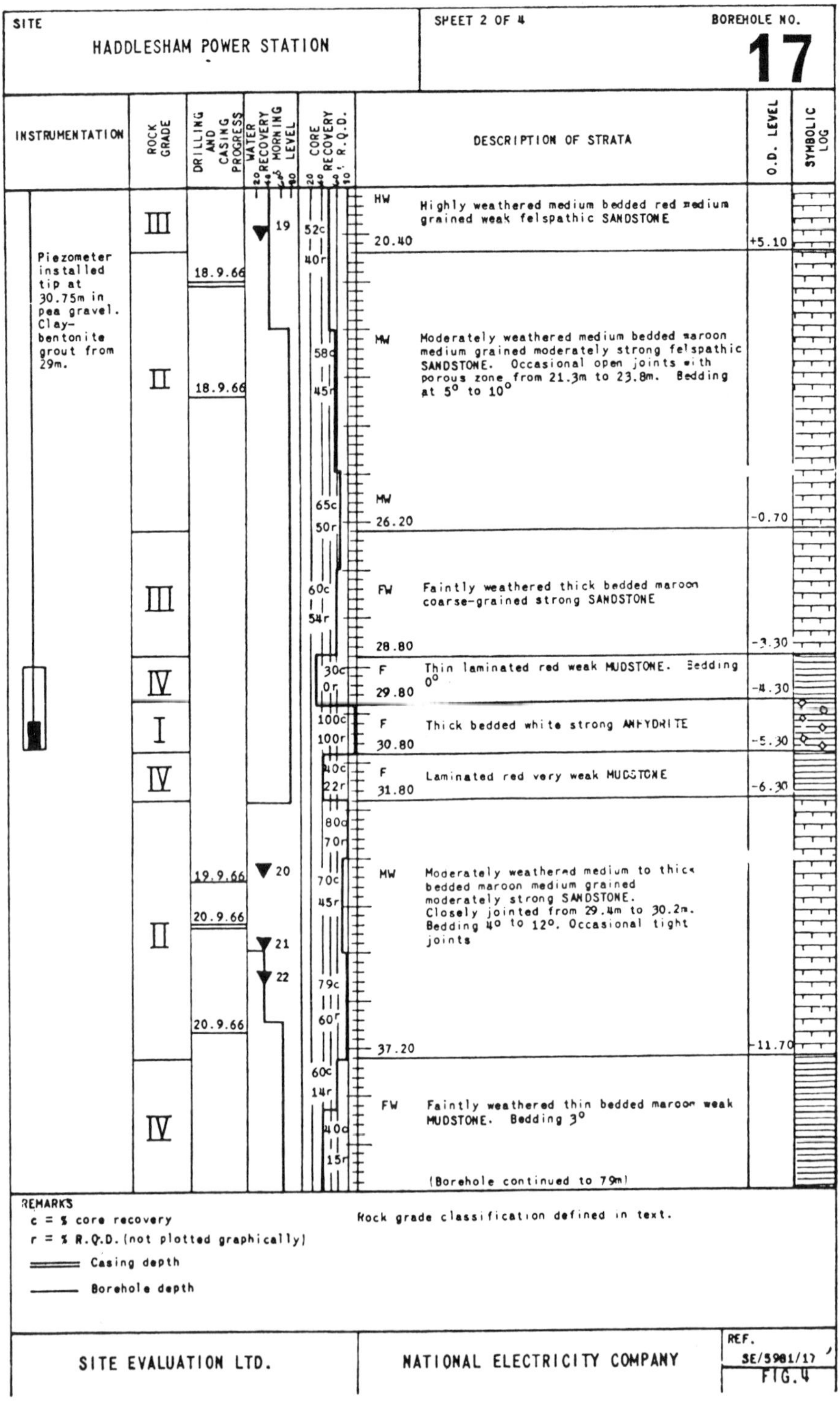

2.7cm binding margin

5.3.2. *Borehole number*

This is placed on the top right hand corner of the log in large lettering for easy identification and retrieval of the record when inspecting a report or sheaf of records.

5.3.3. *Location*

The necessary details of location (site name, grid reference, elevation and orientation) are grouped in the top section of the log adjacent to the borehole number, for easy reference and identification. The site name is in large clear print, should be kept as succinct as possible and is enclosed together with the borehole number in a boldly outlined box.

5.3.4. *Drilling technique*

Details of the drilling method, machine type core barrel and bit design are given at the top left side of the log.

5.3.5. *Contract details*

Details of the drilling or site investigation contractor or both, client and job reference number are given at the bottom of the log in bold legible type.

5.3.6. *Miscellaneous information*

This is located in the blocks between the contract details and the central part of the log and includes the name or cypher of the logger. A key to symbols used in the log, the scale of the log and any pertinent information revealed during the boring may also be included in this space.

5.4. DRILLING INFORMATION

5.4.1. *Arrangement*

The central part of the log includes information on drilling progress, the state of recovery of the core and the descriptive geology displayed in vertical columns. A column for Ordnance Datum levels of strata changes is included together with one for a symbolic log. This last column is placed on the extreme right hand side of the log to facilitate rapid appraisal of bound groups of logs and to allow easy comparison of overlapped loose log sheets for correlation purposes. A wide column is provided on the left side of the log to describe the depth and type of samples taken by methods other than rotary core drilling and the results of any tests conducted either *in-situ* or in the laboratory (figs. 1, 2, 3 and 4).

5.4.2. *Progress and rate of drilling*

The depth of the borehole at the end of each day or shift is displayed in a column on the left side of the record. The depth of the hole is given by a line opposite the appropriate depth on a central divided metric scale; the date or time at which the depth was reached is written above the scale line (Figs. 1 and 3). The ends of each core run may be indicated by scaled horizontal lines or if appropriate by breaks in the core recovery graph (Fig. 1). Records of actual penetration rate may be given as a simple number, possibly emphasized by a symbol (Fig. 1) and placed adjacent to the appropriate core run or, if sufficient detail is available, as a graphic log.

5.4.3. *Core diameter*

This is recorded by index letters or numbers located at the top of the core run (relative to the depth scale) below which this diameter was recovered (Fig. 1).

5.4.4. *Casing*

Casing depth may be shown in the same column as drilling progress by a line symbol (relative to the depth scale) and a date (Fig. 4). Changes in casing size may be recorded numerically above the line symbol.

5.4.5. *Flush returns*

These are given as a percentage recovery for each core run or arbitrary depth unit and may be expressed numerically or graphically (Fig. 1). Any change in the character of the returning circulatory fluid may be noted in the 'Remarks' space.

5.4.6. *Water levels*

Standing water levels and the depth at which ground-water was encountered may be recorded in the flush recovery column by symbols incorporating a date and, if necessary, a time of observation positioned relative to the depth scales (Fig. 1).

5.4.7. *State of recovery of core*

This is provided by a core recovery log (covering RQD and total and solid recovery) and a fracture log and may be expressed graphically or numerically, to give greater precision (Figs. 1, 2 and 4).

5.5. GEOLOGICAL DESCRIPTION

5.5.1. *Presentation*

The greater part of the central portion of the log is occupied by one or more columns devoted to geological description. In the systematic description the rock name should be written in capitals for emphasis. Strata boundaries are drawn as solid lines relative to the central depth scale; the actual depth below surface is written above boundaries adjacent to the depth scale. All the information can be provided in one column (Fig. 1) or separated into two columns (Fig. 3) or more to allow for detailed variations within one lithological unit.

5.5.2. *Dimensions*

The dimensions of the description column may be amended to suit the type of rock described. In the case of a borehole penetrating similar rock throughout its whole depth the rock type description may be short and additional columns inserted to indicate symbolically or graphically changes in weathering state (Fig. 4), joint orientation (Fig. 3), etc. The style of log sheet can be designed according to specific requirements.

5.5.3. *Grade scales*

Grade scales may also be introduced either in an additional column or as a notation to the systematic description (Fig. 4).

5.6. CONTINUATION SHEETS

These need not duplicate the locational and drilling information given in the top section of the first sheet of the record. The continuation sheets illustrated by Figs. 2 and 4 show in the top and bottom blocks of the record only sufficient information to link the

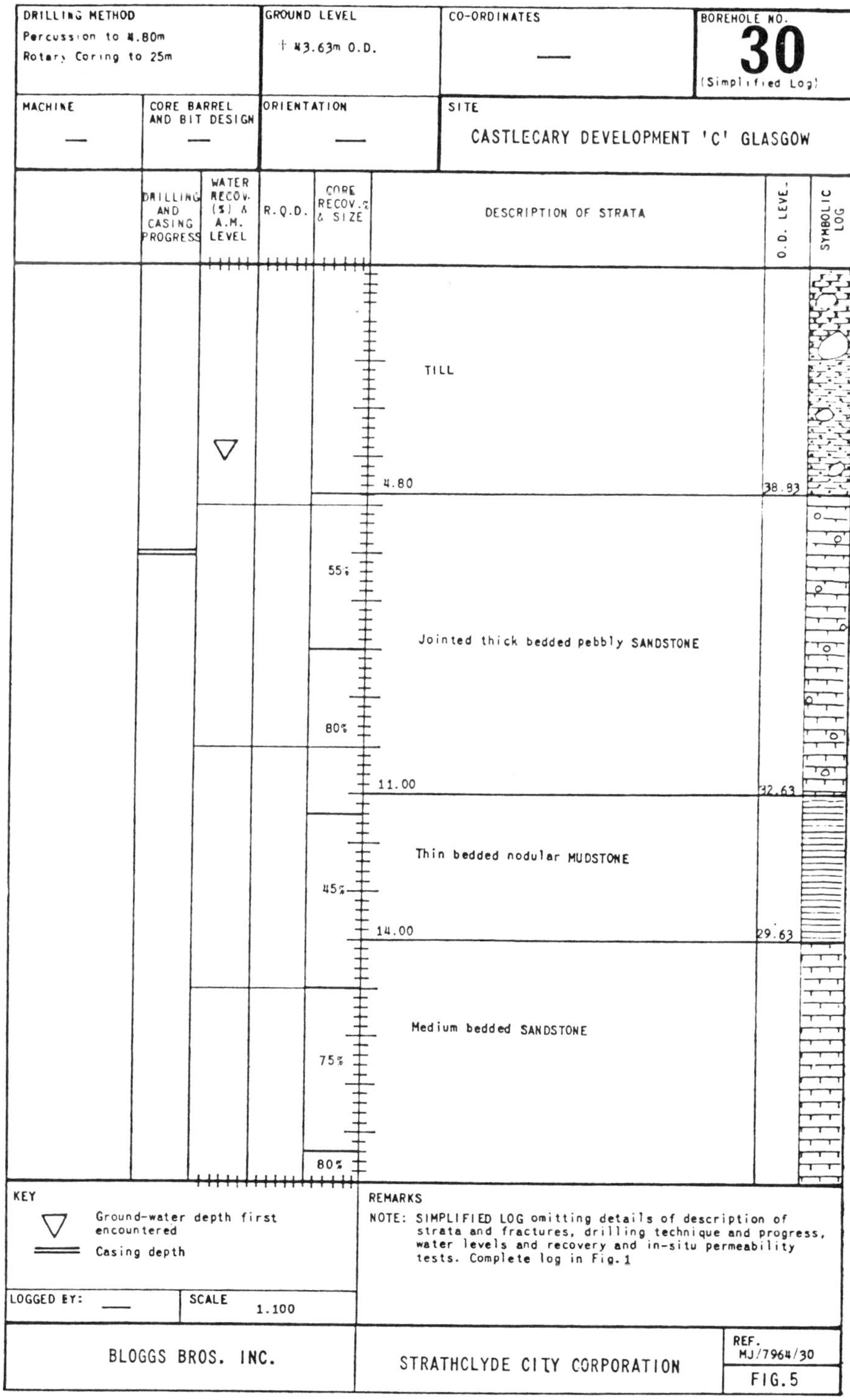

2.7cm binding margin

continuation sheet(s) with the first sheet of the log. The reduction of ancillary information permits a slight expansion of the descriptive part of the log; the continuation sheet in Fig. 2 allows, at a scale of 1 : 100, the description of 21 m of borehole as opposed to 19 m on the first sheet.

5.7. SIMPLIFIED LOGS

In those cases where it is decided to supplement the main log by a simplified log it is suggested that a normal log sheet is used (Fig. 5). By this means it is possible to recognize immediately that certain information has been omitted.

6. Specified requirements

In order to assist in the preparation of Specifications for borehole logging, the following essential and optional information for logging is listed below.

6.1. ESSENTIAL DATA

Borehole number	(4.1.1)
Location	(4.1.2)
Drilling technique	(4.1.3)
Contract details	(4.1.4)
Drilling progress	(4.2 (ii), (iii), (iv), (viii) and (x))
State of recovery of cores	
Total core recovery	(4.3.2)
Descriptive geology	
Systematic description	(4.4.2)
Alteration/weathering state	(4.4.3)
Structure and discontinuities	(4.4.4, 4.4.5)
Assessment of rock material strength	(4.4.6)
Ancillary geological information	(4.4.7)

6.2. OPTIONAL DATA

Drilling progress	(4.2 (vi), (vii), (xi) and (xiii))
State of recovery of cores	
Solid core recovery	(4.3.2)
Fracture log	(4.3.3)
R.Q.D.	(4.3.3)
Stability Index	(4.3.4)

7. References

BRITISH STANDARDS INSTITUTION, 1957. *Site investigations.* C.P. 2001. London.

—— 1966. *Rotary core drilling equipment.* Part I: *Basic equipment.* B.S. 4019. London.

BURTON, A. N. 1965. Classification of rocks for rock mechanics. Correspondence, *Int. Jl Rock Mech. Min. Sci.* **2**, 105.

CHANDLER, R. J. 1969. The effect of weathering on the shear strength properties of Keuper Marl. *Géotechnique* **19**, 321–34.

COATES, D. J. 1964. Classification of rocks for rock mechanics. *Int. Jl Rock Mech. Min. Sci.* **1**, 421–9.

DEERE, D. U. 1964. Technical description of rock cores. *Rock Mech. Engng Geol.* **1**, 16–22.

——& MILLER, R. P. 1966. *Engineering classification and index properties for intact rock.* Tech. Rept. no. AFWL–TR–65–116, Air Force Weapons Lab., Kirtland Air Force Base, New Mexico.

——, HENDRON, A. J., PATTON, F. D. & CORDING, E. J. 1967. Design of surface and near-surface construction in rock. *Proceedings 8th Symposium Rock Mechanics, Minnesota*, 237–302.

——, MERRITT, A. H. & COON, R. F. 1969. *Engineering classification of in-situ rock.* Tech. Rept. no. AFWL–67–144, Air Force Systems Command, Kirtland Air Force Base, New Mexico.

EGE, J. R. 1968. Stability index for underground structures in granitic rock, in Nevada Test Site. *Mem. geol. Soc. Am.* no. 110, 185–98.

KIERSCH, G. A. & TREASHER, R. J. 1955. Investigation; areal and engineering geology—Folsom Dam Project, Central California. *Econ. Geol.* **50**, 271–310.

KNILL, J. L. & JONES, K. S. 1965. The recording and interpretation of geological conditions in the foundations of the Roseires, Kariba and Latiyan dams. *Géotechnique* **15**, 94–124.

MOYE, G. D. 1955. Engineering geology for the Snowy Mountains Scheme. *J. Instn Engrs Aust.* **27**, 281–99.

—— 1967. Diamond drilling for foundation exploration. *J. Instn Engrs Aust.* **CE9**, 95–100.

WARD, W. H., BURLAND, J. B. & GALLOIS, R. W. 1968. Geotechnical assessment of a site at Mundford, Norfolk, for a large proton accelerator. *Géotechnique* **18**, 399–431.

Submitted 7 December 1969.

MEMBERSHIP OF WORKING PARTY

J. L. Knill	Imperial College, London S W 7. (Chairman)
C. R. Cratchley	Institute of Geological Sciences, London S W 7.
K. R. Early	Soil Mechanics Ltd., London S W 3.
R. W. Gallois	Institute of Geological Sciences, London S W 7.
J. D. Humphreys	W. S. Atkins & Partners, Epsom.
J. Newbery	Montrose, Broomfield Ride, Oxshott.
D. G. Price	Central Laboratory, George Wimpey & Company Limited, Hayes.
R. G. Thurrell	Institute of Geological Sciences, London S W 7.

[*Editor's Note:* The Discussion and Plate 1 have been omitted.]

34

Reprinted from *Mining Eng.* **26:**37–40 (1974)

Geotechnical Investigations for Mine Shafts

By James R. Swaisgood & Ronald E. Versaw

Hundreds of shafts have been sunk in the United States in the past. Most of these have been successful; however, in some instances thousands of dollars have been lost due to work stoppage while emergency measures were taken to control unexpected water flows. In cases where it was not possible to control the water, the shafts were completely lost.

Many thousands of dollars have also been wasted in overconservative support design for the successful shafts. These "hell for stout" designs were necessary because of the importance of the shaft in the operation of the mine. If there was trouble in the shaft, all underground operations would come to a halt. Over-design was also necessary because no reliable pre-construction investigation methods or rational support design methods were available.

As time passes, the extra costs for control of unexpected water conditions and for excessive support systems are becoming less acceptable as labor and material prices rise and as shafts go deeper. Because these costs can no longer be afforded, shaft investigation techniques are being perfected and support design methods are being developed. On a few projects, these new techniques and methods have already been successfully used to reduce construction costs. Some of the companies mining the deep potash deposits in Canada have performed pre-construction investigations to evaluate rock and water conditions.[1] At its Henderson Mine in Colorado, Amax drilled and tested small-diameter pilot holes at two shaft sites.[2] Discussion with the Amax engineers revealed that these investigations were extremely useful in predicting the general rock conditions and the exact locations of high water inflow. Other companies engaged in mining have had extensive investigations performed for shafts to be sunk in New Mexico.

The types of geotechnical investigations for shafts can be grouped into several broad categories, including preliminary studies, pilot hole drilling and testing, laboratory testing, and construction monitoring.

Preliminary Studies

Preliminary studies refers to work which should be completed prior to performing any detailed studies at the site. These studies include air photo interpretation, literature search, case history study, and review of exploration boring logs and cores. This information is helpful in planning the more detailed portions of the geotechnical investigation and in providing a reference for comparing the results of the geotechnical studies with the results obtained from previous work in the area.

Air photos are useful in several ways to the geotechnical engineer interested in the siting and construction of a shaft. From air photo studies, the engineer can gain information which is useful in locating shaft sites, observing regional structural geologic trends, and in locating adverse surface features such as landslides, high groundwater areas, karst topography, or subsidence areas. If practical, these adverse surface features should be avoided when locating the shaft collar.

Geological Mapping of the Surface is Equally as Important

Another very important part of the preliminary studies is the literature search. This includes locating and reviewing all available geotechnical information relative to the site. During this phase of work, particular attention should be paid to gathering and reviewing data available on the stratigraphy, physical rock properties, in-situ rock stress, and data relating to groundwater flow and rock permeability. Particularly valuable are accurate geologic maps and cross sections. Generally, most geologic literature is concerned with lithologic rock descriptions and the economic aspects of geology. However, careful review by an experienced person will usually reveal useful engineering information.

Case History Studies

Prior to performing any detailed shaft investigations, as much information as possible should be obtained about previous successful and unsuccessful shafts with similar geologic and hydrologic conditions. Preferably, these shafts should be as close to the proposed shaft site as possible. However, valuable information can be gained from review of other case histories as well.

The information obtained will provide guidelines as to the potential problems which could occur and what has been done in the past to correct them. The case history information will also offer ideas about the type of support and dewatering measures which can be used for shaft completion.

Pilot Hole Drilling and Testing

By far the most significant part of a shaft site investigation is the drilling and testing of a pilot hole. It is the only reliable way to obtain accurate information about the rock and water conditions which will be encountered in the shaft. A pilot hole is a small-diameter borehole (NX to 6-in. diameter) drilled to the total depth of the proposed shaft. To be of most value, the drill hole should be maintained within the proposed shaft diameter. The pilot hole should be directionally surveyed so that its relative position and inclination is known. Cores should be taken and logged by visual examination. Mechanical tests can be conducted on the core in the field. Other mechanical tests and hydrologic tests can be run within the pilot hole at selected depths. Geophysical surveys can be conducted in the pilot hole.

The pilot hole should be cored for its total depth. In some instances, where the rock and water conditions are well known because of nearby exploration and testing, spot coring may be done at critical locations only.

During the coring operation, a complete engineering log should be maintained by the field engineer. The log should contain data and notations relative to: rock type and lithologic description, graphic descriptive log, degree of weathering, relative strength of rock, depth and length of core run, condition of core, core recovery, description of rock defects, joint frequency and spacing, RQD, water level, water or drilling mud loss, casing depth, and drilling rate.

Core orientation during drilling has recently been developed as a practical and useful method of obtaining structural information relating to the attitude of joints, bedding planes, and other rock defects. Although new and cheaper methods are under development, core orientation is presently somewhat expensive and time consuming; and therefore, the necessity of obtaining this information needs to be justified. Generally speaking, core orientation would not be too significant in flat-lying sedimentary rocks with little or no structure. On the other hand, core orientation could be quite useful in more complex geologic situations where the orientation of joint bedding planes, faults, and the like have a significant effect on the support requirements.

An aspect of pilot hole coring which is often neglected is the care and storage of the core. If the rock core is to be of any value to the engineer subsequent to the initial removal from the ground, it must be properly handled and stored. For most rock, this means washing, sealing in plastic, and storing in core boxes in a dry, protected place. Samples which will be used for laboratory testing should be given particular attention so that they will not become broken during handling or transportation. Carefully-stored and preserved core will also be more useful to prospective contractors who inspect it to evaluate the rock conditions.

Field Mechanical Tests

Field mechanical tests which have proven to be useful in geotechnical shaft investigations can be grouped into two categories. These are the tests performed on extracted core and the down-hole tests.

The more useful core tests are the Schmidt hammer rebound and the point load or Brazilian tensile tests. These tests are used to evaluate the strength and elastic properties of the rock. The results of these tests can be included on the engineering log and are useful for correlating with laboratory test data and geophysical well-log data. If done properly, these tests will allow a reduction in the total number of the much more expensive laboratory tests. The particular advantage of field mechanical tests performed on core is that at a very

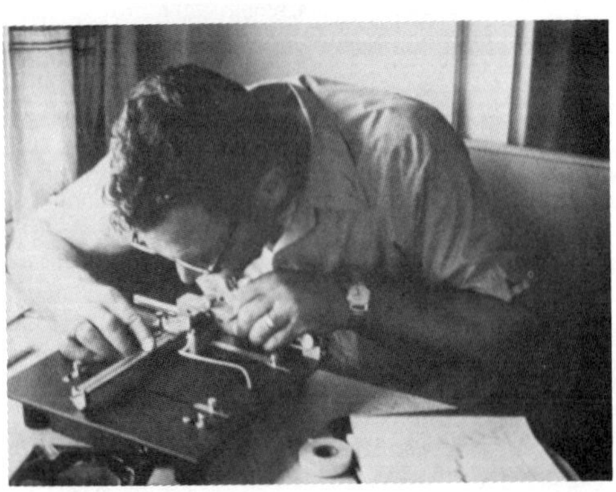

At far left, drilling rig on proposed shaft location near Grants, New Mexico. (Left) Removal of 6-in. core from drill core barrel. (Center) Core being logged for recovery percent, joint spacing, RQD, and lithology. (Bottom) Engineer analyzing results from down-hole pressure record from packer test. test.

low cost, a continuous mechanical log of the core can be obtained.

Limited down-hole tests are performed with borehole jacking tools which are used to measure in-situ modulus in a direction normal to the borehole. Unfortunately, most of these down-hole jacking tests are limited to a few hundred feet in depth; and since most shafts extend in depth well beyon˜ the capability of the jacking tools, their usefulness is somewhat restricted.

Hydrologic Tests

Hydrologic tests performed in the pilot hole are used to determine the rock permeability and the hydrostatic formation pressure from which the anticipated inflow rates into the shaft can be evaluated. The hydrologic tests used in shaft investigations are adaptations of techniques which have been developed and widely used in the water well and petroleum industries. The tests can be performed using the total length of pilot hole. However, the most useful tests for pinpointing exact locations of high inflow involve isolating critical sections such as fault zones or permeable formations with packers. There are two types of tests which can be performed on the isolated zones; drill stem tests and injection tests.

In drill stem testing, the hydrostatic formation pressure and aquifer permeability are determined by isolating the critical zone with packers, reducing the hydrostatic pressure to near zero by swabbing the drill pipe, and then allowing the water to flow back into the test section and up the drill pipe until it stabilizes. The pressure buildup is recorded on a down-hole time recorder to provide a recovery rate curve. The permeability of the test section is calculated from the recovery rate curve. This method has a definite advantage by allowing the walls of the test section to be cleaned of drill mud as the water flows from the rock mass into the pilot hole.

Injection tests are essentially the reverse of drill

stem tests. In this method, the test zone is isolated with packers and water is forced into the formation at a known pressure and rate. The permeability is then determined from the pressure and flow rate data.

Information regarding in-situ rock stresses can be obtained with a variation of the injection test, known as hydrofracturing, wherein the water in the test section is pressurized until the rock is fractured or until existing fractures are forced open as the water pressure equalizes the rock pressure.

Geophysical Well-Logging

Most of the geophysical well-logging techniques were originally developed for use in petroleum and mining exploration. However, they are very useful in geotechnical investigations for shafts because they provide a considerable amount of mechanical information at a relatively low cost. In addition, certain geophysical survey methods have now been developed specifically to obtain engineering characteristics of rock masses.

There are several types of geophysical well-logging techniques, including 3-D velocity, resistivity, self-potential, gamma-ray, neutron, caliper, density, temperature, and seisviewer. Each of these investigate a specific property of the rock mass from which estimations of numerous mechanical and engineering rock properties can be made. All of the geophysical well-logging data can be correlated with laboratory test results so that fewer of the more expensive laboratory tests are required.

Laboratory Testing

To supplement the field data and to provide information about the engineering properties of the rock through which the shaft will be sunk, laboratory tests can be performed on representative samples of the core obtained from the pilot boring. This test data is essential to all design calculations regarding shaft support and dewatering. Unfortunately, laboratory tests can be performed only on unbroken rock core samples; and as a result, the test data often represents only the stronger rock conditions. Allowances for these differences must be made by an experienced engineer during the design process.

The laboratory engineering rock property tests, used under spacial conditions, can be grouped into several categories: index properties (moisture content, density, porosity, and grain size), permeability, compressive strength, tensile strength, shearing resistance, modulus of elasticity, and Poison's ratio.

Construction Monitoring

An important part of geotechnical investigation often overlooked is construction monitoring. During this phase, a close watch is kept on the shaft excavation to make sure that the actual conditions encountered correspond to those anticipated from the field and laboratory investigations, and that they were adequately accounted for in the design of the support and water-control systems. If deviations occur from those conditions which were anticipated, the construction crew can be alerted to possible problems and a redesign can be initiated if necessary. Otherwise, an unsafe condition may go undetected and total construction costs may escalate.

Drilling a series of probe holes at the bottom of the shaft is a common type of monitoring technique used by almost all shaft-sinking contractors today. These probes detect the presence of water-bearing ground just ahead of the shaft. They also can be used to check the effectiveness of previous grouting programs.

Benefits From a Geotechnical Investigation

The benefits which accrue from thorough investigations are considerable in terms of reduced cost and increased safety. They more than offset the total investigation costs, which are on the order of $15 to $20 per ft. Some of the ways in which geotechnical investigations can reduce cost and increase safety during and after construction of a shaft are:

1. Lower bid price from contractor because more information is available prior to bidding, and he does not have to make allowances for "unknowns." Also, there is less chance the contractor will have a need to claim "extras" if the conditions are known in advance of construction.
2. Down time during construction is reduced because adverse conditions ahead of shaft bottoms are anticipated and prepared for in advance.
3. Reduced support costs because the investigations indicate areas where minimal or no support is required, and allowances may be made for these areas in the design.
4. Post-construction maintenance is minimized because support and dewatering requirements are accounted for in the original design.
5. During construction, adverse rock and groundwater conditions can be predicted in advance of shaft bottom and adjustments can be made to insure the safety of the mining personnel.
6. Since the unstable ground and adverse groundwater conditions have been considered in the design, failures in the lining or dewatering system which could be harmful to personnel are minimized.

References

[1] Adamson, J. N. and Scott, R. A., "Borehole Investigations and Logging Methods in Shaft Sinking," *Mining Engineering*, January, 1973, pp. 181-189.

[2] Sharp, J. E., "Engineering Geology Studies of the Henderson No. 1 Shaft, Henderson Mine, Climax Molybdenum Company, Empire, Colorado," June 1969.

35

Paper presented at the *Joint Meeting, MMIJ-AIME, Tokyo, 1972, Print T II e 2,* 14p.

ROCK MECHANICS IN THE PLANNING AND CONTROL OF BLOCK CAVING

J. R. Swaisgood, B. K. McMahon, and L. J. West

INTRODUCTION

Block caving is distinguished from all other mining methods by the use of the natural stresses in the rock mass as the primary agent in rock fragmentation. Ideally, these stresses are concentrated by means of undercuts and boundary stopes until they exceed the shear or tensile strength of the jointed rock mass. A progressive failure then takes place and the broken ore collapses into the undercut to be drawn off through finger and transfer raises.

Problems occur when the ore body is too narrow for the width of undercut required to induce caving; when the ore body is exceptionally strong; when the block width and associated undercut width is too short to maintain sustained caving; when the rock breaks into fragments that are too large to pass through the transfer raises and require secondary fragmentation; and when the concentration of stresses around the sublevel workings results in squeezing ground and the collapse of the openings.

Under these conditions, the additional costs involved in boundary weakening and assistance blasting to induce collapse, the costs resulting from secondary blasting to reduce the size of the blocks, and the costs of supporting and repairing the sublevel workings may substantially reduce profits or even make marginal ore bodies uneconomic. Typically, the total of these expenses will account for approximately 25 to 35 percent of the total mining costs in block caving operations (1,2,3). Accordingly, one objective of the prudent mine manager is to design or modify the mining system and layout to minimize these expenses. The art-science of rock mechanics is used as an aid to achieve this goal.

In the past, an emperical form of rock mechanics has been applied through trial-and-error methods to solve the rock behavior problems related to block-cave operations. If poor caving characteristics occurred initially, mining operations were stopped while additional undercutting or boundary weakening was applied. In contrast, if good caving occurred in the initial block pulled, the development and preparation work for subsequent blocks was continuously reduced until the minimum amount required was found. Changes in the sub-block opening support systems were made using only past experience in the mine as a basis. Using these cut-and-try processes, the optimum mine design features were sometimes never quite found and the minimum costs never totally realized.

Today, the mine manager is faced with high costs of labor and equipment, relatively low-grade ore bodies, and an extremely competitive market. Accordingly, he will rarely take the time or spend the money required for loosely-directed, trial-and-error procedures. Instead, he now finds the optimum design features and minimum costs at the outset of mining operations or shortly thereafter with assistance from today's more sophisticated form of rock mechanics (a blend of past pragmatic knowledge, newly applied theoretical concepts, and reliable measurement techniques). Consequently, he is able to accurately predict how the rock will respond to various mine layouts prior to the time that any underground work is started. He also is able to forecast the effects that any modifications in the basic design will have on rock behavior prior to the time that such modifications are made.

The purpose of this paper is to outline and describe how the present-day rock mechanics tools and techniques are being used in the planning and control of block-cave mining. To serve as background, a description of the rock behavior associated with block caving is also presented.

ROCK BEHAVIOR IN BLOCK-CAVING OPERATIONS

General Concepts

The general stress conditions common to all block-caving operations are illustrated in Figure 1. These have been estimated by combining information from applicable theoretical concepts (4,5,6,14) from past rock mechanics studies of block-caving operations (7), and from a review of actual rock behavior in existing mines (2,3,8,9,10,11,12,13). The pertinent characteristics of the mines which were reviewed are summarized in Table I.

Ultimate rock behavior will be dependent upon the relationship between the stresses existing within the rock mass and the strength and elastic properties of the rock mass. In simple terms, if the stresses in the rock mass exceed the strength or elastic limits, failure or excessive movement will occur. On the other hand, if the strength and elastic limits of the rock are greater than the stress level, then stable conditions will exist.

As shown in Figure 2, successful block-caving operations require a paradoxical situation in which both failure and stable conditions exist simultaneously within the same general area:

1) Above the undercut level, the stresses must exceed the rock mass strength so that failure occurs. This relationship between stresses and strengths is maintained by increasing the stress levels and/or decreasing the rock strengths.
2) Below the undercut level, openings are located and designed so that strengths exceed stresses and stable conditions are developed. In most cases, opening support systems are required to increase the rock mass strength and to counteract excessive stresses.

Maintaining Failure Conditions

The main method used to maintain failure conditions in block caving is to increase or concentrate the existing native stresses by undercutting. As shown in Figure 1, high compressive and shear stresses are developed in the abutments or outer walls of the undercut opening. Sustained caving of the entire block will continue only if these stresses are great enough to exceed the rock strength and cause failure in the corner regions. Otherwise, in instances where the native horizontal stresses are less than the vertical stresses, caving would be arrested after a stable tension arch developed; and in cases where the horizontal stresses equaled or exceeded the vertical stresses, the entire opening would remain stable without any arching occurring.

The magnitude of the stresses developed in the abutment region is dependent upon the native horizontal and vertical stress values existing prior to undercutting; it is also directly proportional to the shortest horizontal dimension of the undercut opening (4,5). As this width increases, the stresses increase. According to this concept (which is based on well-established theories and substan i-ated by actual case histories), it is becoming increasingly apparent that caving behavior is not as dependent on the total area undercut as believed in the past, but more dependent on the relative value of the width of the undercut opening.

Another method used to increase stresses in the ore body is to develop boundary openings around the block. In the vicinity of boundary drifts or corner raises, the stresses are increased to two or three times those existing prior to the development of the openings. These openings also serve to weaken the overall strength of the rock mass in the vicinity of the block boundaries.

Decreasing the rock mass strength in the central portion of the ore block also helps to maintain failure conditions. This is accomplished by fracturing the rock with explosives loaded into blast holes drilled from corner raises and boundary drifts.

Maintaining Stable Conditions

The network of accessways, haulageways, and other sub-block openings should be located and supported so that the adjacent rock strengths are greater than the stresses and relatively stable conditions will exist. This will serve to reduce drift repair costs and increase safety for men and equipment.

One method to improve stability conditions is to locate the sublevel workings in low-stress areas. An example of this is the El Teniente Mine in Chile (15). As shown in Figure 3a, service crosscuts at the grizzly level were initially located directly under the block boundary lines in zones of high stresses. These crosscuts had to be heavily timbered and maintenance costs were high while the crosscuts were in use. Eventually, these crosscuts were relocated to a position outside of the boundary, as illustrated in Figure 3b. This new position was in a region where the stresses were much smaller than those in the initial zone. As a result, only light support was required and maintenance costs were reduced considerably.

Another method used to improve stability is to place the sub-block openings in stronger rock. This method was used in the Jennifer Mine in Kern County, California (10). In this mine, the ore zone was underlain by a low strength, plastic, yielding shale. To minimize the support and maintenance costs, the sublevel development workings were placed in the bottom of the much stronger ore rock zone just above this shale.

Many times it is not possible to locate sub-block workings in low-stress regions or in strong rock areas. In these cases, a support system must be designed to counteract the stresses or to strengthen the surrounding rock mass. A proper choice of a support system will result in an optimum balance of initial costs and future repair expenses. By changing from timber to concrete supports in the grizzly drifts, the cost of block repairs was reduced by 76 percent in Anaconda's Kelly Mine in Butte, Montana (13). In the Miami Copper Company mine in Miami, Arizona, scraper drifts were eventually supported by circular steel sets (2). This system replaced the initial system of conventional square-shaped timber and steel supports. The initial supports were distorted and failed quickly under the high ground stresses. Repair costs were high. The circular-shaped steel sets were more compatible with the stress conditions and usually did not distort. Repair costs were reduced and most sets were reuseable.

APPLYING ROCK MECHANICS TO PREDICT ROCK BEHAVIOR IN BLOCK CAVING

Present-day rock mechanics tools and techniques are used to accurately measure and estimate rock strengths and stresses and to predict rock behavior in advance of actual mining operations. Table II summarizes how these methods are applied in the exploration, development, and production stages to aid in the planning and control of block-caving operations.

The initial step in a rock mechanics investigation for block-cave mining consists of measuring and/or estimating the physical properties of the rock mass and the native stresses within the rock mass. The areas of strong and weak rock are located and defined in this step. The next phase involves determining what the changes in the native stress field will be as a result of undercutting and boundary stoping. The regions of concentrated high stresses and the low-stress zones are defined in this phase. These expected imposed stresses are then compared to the strength and elastic properties of the rock mass and the ultimate rock behavior is predicted. Finally, the actual rock behavior is monitored to check the calculations and to control the caving operations.

Measuring and Estimating Native Stresses and Rock Properties

A great variety of tools and techniques have been developed and are now used to measure and estimate stresses and rock properties. The better-known ones are listed in Table II. The actual selection of the specific tool or technique to be used is dependent on what feature is to be measured and on the unique conditions of the site to be investigated.

One of the most difficult tasks in rock mechanics investigations is to estimate the in-situ rock mass properties from measurements taken in the laboratory on small core samples or in a small-diameter borehole where the rock is relatively intact. In general, the strength and elastic properties of the rock mass are lower than those of intact rock and are related to the number of weak fractures present in the rock mass. When the degree of fracturing is high, the strength and elastic properties of a rock mass are very low when compared to those of an intact rock core. Conversely, when the relative amount of fractures is low, then the rock mass properties can be expected to be only slightly less than those of the intact core sample.

A method has been developed for quantitatively estimating the factor to be used to reduce the properties of the intact rock to those associated with the rock mass. This method involves the use of the RQD (rock quality designation) developed by Deere (16). The RQD is the ratio of the total length of hard sound sticks of borehole core greater than 4 inches in length as compared to the total length of drill hole. An example of the relationship existing between RQD values and the ratio of rock mass elastic moduli to the intact rock elastic moduli is shown on Figure 4. In other studies (17), the RQD has also been shown to be directly related to the relative "cavability" of a rock mass and to the amount of secondary blasting required in block-caving operations.

Determining Stress Changes and Predicting Rock Behavior

Several techniques are used to determine the stress changes which will be imposed by subsurface openings and to predict rock behavior by comparing the resulting stress levels with the in-situ rock mass properties. One type of analysis consists of estimating the changes in stresses caused by the undercut and development openings by using equations based upon elastic and plastic theory. Several publications contain solutions to these equations in easy-to-read charts and graphs (4,5,6,14). Once the stress levels are estimated, they are compared arithmetically to the measured or estimated physical properties of the rock mass and ultimate rock behavior is then predicted.

Scale models can be built with materials simulating the rock mass properties determined by earlier investigations (17) to predict behavior in block caving. In these studies, extreme care must be taken to properly proportion the stresses to the simulated properties.

The most efficient and powerful analytical technique used today is the computerized finite element method. By using it, stress changes and resulting rock behavior are quickly and accurately predicted for any shape and number of openings in complex geological conditions. The advantage of the finite element method over other analytical methods is that the rock mass, which most generally is inhomogeneous, can be modeled as a series of very small elements, each which may be considered to be homogeneous in itself with respect to rock type, structure, and physical properties. The elements can be constructed to represent natural rock materials which cannot resist tension stresses, (18) or materials which do not behave elastically (19). Elements are then assembled mathematically on a computer to form a model which can be every bit as good as the geological knowledge allows, but of course no better.

Alternative mining sequences are modeled in the finite element method by progressively removing elements to represent the openings and by reducing the modulus of deformation to represent broken rock. The calculated stress levels,rock movements, and failures are output from the computer. This method has already been used successfully in a block-caving study (7,20). Recent modifications have now been developed to analyze the interaction between various support systems and the adjacent rock mass as an aid to selecting the most appropriate support method (21).

SUMMARY

A variety of rock mechanics tools and techniques are used by the mine manager responsible for a block-caving operation. Using these rock mechanics methods to accurately predict rock behavior, he now avoids the old expensive and time-consuming, trial-and-error methods and rapidly optimizes the mine design and minimizes block development, preparation, and repair costs.

REFERENCES

1) Peele, Robert: Mining Engineers Handbook. 3rd ed, 3rd printing, p. 10-353, John Wiley & Sons, 1947.

2) Hardwick, W. R.: Block-Caving Copper Mining Methods and Costs at the Miami Mine, Miami Copper Company, Gila County, Arizona. U.S. Bureau of Mines, I.C. 8271, p. 165.

3) Hardwick, W. R.: Block Caving Methods and Costs, Bagdad Mine, Bagdad Copper Corp, Yavapai County, Arizona. U.S. Bureau of Mines, I.C. 7890, 1959.

4) Obert, Leonard; Duvall, Wilbur I; and Merrill, Robert H.: Design of Underground Openings in Competent Rock. U.S. Bureau of Mines Bull. 587, 1960.

5) Coates, D. F.: Rock Mechanics Principles. Canadian Dept. of Mines & Tech. Survey, Mines Br. Monograph 874, 1965.

6) Isaacson, E. de st. Q.: Rock Pressure in Mines. 2nd ed., Mining Pub. Ltd., London, 1962.

7) Dames & Moore: Report on Finite Element Analysis, Cavability Study for the Climax Molybdenum Company. Unpublished report, 1969.

8) Fletcher, J. B.: Ground Movement and Subsidence from Block Caving at Miami Mine. SME of AIME, Preprint No. 59AU27, presented at the annual meeting, Feb. 1959. (Trans. SME of AIME, Vol 217, p. 413-421, 1960)

9) Hardwick, W. R.: Mining Methods and Costs, Inspiration Consolidated Copper Company Open-Pit Mine, Gila County, Arizona. U.S. Bureau of Mines, I.C. 8154, 1963.

10) Obert, L. and Long, A. E.: Underground Borate Mining, Kern County, California. U.S. Bureau of Mines, R.I. 6110, 1962.

11) Long, A. E. and Obert, Leonard: Block Caving in Limestone at the Crestmore Mine, Riverside Cement Company, Riverside, California. U.S. Bureau of Mines, I.C. 7838, 1958.

12) Bucky, P. B.: Fundamental Considerations in Block Caving. Proc. of 1st Symposium on Rock Mechanics, Quarterly of Colo. School of Mines, V. 51, No. 3, 1956.

13) Hannifan, Martin: Concrete for Ground Support in the Kelley Mine. Proc. of 1st Symposium on Rock Mechanics, Quarterly of Colo. School of Mines, V. 51, No. 3, 1956.

14) Woodruff, Seth D.: Methods of Working Coal and Metal Mines. Vol. 1, Pergamon Press, London, 1966.

15) Selters, J. J.: Personal communication, 1971.

16) Deere, D. U.; Hendron, A. J., Jr; Patton, F. D.; and Cording, E. J.: Design of Surface and Near-Surface Construction in Rock. Proc. of 8th Symposium on Rock Mechanics, Minneapolis, Minn., 1966.

17) Brown, E. T. and Trollope, D. H.: Strength of a Model of Jointed Rock. Jour. of Soil Mech and Foundation Div., Proc. of ASCE, Vol SM2, pp 685-704, March 1970.

18) Zienkiewicz, O. C.; Valliappan, S.; and King, I. P.: Stress Analysis of Rock as a "No Tension" Material. Geotechnique, Vol 18, pp 56-66, 1968.

19) Weber, Paul R. and West, Lawrence J.: Nonlinear Finite Element Analysis of Open-Pit Slopes in Incompetent Rock. SME of AIME preprint No. 71-AO-334, presented at Fall Meeting, 1971.

20) McMahon, B. K. and Kendrich, R. F.: Predicting the Block Caving Behavior of Ore Bodies. SME of AIME preprint No. 69-AU-51, presented at the annual meeting, 1969.

21) Dixon, J. D.: Analysis of Tunnel Support Structure with Consideration of Support-Rock Interaction. U.S. Bureau of Mines, RI 7526, 1971.

MINE AND REFERENCE	DEPTH FROM GROUND SURFACE TO UNDERCUT LEVEL	BLOCK PLAN DIMENSIONS	AVERAGE COMPRESSIVE STRESS AT WALL FACE IN UNDERCUT CORNERS (A)	TYPE OF ORE ROCK	AVERAGE UNCONFINED COMPRESSIVE STRENGTH OF ROCK CORE	ESTIMATED RQD (B)	ESTIMATED ROCK MASS COMPRESSIVE STRENGTH (C)
MIAMI COPPER (2,8)	600 FT TO 800 FT	150 FT X 300 FT	4500 PSI TO 6000 PSI	JOINTED SCHIST	2000 TO 5000 PSI (D)	VERY LOW (D)	LESS THAN 500 PSI
INSPIRATION (9) PRODUCTION BLOCKS	600 FT TO 800 FT	100 FT X 100 FT	3000 PSI TO 4000 PSI	JOINTED GRANITE	2000 TO 5000 PSI (D)	VERY LOW (D)	LESS THAN 500 PSI
TRANSFER BLOCK	230 FT	60 FT X 100 FT	1000 PSI	JOINTED GRANITE	1000 TO 5000 PSI (D)	VERY LOW (D)	LESS THAN 500 PSI
(G)	1100 FT	140 FT X 210 FT	7000 PSI TO 8000 PSI	JOINTED MONZONITE	2000 TO 5000 PSI (E)	LOW (E)	1000 TO 1500 PSI (F)
KELLEY (13)	250 FT TO 1000+ FT	80 FT X 120 FT	1300 PSI TO 4500 PSI	JOINTED MONZONITE	2000 TO 5000 PSI (D)	VERY LOW (D)	LESS THAN 500 PSI
JEFFREY (12)	400 FT	200 FT X 200 FT	4000 PSI	MASSIVE ASBESTOS	1650 PSI (F)	HIGH (E)	1000 PSI
BAGDAD (3)	870 FT	100 FT X 100 FT	4000 PSI	JOINTED QUARTZ MONZONITE	5000 TO 10,000 PSI (E)	LOW (E)	1500 TO 2000 PSI
JENIFER (10)	450 FT TO 500 FT	125 FT X 270 FT	2500 PSI	MASSIVE CLAYSTONE & SHALE	3200 PSI (F)	VERY HIGH (E)	2500 PSI TO 3000 PSI
CRESTMORE (11) BLOCK 2B	300 FT	150 FT X 190 FT	2000 PSI TO 2500 PSI	MASSIVE LIMESTONE	11,000 PSI (F)	HIGH (E)	7000 PSI
BLOCK 2C	330 FT	50 FT X 190 FT	1000 PSI TO 1200 PSI	MASSIVE LIMESTONE	11,000 PSI (F)	HIGH (E)	7000 PSI

NOTES:
- (A) CALCULATED AS FUNCTION OF STRESSES DUE TO DEPTH BELOW GROUND SURFACE AND STRESS CONCENTRATION FACTOR RELATED TO MINIMUM PLAN DIMENSION OF UNDERCUT
- (B) ROCK QUALITY DESIGNATION A MEASURE OF FRACTURE STRENGTH AND CORE RECOVERY
- (C) REDUCED FROM INTACT ROCK CORE STRENGTHS ACCORDING TO RQD RATING
- (D) ESTIMATED FROM VISUAL INSPECTIONS BY AUTHORS
- (E) ESTIMATED FROM DESCRIPTIONS IN REFERENCES
- (F) MEASURED VALUES
- (G) PROPRIETARY INFORMATION

RATIO OF AVERAGE STRESS IN UNDERCUT CORNERS TO ROCK MASS STRENGTH	BOUNDARY AND BLOCK WEAK-ENING FEATURES	SECONDARY BLASTING	CAVING CHARACTER-ISTICS
9.1 TO 12.5	CORNER RAISES AND BOUNDARY STOPES REDUCING BOUNDARY STRENGTH BY 25%	MINOR AMOUNTS	GOOD TO EXCELLENT -- BOUNDARY WORK DISCONTINUED FOR SHORT TIME WITH NO CHANGE IN DRAW CHARACTERISTICS
5.9 TO 8.3	CORNER RAISES AND BOUNDARY DRIFTS SOME SHRINKAGE STOPES	MINOR AMOUNTS	GOOD TO EXCELLENT -- AMOUNT OF BOUNDARY WEAKENING WAS LESSENED AND SOMETIMES OMITTED AS EX-PERIENCE WAS GAINED
2.0 TO 2.5	NONE	MINOR	GOOD
4.5 TO 8.3	CORNER RAISES ONLY	MINOR	GOOD TO EXCELLENT
2.6 TO 9.1	MINOR -- ONLY IN AREAS OF HARDER ROCK	MINOR AMOUNTS	GOOD TO EXCELLENT
4.0	UNKNOWN	UNKNOWN	GOOD -- UNIFORM
2.0 TO 2.6	CORNER RAISES AND BOUNDARY SLICING REDUCING BOUNDARY STRENGTH BY 40%	FREQUENT	AVERAGE -- ROCK BROKE COARSE AND BLOCKY
0.83 TO 1.0	BOUNDARY DRIFTS AND EXTEN-SIVE LONGHOLE BLASTING OF ENTIRE ORE BLOCK.	30% OF TOTAL ORE	POOR -- STABLE ARCH FORMED RE-QUIRING LONGHOLE BLASTING TO IN-DUCE CONTINUOUS CAVING
0.29 TO 0.36	COMPLETE CUTOFF ON ALL 4 SIDES BY SHRINKAGE STOPES	30% OF TOTAL ORE	POOR -- PROGRESSIVE CAVING STARTED AFTER 75-FT WIDE UNDER-CUT. REQUIRED UNEVEN DRAW TO CONCENTRATE STRESSES TO FRAG-MENT BROKEN ORE
0.14 TO 0.17	COMPLETE CUTOFF ON 2 SIDES BY VERTICAL SHRINKAGE STOPES	30% OF TOTAL ORE	EXTREMELY POOR -- BLOCK DROPPED "EN MASSE" 6 MONTHS AFTER UNDER-CUT. DRAWING MORE DIFFICULT THAN IN BLOCK 2B

TABLE 1, COMPARISON OF PHYSICAL FEATURES OF SELECTED BLOCK-CAVE MINES

	TYPE OF INVESTIGATION, TOOLS, AND TECHNIQUES
SURFACE EXPLORATION STAGE	STRESS MEASUREMENTS AND ESTIMATES Review of local mining experience. Geological interpretation of structural setting, hydraulic fracturing in exploration holes, analysis of core "discing." MEASUREMENT OF PHYSICAL PROPERTIES Review of available data on local conditions. Geological interpretation of surface and borehole data, surface and borehole geophysical surveys, RQD survey of core from exploration holes, borehole jacking tests for moduli determination in exploration holes. Laboratory tests on samples of rock core. ANALYSIS METHODS Review of local mining experience. Theoretical elastic and plastic calculations. Preliminary finite element modeling.
UNDERGROUND EXPLORATION AND DEVELOPMENT STAGE	STRESS MEASUREMENTS AND ESTIMATES Borehole overcoring or hydraulic fracturing. Flat-jack tests. Load cells in support systems -- all performed in exploration drifts and development openings. MEASUREMENT OF PHYSICAL PROPERTIES Geological surveys of drift walls for lithology; fracture and fault orientation, character, and frequency. Seismic surveys of drift walls. Borehole and plate jacking tests for moduli determination. Additional laboratory tests on rock cores. ANALYSIS METHODS Refined theoretical plastic and elastic calculations. Detailed finite element modeling. Scale model studies. MEASUREMENT OF ROCK RESPONSE Installation of extensometers around openings. Convergence surveys of exploration and development openings.
PRODUCTION STAGE	STRESS MEASUREMENTS AND ESTIMATES Additional use of techniques used in development stage. Installation of elastic stress meters into rock surrounding sub-block openings. MEASUREMENT OF PHYSICAL PROPERTIES Additional use of techniques used in development stage. ANALYSIS METHODS Modifications of finite element model or scale model constructed in development stage. Additional theoretical plastic and elastic calculations. MEASUREMENT OF ROCK RESPONSE Additional use of techniques used in development stage. Installation of extensometers and inclinometers from surface or boundary drifts into block areas. Precise level surveys of surface subsidence.

PURPOSES	
A. To predict the rock behavior in response to various possible arrangements of mine openings as an aid in: 1) Selecting the most appropriate block dimensions; 2) Selecting the types of general amounts of boundary weakening and block weakening required to produce sustained caving; 3) Selecting the most favorable locations and general dimensions for sub-block workings for minimum support requirements and repair work; 4) Estimating the most favorable mining sequence. B. To provide data for cost estimates for feasibility studies.	SURFACE EXPLORATION STAGE
To predict rock behavior in response to the proposed mine layout as an aid in: 1) Determining the exact dimensions of the initial block; 2) Determining the specific types and exact amount of boundary and block weakening required for the initial block; 3) Determining the specific locations and dimensions of sub-block openings; 4) Determining the specific support systems to be used initially.	UNDERGROUND EXPLORATION AND DEVELOPMENT STAGE
A. To predict rock behavior in response to proposed modifications of initial block design as an aid in: 1) Optimizing the mining system to minimize block development, preparation, and repair costs; 2) Selecting the most effective modifications to counterbalance changes in rock conditions. B. To monitor actual rock behavior as an aid in controlling draw.	PRODUCTION STAGE

TABLE II, ROCK MECHANICS TOOLS AND TECHNIQUES AND THEIR APPLICATION TO BLOCK-CAVE MINING

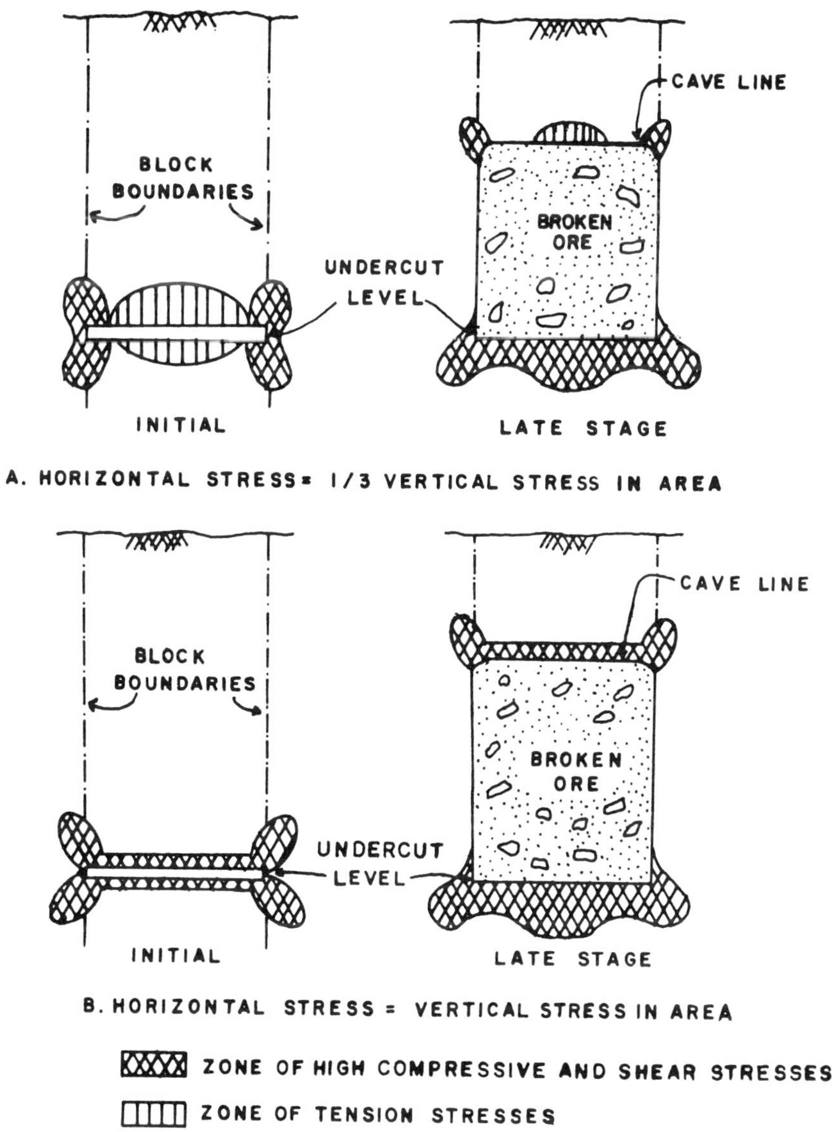

FIGURE 1: STRESS CONDITIONS IN BLOCK CAVING

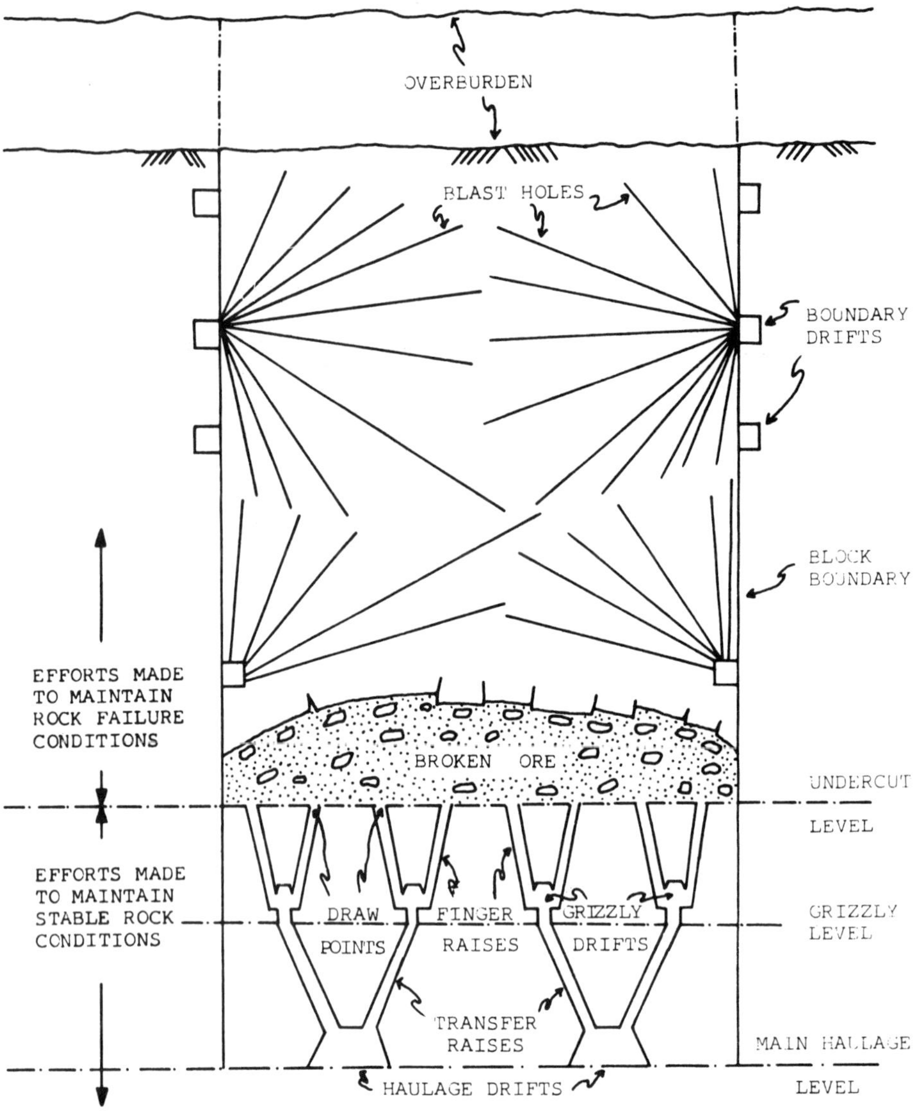

FIGURE 2: FAILURE AND STABLE CONDITIONS PREFERRED IN BLOCK CAVE MINES

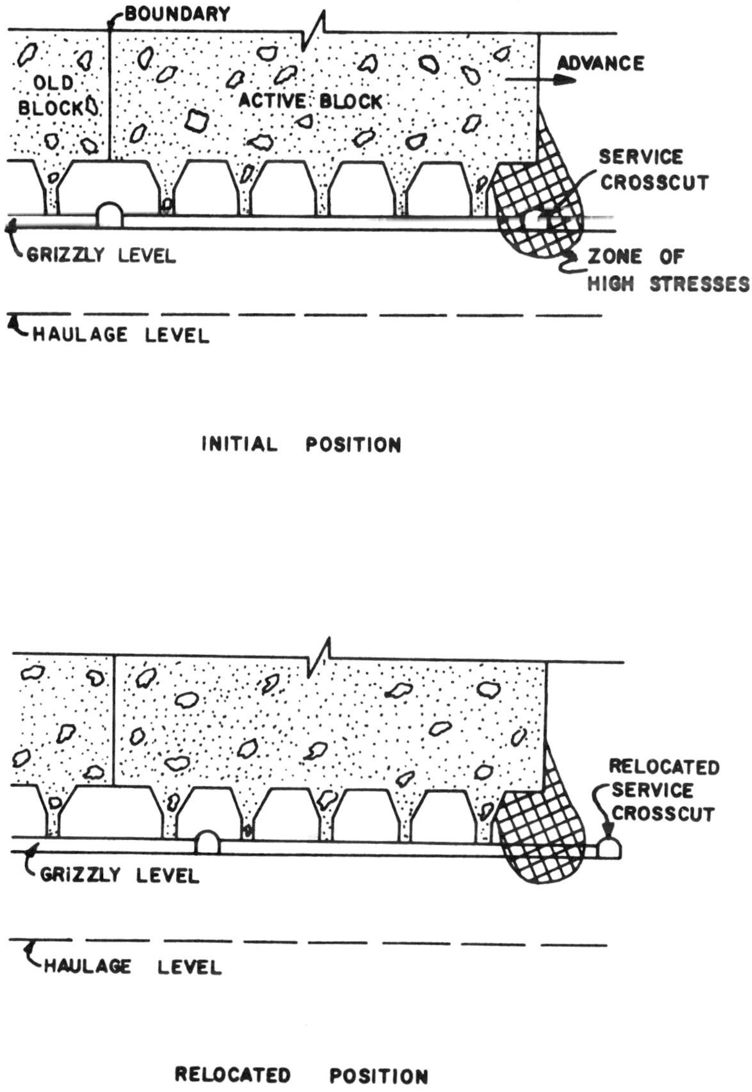

FIGURE 3: EL TENIENTE SERVICE CROSSCUT RELOCATION

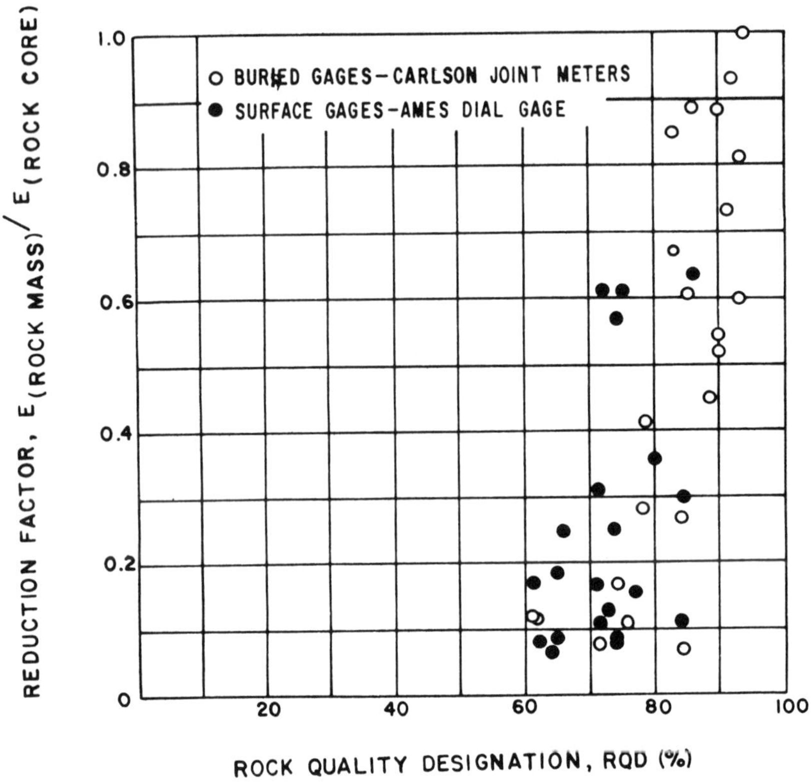

FROM REFERENCE (16)

FIG. 4 VARIATION OF REDUCTION FACTOR WITH ROCK QUALITY FROM PLATE JACKING TESTS, DWORSHAK DAM

36

Reprinted from *Dames & Moore Eng. Bull. No. 39,* pp. 1-9 (1971)

A STATISTICAL METHOD FOR THE DESIGN OF ROCK SLOPES

Barry K. McMahon

INTRODUCTION

The principal sources of uncertainty in the design of a rock slope are usually the locations and orientations of the joints and other rock defects in the rock mass behind the slope. Steep rock slopes have a higher risk of failure than flatter ones mainly because of the greater probability of undercutting joints, or combinations of joints, in unstable orientations.

This article introduces graphic procedures for determining the probability of failure of a rock slope by statistical analysis of joint orientations in conjunction with graphic kinematic and kinetic analyses of stability over a range of possible slope angles. The design slope is then selected on the basis of one of the following criteria:

1. Where safety is the overriding consideration, the design slope is the slope angle at which the probability of failure approaches zero.

2. Where partial failure of the slope can be allowed, the design slope is the slope with the minimum total cost consisting of the initial cost of excavation plus the costs resulting from partial slope failure.

3. Where the slope with minimum total cost has an unacceptably high probability of failure, the slope angle may be decreased to an acceptable compromise between the above two criteria.

In the following sections the design procedure is first demonstrated by reference to a simple example involving sliding along a single joint. Extension of the procedure towards the design of slopes where the failure occurs along combinations of joints is then discussed.

NOTATION

\ Area of Schmidt Projection covered by unsafe orientations of joints.

\ Total area of a Schmidt Projection.

c	Cohesion
C_a	Portion of costs of slope failure that are independent of volume of slide material.
C_b	Unit cost of slope failure where costs are related to volume of slide material.
C_o	Cost of initial excavation.
co	Unit cost of bulk excavation.
C_f	Cost of slope failure.
C_t	Total cost of a rock slope.
i	Angle of roughness.
M	Multiple of the standard deviation being contoured.
N_a	Average number of joint directions observed in a small element of the rock mass.
N_u	Number of joint poles in unsafe orientations.
N_t	Total number of joints in sample.
n	Number of small elements in the rock mass behind the slope.
P_c	Percentage of joints in a set that are judged to be sufficiently extensive and continuous to result in a slope failure.
P_f	Probability of slope failure.
P_i	Probability that the ith element will be undercut by joints in unstable orientations.
P_j	Probability of a joint occurring in an unsafe orientation.
$\varnothing$	Angle of surface friction.
$\varnothing_a$	Apparent angle of friction.
S_x	Sample standard deviation in x-direction.
S_y	Sample standard deviation in y-direction.
σ	Normal stress at failure.
τ	Shearing stress at failure.
u	Hydrostatic uplift.
V_o	Volume of initial excavation.
V_f	Expected volume of material involved in slope failure.
V_i	Volume of the ith element.
x	X-co-ordinate
y	Y-co-ordinate

THE EXAMPLE

The following example will be used to illustrate the design procedure:

A rock slope 100 feet high and facing due west is to be excavated for a highway running N-S. The natural ground surface slopes at 27 degrees and the rock consists of a gneissic granite, with a well-developed foliation. Outcrops are scattered and the foliation is observed to change direction in complex fashion between outcrops. Joints parallel to the foliation extend several hundred feet and are termed joint set 1. Other joints extend less than 10 feet. The groundwater table is below the base of the slope.

Observations on several existing slope failures in the area indicate that failure invariably occurs whenever the true dip of foliation is undercut by a slope and exceeds 34 degrees. It is also observed that sliding on a serious scale only occurs along the foliation joints but that tensile failure can occur in almost any direction in the rock mass, due to separation along combinations of minor joints, whenever necessary to permit sliding on the major joints.

The slope will be allowed to stand for 18 months after excavation before the highway is opened for traffic. Initial bulk excavation is estimated to be $1.00 per cubic yard (in-situ). Clean-up of partial slope failures is estimated to be $5.00 per cubic yard including allowances for scaling down, repair of pavement, and general inconvenience.

DESIGN PROCEDURE

The following steps are necessary for the design of a rock slope by the proposed method:

Geological Mapping

Regardless of whether the problem is to design a highway slope, or the slope of an open-pit mine, the first step must be geological mapping to divide the area into structural regions that are statistically homogeneous with respect to jointing and other rock defects (Ref. 1). This mapping also serves to locate major features such as wide fault zones which must be treated as individuals rather than as statistics for the purpose of rock slope design.

Statistical Sampling

An adequately representative sample of joint orientation must be obtained from each structural region included in the slope. Since only joints exposed on rock outcrops, excavated surfaces or a limited number of bore holes are available for measurement, a perfectly random sample, which would imply that every individual in the population has an equal chance of being chosen for the sample, is clearly impossible. When extensive outcrops are available, formal sampling schemes, which are carefully designed to minimize observer bias, may be applied (Ref. 2). Where only scattered exposures are available, an adequate sample can often be obtained by measuring the three to five most prominent joint directions at each outcrop visited (Ref. 3).

Statistical Analysis

The purpose of taking any sample is to reach conclusions regarding the properties of the population (i.e., the totality of all things that the sample is designed to represent).

Graphic statistical analysis of the joint orientation data in three dimensions is made possible by the Schmidt method of representing the orientation of each joint as a point (pole) on a Lambert Equal-Area Projection* of a sphere, as shown in Fig. 1. Detailed procedures for working with this projection are given in Refs. 1 and 4.

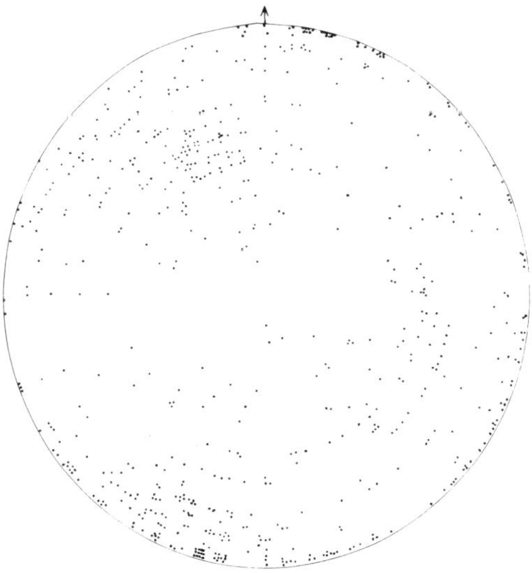

Fig. 1: Representation of joint orientations as poles on a Lambert Equal-Area Projection (631 poles to joints from Geehi Gorge, N.S.W.)

Fig. 2: Conventional contours of the density of poles shown in Fig. 1. Contours ½, 1, 3, 5, 7% per 1% area.

* Throughout this article the lower hemisphere of the Lambert Equal-Area Projection is used.

The traditional geologic method of contouring Schmidt Diagrams in terms of the density of poles as shown in Fig. 2 has statistical usefulness only when the underlying population is suspected of being uniform as demonstrated by Winchell (Ref. 5) and Flynn (Ref. 6). In most actual cases, however, preferred orientations in geologic data are self-evident and the poles are seen to be clustered into groups which, in the case of joints, are called sets.

If it can be considered that the centre of the set represents the ideal orientation of the joints and that the scatter of joints around this ideal orientation is due to chance causes then, as shown by Vistelius (Ref. 7) and subject to some additional limitations discussed by him, the population distribution along any cross section through the joint set may be approximated by the normal distribution.

Extending this concept to three dimensions, if the probability distributions along any two cross sections at right angles through the population are normal and mutually independent, the population follows a two-dimensional normal distribution as shown in Fig. 3.

Fitting two-dimensional normal distributions to joint data is complicated by the presence of several joint sets, of which some may be overlapping. The following graphic procedure makes use of the powerful human ability for pattern recognition and, therefore in the author's opinion, has considerable practical advantages over methods based on vector procedures which are purely mathematical and difficult to envisage.

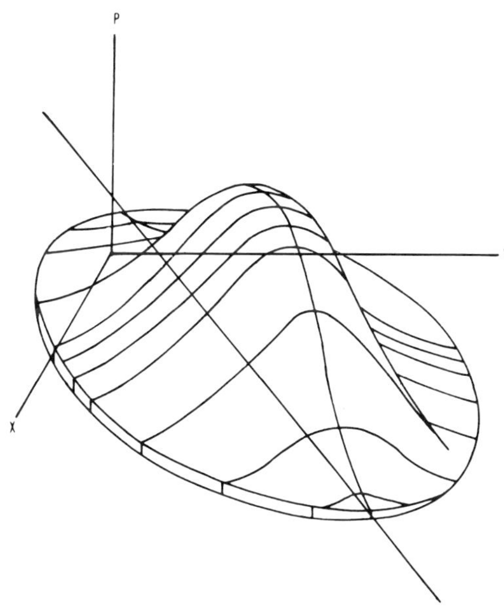

Fig. 3: Representation of a two-dimensional normal distribution (From Crow, Davis and Maxfield Ref. 8).

The first step is to divide the joint system into joint sets using the density contours as a guide. Two great circles, mutually at right angles, are then drawn along the directions of near symmetry through the centre of each joint set. Fitting great circles in this manner is a standard procedure in structural geology (Ref. 1). Where the joint sets are elongated, the correct position of the great circles is obvious. Where the joint sets are near equant, the great circles may be placed arbitrarily.

The standard deviation of the angular distances of the joint poles is then determined along each great circle in turn. Where two joint sets overlap, the standard deviations are computed from the non-overlapping portions.

Elliptical contours, distorted by the properties of the projection, are then drawn with the great circles as principal axes for various multiples of the standard deviation as shown in Fig. 4. These contours define the estimated probability distribution of the population represented by the joint set. They are smooth in contrast to the density contours, whose minor irregularities are usually not reproducible for different samples, and are therefore meaningless in terms of the population.

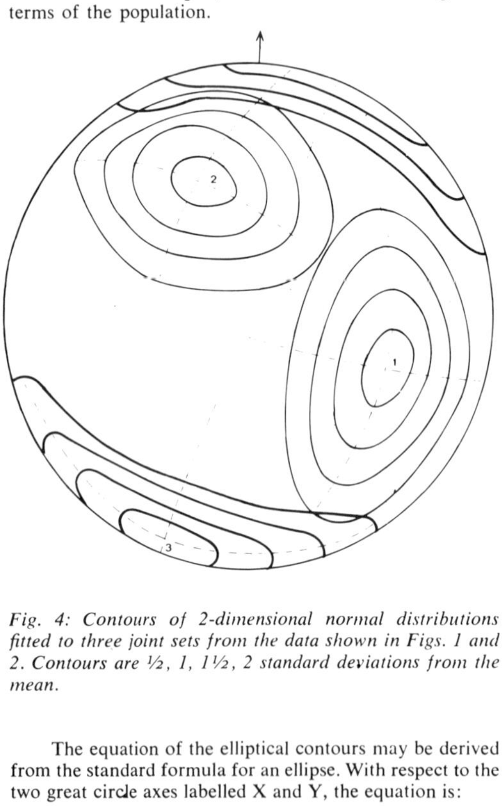

Fig. 4: Contours of 2-dimensional normal distributions fitted to three joint sets from the data shown in Figs. 1 and 2. Contours are ½, 1, 1½, 2 standard deviations from the mean.

The equation of the elliptical contours may be derived from the standard formula for an ellipse. With respect to the two great circle axes labelled X and Y, the equation is:

$$y = \pm M^2 S_y^2 \left(1 - \frac{x^2}{M^2 S_x^2}\right)^{1/2} \quad \text{---- (1)}$$

Using the properties of the two-dimensional normal distribution given by Crow, Davis and Maxfield (Ref. 8), the theoretical probability of a joint falling within any part of the distribution can be computed. The goodness-of-fit of the distribution can therefore be checked by means of the Chi-Square Test (Ref. 8).

Although the normal distribution has the practical advantages that its properties are well known and easily accessible in tabulated form, its theoretical development assumes distribution on a plane, and it is therefore an approximation when used to represent the distribution of joints on a projection of a sphere. The approximation is best when the dispersion is small. If required, more precise representation such as Fisher's distribution on a sphere could be used (Ref. 15).

Although the author believes that the two-dimensional normal distribution is probably the best estimation of the probability distribution of most natural joint sets, it is emphasized that the proposed design procedure does not necessarily depend on this assumption. For example, in highly irregular joint systems it may be suspected that the underlying probability distribution is uniform. In this case the area enclosed by any density contour may be predicted from the Poisson Distribution (Refs. 3 & 5) and the goodness-of-fit determined by means of a Chi-Square Test.

Where no theoretical distribution appears applicable, the sample frequency distribution may be taken as the best estimate of the population probability distribution. However, in this case, a much larger sample is normally required.

Kinematic Testing

The purpose of kinematic testing is to determine the range of joint orientations for which movement is possible. If the action of the resultant force is to move the block into the slope, then failure is clearly impossible, and further consideration of the shear strength of the joint is unnecessary. Extensive discussions on kinematic tests for blocks bounded by two or three joints are given in Refs. 4, 9, 10 and 11. The simple tests given by Panet (Ref. 9) are particularly useful.

Panet's tests distinguish between wedge-type failure where the sliding occurs along the intersection of two joints and block-type failure where sliding occurs along a single joint. Wedge-type failure is kinematically possible only if the line of joint intersection is undercut by the slope. Block-type failure can occur only if the line of the direction of movement on the joint is undercut by the slope. Panet examines only the case where the direction of movement is along the true dip of the sliding joint, which is the case where gravity and hydrostatic uplift are the only active forces. However, the method can be generalized to include the effects of earthquakes and any external forces by using the methods described by John (Ref. 4) to determine the direction of the resultant force. The direction of movement is then given on the equal-area projection by the intersection of the great circle representing the joint plane and the vertical great circle containing the direction of the resultant force.

The problem of kinematic analysis is reduced to its simplest form in the situation where gravity and hydrostatic uplift are the only forces involved. Sliding can occur only along one set of joints, and tensile separation can occur in many directions in the rock mass. Failure in this case is kinematically possible only when the true dip of a foliation joint is undercut by the slope.

On Fig. 5 the great circles on the right hand side of the diagram represent various possibilities for the design rock slope. The closed curves on the left hand side are the loci of the poles of the joint whose true dips lie on the corresponding rock slope (e.g. joint 3 in Fig. 4). These curves enclose the poles of all joints whose true dips are undercut by the indicated slope; that is, all joints that are kinematically unstable with respect to the slope.

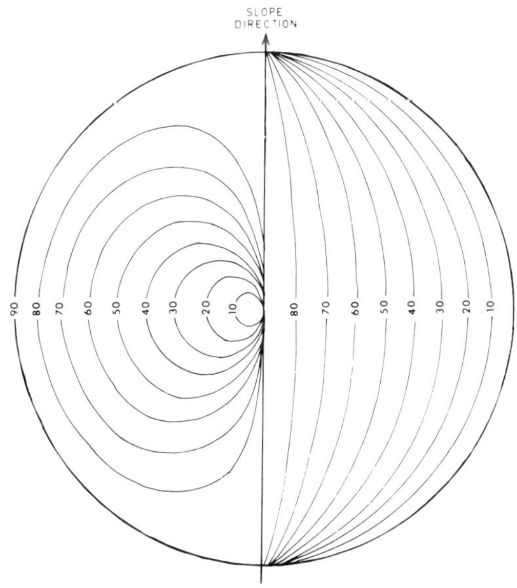

Fig. 5: Kinematic Test for sliding along a single joint. Use is explained in text.

Limiting Equilibrium Analysis

A review of limiting equilibrium methods by Hoek (Ref. 12) shows that the relationship between shear stress τ and normal stress σ at failure can be adequately approximated by a straight line relationship:

$$\tau = c + (\sigma - u) \tan (\varnothing + i) \text{ ---- } (2)$$

In rock slopes the cohesion c is primarily due to the presence of rock bridges between joints as discussed by Jennings (Ref. 13) and Robertson (Ref. 2). The surface friction $\varnothing$ is determined by the nature of the rock material and probably by the effects of minor undulations on the joint surface.

The angle of roughness (i) is defined by Patton (Ref. 12) as the average angle between the major undulations on the joint surface and the direction of sliding along the joint.

In the author's opinion, the best estimates of the shear strength parameters are those obtained by back calculation of existing slope failures using the same analytical method that will be used to design the planned slopes. Assuming that the sliding blocks can be regarded as rigid and as long as the old failures are in slopes of approximately the same height as the planned slopes and have similarly continuous joints, the cohesion and friction can be treated as a single angular quantity ($\varnothing_a$) computed as follows:

$$\varnothing_a = \tan^{-1}\left(\frac{\tau}{\sigma}\right) \text{ ---- } (3)$$

Usually the least satisfactory aspect of back-calculating old failures is that the groundwater conditions at failure are poorly known. However, a realistic approach is to use the same assumptions regarding the height of the water table in both the back calculation and the design analysis. A conservative approach is to assume a higher level for the water table than that used in the back calculation.

In areas where old landslides are not available, the shear strength parameters must be estimated from laboratory and field tests and by judgment based on first principles. In this particular case, careful observation of the slope behavior in the early stages of construction, followed by redesign if necessary, should be considered an integral part of the design process.

The designer has the option of applying a safety factor during the limiting equilibrium analysis (Ref. 4). The value assigned will naturally depend on the purpose of the slope and the possible consequences of failure.

In most cases, however, where partial failures are considered acceptable, the author prefers an approach whereby the limiting equilibrium analysis is carried out for a safety factor of one and judgment applied to the consideration of an acceptable level for the probability of failure.

In the case of a simple block slide, an area of kinetic stability is defined as a small circle of angular radius ($\varnothing_a$) centered on the piercement point of the resultant force. For a dry rock slope affected only by gravity, this small circle is centered on the centre of the projection as shown in Fig. 6 for the example where $\varnothing_a$ is 34°.

Superposition of the small circle defining the area of kinetic stability and the curves defining areas of kinematic instability for various design rock slope angles on the probability distribution of the joint set 1 (from Fig. 4) is shown in Fig. 6. Joints with poles lying inside the area of kinematic instability and outside the area of kinetic stability are in unsafe orientations with respect to the indicated slopes.

Determination of the Probability of Joints Occurring in Unsafe Orientations

In any small portion of a rock mass, joints are usually found to occur in between three and six directions. Where the joint system can be divided into sets, it can be assumed for the purposes of slope design that one member of each set is present in any small element of the rock mass.

For any specific rock slope, it is usually found that sliding can occur only along joints belonging to one or two of the joint sets. These joint sets, and the safe and unsafe orientations within them, are determined by superimposing the results of the kinematic and limiting equilibrium analyses on the contours of the joint set. The probability that a joint will occur in an unsafe orientation behind the slope is

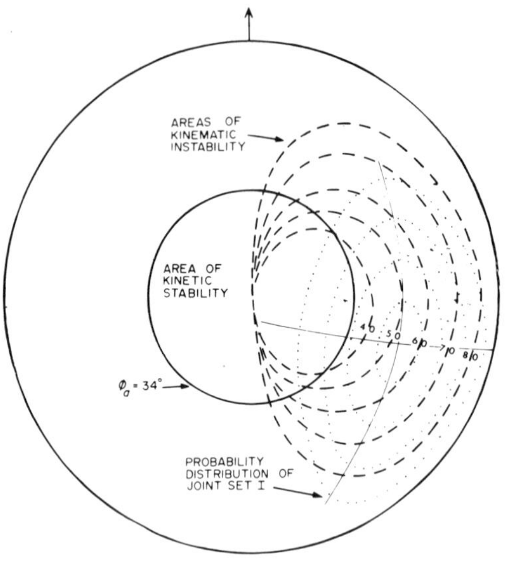

Fig. 6: Superposition of area of kinetic stability and areas of kinematic instability on the probability distribution of a joint set. Contours of the joint set are ½, 1, 1½, 2 standard deviations from the mean.

obtained in one of the following three ways:

(1) If the joint set fits a two-dimensional normal distribution, the probability that the joint will occur in an unstable orientation can be determined by the graphic transformation (e.g. Fig. 7 shows a graphic transformation of Fig. 6) which is superimposed one quadrant at a time over the Rectangular Normal Probability Chart published by Crow, Davis & Maxfield (Ref. 8) and reproduced in Fig. 8. The probability of an event occurring in any portion of this diagram can then be determined by counting rectangles.

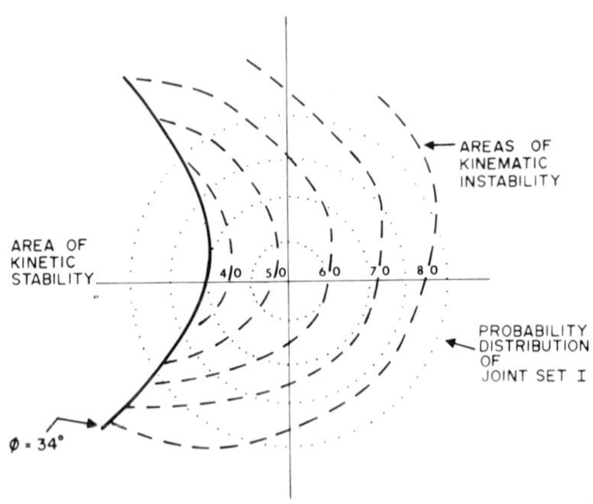

Fig. 7: Graphic transformation of the data on Fig. 6. The elliptical contours defining the probability distribution of the joints in Fig. 6 are transformed to circles of radius suitable for superposition on Fig. 8.

(2) If the joint system has a uniform population distribution, the probability (P_j) that a joint will occur in an unstable orientation is given by:

$$P_j = N_a (A_u / A_t) \text{ ---- } (4)$$

(3) If the assumption is made that the sample distribution represents the best estimate of the population probability distribution, the probability that a joint will occur in an unstable orientation is given by:

$$P_j = N_a (N_u / N_t) \text{ ---- } (5)$$

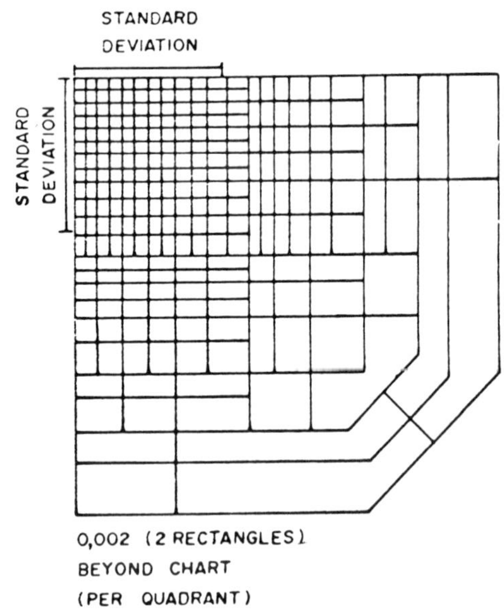

Fig. 8: Rectangular Normal Probability Chart reproduced from Crow, Davis and Maxfield (Ref. 8) and attributed by them to A. D. Sprague. Each rectangular area represents a probability of 0.001.

Determination of the Probability of Failure for a Range of Slope Angles.

In simple cases where the joints are uniformly continuous for distances that are large compared to the size of the slope, the probability that a joint occurs in an unstable orientation (P_j) can be taken as the probability of failure. Where only a proportion (P_c) of the joints of any set are sufficiently continuous to result in a slope failure the probability of failure (P_f) is given by:

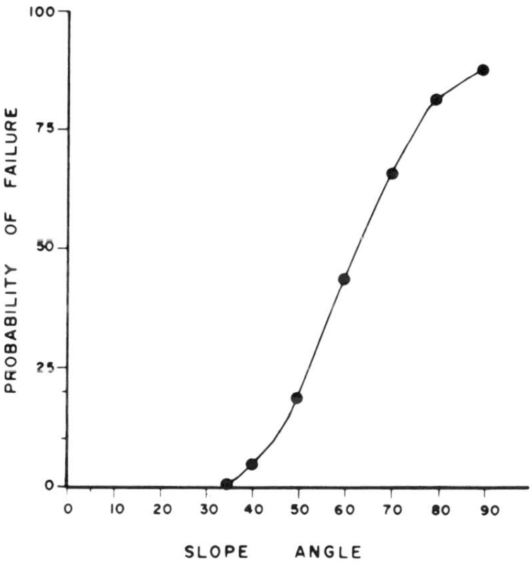

Fig. 9: Curve of Probability of Failure vs. Slope Angle for the example shown in Figs. 6 and 7. (In this case $P_c = 1$.)

$P_f = P_j . P_c$ ---- (6)

The probability of failure is then computed for a range of slope angles as shown in Fig. 9.

Where a slope is to be designed with safety as the overriding consideration, the slope angle, at which the curve becomes asymptotic to zero, would be the appropriate design angle. Where partial slope failure can be contemplated, the design slope angle is determined as described in the following section.

Determination of the Most Economic Slope Angles

The total cost of a rock slope is given by:

$C_t = C_o + C_f$ ---- (7)

where:

C_t = the total cost of the slope

C_o = the initial cost of excavation

C_f = the cost of failure

For any rock slope the initial cost of excavation can be considered to be the cost of excavation of the volume of rock in excess of that required for a vertical rock slope, therefore:

C_o = co Vo ---- (8)

where:

co = the unit cost of bulk rock excavation

Vo = the volume of rock excavated in excess of that required for a vertical slope

The cost of slope failure is given by:

$C_f = P_f (C_a + C_b V_f)$ ---- (9)

Where C_a represents those costs that are independent of the volume of material in the slide, such as part of the costs for loss of production, mobilization of equipment, and the costs of slope monitoring.

Determination of the portion of the cost of failure that is proportional to slide volume is complicated by the fact that the probability of failure and the expected volume of failure are related quantities and are both determined by the probability distribution of the joints behind the slope. A steep slope will fail back to the plane of the undercutting joint which can lie anywhere between the plane of zero probability of failure and the excavated slope surface. The product V_f P_f is therefore best determined by an incremental procedure as follows:

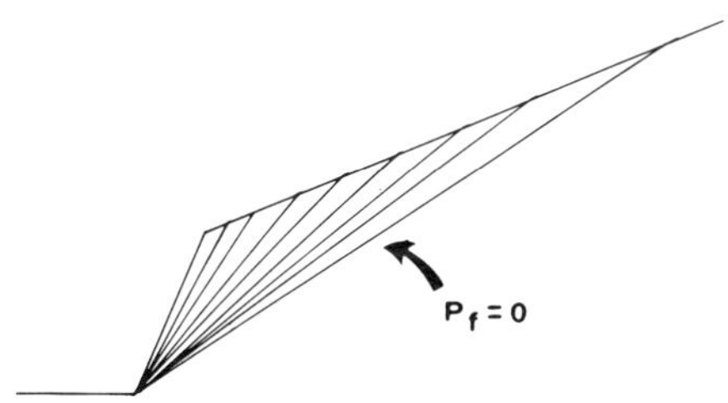

Fig. 10: Slope divided into a series of n elements.

If the slope is considered to be divided into a series of n elements as shown in Fig. 10, such that the base of the first element is along the plane with zero probability of failure, the top of the nth element is along the proposed slope surface, the volume of the ith element is Vi, and the probability that a joint undercuts the base of the ith element is Pi then:

$$P_f V_f = \sum_{i = 1, n} \text{Pi Vi} \quad \text{---- (10)}$$

Fig. 11 shows the curve of total cost per foot of slope vs. slope angle for the example (assuming $C_a = 0$ and the slope direction and probability distribution of the joints are as shown in Fig. 6 and the probability of failure is as shown in Fig. 9).

The following conclusions may be drawn from Figs. 9 and 11:

(1) The slope with the least total cost ($600 per foot of slope length) is between 45 and 50°. The probability of failure for a slope of 45° is 10% compared to 19% for a slope of 50°.

(2) The slope with zero probability of failure is 34°. This slope has a total cost of $1,130 per foot.

It is interesting to note that a safety factor of 1.5, applied to the slope of zero probability of failure in this case, would result in a cut slope of 24° which would be 3° flatter than the natural slope. This illustrates the absurdity of applying arbitrary safety factors to slope designs already based on the worst likely conditions.

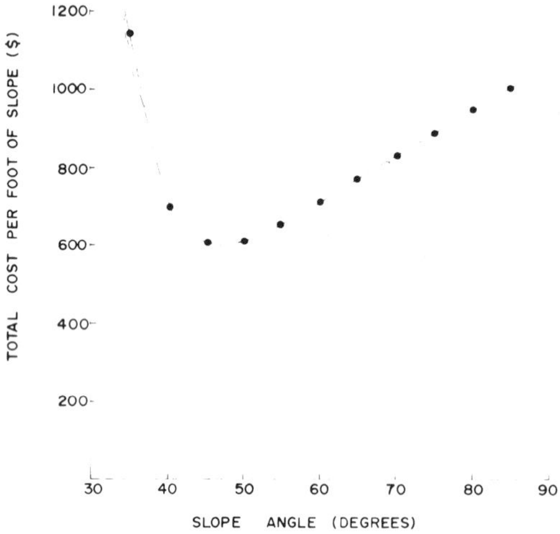

Fig. 11: Plot of total cost of slope vs. slope angle.

EXTENSION OF THE PROCEDURE TO SLIDING ON COMBINATIONS OF JOINTS

When more than one joint set is involved in the formation of the sliding surface the problem is complicated by the fact that the joint sets are rarely independent in their orientations. A conservative method is to assume that the joint sets are completely dependent so that any joint in the first set will be combined with a joint in the second set that is orientated to form the least stable combination. The methods of John (Ref. 4) can then be used to define the areas of kinematically and kinetically unstable orientations for joint set 1. The slope design can then be completed as described in the previous section.

The least conservative approach is to consider that the joint sets are independent so that the probability of finding any combination of orientations is the product of the probabilities of finding each orientation alone. An intermediate approach, which the author feels is probably the most realistic in many bases, is to assume that any joint in one set is combined with the mean joint orientation in the other set.

In open pit mines large-scale failures occur along complex failure surfaces which may consist partly of major weaknesses such as faults and partly of step-like surfaces composed of many joints. If as shown in Fig. 12 the failure is rotational, it may be analyzed by the methods of soil mechanics as suggested by Wilson (Ref. 14) using the failure criterion given in equation 2. In this case one of the principal variables is the roughness of the step-like series of joints. The probability distribution of the roughness can be determined for any small portion of the failure surface by the following method:

(1) Determine the probability distribution of the joint set forming the flatter portion of the step.

(2) Plot the direction of the average failure surface as a pole. The angle between this pole and a given joint orientation is the angle of roughness where this joint direction prevails.

(3) Determine the probability that a joint pole lies within any required radius of the pole to the failure surface by constructing inclined small circles around the pole and analyzing by the graphic procedure described in the preceding section.

(4) The probability of failure is then given by the probability that the roughness is less than that required to provide the required safety factor.

If as shown in Fig. 12 the failure takes place by sliding along the flatter joints and separation along the steeper joints as suggested by Jennings (Ref. 14), the problem is simply an extension of the problem of sliding along a single joint as described in the preceding example.

In actual slopes, the sliding mass is, of course, non-rigid and the initial movement is probably partly rotational and partly translational.

If the step-like failure surface is considered to be very large with respect to the joint spacing so that it contains a large number of joints, then the mean direction of the joints, in the direction of sliding, can be used in the analysis as suggested by Jennings (Ref. 14). In this case the probability that the joint continuity is greater than the value required for stability may become the principal basis for design.

For very large slopes the probability that the slope will fail can be interpreted as the percentage of the slope that can be expected to fail. Estimation of this percentage is the principal problem in open-pit design.

CONCLUSIONS

The probability of failure provides a realistic basis for the design of rock slopes where the locations and orientations of the rock defects behind the slope are not known in detail.

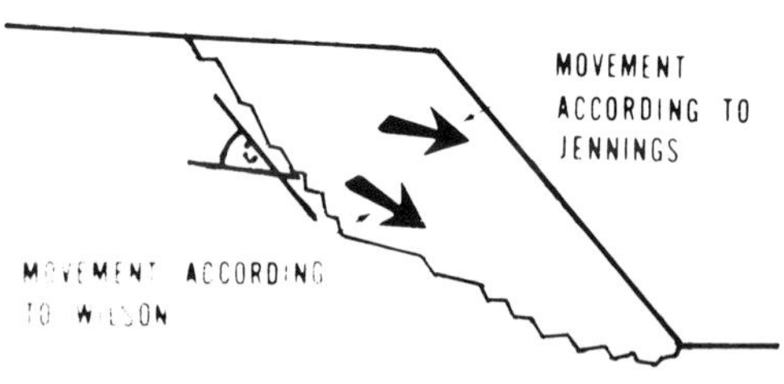

Fig. 12: Rotational failure along step-like patterns of joints.

ACKNOWLEDGMENTS

The author is particularly grateful for the encouragement by Dr. Ing. Klaus John of Karlsruhr whose pioneering work in the graphic analysis of rock slopes provided the inspiration for this study. He is also grateful to Mr. John Herington of the Sydney office of Dames & Moore for his assistance and constructive criticism during the preparation of this article.

References

[1] Turner, F. J. and Wiess, L. E. – *Structural Analysis of the Metamorphic Tectorites*, especially Chapter 3, Graphic Treatment of Fabric Data, New York, McGraw-Hill, 1963 pp 46-75

[2] Robertson, A. MacG. – The Interpretation of Geological Factors for Use in Slope Theory, *Preprints South African Symposium on Slope Stability in Open-Pit Mines*, Sept. 1970.

[3] McMahon, B. K. – Indices related to the mechanical Properties of Jointed Rock, in Grosvenor N. E. and Paulding B. W. Ed., *Status of Practical Rock Mechanics*, New York, American Institute of Mining, Metallurgical and Petroleum Engineers Inc., 1968 pp 117-133.

[4] John, K. W. – Graphical Stability Analysis of Slopes in Jointed Rock. *J. Soil Mech. and Found. Div. Proc. A.S.C.E. Vol. 94, SM2.* March 1968 pp 497-526 with discussions and closure *Vol. 59 SM6* Nov. 1969 pp 1541-1546.

[5] Winchell, M. – A New Method for the Interpretation of Fabric Diagrams, *American Mineralogist* 1937, Vol. 22, pp 15-36.

[6] Flynn, D. – On Tests of Significance of Preferred Orientation in Three-dimensional Fabric Diagrams, *J. Geol.*, 1958, Vol. 66, pp 526-539.

[7] Vistelius, A. B. – *Structural Diagrams*, Pergamon 1966, 176 pp.

[8] Crow, E. L., Davis, F. A., Maxfield, M. W. – *Statistics Manual*, New York, Dover Publications, 1960, 288 pp.

[9] Panet, M. – Discussion on Graphical Stability Analysis of Slopes in Jointed Rock by K. W. John, *J. Soil Mech. and Found. Div. Proc. A.S.C.E., Vol. 95, SM2*, March 1969, pp 685-686.

[10] Wittke, W. – Verfahren zur Berechnung der Standsich er heit belasteter und un belasteter Felsboschungen (*A Numerical Method of Calculating the Stability of Loaded and Unloaded Rock Slopes*) *Felsmechanic und Ingenieurgeologie* Supp. 2, Safety in Rock Engineering, Springer, 1963 pp 52-79.

[11] Londe, P., Vigier, G. and Vormeringer, R. – Stability of Rock Slopes, A Three-Dimensionaly Study, *J. Soil Mech. and Found. Div. Proc. A.S.C.E.* Vol. 95 SM1, January 1969, pp 235-261.

[12] Hoek, E., 1970, Rock Slope Stability in Open-Cast Mining, Rock Mechanics Progress Report, Imperial College, No. 4, July 1970, 71 pp.

[13] Jennings, J. E. – A Mathematical Theory for the Calculation of the Stability of Slopes in Open-Cast Mines, *Preprints, South African Symposium on Slope Stability in Open-Pit Mines*, Sept. 1970.

[14] Wilson, S. D. – The Application of Soil Mechanics to the Stability of Open-Pit Mines. *Third Symposium on Rock Mechanics, Colo. School of Mines Quarterly Vol. 54, No. 3*, 1959, pp 94-113.

[15] Fisher, R. A. – Dispersion on a Sphere. *Proc. of Royal Soc. of London, 1953, Series A, Vol. 217*, 295 pp.

ERRATA

On page 1, paragraph 3, line 1, "Where safety is the overriding consideration" should read "Where stability is the overriding consideration." The article is based on the consideration that in many situations unstable slopes that are properly monitored and inspected are safe.

On page 7, Equation 10 is an error although it gives approximately correct results at low probabilities of failure. The correct relationship is given in: McMahon, B. K., 1975, Probability of Failure and Expected Volume of Failure in High Rock Slopes, in *Australian-New Zealand Conference in Geomechanics (2nd) Proceedings*, pp. 308–313.

37

Reprinted from *Soc. Mining Engineers Preprint 68-AM-85,* 1968, 29p.

GEOLOGIC STRUCTURAL ANALYSIS FOR OPEN PIT SLOPE DESIGN, KIMBLEY PIT, ELY, NEVADA

by

Perry N. Halstead[1/], Richard D. Call[2/], and Kenneth H. Rippere[3/]

ABSTRACT

The primary objective of the structural geology portion of the Kennecott Copper Corporation-U. S. Bureau of Mines Pit Slope Stability Study was to determine the attitude, geometry, and spatial distribution of geologic structures of the west wall of the Kimbley Pit for use in designing a maximum stable slope for the pit. A corollary objective was to develop and refine techniques for collecting and analyzing geologic structural data with emphasis on numerical methods that could be integrated with other aspects of pit design. Surface mapping techniques used were a modification of conventional joint set mapping and a detailed linear sampling method. For subsurface investigations both oriented core and a borehole camera were used.

It was determined that the major faults in the area were favorably oriented for mining a steep stable slope. The jointing or natural fracturing, although complex, was not random in attitude and consisted of several persistent fracture sets and a number of secondary sets. The attitude of the persistent fracture sets was favorable for steeper slope angles, and

[1/]Geologist, U. S. Atomic Energy Commission, Las Vegas, Nevada (formerly Geologist, Bureau of Mines, U. S. Department of the Interior).
[2/]Geologist, Bureau of Mines, U. S. Department of the Interior, University of Arizona, Tucson, Arizona (formerly Geologist, Kennecott Copper Corporation).
[3/]Geologist, Kennecott Copper Corporation, Western Mines Division, Ruth, Nevada.

these attitudes were used in the design considerations.

There was no significant difference between the attitude of the fracture sets determined from the former and new slope face.

INTRODUCTION

The rock of an open pit slope is structurally complex, containing many faults, joints, and lithologic contacts. These structures have appreciably lower strengths than the intact rock and constitute mechanical discontinuities in the rock mass. Thus, a rock mass with many discontinuities should fail at a lower applied stress than rock with fewer discontinuities and failure will typically occur along discontinuities that are critically oriented to the pit wall and the applied stress. Therefore, one part of the Kennecott Copper Corporation-U. S. Bureau of Mines slope stability research project, conducted at the Kimbley Pit, Ely, Nevada, was devoted to an analysis of the geologic structure. The project also consisted of studies of the stress, strain, displacement, and strength of the slope wall with the overall objective of improving methods of designing open pit slopes. This paper covers only the structural geology portion of the study.

The primary objective of the structural analysis study was to determine the attitude, geometry, and spatial distribution of geologic structures in the west wall of the Kimbley Pit. A corollary objective was to develop and refine techniques for collecting and analyzing geologic structural data with emphasis on numerical, descriptive methods. These techniques could then be applied at other pits to provide data for slope design.

The program was divided into two parts: the first, or pre-mining phase, during which structural data were collected, analyzed, and used in the design of a steeper slope angle, and the second, or mining phase, during which additional structural data were collected to verify the results of Phase I.

GEOLOGIC SETTING

The Kimbley Pit is located in the Egan Range in east-central Nevada. The Egan Range is a typical fault block mountain of the Basin and Range structural province. Cretaceous monzonite porphyry stocks intruded and, in part, metamorphosed older, Paleozic sedimentary rock in an east-west trending zone through the central portion of the range; Tertiary volcanics have intruded and partially covered the older rocks.

The west end of the Kimbley Pit, the area studied in detail, is in monzonite porphyry and granitized shale complex. The rock is hydrothermally altered and contains disseminated copper mineralization. An isometric view of the area is shown in figure 1, and a photograph of the west slope of the pit is shown in figure 2.

DESCRIPTION OF THE PHYSICAL NATURE OF THE ROCK

The rock forming the upper 200 feet of the Kimbley Pit is heavily oxidized. It is a soft, porous material high in clay content and has a specific gravity of 2.3. Seismic velocities range from 1,500 to 4,500 fps.

The laboratory determined unconfined compressive strength is about 1,000 to 5,000 psi[4/]. Small-scale fracturing is very intense, resulting in 20 to 30 fractures per foot. The rocks were excavated with tractor-mounted rippers and a complementary fleet of scrapers. No blasting was required except a few zones of residual silication.

Below the oxidized zones is a hard, silicified porphyry-granitized shale complex. The contact marks a former water table and is the oxide-sulfide interface in the ore body. The rocks have a specific gravity of 2.8 and seismic velocities range from 4,500 to 8,000 fps. Laboratory determined unconfined compressive strengths vary between 6,000 and 10,000 psi. The fracture intensity is on the order of two or three per foot and the rock breaks into angular fragments, generally less than 2 feet in size. Owing to the hardness of the rock, conventional blasting techniques were employed to excavate this portion of the pit.

PHASE I STUDIES

Geologic Model

Most structural discontinuities are essentially planar and may be described by their attitude and position in space. Conventional geologic nomenclature is used to define fracture attitude by strike and dip, while its position in space is determined by reference to some fixed point. Since faults, contacts, and other structures range from a few inches to several miles in extent, a segregation by size is necessary. In this study structures were divided into major

---

[4/] The results of the laboratory tests on the rock are discussed in a companion paper "Bureau Contribution to Slope Angle Research at the Kimbley Pit, Ely, Nevada" by Robert H. Merrill.

and minor structures based on their extent relative to the pit. Major structures are those single, continuous features which have an extent in the same order of magnitude as the pit dimensions, in this case about 500 feet or greater. Large faults and lithologic contacts would fall in this category. Minor structures have an extent of less than 500 feet, and are typified by joints, fractures, and localized faults.

The rock mass between major structures was designated a structural domain. Within each structural domain there is a population of minor structures varying in attitude and extent but usually grouped into sets by parallel attitudes rather than being randomly oriented. All of the minor structures, including joints, small faults, and fissures, were referred to as fractures to avoid any generic implication. A series of two or more parallel fracture planes was called a fracture set. Within any structural domain a number of fracture sets could exist.

The structural domain concept was utilized because two rock masses with different structural patterns may be juxtaposed along a major fault; or, displacement of similar rock masses on a fault may result in rotation of one mass relative to the other. By sampling and analyzing each mass separately, the structural patterns for each domain can be compared and, if similar, combined.

Since the large number of minor structures constitutes, for all practical purposes, an infinite population, they can best be sampled and described statistically. Certain geometric and physical properties of fractures and fracture sets can similarly be assigned a numerical value and be treated statistically. These properties such as (1) continuity, (2) planarity, (3) intensity, and (4) fracture filling are considered to affect the strength of the fracture and, consequently, the rock mass, and are described as follows:

Continuity.- Continuity is the two-dimensional extent of a fracture in the plane of its attitude. The greater the continuity of a fracture the lower its effective strength, other factors being equal. The true three-dimensional continuity cannot be measured because the total fracture surface cannot be observed; therefore, the length of the trace of the fracture at its point of observation must be used as an approximation.

Planarity.- Deviations of a real fracture from a true plane surface also affect the strength of a fracture. A planar structure will offer less resistance to sliding than an irregular structure. A fracture trace can be classified as planar, wavey, or irregular.

Intensity.- The intensity is the number of fractures per unit distance along the normal to the fracture set.

Fracture Filling.- The composition and quantity of material such as clay, quartz, or pyrite in a fracture affects the strength of the fracture.

Surface Mapping

Conventional geologic mapping was performed to determine rock types, locate major structures, and delineate structural domains. The area has been mapped by the Geology Department of the Nevada Mines Division of Kennecott Copper Corporation; therefore, the mapping was updated by adding detail and revising the original map based on new information obtained in this study. Special emphasis was placed on the location and physical characteristics of the major structures. Both aerial photographic and plane table techniques were utilized.

Fracture Set Sampling

The fracture sets in the pit slope were visually observed and the attitude measured with a Brunton compass. Spacing, continuity, and area over which the set could be observed were estimated and recorded. This was done by a two-man team systematically observing rock exposed by the pit benches. The location of the points of observation was also recorded. Individual faults were recorded as well as fracture sets. A total of 343 fracture set observations were made by this method. The east-west striking fracture sets, mapped by this method, are shown in figure 3.

Detail Line Sampling

To provide more precise data on fracture attitude and spacing, a detail line technique was utilized. This consisted of stretching a tape along the pit face and recording the attitude and location of each fracture. All the fractures in a zone about 3 feet above and 3 feet below the tape were included. Originally it was intended that planarity, fracture filling, and surface roughness should be recorded; however, most fractures exhibited nearly identical properties. Almost all were wavey and contained some silica and sulfide fracture filling. Therefore, unless unusual conditions were encountered, only the attitude of the fracture set was recorded.

Sample lines were nominally 20 feet long and spaced about 100 feet apart horizontally. The number of fractures observed per sample line ranged from 30 to 140 with an average of 75 . A total of 29 detail-line samples were taken on all four levels (roadways - see fig. 2) in the west end of the pit.

Subsurface Sampling

A major element of the Phase 1 Geologic Study was the investigation of the rock behind the pit face by drill-hole techniques. This portion of the study was considered important since it provided direct sampling of geologic structures in the pit wall rather than extrapolating surface data. Also, development of subsurface techniques was considered necessary for future application to new mines where drilling would be the only means of exposing data.

The drilling consisted of 21, NC-size (3.6") core holes totaling 9,078 feet. Ten of the holes were inclined 20° from the vertical; the other 11 holes were vertical. In addition six 8-inch holes were vertically drilled for a total of 2,201 feet; about half of this was core drilled. The location of the drill holes is presented in figure 4. The NC holes were for structure and rock-type information. Core from the 8-inch holes was sent to the Denver Mining Research Center, Denver, Colorado, for laboratory tests to obtain estimates of the strength of the rocks in the slope wall.

Core Logging

The NC core was logged for rock type, core recovery, nature of the core breaks, and type of fracturing. The nature of the core breaks was divided into three categories on the basis of fragment size. Core which was intact (except for occasional breaks) was classified as broken; core pieces whose length was about equal to the core diameter were classified as fragmented; material less than 1/2 inch was considered as rubble. Wherever possible, planarity of fracture surfaces was recorded as planar, wavey, or irregular.

The dip of the fracture planes forming the breaks in a 130-foot section of core from a vertical drill hole was obtained by measuring the angle between the fracture and the axis of the core. A total of 790 fracture planes were measured. The number of fractures intercepted by the drill core is dependent upon the angle of the fractures with respect to the drill hole. The frequency distribution was corrected by the following relationship:

$$n = \frac{n_o}{\cos \theta},$$

where n = corrected number of fractures

n_o = recorded number of fractures

θ = dip of fractures.

The frequency distribution in the drill core showed an almost linear increase in number of fractures from 0° dip to 30° dip, and an almost uniform concentration of fractures from 30° to 90°. This test indicated that the information gained did not warrant the effort involved, particularly as the core was not oriented; therefore, fracture dips were not recorded for subsequent holes.

Borehole Camera

An NX-size borehole camera was also used to obtain oriented photographs of the walls of the drill holes, and techniques were developed to obtain the attitude of fractures observed in the photographs (2)[5/]. Difficulties in maintaining an open hole and in obtaining satisfactory photographs limited the amount of useable data obtained by this method. Suitable photographs were obtained from an average of only 34% of each NC-size hole, resulting in 801 measured fracture attitudes. No useable photographs were obtained in the large diameter holes.

Oriented Core

Of the 1,116 feet of core obtained from the 8" holes only 37 feet was ultimately oriented by use of a television borehole camera borrowed from the U. S. Bureau of Reclamation. The attitudes of 668 fractures were obtained from this core.

DATA ANALYSIS AND RESULTS OF PHASE I

Major Structures

Structures in the west end of the pit consisted of the Kimbley and the Jupiter faults, and a number of smaller unnamed faults. The Jupiter fault strikes north-south across the west pit face, intersecting the slope just below the rim, and dips 25° to the west away from the structure of

---

[5/] Underlined numbers in parentheses refer to items in the list of references at the end of this report.

the pit. It forms the contact between a sandstone (above) and the quartz monzonite, granitized shale (below). The Kimbley fault strikes northwest and parallels the south slope of the pit. It intersects the pit face about one-quarter of the way up the south side of the pit and projects across the west face of the pit. It dips to the south away from the pit at about 55°. A limestone forms the hanging wall and the quartz monzonite, granitized shale is in the footwall. The unnamed faults strike generally northeast and are either near vertical or dip west away from the pit. These structures formed the boundaries of the structural domains.

Comparisons of fracture patterns and rock types showed little difference between structural domains, with the exception of the areas in the hanging wall of the Jupiter and Kimbley faults, and, therefore, these data were treated as one structural domain. Since the south wall of the pit was not part of the experiment and the rocks above the Jupiter fault represent a very small portion of the west slope, data from these areas were not used.

Fracture Patterns

The complex fracturing of a typical porphyry copper deposit often gives the initial appearance that the fracturing is random. It was established that this was not the case for the Kimbley Pit. A test for randomness for fabric diagrams, utilizing the Poisson exponential-binominal limit, was used to verify the nonrandomness of project data (1).

The fabric diagram is a standard method of portraying structural orientation data and is widely used in structural geology texts (3, 4). The poles of the fracture planes are plotted on a circular grid which is a projection of a reference hemisphere to a horizontal plane. By convention, the lower hemisphere is used in a fabric diagram, and the grid is constructed such that an area on the projection is equivalent to the corresponding area on the reference hemisphere (Schmidt or Lambert equal-area azimuthal). When plotted, the pole of a fracture will be 180° from the dip direction and 90° from the strike direction. The concentration of poles is contoured using a counting circle which represents 1% of the total area of the diagram. A rectangular grid pattern, based on the radius of the counting circle, is superimposed on the fabric diagram and the counting circle is moved systematically over the grid. All fracture poles within the circle are totaled and this number is recorded on the grid intercept. This matrix of numbers can be contoured to provide a diagram to visually inspect the concentration and attitude of the fracture sets.

According to the Poisson limit for the sample of 100 fractures the probability of finding at least x number of fractures in a 1 percent area (with random samples) is as follows:

x	probability
0	1.00
1	.63
2	.26
3	.08
4	.02
5	.004
6	.0006
7	.0001

Thus, a 3% concentration (three fractures) on a fabric diagram would occur only eight times in 100 samples, and a 7% concentration only once in 10,000 samples. Taking a probability of 0.02 as a level of significance (one chance in 50 of being wrong), for any sample of 100 fractures, a concentration of four or more fractures per 1% area would be nonrandom. Because all of the individual detail-line samples of 100 or more fractures had concentrations of greater than four fractures per 1% area, as did the total detail line and the fracture set mapping, the fractures appear to be nonrandom.

However, apparent nonrandom patterns could also be introduced in a plot of a sample because of sample technique bias. Therefore, the statistical test does not prove conclusively that the true population is nonrandom. More significant is the coincidence of concentrations on the plots regardless of the sampling methods. Several of the strong fracture sets reappeared fairly consistently in the individual plots of each detail line, the total detail line, and the plots of the fracture set mapping. Borehole camera data and oriented core data also showed similar concentrations although there was considerable bias in the sampling technique. The stereoplots of the data from the Phase I geologic investigations are given in figure 5. The regularly shaped segments denote areas in which fracture concentrations appear in one, or more, stereoplots.

Three fracture sets appeared to be quite persistent: one striking east-west and dipping 60° to 70° to the south, the second striking north-east and dipping 40° to 55° to the northwest, and the third striking northwest and dipping 75° to 85° to the southwest. A number of other fracture sets are present but are not as clearly defined. In particular, there are two sets striking northwest, one dipping east 70° to 80° and one west 70° to 80° respectively; and two fracture sets striking northeast, one dipping between 30° and 40° to the east and one dipping 50° to 60° to the east, respectively.

Sample Bias

Apparent differences in fracture intensity between samples can result from (1) orientation of linear samples, (2) the measuring technique, and (3) actual differences of fracture intensity from sample point to sample point. To illustrate, a vertical linear sample, such as a borehole, would not intersect a vertical fracture set unless drilled exactly along a fracture. Also a fracture set at a high dip angle would show fewer intercepts than a fracture set with the same spacing but at a lower dip angle. The geometric relationship is

$$I = \frac{n}{L \cos \theta}$$

where I = intensity
L = length of sample
n = number of fractures observed
θ = dip of fractures.

A horizontal sample (such as a detail line) would intersect very few horizontal fractures or fractures striking parallel to the line. The geometric correction is:

$$I = \frac{n}{L \sin \theta \sin b}$$

where b is the acute angle between the line bearing and the fracture set bearing. The uncorrected plot of the oriented core data (figure 5D) shows the effect of this bias. Note that the uncorrected plot shows fewer high-angle fractures.

Sampling techniques may also introduce a bias. For example, because it is more difficult to log flat fractures in the borehole camera photographs, fewer low-angle fractures appear in the plot (figure 5C). A study is now in progress to determine the optimum bias corrections to be applied to the data.

The 1% counting circle of the contouring method covers a 20° segment on the projection, resulting in a rounding effect when data are combined for the contoured plot. The actual mean attitude of a fracture set could be ± 10° away from a grid point where the summation is recorded; also, two concentrations which are 10° apart may be combined in a single high value and recorded at a grid point. Thus, a more accurate representation of the attitudes of a fracture set results from a plot of poles or from the computed, mean value of the actual poles.

Fracture Prediction Program

A computer program was developed to calculate the variation of fracture intensity, with respect to the position of the sample location, and estimate the intensity at points where samples could not be obtained (5). Fracture intensity values are predicted for every 50-foot-square block on the basis of a weighted average of the neighboring samples.

The method of weighing and the attitude range to be predicted are the input data. A predictability ratio is also computed for each block based on agreement between predicted values and actual observations at sample points as well as distance from the nearest sample.

A statistical test of Phase I data gave an average of 43% agreement between predicted and measured intensities, 35% agreement for detail-line data, and 14% agreement for borehole camera data. The three persistent fracture sets gave significantly higher agreement than the others.

Determination of Pit Slope Angles

The fracture set data and major structural mapping were used together with the stress and strength determinations to select new slope angles. The major structure mapping showed that no critical failure conditions such as a major structure or an intersection of major structures dipping into the pit was present in the wall of the Kimbley Pit. On the basis that a slope angle which exceeded the dip of a prominent fracture set would leave unsupported wedges of rock which could slide into the pit, the strong fracture set striking east-west and dipping south into the pit at 60° to 70° was considered to be the maximum slope angle along the north wall. (This limit was substantiated during the mining phase where a small wedge failed when inadvertently undercut.) The west wall did not have as clear cut a limit, but an overall slope angle of 57° was chosen on the basis of the fracture set which strikes north-south and dips 50° to 60° to the east.

Several lesser fracture sets strike about north-south and dip 30° to 40° to the east; however, because these sets had not failed in the original pit wall, and the specimens of rock tested in the laboratory did not fail along these sets, there was reason to believe that the lower angle sets would not create slope instability. This result illustrates that fracture strength data are necessary.

PHASE II STUDIES

Following the development of a preliminary pit design, two adits were driven into the west wall of the Kimbley Pit. These were at the 6725 and 6515 elevations. The primary purpose of the adits was to provide a site for strain and displacement instrumentation and to monitor stress changes in the slope behind the pit. In addition, they provided access to a portion of the rock mass that had been previously sampled only by borehole techniques. This information had been used in the structure prediction program and had been part of the basis for the determination of the probable structural fabric existing throughout the rock mass. Direct sampling of visible structures in the adits permitted an evaluation of the borehole sampling technique and formed the basis to prove the ability of the prediction program to provide detailed structural information for inaccessible portions of the rock mass. A section of the pit wall, which shows the relationship of the old and new pit faces, the adits, and some of the boreholes, is given in figure 6.

Sampling Methods

The detail line sampling technique was selected to sample the adit fractures. A line was painted at waist height along both ribs of both adits. A tape measure was strung along the line and the attitude of every fracture that intersected the line was measured and logged together with the tape intercept. A notation as to the visible length (extent) of the fracture trace was also made. In addition, vertical sample lines were run from the floor to the back at 25-foot intervals along both ribs. The data from the horizontal lines and the vertical lines were compiled separately in order to permit an evaluation of the linear sampling technique along a blasted rock face. The fabric diagrams of the data from the Phase II geologic investigations are given in figure 7.

The data were used to refine the detail-line sampling technique. The following parameters were developed to define the requirements of an adequate structural sample:

1. To insure an adequate sample, a minimum of 100 fractures should be recorded.

2. Maximum permissible sample separation to insure continuity between samples is 100 feet.

3. Fractures less than 2 feet need not be included in a sample since they do not contribute significantly to the overall pattern.

It must be emphasized that the quantitative aspects of these parameters will have to be determined anew for each different geologic environment. However, these criteria indicate the approximate magnitudes to be expected in a typical blasted rock mass.

Based on this information, a sampling program was outlined for use on the new pit face as it was exposed by mining. To insure complete coverage, the sample spacing was reduced from the recommended 100 feet to 40 feet and an attempt was made not to exceed this interval. The sampling program produced 35 detail-line samples consisting of about 100 fractures each.

Comparison With Phase I

The general agreement between the results of the Phase I and Phase II was very good, see figures 5 and 7. The three strong fracture sets and most of the lesser sets shown in the Phase I data appear in the plots of the Phase II data. However, the relative magnitude of the concentrations varied considerably, especially in the plots of the data from the adits, see figures 7B and 7C. This difference is probably due to the fact that the sample area in the adits is small compared to the sample area in the slope wall, and to the fact that the adit data is subject to linear bias.

Vertical sampling in the lower adit showed a fracture set that dips into the pit at 20° or less. Although this fracture set appeared in the data from the Phase I investigations, its concentration was so weak that it was not considered significant, see figures 5A and 7A.

Horizontal sampling in the adits (in 25-foot lengths) showed large variations in fracture intensities between the segments and also between opposite ribs of the same adit. It was also observed on the pit face that fracture sets would be pronounced and well developed in one area of the pit and would be less pronounced elsewhere. B. K. McMahon of the Colorado School of Mines made similar observations concerning the metamorphosis of the Colorado Front Range.

The fracture prediction program, which used Phase I data, gave predicted fracture intensities at points along the adit which were in the same range of agreement as the rib-to-rib comparisons. Although the program produced acceptable data, it is being modified to account for the observed variations in fracture intensities over short distances.

Evaluation of Sample Techniques

In regard to future pit slope design investigations, the following comments on sampling techniques are pertinent.

No single sampling technique is clearly superior to the others, in many respects they are complementary.

Fracture set mapping is the most comprehensive technique in that a large amount of information is assimilated by the geologist in choosing fracture sets. It is the fastest and lowest cost technique but relies heavily upon the judgment of the observer. Each fracture set observation has only one attitude measurement, which may not reflect the true mean attitude of the fracture set. In addition, spacing is estimated rather than measured.

Detail-line sampling gives more precise data on fracture set intensity and attitude variability at the sample point. It is more time consuming and thereby more expensive. Linear bias must also be considered.

Ideally, a sample program should consist of a combination of sampling methods, as follows:

1. A conventional geologic mapping program to locate major structures and outline structural domains.

2. A fracture set mapping program to determine the well-developed fracture sets.

3. A detail-line program starting with widely spaced samples. Sequential sampling should be continued between the original sample points until the last few samples make no significant change in the fracture pattern for the structural domain.

4. Oriented core and/or borehole camera techniques should be used to obtain subsurface information. Hole locations and orientations should be laid out so that all fracture attitudes will be sampled.

REFERENCES

1. Friedman, M., Petrofabric Techniques for Determination of the Principal Stress in Rock, International Conf. on the State of Stress in the Earth's Crust, Santa Monica, Calif., June 1963: American Elsevier, New York, 1964, pp. 451-550.

2. Halstead, P. N., R. D. Call, and S. J. Hubbard, Two Borehole Photograph Goniometers: BuMines Rept. of Inv. 7097, 1968.

3. Phillips, F. C., The Use of the Stereographic Projection in Structural Geology: Arnold Press, London, England, 1954, 86 pp.

4. Terzaghi, Ruth D., Sources of Error in Joint Surveys: Geotechnique, v. 15, No. 3, 1965, pp. 287-304.

5. Weaver, R., and R. D. Call, Computer Estimation of Oriented Fracture Set Intensity: Symposium of Computers in Mining and Exploration, March 1965; College of Mines, University of Arizona, Tucson, Arizona, 1965, 16 pp.

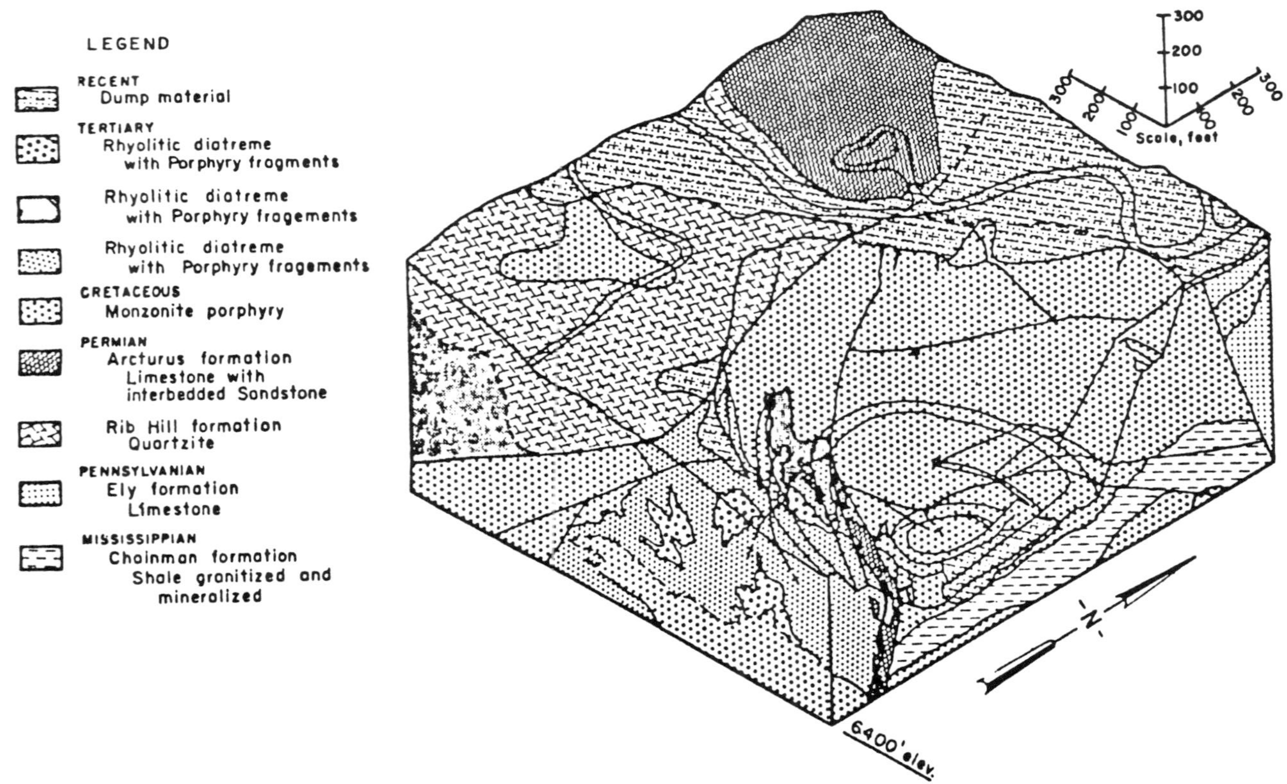

FIGURE 1.—Isometric View of West End of the Kimbley Pit, Showing the Geology

Figure 2.- View of Kimbley Pit

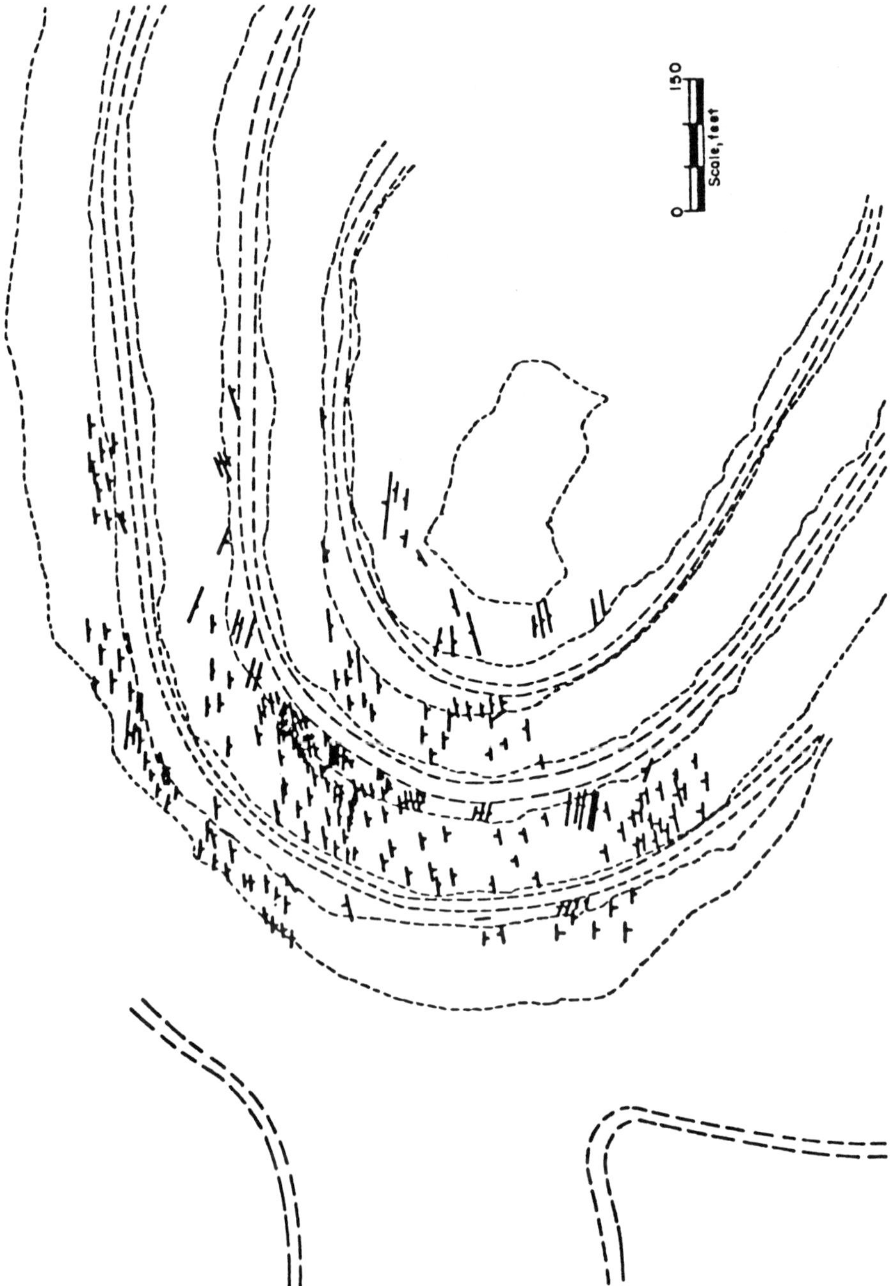

FIGURE 3.-East-West Fracture Sets

FIGURE 4—Plan Map Showing Locations of Drill Holes, used for Borehole Investigation

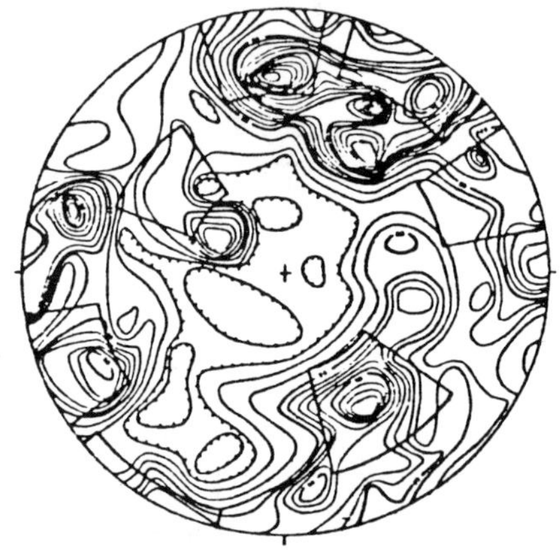

A.–Plot of 211 Fracture Sets, Macro Data; Contour Interval Equals 100 Fractures.

B.–Plot of Detail Line Data; 2,178 Fractures, Contour Interval Equals 5 Fractures.

C.–Plot of Borehole Camera Data; 801 Fractures, Contour Interval Equals 2 Fractures.

D.–Plot of Oriented Core Data; 662 Fractures, Contour Interval Equals 1 Fracture.

FIGURE 5.–Plots of Data from Structure Examinations in Phase I.

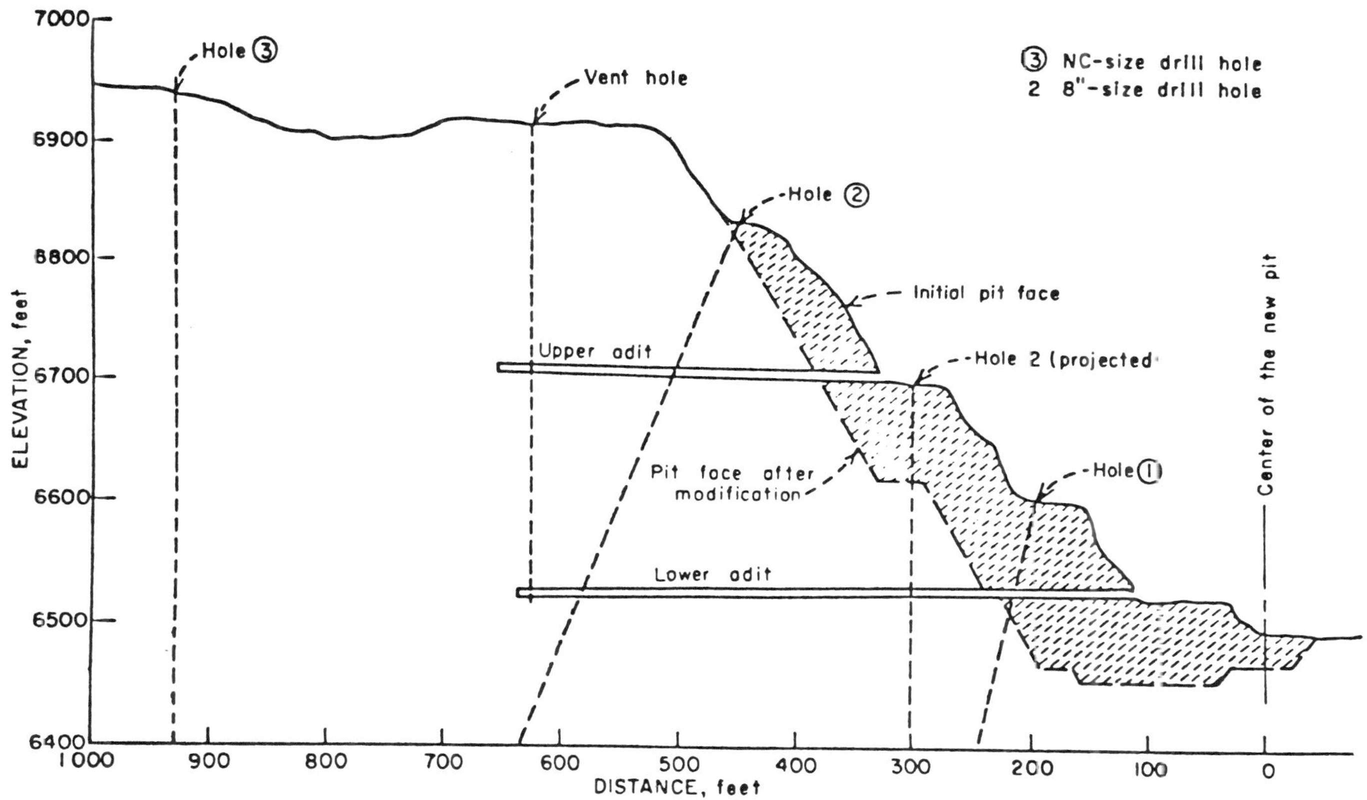

FIGURE 6.—Section Showing the Spatial Relationship Between the Two Pit Faces, the Adits, and Four of the Bore Holes

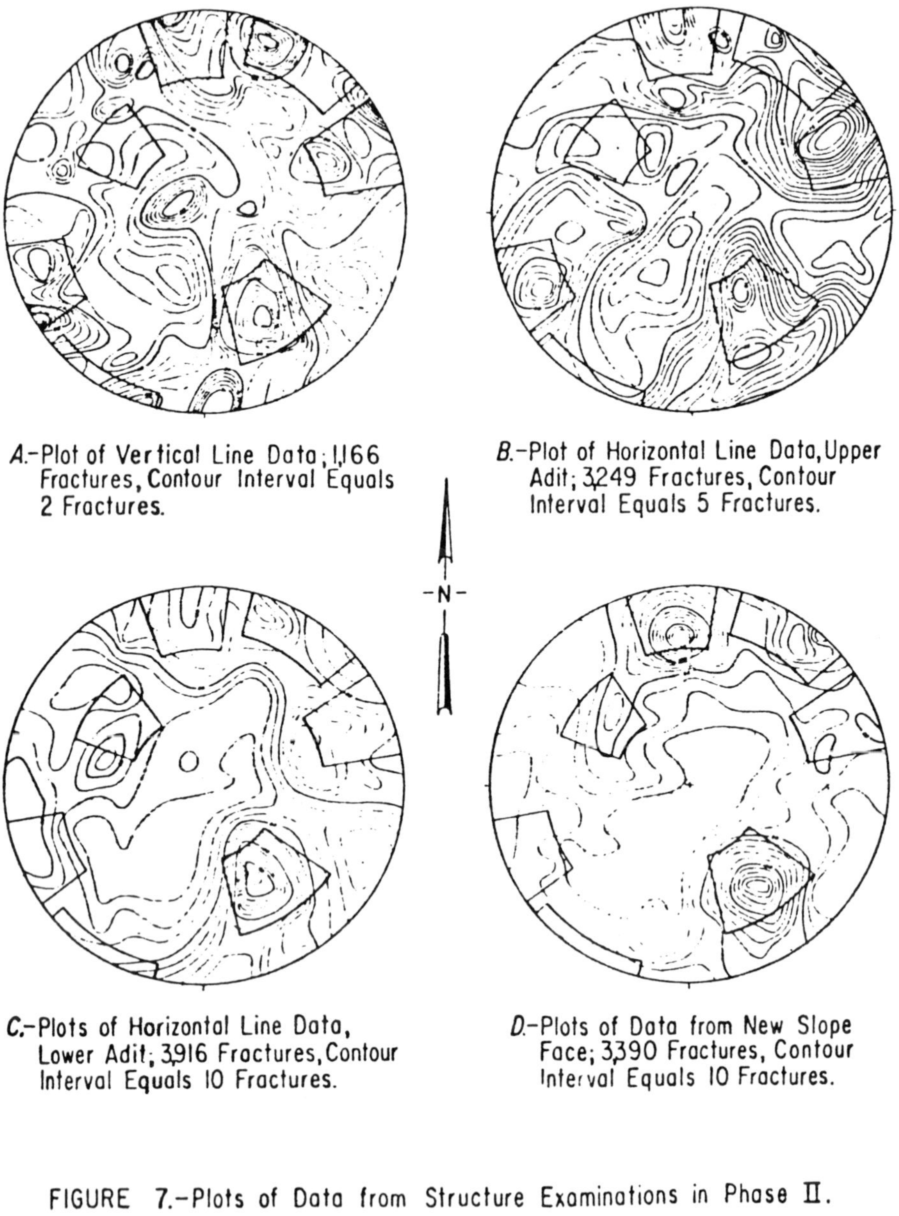

A.–Plot of Vertical Line Data; 1,166 Fractures, Contour Interval Equals 2 Fractures.

B.–Plot of Horizontal Line Data, Upper Adit; 3,249 Fractures, Contour Interval Equals 5 Fractures.

C.–Plots of Horizontal Line Data, Lower Adit; 3,916 Fractures, Contour Interval Equals 10 Fractures.

D.–Plots of Data from New Slope Face; 3,390 Fractures, Contour Interval Equals 10 Fractures.

FIGURE 7.–Plots of Data from Structure Examinations in Phase II.

38

Reprinted from *South African Inst. Mining and Metallurgy Jour.* **64**:310-327 (1964)

CRITICAL REVIEW OF STRATA MOVEMENT THEORIES AND THEIR APPLICATION TO PRACTICAL PROBLEMS

By **H. G. Denkhaus***, Dr.-Ing. (Member)

INTRODUCTION

Rock mechanics is the science of movement, stress and failure in the ground around excavations at depth. Its objective is to relate these three groups of phenomena to factors describing the geological and physical conditions of the ground, such as stratification, fissurization, strength and brittleness, and to factors describing the mining operations such as depth, stoping width, support and rate of face advance. Once such relationships are established and systematically interpreted in what is called a 'theory', it is relatively easy to derive therefrom practical rules for the design of layouts of stoping.

While the formulation of the objectives of rock mechanics is a relatively simple matter, its implementation requires long-term research. The lines of approach towards a solution of the problem are firstly observations, secondly experiments underground and in the laboratory and thirdly theoretical analyses. While practical observations and experiments constitute the indispensible foundations of the body of knowledge in rock mechanics, a theoretical concept, however simple, enables an analysis and assessment to be made of the significance of the results of observations and experiments.

When establishing a theory in the above sense, one is faced with the discrepancy between the complexity of the behaviour of Nature and the need for clear, systematic and simple thinking.

A theory is therefore nothing other than an image, a model, of the actual natural system. Such a model must contain all the essential features of the prototype, but must also be simple enough to lend itself to logical reasoning, that is, in most cases to mathematical treatment. This requires certain assumptions.

In rock mechanics, as in all branches of engineering mechanics, the most important assumptions refer to the relationship between movement (deformation, displacement, strain) and stress (load, force, pressure) and to the relationship between failure (gradual fracturing, bursting, yielding) and stress. One such assumption, for instance, would be that the displacement is reversible and proportional to the load; this would lead to an elastic strata movement theory. Another assumption could postulate proportionality between stress and rate of displacement (straining velocity) and would lead to a rheological strata movement theory.

Phenomena such as rockbursts and water influx may be regarded as special problems of strata movement and control.

This paper is an attempt to review the various theories of strata movement and stability that have been propounded and to assess their practical significance. With a view to the variety of such theories the survey cannot claim completeness, but the

---

*Director, National Mechanical Engineering Research Institute of the South African Council for Scientific and Industrial Research.

author is of the opinion that all the theories can be grouped into three major categories, namely the dome theories, the trough theories and the continuum theories. These will be dealt with and compared with each other with a view to their practical significance.

DOME OR ARCH THEORIES

The rigid dome

The observation that cavities and tunnels in soil or rock often remain stable without artificial support if they have assumed shapes of domes or arches, has led to certain theories. The builders of early Roman or Gothic arches on bridges and edifices made intuitive use of these ideas.

The simplest concept is based on the equilibrium of forces at the dome boundary. It does not account for deformation but for failure which occurs when the load attains a critical value, the failure load.

In the general case, the dome above an excavation is three-dimensional and deriving a formula for its shape is difficult. In tunnels or longwalls remote from their ends, however, the dome shape does not curve in the longitudinal direction and the problem simplifies to the two-dimensional one of determining the shape of the dome cross-section normal to the longitudinal axies of the tunnel or longwall, as depicted in Fig. 1.

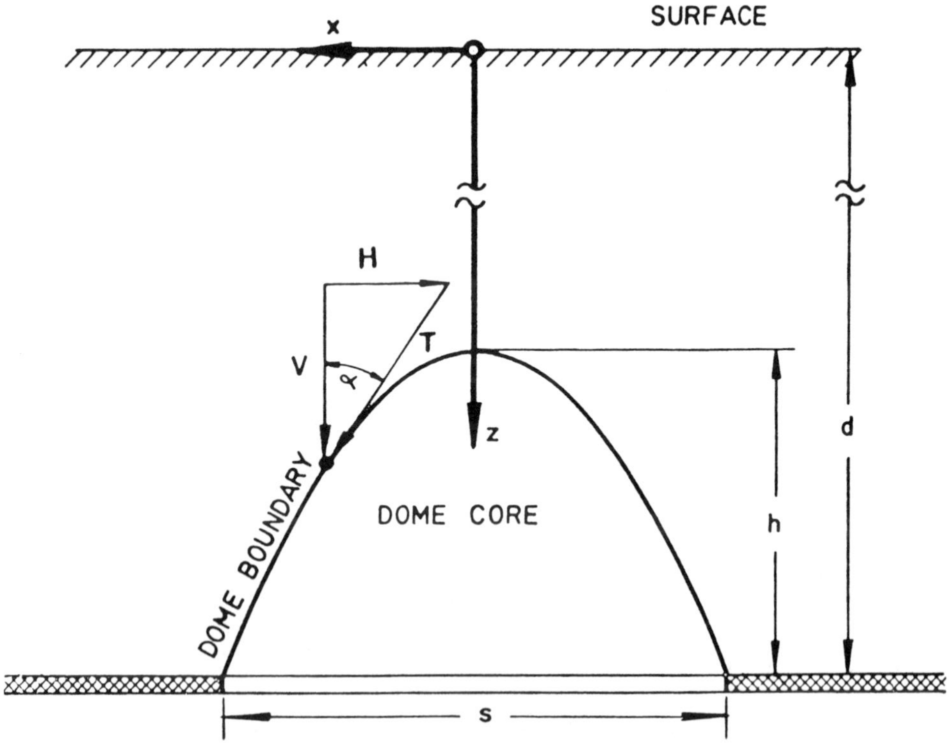

Fig. 1—Forces at dome boundary

In treating the problem distinction is made between the two cases of sufficiently cohesive rock and insufficiently cohesive rock.

If the rock is *sufficiently cohesive*, the core of the dome (intradosal ground) will not separate from the dome boundary which thus must carry the weight of the core. If the span of the dome becomes too large, the weight of the core exceeds the cohesive resistance and sudden collapse of the core or major portions of it may occur. Support under the core then has to take a large weight suddenly and this may be disastrous.

If the rock is *insufficiently cohesive*, portions of the core will separate from the dome boundary gradually or at short intervals while the span is increased (the faces advance). This seems to be the more favourable case with regard to ground control; the core has to find support partly by resting on artificial support entities or on natural 'intradosal' abutments at the faces. With increasing span the core may support itself by forming smaller domes within the main dome; the abutments of the sub-domes may be found in areas where the hanging rests on the footwall or regional support is taking full weight.

The actual condition will usually be one between these two extreme cases. The problem is to determine the shape and size of a dome whose boundary remains stable, as depicted in Fig. 1. The analysis is given in Appendix I and it is found that in the case of insufficient cohesion, the relationship between span and height of the dome is given by

$$s = \sqrt{\frac{8\,\sigma_c d}{w}(1 - h/d)\log(1 - h/d)} \qquad \text{(1a)}$$

where s is the span and h the height of the dome, d the depth below surface, w the specific weight, and σ_c the uniaxial compressive strength of the rock. The largest possible height for the dome to remain stable and the corresponding span are then

$$h_{max} = 0{\cdot}63d \qquad \text{(2a)}$$

$$s_{max} = \sqrt{\frac{2{\cdot}96\sigma_c d}{w}} \qquad \text{(3a)}$$

For a dome with sufficient cohesion the corresponding formulae read:

$$s = \sqrt{\frac{8\,\sigma_c h}{w}(1 - h/d)} \qquad \text{(1b)}$$

$$h_{max} = 0{\cdot}5d \qquad \text{(2b)}$$

$$s_{max} = \sqrt{\frac{2\,\sigma_c d}{w}} \qquad \text{(3b)}$$

In Fig. 2, the shape of the dome is depicted for both cases. In the case of sufficient cohesion, when the dome core remains *in situ*, the shape is parabolic. It will be noted that the dome whose core does not separate from the boundary, is the steeper one of the two, since it has to carry the additional weight of the core.

In Fig. 3, the relationship between height and span is graphically presented according to eqns. (1a) and (1b) and the limits of span and height according to eqns. (2a), (2b), (3a) and (3b) are also shown.

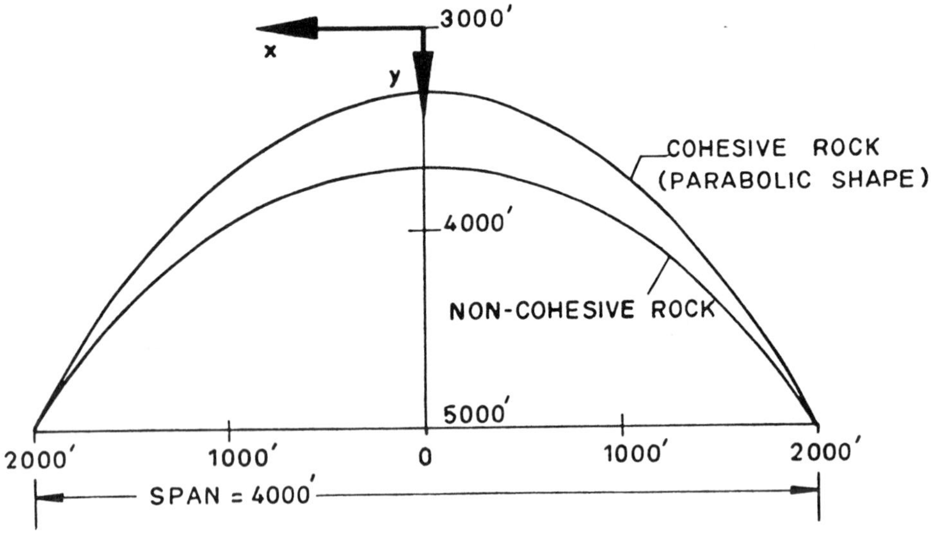

Fig. 2—Shapes of rigid domes

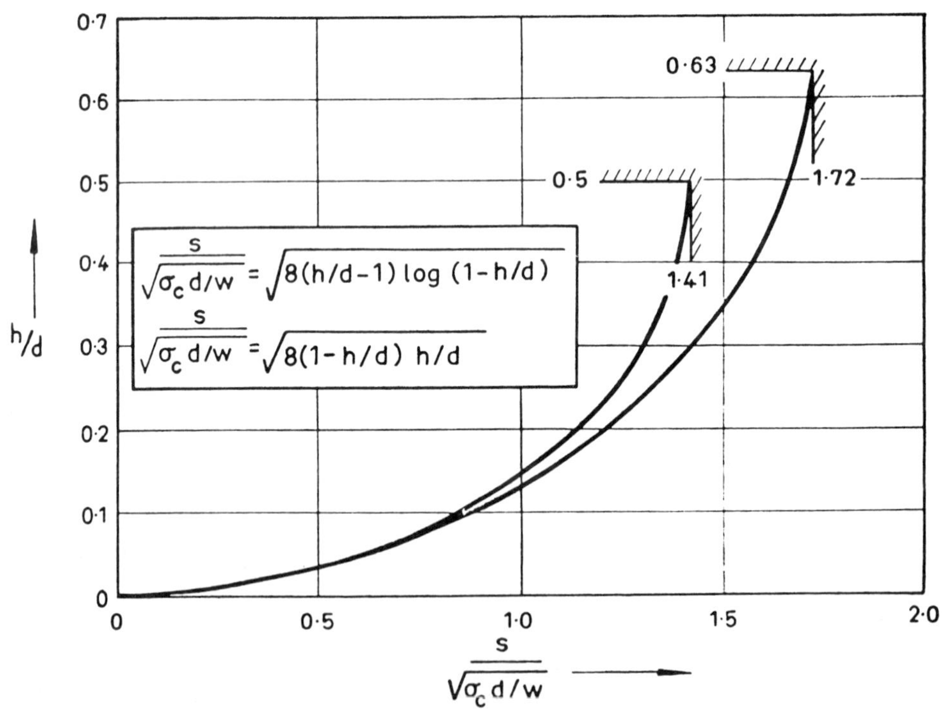

Fig. 3—Relationship between height and span for rigid domes

If the span exceeds the limits given by eqns. (3a) or (3b) and shown in Fig. 3, the dome fails; it may break through to surface. The failure may be accompanied by a major shock or a series of shocks.

It must, however, always be remembered that the concept is based on the assumption of a homogeneous isotropic ground and on a number of additional simplifying assumptions mentioned in Appendix I. With increasing height and span the probability increases that non-homogeneities and anisotropics such as dykes, faults and bed separation planes weaken the actual strength of the rock system. In practice, the maximum possible span must therefore be expected to be smaller than given by eqns. (3a) or (3b).

The concept does not yield information with regard to stress or movement ahead of the faces, around or within the dome, nor does it permit any statements with regard to the conditions in the footwall. It rather furnishes limits of a zone of movement: displacement does not occur outside the dome but is possible inside the dome. This is due to the fact that the concept does not account for deformation of the rock, i.e. considers the rock as a rigid medium.

The fracture dome in elastic rock

If the deformability of the rock is to be accounted for in a dome theory, the deformable medium, that lends itself most easily to theoretical treatment, is the homogeneous isotropic elastic body. Although this concept still constitutes an idealization of the behaviour of rock, it is a better approximation than the previous one. It permits determining stress and displacement not only in the hanging but also in the footwall of an excavation.

In an earlier paper[2] the author has dealt with the assumptions on which the theory of elasticity is based and summarized the results of analyses of the stress distribution around excavations of simple shape. If the shape becomes more complicated, experimental methods using the photo-elastic and similar analogue model techniques can be resorted to. During recent years these techniques have been developed to most efficient tools for determining stress and displacement even around excavations of complicated three-dimensional shapes.

If a failure criterion for rock, that is the condition under which rock fails for a given state of stress, is applied to the stress distribution around an excavation, it is possible to predict the shape and size of the fracture zone that may surround the excavation.

In his abovementioned earlier paper the author has modified a concept propounded by Fenner[3] and assumed that the fracture zone will assume such a shape and size that at the boundary neither the tensile stress exceeds the uniaxial tensile strength nor the compressive stress exceeds the uniaxial compressive strength.

The axis ratio $2\,h/s$ can be obtained from the following two formulae

$$2\,h/s = (1 + \sigma_t/wd)/2k - 0{\cdot}5 \quad \text{(4a)}$$

$$2\,h/s = 2/(k - 1 + \sigma_c/wd) \quad \text{(4b)}$$

where k is the ratio of lateral to vertical initial rock pressure, σ_t the uniaxial tensile strength and the other symbols as has previously been specified. The ellipse will adopt the greatest of the two ratios obtained from these two equations.

Obviously, these formulae are of limited validity since eqn. (4a) would furnish $2\,h/s = \infty$ for $k = 0$.

The concept assumes that the ground inside the ellipse (intradosal ground) is fractured and therefore cannot be regarded as a homogeneous elastic any more, while the extradosal ground is still a homogeneous elastic medium whose stress and displacement distribution can be predicted by the theory of elasticity.

Practical observations

In 1955, strata movement measurements were started in the ventilation shaft of Harmony Gold Mine, when the mine was still in the early stage of opening up. The upper limit of movement in the shaft with increasing stoping span has been recorded[4] and this provides an opportunity for comparing the two concepts of a rigid and elastic dome with practical observations. In Table I are listed the stoping spans and the heights to which movement extended at various dates. If one applies eqns. (1a) and (1b) for a rigid dome to these figures, observing that the depth of the excavation below surface was $d = 4{,}360$ ft, one obtains values σ_c/w for the various span to height ratios. It will be noted from Table I that the values for σ_c/w are not inconsistent.

If one introduces $w = 165$ lb/cu.ft, one finds values for σ_c of the order of 950 to 1,450 lb/sq.in. These values are low compared with $\sigma_c = 40{,}000$ lb/sq.in. obtained on laboratory specimens. However, if one realizes that rock *in situ* contains fissures, the low strength values may represent the actual condition. The strength of a brittle material is proportional to the square root of incipient cracks. The length of incipient cracks in a laboratory specimen is of the order of five thousands of a foot, while the length of cracks and fissures in rock *in situ* may be of the order of 10 ft. Hence, the structural strength may be $\sqrt{0{\cdot}005/10} \times 40{,}000 = 900$ lb/sq.in.

The fact that no appreciable surface subsidence was measured at Harmony before the span reached about 2,000 ft speaks for the qualitative validity of a dome concept. Nevertheless, the rigid dome concept contains too many oversimplifications to lend itself to quantitative analyses.

In Table I the ratios 2 h/s are also listed. The application of eqn. (4a) and (4b) is ruled out by the inconsistency of these values. It can therefore not be concluded that the domes had an elliptic shape.

TABLE I

POSSIBLE STRATA MOVEMENT DOME AT HARMONY GOLD MINE

Depth $d = 4{,}360$ ft

	Measured*		Eqn. (1a)	Eqn. (1b)	
Date	Span s (ft)	Height h (ft)	σ_c/w (ft)	σ_c/w (ft)	$2h/s$ (—)
June, 1955	251	96	83	84	0·76
Sep., 1955	306	143	83	85	0·93
Dec., 1955	436	194	125	128	0·89
Mar., 1956	468	309	92	95	1·32
June, 1956	577	410	107	112	1·42
Sep., 1956	652	596	96	103	1·83
Dec., 1956	711	598	113	122	1·68

*See Barcza and Von Willich[4]—Fig. 326.

Modification of the elastic dome concept

The weakness of the elastic dome concept lies in the fact that it is based on over-simplifications of the conditions under which rock fails. It is therefore an important objective of fundamental research in rock mechanics to find a failure theory for rock. It appears that the so-called modified Griffith failure criterion[5] is applicable to rock. Hoek[6] has dealt with this problem and it appears that application of this theory to the distribution of elastic stress around excavations may lead to a more realistic picture of the development of fracture domes around excavations.

Even with these refinements, however, two major factors of influence, that have been neglected in the first elastic dome concept, must be taken into account.

Firstly, whenever failure is initiated, the stress distribution around the excavation changes. The new stress distribution for the excavation of changed geometry must then be determined and the failure hypothesis re-applied. The procedure must be repeated until a stage is reached where the application of the failure hypothesis indicates no further propagation of fracture. The implementation of this approach is a major problem in fundamental rock mechanics research and is receiving attention by research workers.

Secondly, forces exerted from the fractured rock mass inside the dome onto the boundary of the dome, must be taken into account. Due to its own weight the fractured rock mass in the hanging tends to sag until the irregularly shaped blocks of which it consists wedge between each other and against the solid boundary of the dome. Consequently the vertical downward forces of the weight of an individual block may be resolved into components causing thrust between the blocks and against the dome boundary and components tending to shear blocks out of the arch system formed by the thrust. The shear is resisted by the friction between the blocks which in turn is increased by the thrust so that the system remains stable. After the French word 'voussoir' for 'brick' such systems are called voussoir arches or voussoir beams or voussoir domes.

The thrust and the shear forces exerted from the voussoir system inside the dome onto the dome boundary convert the state of stress, that would be uniaxial in the case of a force-free boundary, into a multi-axial boundary stress. Any brittle material is stronger in multi-axial compression than in uniaxial compression. The thrust and shear forces from the voussoir system are therefore an important factor controlling the propagation of fracture at the face and at the other boundaries of the dome. Depending upon their magnitude and distribution the fracture zone will assume a shape that differs from the elliptical one that was based on a force-free boundary. A hypothetical shape of the fracture zone accounting for the influence of boundary forces, is sketched in Fig. 4. In this connection it should also be mentioned that the dome boundary may not be a smooth plane but rather a transition zone from completely separated rock pieces and slabs over cracks penetrating into rock to the solid rock mass.

Obviously, the thrust from the fractured rock in the footwall onto the dome boundary in the footwall is particularly great. Its strengthening effect results in the fracture zone extending less far into the footwall than into the hanging wall. The dome may break through to surface depending upon the conditions prevailing in the upper strata.

No quantitative analyses using the modified Griffith criterion and accounting for the effect of boundary forces have been carried out yet. Knowledge of the magnitude and distribution of forces in the voussoir system is a prerequisite for such analyses.

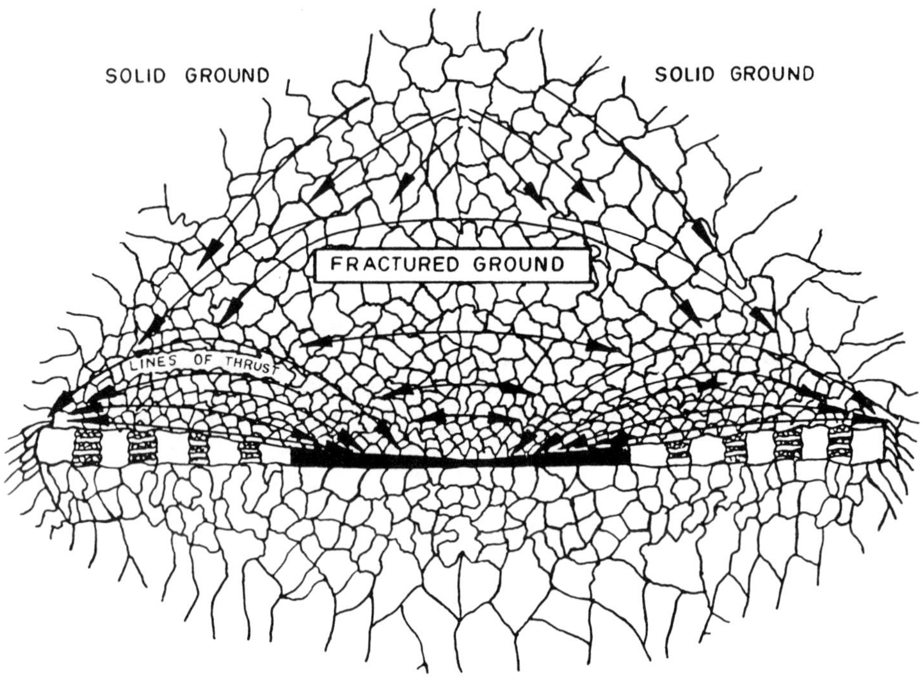

Fig. 4—Hypothetical shape of dome with intradosal voussoir systems

Voussoir systems

The voussoir arches are abutted at the boundary of the fracture zone and, if the stoping span is large, within the stope on areas where the hanging has settled on the footwall or regional support is taking weight. If these abutments are mined out, e.g. in reclamation operations or if the settlement is disturbed, say by sudden changes in face direction, too fast a rate of face advance or exposing dykes or faults, sudden adjustment of the voussoir system may occur and manifest itself as intradosal bursting. Such bursts may also occur if the settlement of the voussoir system is too slow due to delayed fracture so that a voussoir arch has not yet found an intradosal (back) abutment while its span is increased due to the face advance. It is therefore of great advantage for the operational planning to establish a theory of the behaviour of voussoir systems in rock.

Such an attempt has been made by Evans[7] but has not led to quantitative results. It is therefore a field for further research. Such a theory should be able to predict the position and size of intradosal abutments, the magnitude and direction of forces and movements in the system and the rate of settlement. Most probably such a theory must be of a statistical nature since the pattern of distribution and orientation of the fracture planes forming the voussoirs is a matter of the law of probability.

Even the existence of back abutments (intradosal abutments) is a matter of divergent opinion. Haack and Spruth[8] postulated the existence of a back abutment in coal mines, as shown in Fig. 5a, while pressure measurements at Neumühl Colliery[9] did not yield evidence of a back abutment for a distance of 1,400 ft from the face.

The author is of the opinion that back abutments must exist, if the stoping span is large. Obviously, however, conditions may often occur where the voussoir arch spans from face to face without an abutment in between. This may have been the case at Neumühl.

A voussoir arch need not have the geometrical shape of a concave arch, it can bc stragith or cvcn convcx; it is an arch of forces rather than an arch of material.

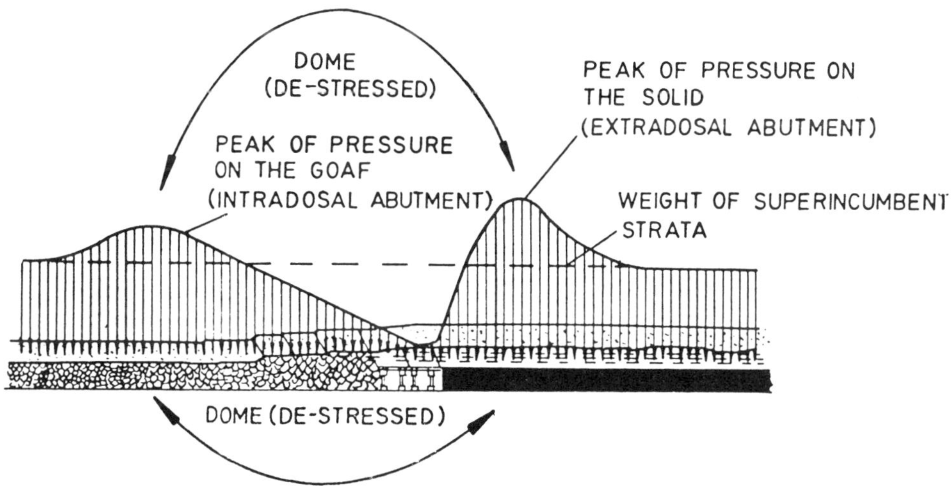

Fig. 5a—Illustration to the theory of Haack modified by Spruth

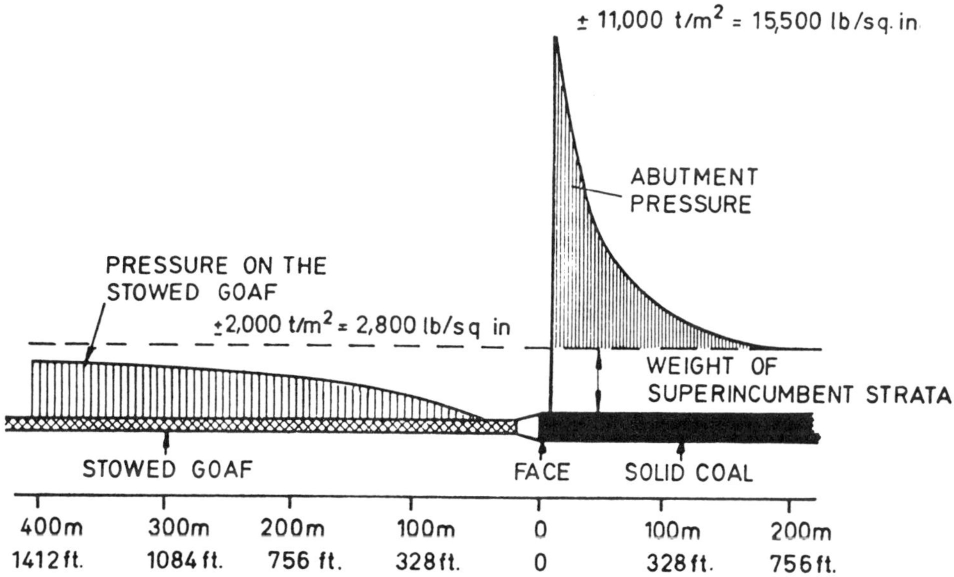

Fig. 5b—Results of pressure measurements in the abutment and in the stowed goaf during removal of a pillar in the Girondelle seam Neumühl Colliery

Beam or plate theories

In the rigid and elastic dome concepts the virgin ground was considered as a homogeneous* and isotropic* medium. No doubt exists that most mines are in rock that is non-homogeneous and anisotropic to various degrees. A theory dealing with such disturbances does not exist and would be extremely difficult to evolve. All that can be hoped for is that research succeeds in establishing a theory that would be statistical in nature in the sense that it deals with the effects of the probability of the spacing and orientation of non-homogeneities and anisotropies in a given type of ground. Furthermore, it may be possible to establish a theory describing the influence of, say, a major dyke or fault of known size, shape and orientation on the behaviour of an otherwise undisturbed rock system.

One type of non-homogeneity and anisotropy lends itself more easily to theoretical treatment, namely the stratification of a rock system. This has led to the concept of treating, in the first approach, the strata as a series of plates or, if the problem is simplified to a two-dimensional one, as a series of beams overlying each other.

The theory of beams is well known and the methods for determining the stresses in and the deflections of the beam are found in any engineering handbook. In mining, one usually considers the beams as built-in at both ends, i.e. at the faces. The distribution of stress and deflection (sag) of such a beam loaded by its own weight is shown in Fig. 6a. The maximum stress occurs at the face and if the stress exceeds the strength failure will occur there. This is, however, only valid for rigid abutments. If the abutment is elastic, the condition changes and, according to model tests by Lehr[10] the point of maximum stress and possible failure shifts towards midspan. Moreover, the points of inflection of the sag curve also shift inwards towards midspan.

Once the immediate hanging wall beam has been treated in this manner, the consideration is repeated for the next overlying beam. Due to the bending of the lower beam the span of the second beam will be smaller than that of the first one. The deflection of each beam depends upon its thickness and its elastic deformation modulus. Hence, it is possible that a lower beam deflects more than the overlying one so that a gap (termed the Weber cavity) occurs between the two beams. If one joins the abutments of all beams, one obtains a dome-shaped curve, as depicted in Fig. 6b.

This dome is one within which the layers deflect and possibly fail and Weber cavities may occur, while no deflection or failure occurs in the ground outside the dome. This dome, too, can break through to surface depending upon depth, span and rock properties.

If failure occurs, this may lead to visible arching over portions of the span or over tunnels, as depicted in Fig. 6c. Such occurrences obviously cause the remainder of the broken beam to act as a cantilever beam. If more than one cross-section fails, the beam ceases to be a continuous elastic beam and remains *in situ* only if it becomes a voussoir beam by friction and lateral thrust.

---

*Definition: A medium is homogeneous with respect to a certain property (e.g. strength, deformation moduli), if this property has the same magnitude if measured at any point.

A medium is isotropic with respect to a certain property, if this property has the same magnitude if measured in any direction.

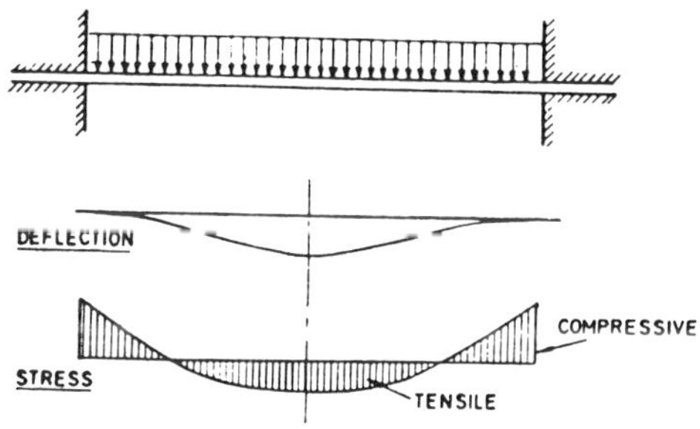

Fig. 6a—Deflection and stress of built-in beam loaded by its own weight

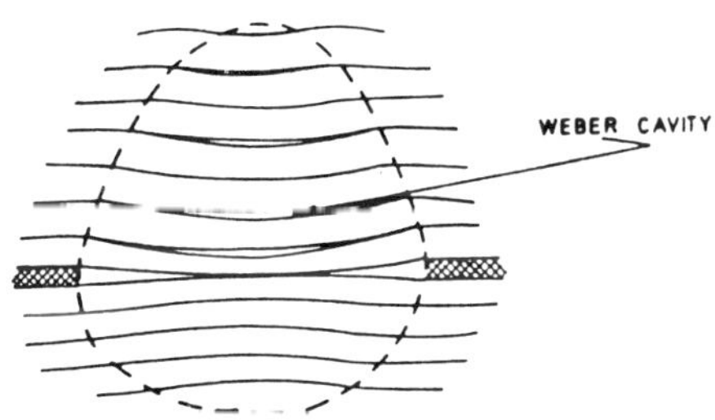

Fig. 6b—Beam movement dome

Fig. 6c—Beam fracture arch

The above concept is based on the assumption that no bond exists between successive layers so that they can deflect freely. If such a bond exists the two layers behave like one thick beam. If, however, the shear forces caused by the deflection loosen the bond, the deflection of both beams suddenly increases and this may manifest itself in a burst-like fall of hanging over a large area.

The concept could be refined by accounting for the lateral rock pressure which increases the deflection and also causes upward bulging of footwall layers. Accordingly, a footwall dome can be visualized as shown in Fig. 6b.

The theory of elastic beams is part of the theory of elasticity, but it is an approximation only, since it neglects the influence of shear stresses in the beam. It is therefore valid for long, thin beams only. Moreover, it is a two-dimensional theory. Refining the theory to a three-dimensional plate theory and treating thick plates would be cumbersome and it is then better to treat the whole rock system as a 'transversely isotropic elastic medium'. This theory has been referred to by Salamon[11] in a recent paper.

TROUGH THEORIES

Apart from the dome theories a concept has been propounded that considers a zone of influence in which movement occurs and which spreads from the excavation to the surface, forming what may be called a movement trough. This concept considers the ground as loose granular matter.

If an excavation is made at depth, movement of the strata extends to surface and appears as subsidence (vertical displacement) as well as horizontal displacement within a zone of influence. This zone is bounded by a three-dimensional plane joining the upper edges of the faces or sidewalls with the perimeter line on surface where movement ceases. On a vertical cross-section this *limit plane of influence* appears as a line. In the first approach this line is assumed to be straight as depicted as *AB* in Fig. 7. The dip of the line is called the *limit angle of influence*, β, and its complementary angle the *angle of draw*, γ.

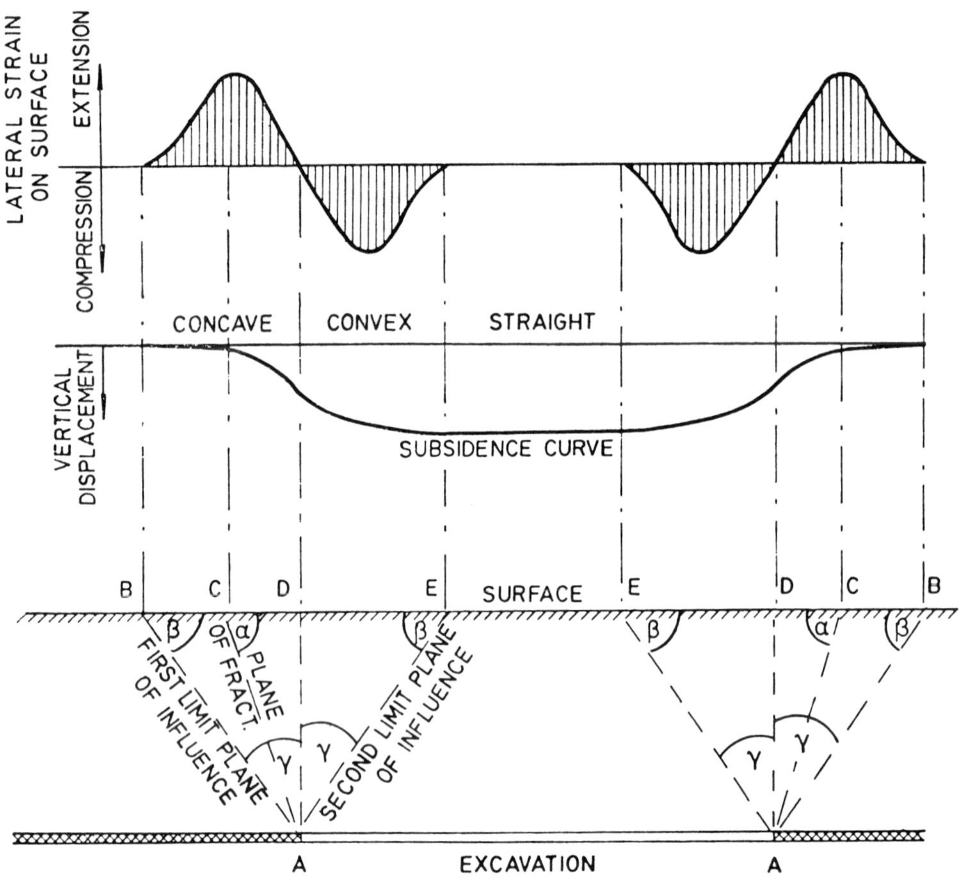

Fig. 7—Movement and fracture troughs

Within the zone of influence *ABBA* various sub-zones with different movement patterns are visualized. From *B* to *C* the vertical displacement increases gradually and the lateral movement, which is tensile in character, also increases.

At point *C* the vertical displacement increases sharply and the lateral tensile strain attains a maximum and begins to decrease again. It is assumed that this is due to downward sliding of the ground along *CA*, which lines are the traces of what therefore is called the *plane of fracture*. Its dip is the *angle of slide*, α.

In the regions *CD* the vertical displacement curve is still concave downwards as in *BC* but with sharper curvature and the lateral tensile strain decreases towards zero.

At point D vertically above the face A, the curvature of the subsidence line changes from concave to convex downwards. This point is therefore called the *point of inflection*. The lateral strain there is zero.

From D to E the vertical displacement increases further, the lateral strain, which is compressive, increases to a maximum and then decreases to zero.

At point E the subsidence has attained its maximum value, δ_{max}, and the curve of subsidence becomes a straight line; the lateral strain is zero. The lines EA are traces of what is called the '*Second limit plane of influence*'. Its dip (or its angle of draw, for this matter) is the same as that of the first limit plane of influence BA.

These explanations apply to the surface as well as to any horizon between excavation and surface. The movement obviously depends upon the depth, d, and the span, s, of the excavation.

If the geometry of Fig. 7 is applied to excavations with various spans at various depths, three stages of subsidence, as illustrated in Fig. 8, may be distinguished.

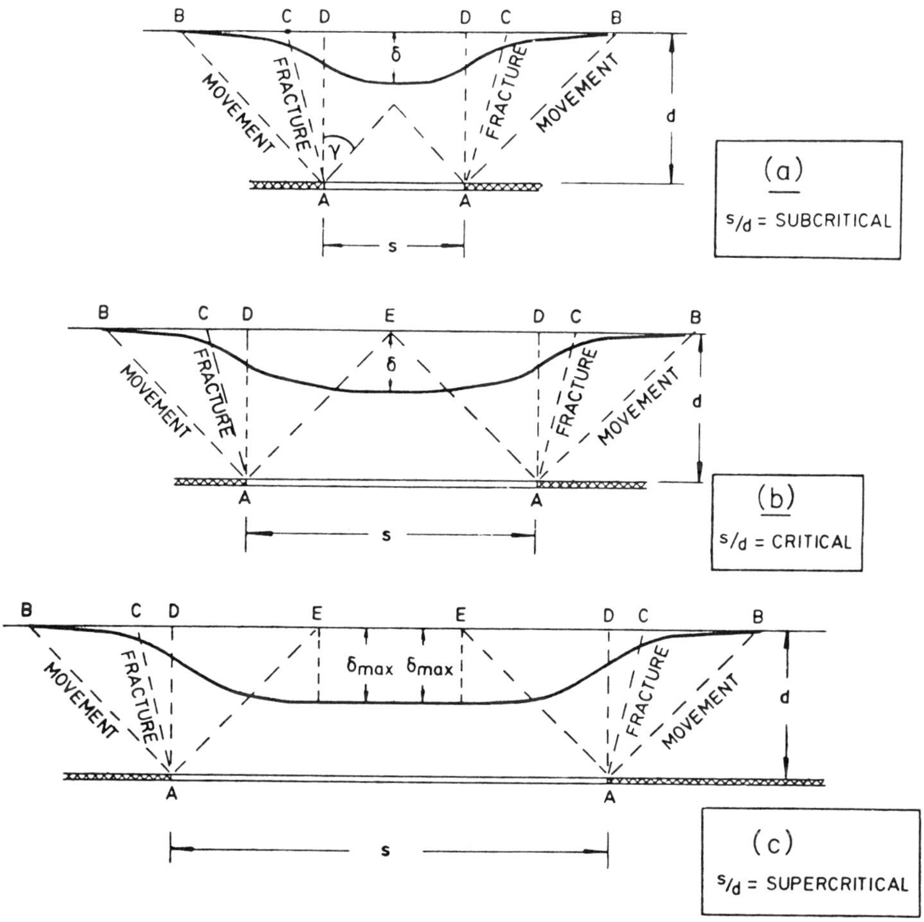

Fig. 8—The three stages of subsidence

If the ratio s/d is such that the second limit planes of influence (EA) from the two opposite faces intersect below the surface horizon, the stage is called *subscritical.* The greatest displacement δ of the subsidence curve increases with increasing stoping span.

If the ratio s/d has increased to a value where the second limit planes intersect at surface (E) the *critical* stage has been reached. The vertical displacement has attained its maximum value δ_{max}.

If the ratio s/d is such that the second limit planes intersect above surface, the stage of subsidence is called *super-critical.* With increasing span, s, no further movement occurs in the zone EE which spreads out with increasing span. In this zone the vertical displacement remains at the value δ_{max} and there is no lateral strain. This is attributed to completed settlement (closure) in the central region of the stope. The practical implication is that buildings on surface or shafts in the region EE are not affected by movement any more as was the case in the sub-critical and critical stages.

The concept does not readily facilitate calculating the magnitudes of the movements. These can only be determined by field measurements. In Fig. 9, for instance, the maximum subsidence is plotted versus depth of the excavation, as derived by Seldenrath[12] from subsidence measurements in Dutch collieries.

The concept limits itself to defining the areas in which movement occurs as a result of excavating. Hence, it is of practical value for planning purposes only if the extent of these areas, depending upon the angle of slide and the angle of fracture, can be predicted.

It is not known yet how the angle of draw (or the limit angle of influence, for that matter) depends upon the properties of the strata. It is derived from analyses of actual subsidence measurements. Some figures obtained for various European coal mining areas are given in Table II. Wardell[13] has pointed out that the limit planes of influence most probably do not start at the face. Due to weakening of the solid abutment the first limit plane may start at a point ahead of the face and the second limit plane may start at a point back from the face where the support starts to take weight.

TABLE II

ANGLES OF DRAW DETERMINED FOR EUROPEAN COAL MINING AREAS

Author	Area	Limit angle, β	Angle of draw, γ
Drent	Netherlands	45°-55°	35°-45°
Wardell, Orchard ..	Britain	50°-58°	32°-40°
Fläschenträger	Lower Rhine	51°-61°	29°-39°
Knothe	Poland	56°-71°	19°-34°

It must also be expected that the limit planes of influence are not straight lines but, depending upon the strata, either curves or stepped lines following the bedding planes. Mohr[14] concludes this from observations in shafts.

The angle of slide may be predicted on the basis of a failure hypothesis for loose granular masses (soil), which is dealt with in Appendix II. According to this hypothesis

$$\alpha = 45° + \rho/2 \qquad \text{......................(5)}$$

where ρ is the angle of internal friction of the material. It can be determined as the natural angle of slope of the granular material. Values of ρ and α for various types of rock are given in Table III.

TABLE III

ANGLES OF INTERNAL FRICTION AND ANGLES OF SLIDE FOR VARIOUS TYPES OF ROCK (AFTER SELDENRATH AND LABASSE—1951)

Rock	Angle of internal friction, ρ	Angle of slide, α
Clay	15°-20°	52·5°-55°
Sand	35°-45°	62·5°-67·5°
Moderate shale ..	37°	63·5°
Hard shale ..	45°	67·5°
Sandstone	50°-70°	70°-80°
Coal average ..	45°	67·5°

Seldenrath[12] has measured subsidence in drifts about 250 ft above the advancing face in Dutch collieries and found the point of inflection C not above the solid as depicted in Figs. 7 and 8, but above the goaf as sketched in Fig. 10. This points towards the existence of a fracture dome (the concept of the *fracture* trough depicted in Fig. 7 would then not be correct) but still proves the existence of a trough shape of the zone of influence.

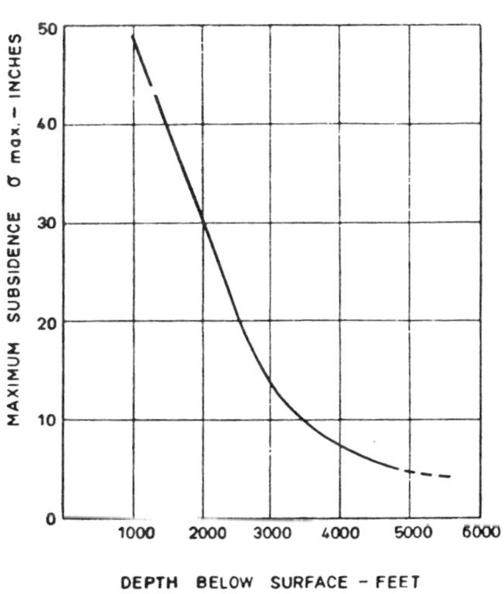

Fig. 9—Maximum subsidence as a function of the depth of an excavation in collieries (after Seldenrath)

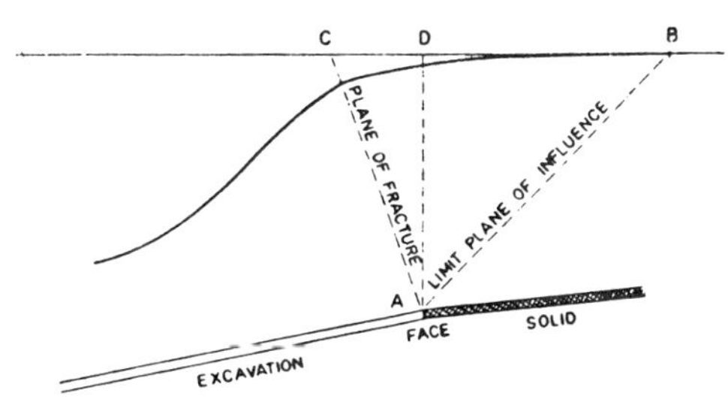

Fig. 10—Subsidence curve and fracture plane as determined by Seldenrath

CONTINUUM THEORIES

The dome theories as well as the trough theories deal with the limits of zones of different movement and fracture patterns rather than with the magnitude of the movement. If one sets out to predict the actual displacement for a given stress field in rock with given properties, one will have to treat the ground, in the first approach, as a continuous body bounded by the surface above and the excavation within.

As already mentioned, the type of idealized substance that offers itself readily for such treatment is the perfectly linear elastic medium whose theory, the well-known theory of elasticity, has been developed to a most advanced stage. In its concept it is assumed that the displacement at any point in the medium is proportional to the stress at that point and that it is reversible (i.e. it vanishes if the stress vanishes) and does not depend upon time. Hence, if the elastic moduli, the initial state of stress in the ground and the boundary conditions in terms of displacement or stress on surface, on the roof or on the floor are given, it is possible to predict stress and displacement at any point of the medium. Such a strata movement theory has been developed by Salamon[11] and presented in a paper of this Symposium. Ryder and Officer[15] have applied this theory to practical measurements in a Witwatersrand gold mine and found good agreement.

The concept of an elastic medium has already been referred to elsewhere in the present paper, but there it was used in an attempt to predict the shape and size of the fracture zone, while the theory described by Salamon mainly deals with the elastic displacement. Both approaches do not exclude but rather supplement each other. The continuum theories can also be applied if the ground exhibits zones with different deformation moduli (e.g. fracture zone and solid zone); then each zone must be treated separately with appropriate boundary conditions at the limiting planes between the zones.

The strata movement theory based on the concept of an elastic medium is a step forward to the establishment of a theory that yields quantitative results. It can even be extended to transversely isotropic (stratified) ground as pointed out by Salamon[11]. It is, of course, still an idealization and the question arises as to whether the actual rock mass can be considered as an elastic medium. The findings of Ryder and Officer[15] and also studies by Cook[16] of the propagation of seismic waves indicate that the idealization is at least a good approximation in the case of quartzite. On the other hand, no doubt exists that certain types of rock are certainly not perfectly elastic. In the general case, it must be expected that the displacement consists of four components, namely the time-independent elastic, the time-dependent elastic, the time-independent plastic* and the time-dependent plastic* components. The above theory caters only for the first component; whether the other three are negligible or not, depends upon the type of ground. In addition, the deformation moduli governing the relationship between displacement and stress may be non-homogeneous and/or anisotropic to various degrees and these properties may vary continuously or discontinuously (stepwise) over a given rock mass, the latter being due to fissurization and other disturbances of the ground.

The ideal strata movement theory would have to account for all these factors and, in addition, for failure criteria specifying the conditions under which the medium

---

*In this context the term 'plastic' is used to denote 'irreversible', irrespective of the physical mechanism of deformation.

or certain zones thereof change properties, e.g. from the continuous to the discontinuous state. This, however, is a target for fundamental research and the present policy to firstly exhaust the possibilities of the concept of a perfectly elastic medium appears to be sound from a practical point of view.

It may be mentioned, however, that Polish researchers[17] are working on a theory which accounts for the time-dependence of the displacement, i.e. for the rheological behaviour of the ground.

The fact that strata movement around excavations continues even after mining has ceased, is a manifestation of rheological behaviour and the development of such a strata movement theory is therefore of practical significance.

CONCLUSION

The more important strata movement theories have been discussed under the three headings of dome, trough and continuum theories. Although at first sight the impression may be gained that they diverge from each other, it will have become clear from the paper that they do not necessarily contradict each other. In fact, they appear as different modifications of one universal theory although this may not be clearly established at the present time. It has been pointed out that, for instance, the trough theory does not exclude the existence of a fracture dome and that the continuum theories may be applied to rock inside as well as outside a dome (intradosal and extradosal ground). In this connection, a diagrammatic sketch by Wiggill[18], which is repeated in a modified form in Fig. 11, is most interesting. Wiggill depicts the

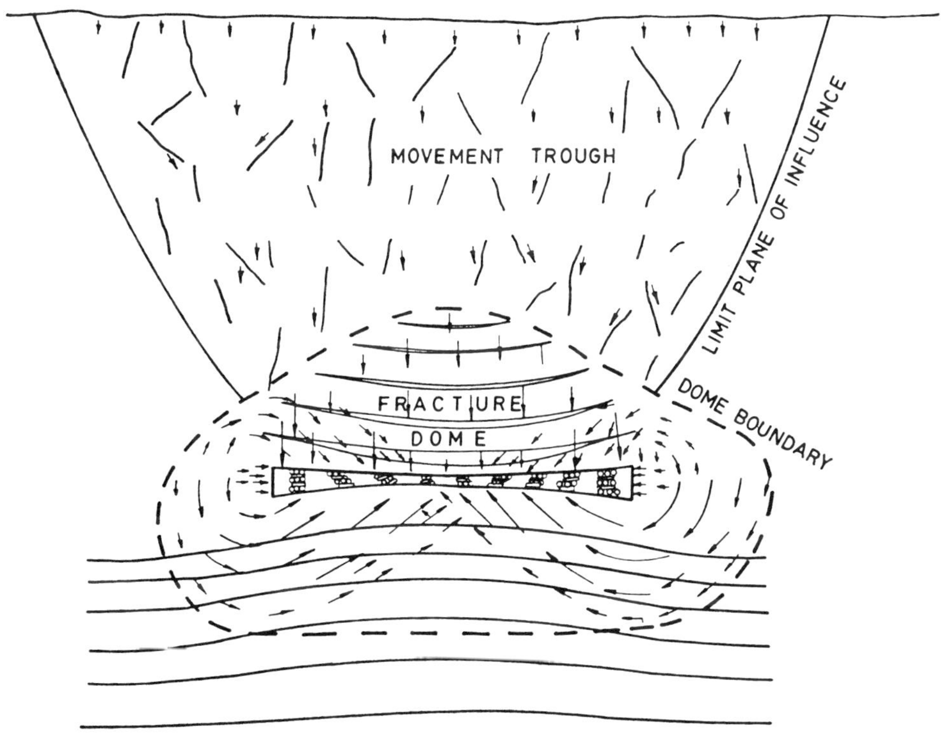

Fig. 11—Dome and movement trough (after Wiggill)

concept of a combined dome and trough theory, where the movement trough does not start from the face but higher up from the dome boundary. It could even be possible that a fracture trough develops on top of the dome within the movement trough (zone of influence). All this will depend upon the local conditions. Needless to say that the continuum theory can be applied to the various zones to determine the magnitude of the movements. The difference between the various concepts appears to be only a difference in emphasis placed on various characteristic properties of the strata. The dome theory has been favoured by mine engineers dealing with hard rock while the trough theory stems from experience in soft rock.

All these considerations point towards the necessity for finding parameters that specify the properties and geological conditions of the ground (strength, deformation moduli, degree of non-homogeneity and anisotropy) and to establish a theory, a 'model', that relates movement and failure to these parameters and to parameters describing the mining operations (depth, stoping width, span, support, rate of face advance). The theory should then be verified by the results of practical strata movement measurements.

This must be the aim of research in rock mechanics.

REFERENCES

1. Börger, H. 'Die Gebirgsdruckhypothesen für elastische Gesteine und ihre Anwendungsmöglichkeiten im Salzbergbau.' *Kali u. Steinsalz*, Heft 6, pp. 10/20, Sep., 1954
2. Denkhaus, H. G. 'The application of the mathematical theory of elasticity to problems of stress in hard rock at great depth.' *Pap. Ass. Min. Mngrs., S.Afr.*, Vol. 1958/59, pp. 271/310, Sep., 1958.
3. Fenner, R. 'Untersuchungen zur Erkenntnis des Gebirgsdruckes.' *Glückauf*, Bd, 74, pp. 681/695, 705/715, 1938.
4. Barcza, M. and Von Willich, G. P. R. 'Strata movement measurements at Harmony Gold Mine.' *Pap. Ass. Min. Mngrs. S.Afr.*, Vol. 1958/59, pp. 447/464, Nov., 1958.
5. McClintock, F. A. and Walsh, J. B. 'Friction on Griffith cracks in rock under pressure.' Proc. Internat. Congr. Appl. Mech., Berkeley, 1962.
6. Hoek, E. 'Rock fracture around mining excavations.' To be published in Proc. Internat. Conf. Strata Control, New York, May, 1964.
7. Evans, W. H. 'The strength of undermined strata.' *Trans. Instn. Min. Metall.*, London, Vol. 50, pp. 475/532, April, 1941.
8. Spruth, F. 'Distribution of pressure up, on and near the coal face.' Proc. Internat. Conf. Rock Pressure, Liege, 1951.
9. Jacobi, O. 'The pressure on seam and goaf.' Proc. Internat. Conf. Strata Control, Essen, 1956.
10. Lehr, E. 'Modellversuche an Balken auf elastischer Unterlage zur Klärung der Spannungsverteilung im Hangenden von Abbauörtern.' VDI-Forschungsheft 372, VDI-Verlag, Berlin, 1935.
11. Salamon, M. D. G. 'Elastic analysis of displacements and stresses induced by mining of seam or reef deposits.' *J. S.Afr. Inst. Min. Metall.*, Vol. 64, pp. 128/149, 197/218, Nov., 1963/Jan., 1964.
12. Sedlenrath, T. R. 'Can coal measures be considered as masses of loose structure to which the laws of soil mechanics may be applied?' Proc. Internat. Conf. Rock Pressure, Liege, 1951.
13. Wardell, K. 'The problem of analyzing and interpreting observed ground movements.' Proc. Internat. Conf. Strata Control, Leipzig, 1958.
14. Mohr, F. 'Observations in shafts on rock movement due to mining.' Proc. Internat. Conf. Strata Control, Leipzig, 1958.
15. Ryder, J. A. and Officer, N. C. 'An elastic analysis of strata movement observed in the vicinity of inclined excavations.' *J. S.Afr. Inst. Min. Metall.*, Vol. 64, pp. 219/244, Jan., 1964.
16. Cook, N. G. W. 'The seismic location of rockbursts.' Proc. Fifth Symp. Rock Mech., Univ. Minnesota, pp. 493/516, May, 1962.
17. Litwiniszyn, J. 'Statistical methods in rock mechanics.' Proc. Internat. Conf. Strata Control, Leipzig, 1958.
18. Wiggill, R. B. 'The effects of different support methods on strata behaviour around stoping excavations.' *J. S.Afr. Inst. Min. Metall.*, Vol. 63, pp. 391/426, April, 1963.

Editor's Comments on Paper 39

39 LeGRAND
Overview of Problems of Mine Hydrology

MINE HYDROLOGY

Mines have been forced to close in part or in entirety, and miners have been trapped by flooding of mine workings. However, drainage of mines has on numerous occasions, had the effect of lowering the water table in the district and drying up wells in adjoining farms. One of the classical near-disaster situations is described in the book *West Driefontein—Ordeal by Water* by A. P. Cartwright (1969). It is recommended to the reader as an example of the need for detailed analysis of mine hydrology because few such documented case histories exist. However LeGrand (Paper 39) outlines the basic principles to avoid operational hazards and regional environmental damage. Safety of mine workers and environmental protection require careful hydrological analysis and a continuing monitoring system (Swaisgood and Toland, 1972). Engineering studies (Kendorski et al., 1979; Loofbourow, 1973) are gradually establishing safety criteria as well as surface and groundwater control techniques.

REFERENCES

Cartwright, A. P., 1969, *West Driefontein—Ordeal by Water,* Ince, Johannesburg, 83p.

Kendorski, F. S., I. Khosla, and M. M. Singh, 1979, *Criteria for Determining When a Body of Surface Water Constitutes a Hazard to Mining,* U.S. Bureau of Mines Contract J0285011, Washington, D.C., 364p.

Loofbourow, R. L., 1973, Ground Water and Gound-water Control, in *SME Mining Engineering Handbook,* vol. 2, I. A. Given, ed., American Institute of Mining, Metallurgical and Petroleum Engineers, New York, pp. 1-55.

Swaisgood, J. R., and G. C. Toland, 1972, Control of Water in Tailings Structures, *Dames & Moore Eng. Bull. No. 41,* pp. 17-28.

39

Reprinted from *Soc. Mining Engineers Trans.* **252**:362-365 (1972)

Overview of Problems of Mine Hydrology

by Harry E. LeGrand

The purpose of this paper is to outline briefly some basic principles of ground-water hydrology that have application in mine hydrology and to suggest that an understanding of these principles can help to forestall problems that are internal as regards mining operations and external as regards damage or harm to surrounding areas. The development of a mine below the water table pleads for accurate values to answers of hydrologic problems. Such internal problems involve both an inflow of water without adequate drainage facilities and a deficiency of water for the required mining or milling use. Methods which give acceptable quantitative values for the solution of many ground-water problems depend on rather simple geologic structures, characterized chiefly by homogeneous rocks of considerable areal extent. Unfortunately, mining operations are likely to be centered where rocks are not homogeneous. Consequently, methods that attempt to show values of the storage and transmitting capacity of rocks in mining areas must be applied with caution. In spite of complex geology and apparent uniqueness of individual mines and quarries, adherence to some concepts of ground-water occurrence and movement should be helpful to mine operators.

Mining operations that extend below the water table disturb the natural hydrologic regimen. Prior to development, the water table slopes away from the site of a mine in at least one direction, but as soon as development begins and water is removed from the mine, the mine becomes the apex of a depressed cone in the water table, causing a convergence in the flow of water toward the mine. The behavior of the cone of depression, as it tends to deepen and to increase in areal extent when the mine is deepened, is related to the rate of withdrawal of water necessary to keep the mine dry. The significance and nature of the cone of depression in the vicinity of pumped wells has been studied by ground-water hydrologists for years, but the cone caused by pumping water from mines has generally been less systematically appraised. The geologic and hydrologic settings of mines range greatly so that a simple standard procedure for dealing with problems of hydrology of mines is not possible. Yet, truly unique hydrologic conditions in mines are rare.

The literature on mine hydrology is widely scattered, most of the publications being restricted to conditions of a particular mine. Because of the differences of emphasis of the individual authors and because of the great range in hydrologic conditions at different mines a direct synthesis of information from publications was not attempted in this paper. Moreover, the number of hydrologists or mine officials who have studied the hydrology of more than a few mines is surprisingly small. R. L. Loofbourow has kindly furnished the author notes which he has compiled on the occurrence of mine water, as well as on the effects and control of mine water.

The Hydrologic Setting Prior to Mining

A great range in capacity to hold and yield water has been noted in granular materials and consolidated rocks. Mines are developed in many kinds of rocks, which differ considerably in the number, size, and shape of their open spaces. Clay and shale, although quite porous, are not considered permeable because the open spaces are quite small and most of the contained water is bound to the surfaces of particles. Thus the flow of water in clay and shale under ordinary hydraulic gradients is very small or negligible. At the other extreme, sand and gravel, which may contain less volume of pore space than clay or shale, are considered permeable because water can move freely through the large interconnecting pore spaces. Igneous and metamorphic rocks, as well as other massive rocks, transmit water only through interconnecting fractures, crevices, or interbed partings. The pore space, no matter what form it has, cannot be logically evaluated apart from the rock material and from the geologic history.

Prior to disturbance by man, such as the development of a mine, the hydraulic system of water-bearing rocks is in dynamic balance: the natural discharge of water, as seepage or spring outflow or as evapotranspiration, is equal to the replenishment or recharge. The water table, therefore, is almost fixed in position as regarded from a long-range view, its position changing in accord with rates of recharge and discharge. Even if mining has started and the natural hydrologic setting has been altered, it may be advantageous to mentally reconstruct the natural flow pattern of the water in the surrounding area prior to pumping from the mine. The main phases of the hydrologic cycle that need to be considered by the mine operator are: recharge, storage and transmission in the rocks, and discharge. The ground-water hydrologist, or person whose responsibility is water in the mine area, might get a good perspective of impending problems if, prior to mining, he would look over the terrain from a vantage point and then reflect on the hydrologic system in the area. Some aspects of groundwater recharge and discharge are observable and are not difficult to evaluate. The important consideration of precipitation does not need elaboration here. If there is recharge there must also be discharge in one form or another. If streams are widely spaced and the climate is arid, the prevalent discharge (if any) may be in a remote area and the movement of water (if any) is by underflow toward that area. In humid regions the streams usually delineate the chief areas of groundwater discharge. Springs and diffuse seepage contribute to the flow of streams at all times and may be the total source of streamflow in long periods of fair weather. This part of ground-water discharge can be measured but that part resulting from evapotranspiration is difficult to determine. Over a long period of time, the discharge of water by outflow in liquid form and by evapotranspiration is equal to the average recharge. Although changeable in rate, discharge is generally continuous. Recharge, on the other hand, is sporadic because it results from precipitation. A period of significant precipitation generally results in a rise of the water table and a consequent increase in the underground storage of water. From the viewpoint of water in the ground, the rock materials are underground reservoirs in which changes in storage are reflected by changes in the position of the water table. To a drop of water, however, storage is only a transient phase of slow movement to a discharge point. The subject of movement of ground water must be tempered by the realization that the character and structure of the rocks are all important. Prior to consideration of specific geologic conditions it can be said that wherever water enters openings in the rocks it moves by gravity and pressure through interconnecting openings to some point of discharge at a lower elevation.

Induced Changes in Hydrology with Mine Development

Starting with the first withdrawal of water from a mine, the movement of some water in the immediate area is changed in direction. The withdrawal of ground water causes a lowering of the water level in the form of an inverted cone, the apex of the cone being at the center of pumping. At first, when mining operations are at shallow levels, the cone of depression may not extend far laterally. As the apex of the cone in the mine deepens, the cone enlarges laterally at a rate that depends on the ability of the rocks to store and transmit water. The more permeable rocks, transmitting water easily, tend to have a shallower but broader cone of depression than do the less permeable rocks, such as clays and unfractured rocks. In the beginning, the water pumped out is drawn from storage in the vicinity of the mine. As pumping continues the cone of depression enlarges until it intercepts additional sources of water equal in amount to that pumped out or reduces the natural discharge of the rock unit by that amount.

In order to see graphically what happens when water is withdrawn from a mine, Fig. 1 shows a hypothetical mining situation in homogeneous and isotropic rocks. Notice that prior to mining, the water table is depressed only toward the streams or other natural discharge areas. As mining operations start and a depth of 600 ft is reached, the volume of material represented by a is unwatered. The cone of depression has extended to the limits of the mine property. Note that the water level in the water-supply well of the mine has lowered and that its yield may be impaired. As the mine workings are lowered to 1500 ft, an additional volume b is unwatered. In order to keep the mine dry at 1500 ft, the rate of pumpage must be greater than at the 600-ft level because at first more water has been drawn from storage and the cone of depression has enlarged. Note that at the 1500-ft pumping level the cone extends beyond the limits of the mine property, the water table has fallen below the bottom of the water-supply well at the mine, and three other wells may decline in yield because of interference from the cone of depression in the mine. In lowering the mine operations to the 2500-ft level, the rate of pumping may have to be increased still further because the cone of depression has enlarged and new sources of water enter the cone that could not enter at higher levels. The cone has intersected a swampy area on the flood plain where, previously, water had discharged as evapotranspiration. As the water table lowers in the flood plain, this natural discharge is diverted into the mine. Note, also, that a ground-water divide no longer exists between the mine and the stream; now a hydraulic gradient has been established from the stream to the mine. and a major—perhaps serious—source of recharge has been intercepted.

The literature on hydrology is replete with intensive studies of the behavior of water levels resulting from pumping wells in homogeneous materials. Theis[1] developed a non-equilibrium equation which set the stage for late refinements in ground-water hydraulics. Some

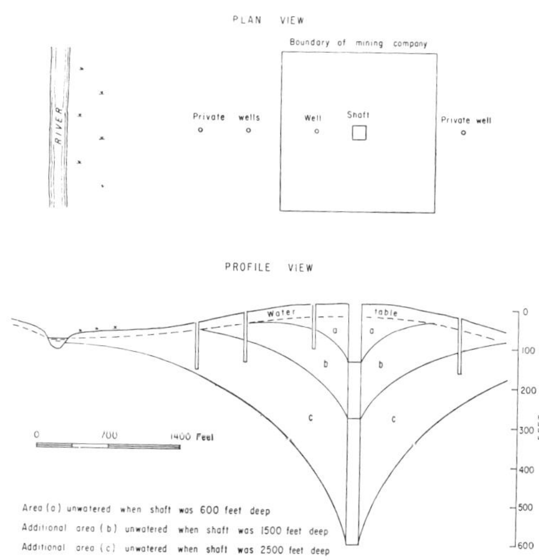

Fig. 1—Hydrologic setting of a mine area during several stages of operation. In order to keep the mine dry, it is necessary to reckon with the water drawn from storage within the cone and also that which moves radially toward the mine through the saturated materials below the cone of depression.

results of earlier workers were modified and applied by Worley[2] to develop a method for estimating ground-water influx from a homogeneous permeable formation into a vertical mine shaft during sinking. Loofbourow and Brittain[3] described the dewatering process through wells before mine development.

Some Effects of Geologic Boundaries on Mine Hydrology

It should be clearly realized that the idealized conditions postulated above are rarely completely satisfied in mining areas. Instead of being centered in homogeneous and isotropic rocks that are capable of conducting water equally well in all directions, mines generally penetrate certain rocks of limited extent in a complicated geological environment. The hydrology of the area will be correspondingly complicated. As the cone of depression enlarges, it reaches geological bodies with greater or less permeability than the rock in the mined area. Boundaries between areas of relatively good and relatively low permeability are both hydrologic and geologic.

Such geologic features as contacts between types of rocks, faults, folds, and bodies of surface water form ground-water boundaries. So complicated is the geology in ordinary mines, compared with the simplified geology referred to in Fig. 1, that a discussion of all possible situations dealing with ground-water boundaries is out of place here. It may suffice to say that boundaries may be classified as either negative or positive, according to their significance in transmitting water to the cone.

If the mine is in a relatively permeable rock that abuts against an impermeable one, a negative boundary is formed along the contact because the replenishment area is restricted on the side of the impermeable rock.

The right half of Fig. 2 shows a cone of depression which intersects an impermeable boundary. Because the impermeable rock cannot contribute much water to the mine, the area of influence is constricted, and interception of recharge is limited by the impermeable rock. The inability of the cone to spread beyond the impermeable boundary and to induce more water to enter the mine results in the need for less pumping from the mine than would otherwise be expected.

A positive boundary is formed where the cone of depression extends to a perennial stream or other body of surface water. An example has already been shown in Fig. 1. The positive boundary there is along the line where the cone has reached the river. So long as the stream has water in it and maintains hydraulic continuity with the mine, the cone in its growth may not cross the stream, but its base may extend linearly to coincide with the course of the stream. Under such conditions a source of recharge is available to the mine

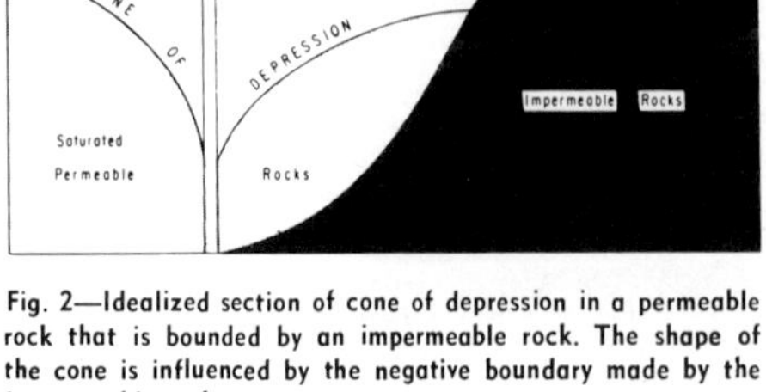

Fig. 2—Idealized section of cone of depression in a permeable rock that is bounded by an impermeable rock. The shape of the cone is influenced by the negative boundary made by the impermeable rock.

that could result in greatly increased pumping rates if the water table must be lowered as the mine deepens.

Some Practical Considerations

The geological work necessary for the operation of a mine is a basis on which evaluation of the ground water can be made. Where mining operations extend below the water table, careful geological interpretations must precede an attempt to solve problems of mine drainage.

A question of importance concerns the rate of pumping at various levels in order to keep the mine dry. Will the rate of pumping be greater with depth? Will it increase directly in proportion to depth? Will there be a sudden, catastrophic inflow of water at any time? What will be the effect of drifts extending outside the cone of depression? Answers to these questions must be based upon an understanding of the local geology and of the principles of ground-water behavior.

There are two tendencies about the pumping level that are favorable to the mine operator. In ground-water terminology, specific capacity of a well is the rate of discharge per unit of drawdown of the water level. It is sometimes assumed to be a constant at differing rates of pumping. For example, a well with a specific capacity of 10 might yield 100 gpm with a drawdown of 10 ft and might be assumed to yield 200 gpm with a drawdown of 20 ft. For water-table conditions, which apply to most mining areas, this direct relationship between withdrawal and drawdown is not accurate because the saturated thickness of the water-bearing material decreases with increased drawdown. Thus with greater drawdown (dewatering at greater depth) the increments of inflow are proportionately smaller. It is interesting to note that, in a homogeneous water-table aquifer of infinite thickness and extent, about 75% of the relative yield of a well comes during the first 50% of the drawdown;[4] conversely, a well whose pumping level is at the bottom of the well will yield water at a rate only about 25% greater than if the pumping level were halfway between the static water level and the bottom of the well.

The other favorable consideration is that in a great majority of rocks the permeability decreases with depth. In many consolidated rocks, particularly igneous and metamorphic rocks, the interconnecting fractures, which represent the gross permeability, decrease in both number and size with depth. In such cases the enlargement of the cone of depression is constricted, and great drawdown can be accomplished with only slightly increased pumping rates. In dense impermeable rocks, the lower part of the main shaft is, of course, the apex of the cone of depression but it may also be considered as a large tubular sump.

Thus far only the unwatered area in the cone has been considered; however, deep mining operations must extend laterally outside the cone. The fact that lateral workings may be dry is fortunate, but a distinction should be made between those appearing to be dry because of the low permeability of the rock and those that are dry because they lie within the cone of depression. The risk of a sudden inflow of water from a permeable zone or abandoned workings must be considered in extending operations in impervious rocks. Fig. 3 shows how workings at great depth may extend in impermeable rocks without increasing the pumping rate. However, risks of flooding arise from extending laterals outside the cone of depression unless the hydrologic properties of the rocks around the drifts are completely known and measures are taken to prevent the risks.

The external aspects of mining are looming ahead, as environmentalists and land owners around the mines

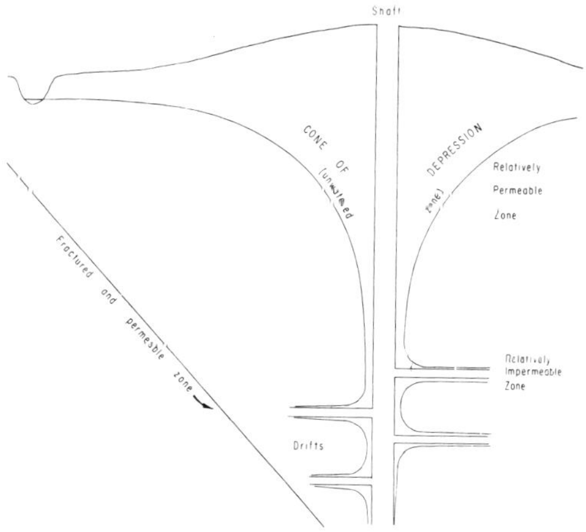

Fig. 3—Section showing drifts extending into dense impermeable rocks outside the cone of depression and the imminent danger of flooding if workings extend to the fractured and permeable zone which is saturated.

are now on guard against possible deterioration of the environment that results from mining. Complaints of polluted water from mine operations are loud and clear. Acid mine water discharged from coal mines is a subtle type of geochemical pollution that causes critical problems to surface waters for great distances from the coal mines; the mines may also pollute the ground water.[5] Where the permeability of the country rock is high the effects of the lowering of water levels due to pumping from a mine can be felt for great distances beyond a mine. Operations of a single phosphate mine in the Coastal Plain of North Carolina resulted in the withdrawal of about 60 million gal of water a day; this withdrawal, according to Peek,[6] caused a lowering of water level from the artesian system of more than 5 ft in an area of about 1,300 sq miles. Subsidence of the land surface as a result of dewatering mines is not common but is a severe problem around some mines. Brink and Partridge[7] report that withdrawal of water from the deep gold mines of Transvaal, South Africa, has caused acceleration of sinkhole development in the overlying dolomite; they report that catastrophic sinkhole development has caused great property damage, the engulfment and collapse of large buildings, and the death of 29 people in 1962 and 5 in 1964. According to Foose,[8] pumping from a limestone quarry in the Hershey Valley of Pennsylvania caused sinkhole development in the surrounding populated area, and withdrawal of ground water in a part of Birmingham, Ala., caused local subsidence.[9] Almost all of the external hydrological effects of dewatering a mine are to some degree harmful rather than beneficial. The potential harm needs to be considered before mining operations begin.

Perhaps the development of hydrologic problems in mining has been due less frequently to their having been overlooked than to the operators having taken a wait-and-see attitude. Of increasing importance is the need for a *mine hydrology framework and plan;* such a framework and plan would be continually upgraded as stages in mining develop. The report or document would (1) itemize the pertinent hydrologic principles, (2) depict the hydrogeologic framework, and (3) express (perhaps as tendencies) the interrelated hydrological and environmental factors for both natural, or pre-mining, setting and for the active changes during mining. It would give hydrologic consideration to anticipated actions, consequences of actions, remedial plans, and contingencies. An early phase of the plan might be the preparation of a matrix, as a chart, including on one axis the actions which cause hydrologic impact and on the other existing environmental or mining conditions that might be affected. A schematic matrix not restricted to mining but showing environmental impact assessment has been devised by Leopold and others.[10] Such matrices should be especially helpful during pre-mining and exploration stages to evaluate the external environmental impact, to facilitate compliance with State environmental regulations, and to provide proper defense in cases of unfounded external complaints concerning changes of the environment.

Neither the internal nor the external hydrologic problems in mining are new, but some of the external ones that were disregarded in the past must now be considered and handled effectively. There can be no simple and inexpensive way to solve many of the hydrology problems related to mining; yet, a program which leads to a *hydrologic framework and plan* will help to prevent some problems, to forewarn and forestall others, and to better manage the inevitable ones.

References

[1] Theis, C.V., "The Relation Between the Lowering of the Piezometric Surface and the Rate and Duration of Discharge of a Well Using Ground-Water Storage," *Transactions*, American Geophysical Union, Vol. 16, Pt. 2, 1935, pp. 519-524.

[2] Worley, M.T., "Groundwater Influx into a Vertical Mine Shaft," *Trans. SME/AIME*, Vol. 223, 1962, pp. 428-431.

[3] Loofbourow, R.L., and Brittain, R.L., "Dewatering Through Wells Before Mine Development," *Mining Congress Journal*, July 1964.

[4] *Ground Water and Wells*, Edward E. Johnson, Inc., St. Paul, Minn., 1966, 440 pp.

[5] Emrich, G.H., and Merritt, G.L., "Effects of Mine Drainage on Ground Water," *Ground Water*, Vol. 7, No. 3, 1969, pp. 27-32.

[6] Peek, H.M., "Effects of Large-Scale Mining Withdrawals of Ground Water," *Ground Water*, Vol. 7, No. 4, 1969, pp. 12-20.

[7] Brink, A.B.A., and Partridge, T.C., "Transvaal Karst: Some Considerations of Development and Morphology with Special Reference to Sinkholes and Subsidence on the Far West Rand," *South African Geographical Journal*, Vol. 47, 1965, pp. 11-34.

[8] Foose, R.M., "Ground-Water Behavior in the Hershey Valley, Pennsylvania," *Bulletin 64*, Geological Society of America, 1953, pp. 623-645.

[9] Newton, J.G., and Hyde, L.W., "Sinkhole Problem in and near Roberts Industrial Subdivision, Birmingham, Alabama, A Reconnaissance," *Circular 68*, Alabama Geological Survey, 1971, 42 pp.

[10] Leopold, L.B., et al., "A Procedure for Evaluating Environmental Impact," *Circular 645*, U.S. Geological Survey, 1971, 13 pp.

Part VI

ASSISTING MINERAL BENEFICIATION

Editor's Comments on Papers 40, 41, and 42

ORE MICROSCOPY

Application of the metallographic reflecting microscope to study opaque ore minerals was introduced in the early part of the twentieth century. This innovation coincided with the development of selective froth flotation as a means of separating complex intergrowths of ore minerals. Thus, identification of the minerals and determination of their intergrowth patterns became an essential adjunct to control of comminution of ores and reagent selection and control.

Evaluation of ore samples for beneficiation entails first the identification of ore and gangue minerals and the determination of their relationships and variations. Thin and polished section microscopy is an essential tool in delineating and solving beneficiation problems. Mineragraphy can establish practical limits to separations, and probable metallurgical recoveries, dependent upon the character of mineral intergrowths and solid solution phenomenon (Edwards, 1960; Groves and Whittle, 1976; Ramdohr, 1969).

Polished section techniques are applied with equal success to coal. Reflectance in oil enables identification of constituents (macerals, ash, and sulphides) and allows prediction of the effects of beneficiation processes for the removal of unwanted ash content and iron sulphide minerals. It also permits prediction of coking, hydrogenation, and liquifaction characteristics that cannot be easily determined from standard analyses, thus saving costly testing (Fisher, 1942; Schapiro and Gray, 1964).

The fundamental principle in minerals beneficiation is first to reduce the ore as mined to a particle size, so as to cause optimum liberation of the valuable minerals, then to separate the valuable from associated valueless minerals. The type of data sought will include the following:

- range of grain sizes of ore and gangue minerals;
- percentage of free ore and gangue minerals at various sizes;
- percentage of various types of composite mineral particles at various sizes;
- states of oxidation, hydration, and replacement in minerals;
- presence of surface coatings on mineral grains; and
- presence of exsolution and/or solid solution phenomena.

Schwartz (Paper 40) was one of the earliest investigators to recognize the importance of this tool to the metallurgist. The geologist at the time was more interested in using the technique to throw light on paragenetic relations and ore genesis problems than in assisting the metallurgist. As mining has progressed to lower-grade ores, metallurgical losses due to entrainment of ore minerals in gangue has become an important factor in property evaluation and in maintaining concentrator output. Likewise, recovery of ore minerals may be severely reduced by the presence of oxide coatings. Certain portions of ore bodies may require specialized treatment. Thus, increasing dependence has been placed upon the geologist/mineralogist to recognize potential metallurgical problems and to alert the metallurgist. Amstutz (Paper 41) illustrates some of the intergrowth textures and defines many of the problems that cause difficulty in effecting mineral liberation.

Agarwal, Schapiro, and Mallio (Paper 42) emphasize that prediction of metallurgical behavior based upon petrographic examination should begin with the exploration phase and that it can be used to select appropriate processing. Petrographic and mineragraphic usage is not confined to sulphide ores but applies equally as well to beach sands, iron ores, manganese nodules, and coals. Prediction and examination are essential tools of the mining and exploration geologist.

REFERENCES

Edwards, A. B., 1960, *Textures of the Ore Minerals,* Australasian Institute of Mining and Metallurgy, Parkville, Victoria, Aust., 242p.

Fisher, C. H., 1942, Hydrogenation and Liquifaction of Coal. *U.S. Bureau of Mines Tech. Paper 642.*

Groves, D. I., and A. W. G. Whittle, 1976, *Mineragraphy Applied to Exploration and Mining,* Australian Mineral Foundation, Adelaide, S. A., 176p.

Ramdohr, P., 1969, *The Ore Minerals and Their Intergrowth,* Akademie-Verlag, Berlin, 1077p.

Schapiro, N., and R. J. Gray, 1964, The Use of Coal Petrography in Coke-Making, *Inst. Fuel Jour.* **6:**234–242.

40

Reprinted from *Eng. and Mining Jour.* **116:**237-238 (1923)

SOLVING METALLURGICAL PROBLEMS WITH THE REFLECTING MICROSCOPE

By G. M. Schwartz

THE RECENT application of the methods of metallography to the study of the polished surfaces of opaque ores has led to a knowledge of ore deposition, enrichment, and alteration which has been of great value to the mining geologist. Descriptions of the geology of ore deposits are now scarcely considered complete without a study of the polished ores. The growth of this method has been rapid, as practically the first paper published was by William Campbell[1] in 1906. The publication of the first comprehensive paper on the copper ores by Graton and Murdoch[2] in 1913 has been followed by a number of papers dealing with many kinds of ores. These papers have dealt principally with the geologic aspects of the work, and the mining and metallurgical engineer is not as familiar with the value of the results as might be wished.

Several instances in which a study of polished surfaces has had a bearing on metallurgical problems have recently come to my attention. These have emphasized the possibilities of the work as an aid to the metallurgist, and it is proposed to indicate, by example, some of them. A detailed description of the methods of polishing and identification would be out of place here. These are treated in some detail in the latest book of determinative tables by Davy and Farnum.[3] Some suggestions on the general problem, more particularly as applied to mining geology, were made in an article by Ray.[4] The only paper dealing with the purely metallurgical side is by Head,[5] who applied the method to a study of head and tailing samples of a mill with special reference to data on treatment of the ore by flotation.

Some of the specific questions of the metallurgist regarding an ore, which may be answered by a study of polished surfaces, are:

1. What minerals carry the values? From this knowledge inferences as to the treatment may be made based on the chemical composition of the minerals.

2. What minerals are associated with the ore-bearing minerals? Several inferences as to the possibility of separation may be made with this data.

3. What is the state of aggregation of the various ore minerals in a complex ore? The mesh to which the ore must be ground may be inferred where differential flotation or other processes are to be used.

4. What is the textural relation of the ore and gangue minerals? From this the possibility of separation by mechanical means may be determined. The three examples described below indicate specific applications of data obtained in this way.

In examining copper-magnetite ores on which extensive magnetic concentration tests had been made with poor results, an unusual copper mineral was found associated with chalcopyrite. By means of the tables mentioned above the mineral was identified as chalmersite ($Cu\ Fe_2S_3$). This mineral is magnetic, and,

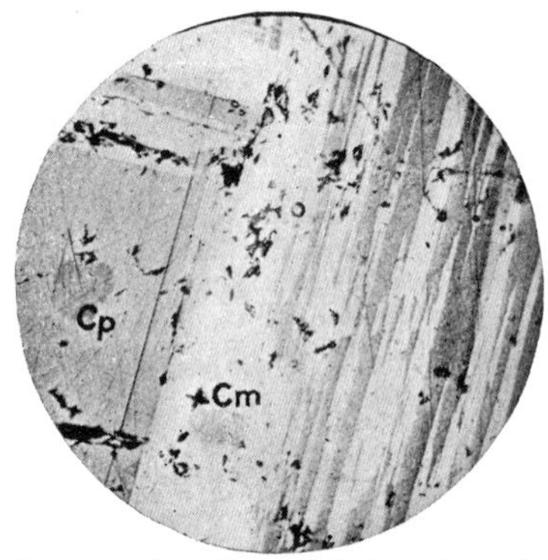

Fig. 1—Micrograph of the polished surface of an intergrowth of chalcopyrite (Cp) and chalmersite (Cm)

Note the fineness of the blades on the right-hand side. Magnification X75.

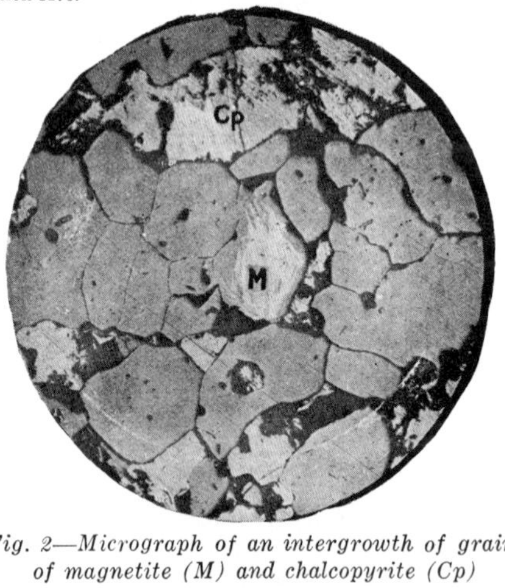

Fig. 2—Micrograph of an intergrowth of grains of magnetite (M) and chalcopyrite (Cp)

Here the texture is coarse and relatively coarse grinding may be expected to allow the separation of the two, but as chalmersite, as shown in Fig. 1, is intergrown with some of the chalcopyrite, magnetic concentration cannot be used. Magnification X70.

furthermore, associated with it and chalcopyrite were considerable amounts of pyrrhotite (Fe_3S_4), which is also magnetic. Neither of these minerals could be seen in the hand specimen with ease, and the chalmersite could not be detected with a hand lens. The chalmersite and chalcopyrite formed a most intimate intergrowth (See Fig. 1). On the contrary, the mixture of copper sulphide and magnetite was rather coarse (See Fig. 2). With the above data at hand it becomes ob-

[1]Campbell, W. "The Microscopic Examination of Opaque Minerals," *Economic Geology*, Vol. I, 1906, p. 751.
[2]Graton, L. and Murdoch, J., "The Sulphide Ores of Copper." *Trans. A.I.M.E.*, Vol. 45, 1913, pp. 26-81.
[3]Davy, W. M. and Farnum, C. M., "Microscopic Examination of the Ore Minerals." McGraw-Hill Book Co., 1920.
[4]Ray, J. C., "The Reflecting Microscope in Mining Geology and Metallurgy." *Min. & Sci. Press*, June 6, 1914.
[5]Head, R. E., "Quantitative Microscopic Determination of Chalcopyrite, Bornite, and Pyrite, in a Porphyry Ore." *Reports of Investigations*. Bureau of Mines, June, 1921.

vious that magnetic concentration tests are unnecessary, as at best this method could give only an imperfect separation of the copper and iron, inasmuch as both chalmersite and pyrrhotite would tend to go with the magnetite. But Fig. 2 strongly indicates that relatively coarse grinding, followed by flotation, would probably give a reasonable separation. Tests could be carried out to determine the commercial possibilities of the process.

In another case the metallurgists desired to know the form in which silver occurred in a disseminated and siliceous lead-silver ore in which silver minerals could not be detected with a hand lens. Specimens assaying as high as 100 oz. of silver per ton apparently contained no silver minerals. Polished surfaces of the galena in the ore, when etched with nitric acid, showed inclusions of argentite (Ag_2S) which remained unetched. (See Fig. 3.) It was not at all certain that all of the silver in this ore occurred as the sulphide, but further work might serve to clear up the point. The knowledge that even part of the silver occurs as definite grains of sulphide should be of value. Even the information that much of the silver occurred so finely disseminated as to defy identification might assist the metallurgist in his problem.

In the third example a low-grade iron ore had defied attempts to concentrate it by the usual washing process, although apparently the material was similar to other ore which had been treated successfully. Examination of the polished surfaces of some of this ore showed fine grains of silica mixed with the iron oxide. (See Fig. 4.) Measurement of these grains showed them to average about 1/200 of an inch in diameter. To separate the silica grains from the iron oxides it would be necessary, presumably, to grind the material to about 200 mesh. This, on such an ore, is commercially impracticable, and there is little need of continuing the tests.

It should be emphasized that the application of the study of polished surfaces to metallurgical problems will not solve the problems, but it should give information invaluable to the metallurgist. In certain cases studies of this sort will indicate clearly that some methods are impracticable and that others are much more likely to succeed, and in any event the information obtained will enable the worker to proceed with a clearer understanding of the problem with which he is confronted. Often much time and money might be saved by a little spent in a preliminary microscopic examination of the ores to be treated.

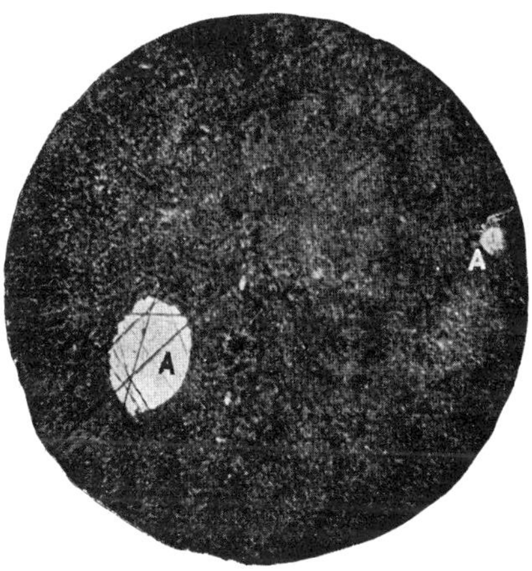

Fig. 3—Micrograph of an etched surface of polished galena showing inclusions of argentite (A)

Argentiferous galena often carries the silver in this form. Magnification X140.

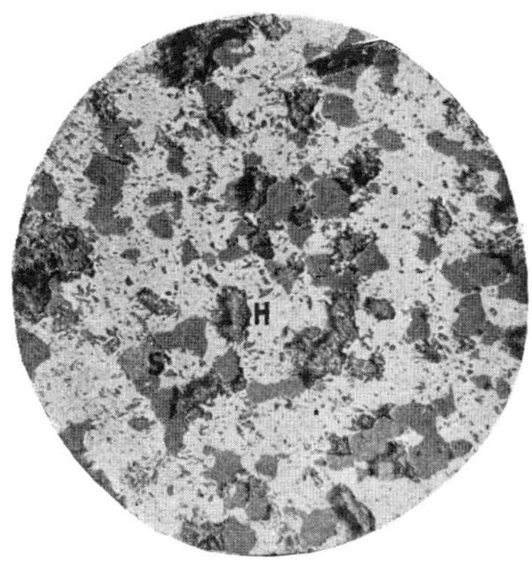

Fig. 4—Micrograph of an etched surface of lean iron ore showing silica grains (S) and iron oxide, chiefly hematite (H)

The silica grains are 1/200 of an inch or less in diameter. Magnification X150.

41

Reprinted from *Mining World* **24:**19–23 (1962)

HOW MICROSCOPY CAN INCREASE RECOVERY IN YOUR MILLING CIRCUIT

By Dr. G. C. Amstutz,
Professor of Economic Geology,
Missouri School of Mines and Metallurgy,
Rolla, Missouri

Geometric Classification of Basic Intergrowth Patterns of Minerals

Type 1a: Simple intergrowth or locking type; rectilinear or gently curved boundaries. Most common type, many examples.

Type 1b: Mottled, spotty, or amoe-[illegible]-type locking or intergrowth. [illegible]ple, common pattern; many examples.

Type 1c: Graphic, myrmekitic, or "eutectic" type. Common; examples: chalcopyrite and stannite; quartz and feldspars; etc.

Type 1d: Disseminated, emulsion-like, drop-like, buckshot or peppered type. Common; examples: chalcopyrite in sphalerite or stannite; sericite, etc. in feldspars; tetrahedrite in galena; etc.

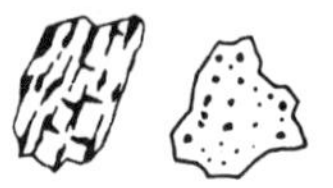

Type 2a: Coated, mantled, enveloped, corona-, rim-, ring-, shell-, or atoll-like. Common; examples: chalcocite or covellite around pyrite, sphalerite, galena; etc.; kelyphite rim, and other rims.

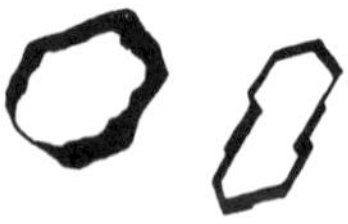

Type 2b: Concentric-spherulitic, or multiple shell-type. Fairly common; examples: uraninite with galena, ch[illegible]yrite, bornite; cerussite-limo[illegible]e; Mn- and Fe-oxides; etc.

Type 3a: Vein-like, stringer-like, or sandwich-type. Common; examples: molybdenite-pyrite; silicates; carbonates; phosphates; etc.

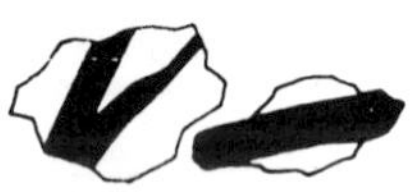

Type 3b: Lamellae-layered, or polysynthetic type. Less common; examples: pyrrhotite-pentlandite; chlorite-clays; etc.

Type 3c: Network, boxwork, or Widmanstätter-type. Less common; examples: hematite-ilmenite-magnetite; bornite or cubanite in chalcopyrite; millerite-linneite; metals; etc.

Between most of these nine common locking types there are naturally gradational transitions with regard to both, pattern and size. Particle or grain size data are a prerequisite of any accurate study of rocks and mineral deposits and enhance the value of this chart.

Forty to 50 years ago the average grade of most milling ores was considerably higher than it is today, and the concentration of minerals, as a whole, was a much smaller section of the mineral industry than today. During the last 50 years the concentrating departments of mining companies have gradually grown to about the same size or even to a larger size than the mining departments. This fundamental change, largely caused by the introduction of flotation, may well be considered a major revolution in the field of the mineral industry.

This change has made the close cooperation between the ore dressing engineer and the geologist-mineralogist more vital than it was before. During the last century, when the concentration methods were almost only megascopic, the geologist and the metallurgist did not find too many technical reasons for close team work.

Today the bulk of almost all concentration processes is so to say "microscopic", because the size of the particles is within the size range of microscopic observation. This is perhaps the major reason why, today, microscopic testing has become an indispensable tool for any ore dressing work. The practical size range in flotation corresponds to the practical size range of microscopy. Although, today 90 to 95 percent of all flotation microscopy is still done with visible light, there is no reason why electron- and X-ray-microscopy should not be included in the term of ore dressing microscopy. But, since this article is dealing mostly with mill control methods available to an average mill, only those microscopic methods using visible light will be mentioned.

It should perhaps be mentioned at this time that someone who is familiar with the principles of microscopic materials testing, may think, that chemical assays are all that is needed for flotation control. However, a chemical analysis indicates only the bulk chemical composition. It does not tell anything about the many mineralogical and geometric properties affecting the choice of reagents, the size of grinding and regrinding, the limitations of liberation due to the size and type of locking, or the distribution of common and rare elements among the various minerals; for example the presence of marmatitic sphalerite.

A microscopic analysis has the important additional advantage over a chemical analysis, that it requires much less time. Whereas it usually takes 12 to 24 hours to obtain chemical results, a mineralogical test can be made in 20 to 60 minutes, depending on the methods used. In some respects, however, chemical and mineralogical analyses can not replace each other. An accurate chemical analysis should of course always be obtained if a new type of ore is investigated. In ores of unknown composition a chemical analysis speeds up the mineralogical testing work considerably. In much of the routine testing, however, microscopic observations are in many cases entirely satisfactory or even superior. Whereas, on one hand, the quantity of lost metal may not be quite as accurately determined with microscopic testing, it shows, on the other hand, normally **what caused the loss.** The reason why some ore dressing engineers still need to be sold the idea of using a microscope for routine testing lies probably in the fact that

Locking pattern sets mill recovery from these ores

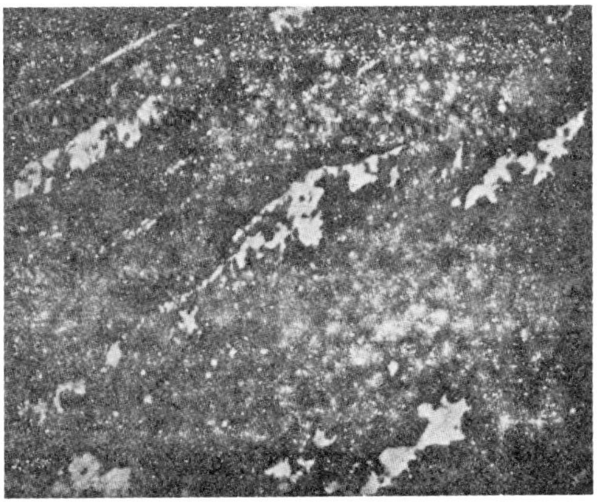

Layered copper-lead-zinc ore from Rammelsberg, Germany. Mineralogic composition, grain size, locking sizes and types can be seen before grinding and testing. – 60 ×.

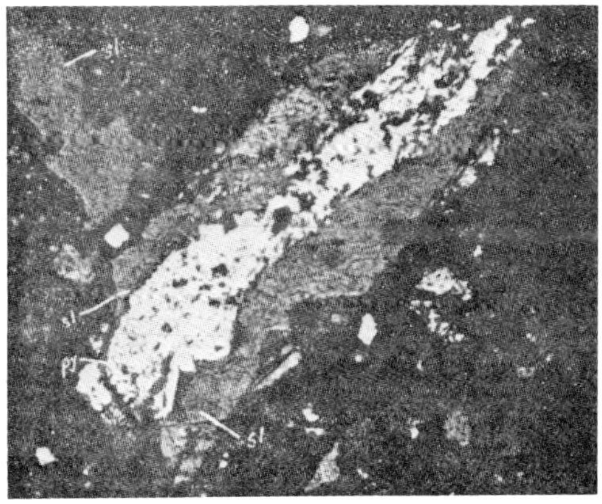

Simple vein-type locking between sphalerite (sl) and py[illegible] (py) in a Cerro zinc concentrate; 48 mesh particle. Regrin[illegible] liberates sphalerite. Bakelite matrix, – 600 ×, oil immersion.

Sphalerite particle with blebs of 5 micron-chalcopyrite which cannot be liberated. Grey corner of galena. Typical locking from Ticlio area, Peru. – 600 x, oil.

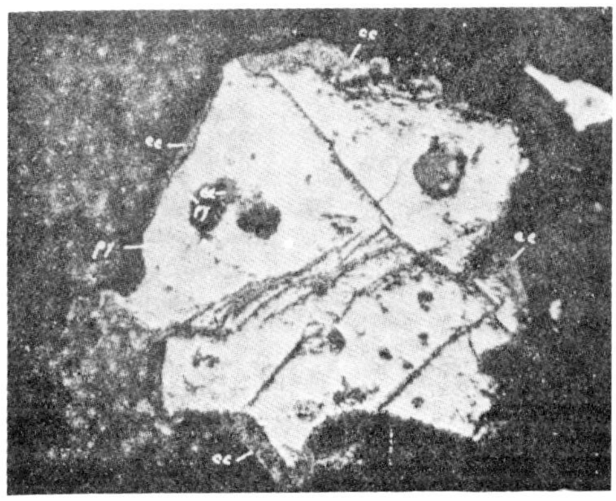

Ring-type locking (type 2a) of chalcocite (cc) coating [illegible]rite (py) causing high iron in copper concentrate. Cuajone porphyry copper, Peru. – 600 ×, oil.

Pyrite particle veined by chalcopyrite. Type 3b locking. Morococha, Peru. Further grinding will not liberate copper; veinlets too small. – 600 x, oil.

Ring-type locking between sphalerite dark grey and chalcopyrite (whitish). Liberation difficult. Morococha, Peru. – 600 ×, oil.

they don't know how simple it is. They do not realize that they can do it themselves and that the few minutes time they may need to look at the products under a microscope comes back manifold in a quicker and better solution to the troubles. As a matter of fact, the microscope tells many fundamental things which create a better basic understanding of what is going on in the different steps of flotation, or any other concentrating method. If the microscope is used the trial and error method does not have to be applied so often.

Microscopy's two key clues: mineralogy and space geometry

The information available from a microscopic study may be subdivided into two distinct groups: 1. mineralogic or compositional information; and 2. space or geometric information. Both these groups may be further subdivided into qualitative and quantitative data.

One or two of the compositional mineralogic properties which can be seen with the microscope will be mentioned and some more are shown on the figures.

An important piece of information is always the degree of oxidation of a sulphide ore. Another vital observation is the presence, or absence, of objectional minerals, such as soluble salts, sheet shaped minerals like clay, graphite, or covellite, or the presence of phosphorous, sulphur, or arsenic in iron oxide ore, or of bismuth in lead ores.

A common problem which can be answered rapidly h a microscope is the form of the iron in zinc concentrates. Whereas in one case the iron may be entirely locked up as marmatitic sphalerite, in another ore it may be present as locked or even as free particles of iron sulphides.

The second type of information, which only the microscope can reveal, consists of the space or geometric properties of the minerals which enter the concentration process. Ores are aggregates and the ore minerals are usually tightly interlocked with each other or with gangue minerals. The liberation of the valuable minerals for flotation is therefore largely a function of the size and geometric type of intergrowth.

Both, the size of locking and the pattern of intergrowth are normally observed without difficulty under the microscope. Locking size and locking pattern information on new ore types can cut down the number and consequently the cost of laboratory and pilot plant testing considerably.

In one case overgrinding may be avoided. In another ore it may be clear from a microscopic study, to begin with, that a separation of two minerals may never be economically feasible. In the case of the copper-nickel-cobalt ore of Fredericktown, Missouri, this was true with regard to the chalcopyrite and the very minutely interlocked nickel-cobalt mineral siegenite.

Since the locking pattern is also essential, a chart of important locking or intergrowth patterns was set up some years ago. This chart is reproduced on the first page of this article.

Perhaps it is in order to explain why different locking types may affect the course of flotation. Let us assume that we have two mineral species, for example 80 percent sphalerite and 20 percent chalcopyrite. If the chalcopyrite occurs as exsolution blebs in sphalerite, which is rather common, the particle will float as sphalerite and carry the copper and iron into the zinc concentrate. If, however, the chalcopyrite surrounds the sphalerite as a

continued on page 22

Who can do the microscopic work for three basic investigations

We may ask now who is performing this microscopic analysis or testing. The study of common minerals and less complex problems can, in general, be performed by the flotation engine' or technician himself. Special problems and minerals requ_ the special training and knowledge of a mineralogist-geologist.

Then, the time or occasion at which microscopic work is needed, can be subdivided into three basic groups:

1. Periodic routine control to insure best performance and recovery.
2. Investigation of new ore types (before concentration).
3. Development of new or improved recovery methods.

The control listed under No. 1. and perhaps occasionally also under No. 2, should be able to be carried out by a person without special scientific training. For case No. 2 and usually No. 3, a specialist has to be hired or consulted.

At the Missouri School of Mines and Metallurgy, the Colorado School of Mines, the Royal School of Mines in London, England, and at various other mining colleges, a course in ore dressing microscopy is taught which intends to prepare the ore dressing engineer for routine control work of his plant.

In addition, at the Missouri School of Mines and Metallurgy and also in a few other colleges, courses are taught which offer special training for more detailed and complex testing and research work on the nature of ores.

It goes of course without special mentioning that for microscopic as well as for chemical analyses proper sampling is a prerequisite.

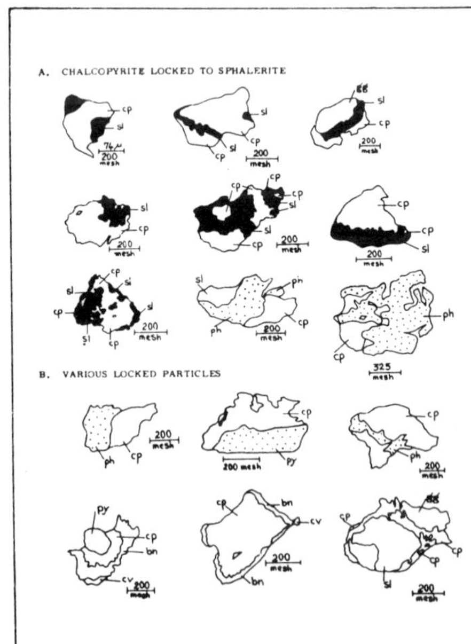

COMMON LOCKING types in Tennessee Copper Company's copper ores from Ducktown, Tennessee mines.

continued from page 21

last generation coating, this particle will of course migrate into the copper concentrate which will then show an excess of zinc.

There are, of course, many more examples of geometric properties of the mineral aggregates as well as of individual free minerals, which affect the flotation properties and results. Some of them are shown in the two pages of photomicrographs.

Both, the compositional as well as the geometric information can be obtained in a purely qualitative way or in a partly or completely quantitative manner. We can measure or estimate the mineralogic and geometric properties just mentioned.

Sample preparation is easy; direct observation often used

The flow sheet of microscopic flotation control is relatively simple. If a specialized microscopist is at hand, the mill man will obtain the carefully selected sample for him. He fills out a problem or question sheet for the microscoper. If available, the results of chemical analyses should be attached.

The microscopist then prepares the samples and makes the necessary observations. One or more conferences between the flotation engineer and the microscopist or consultant may be needed in order to clarify some of the specific details and observed properties of the ore minerals. In general it is of great advantage if the concentration engineer and the microscopist can sit down at the microscope and talk over the specific properties of an ore as revealed by the microscope.

A lot could also be written about instrumentation and preparation of polished section. Only the essential principles will be mentioned.

There are methods and microscopes available for a direct observation of the minerals as removed from the flotation cell. With the aid of a Petri-dish and an immersion system for water, or any other liquid, observations on coatings, grain size, locking, liberation, and composition, etc. can be made without briquetting and polishing. A stereoscopic, or an Ultropak miscroscope can be set up in a clean corner of the mill, perhaps in the control room, and routine observations can be performed within a few minutes.

But even the methods using polished surfaces are not time consuming. With the diamond abrasive method a good polished section can be obtained within 20 to 30 minutes and the sample is also ready for observation.

It is of course clear that the time factor is of prime importance. The flotation engineer often does not have time to wait many hours for an answer on a control problem, because, while he is waiting, a large tonnage of ore may be flowing through the mill and a lot of valuable material may get lost. By the time the results of chemical analysis are received much can go wrong and be lost.

In addition or in connection with the microscopic ing methods there are naturally many auxiliary methods which are not discussed here. Among the most important tools are heavy media, magnetic or electrostatic separations of particles before, or, after microscopic observation. X-ray powder methods or X-ray thin section photography may also be applied for difficult cases.

To make available the many valuable papers on special methods and problems to those interested an extensive bibliography has been prepared and published in the Colorado School of Mines Quarterly, Volume 56, Number 3, July 1961, *Fiftieth Anniversary of Froth Flotation in the U. S. A.* Many Russian papers are lacking which I regret, because they are experts in both, flotation and microscopy.

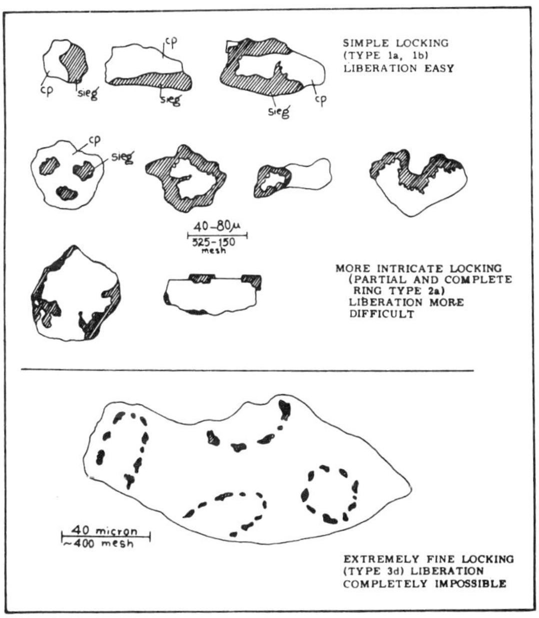

INTIMATE LOCKING between chalcopyrite and siegenite. Ore from National Lead Company's Fredericktown, Missouri mine.

What is the cost of testing, and buying a complete laboratory?

Perhaps, it should be mentioned at this time that the quality of the microscopic testing work in many cases also a function of the understanding the ore dressing engineer or the microscopist has of the geological nature of the deposit, such as the degree and nature of oxidation or brecciation, etc. If a mineralogist-geologist is employed, he ought to work in close touch with the flotation engineer on one hand, and with the mine geologist and engineer on the other hand. He is a most important liaison officer who can save a company a lot of trouble and consequently he can cut the cost of flotation and avoid excessive metal losses in tailing.

The questions are often asked, how much does microscopic testing work cost? What does a microscopic testing laboratory cost? The cost of testing work by a consultant depends very much on the nature of the problem. A simple problem may be solved in a few hours, whereas complete and complicated investigations may take several days.

A complete microscopic laboratory with a stereoscopic and a regular microscope, polishing equipment and materials costs between $4,000 and $8,000.

Typical mineral arrangement of South American ores

Cerro lead-zinc heads (– 200 + 270 fraction). Counting free and locked particles, locking size and type, etc. greatly assists planning and checking. Some minerals are persistently ·, other consistently locked.

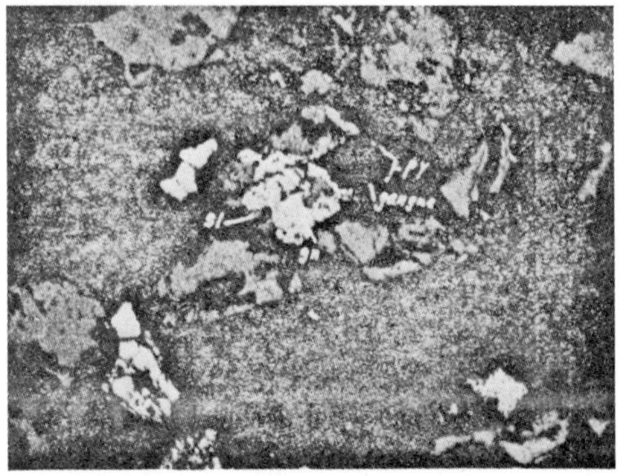

Cerro lead-zinc heads (– 100 + 150 fraction). Some composition as in picture to the left, but more locking present. Finer grinding will liberate many ore minerals. (Gangue always as dark as leucite matrix.)

Irreaular chalcopyrite (cp) particle coated by covellite (cv). Th nd figure below typical for porphyry copper deposits; espe.ially the supergene enrichment zone, 600 ×, oil.

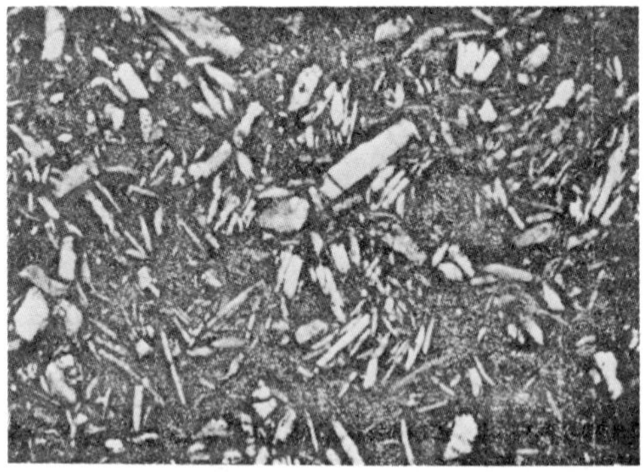

Molybdenite concentrate, El Teniente mine, Chile. High degree of liberation. Only traces of pyrite and chalcopyrite locked to MoS_2. – 180 ×.

Chuquicamata copper concentrate. Replacement of chalcopyrite by covellite typical for the "secondary sulfide" or enrichment zone. Locking is not detrimental. – 400 ×, oil.

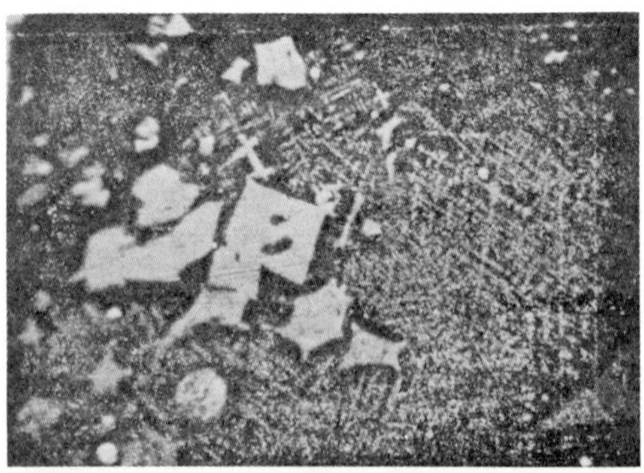

Copper reverb slag with converter and reverb magnetite and copper as delafossite, native and sulfide globules. Liberation and flotation impossible. – 600 ×, oil.

42

Reprinted from *Mining Cong. Jour.* **62**(3):28-35 (1976)

Process Petrography and Ore Deposits

By J. C. AGARWAL
Director, Development

N. SCHAPIRO
Group Leader

and W. J. MALLIO
Geochemist, Applied Mineral Technology Group, Ledgemont Laboratory, Kennecott Copper Corp.

Until recently, it has mostly been possible to determine what constitutes an ore by considering ore grade, extent of reserves, transportation services and the political structure. These considerations were generally made on the assumption that conventional concentration and smelting technology would be utilized.

What constitutes an ore?

Past	*Present*
Grade	Grade
Tonnage	Tonnage
Location	Location
Political structure	Political structure
	Environmental constraints
	• Sulfur disposal
	• Trace elements
	Hydrometallurgical
	Alternative
	Energy requirements

Processing options now available through the development of hydrometallurgical processes, coupled with lower ore grades and new environmental constraints, make it necessary to answer more questions about a deposit before the optimum extraction technology can be selected. In order to meet the needs of selecting the best process scheme for new or unconventional deposits, the materials must be evaluated so that some predictions can be made regarding their behavior in extractive processes.

In this connection, let us look at process petrography as a necessary part of developing an ore deposit into an economic success.

PROCESS PETROGRAPHY

What is process petrography?

On the whole, scientific theory permits us to predict reasonably well any process change which occurs in a liquid melt or in a gas, if we know the chemical composition, temperature and flow conditions. However, theory is as yet unable to predict the behavior of solids under given process conditions, because solids are infinitely more complex than fluids, to the extent that their surface geometry is often as significant as their chemical compositions.

In the mineral industry, most processes occur between solids, or by solids interacting with liquids and gases. Therefore, with the exception of reactions in the molten or gaseous states, the results of mineral processes can only be predicted from empirical data. Lacking an insight into what occurs at the surface and in the interior of particles, empirical data collection has had to follow a "black box" approach, involving costly experimentation on a large scale at several levels of the significant variables and sometimes costly failures.

Modern process petrography supplies such an insight (fig. 1). It permits us to observe specific changes in texture, microstructure, placement and association of substances within aggregates, and physical properties; to relate them to changes in process conditions; and thus, even after only small-scale experiments, to predict process performance and identify optimal conditions.

Petrography per se is not new. Its main application, however, has traditionally been geochemistry, the historical reconstruction of how a given ore has been processed by nature. Process petrography uses the same tools, but in a predictive mode, to identify the impact of individual

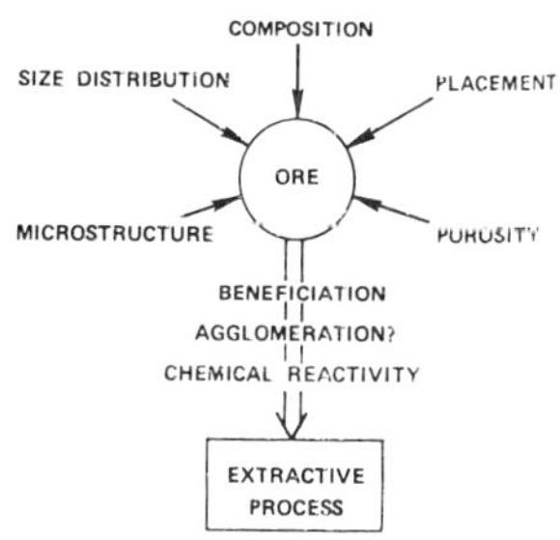

Fig. 1. Petrographic parameters as they relate to important processing parameters

process variables. Used in this fashion it is still quite new, and although the bulk of all mineral development work is still performed with a "black box" approach, process petrography is being increasingly used by the leaders in the industry.

This type of study has been of great value in the pre-pilot plant and pilot plant development work where the identification of unusual behavior of a particular material means substantial savings in money and time in future processing steps. The raw materials appropriate to this type of examination are ores, slags, coal and carbon, fluxes, refractories, portland cement, etc. and even plastics.

Of particular interest is the application of process petrography to the increasingly important field of hydrometallurgical processing. Regardless of how performed—leaching ore in a reactor or in-situ or leaching a concentrate—it involves contacting a heterogeneous mixture of minerals of irregular geometry with a leach solution. Without process petrography it is difficult to gain an understanding of the impact of process changes on the numerous physical phenomena and chemical reactions that occur simultaneously.

Tools of process petrography

Petrography is not limited to optical microscopy, although optical techniques are a primary means of observing changes in materials during processing. Rapid quantitative determinations are made by comparing unknowns with standard reference materials using physical methods of analysis. Optical microscopy is the technique which ties together the small scale variations in chemical distributions, such as those observed using the microprobe with the texture and placement of phases developed in the process. Optical examination must always be the first step in the application of any micro-analytical technique. Skills in the use of simulation studies, x-ray spectroscopy, electromicroprobe, thermal analysis, stress-strain tests, etc. are also necessary. Extensive process experience is, of course, a "must."

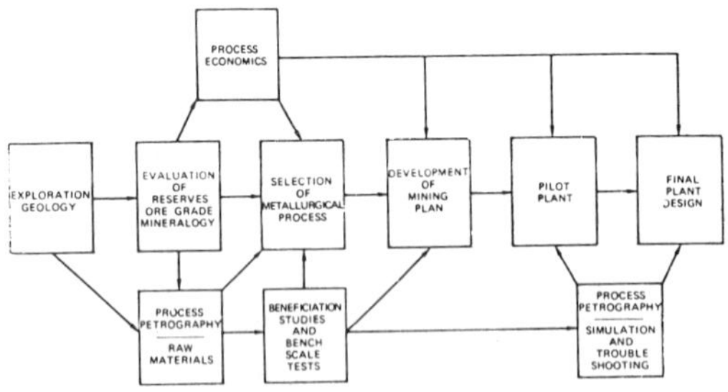

Fig. 2. Development of an ore deposit.

EXPLORATION AND DEVELOPMENT OF A MINERAL PROSPECT

Although the initial examination and preliminary evaluation of deposits is primarily the responsibility of exploration groups, petrographic study should also begin at the same time (fig. 2). During the exploration phase, while the ore reserve is being estimated and a preliminary mineralization inventory established, it is necessary to anticipate potential processing difficulties in order to scope the order of magnitude of the investment and operating cost from ore-in-the-ground to finished product. This requires that various qualitative and quantitative data related to the mineralogy and texture of the deposit be developed in detail. This would include the identification of the ore minerals, the valueless metallic minerals and the gangue minerals, their individual chemical compositions and intergrowths, parameters that can never be established by assay values. Quantitative determination of their abundances would also be made.

The quantitative data would then be combined with the large scale chemical data from assays of drill samples. A parallel effort would be going on with process engineers in selecting the best process for extracting metal values from the ore and starting beneficiation studies, as indicated in fig. 2.

For example, questions asked by process engineers which must be addressed include:

1) How is the element of interest partitioned among the phases?
2) What is the particle size of the ore minerals and how are they intergrown?
3) What trace elements are present that might be environmentally damaging and how can they be disposed of?
4) What will the chemical and physical nature of the residue be after processing?
5) What gangue minerals would react in hydrometallurgical processes to consume acids or other active chemicals?
6) What reactions between the gangue and pregnant solutions could occur in hydrometallurgical treatment to remove a valuable element after it has gone into solution?
7) Can certain process steps be simulated to gain engineering information and anticipate trouble spots?

Since process petrography is oriented to engineering, these questions can now be answered for each deposit at an early stage in the course of development, instead of recognizing the problems piecemeal as they arise, and different process options can be evaluated.

The assistance of process petrography in benefication is invaluable. Petrography supplies an understanding of how an ore fractures when comminuted in different types of equipment, how many distinct kinds of particles form, and how they may consequently be best separated. Working closely with beneficiation engineers can cut down the number of tests by a whole order of magnitude. Yet it is surprising how many beneficiation studies still follow the "black box" approach, in which many combinations of grinds and of reagents are tested in order to arrive at the optimum process.

THE USE OF PROCESS PETROGRAPHY

We will now cite several examples which illustrate the use and importance of process petrography in different stages of development of ore deposits from exploration to final processing.

Copper ore deposits

To show how petrographic information about a deposit can be used, we will discuss two examples. These descriptions are of actual deposits but not necessarily new deposits.

Deposit A. This deposit has the petrographic characteristics specified below:

Grade	0.9% Cu
Mineralogy	chalcopyrite, bornite, pyrite and molybdenite
Grain size	150 microns average for sulfides
Texture	disseminated
Gangue	plagioclase, biotite, chlorite, iron-titanium oxides, calcite, quartz

The photomicrograph displayed as fig. 3 shows chalcopyrite and pyrite as disseminated blebs. Gold and silver are present at levels ~100 ppm.

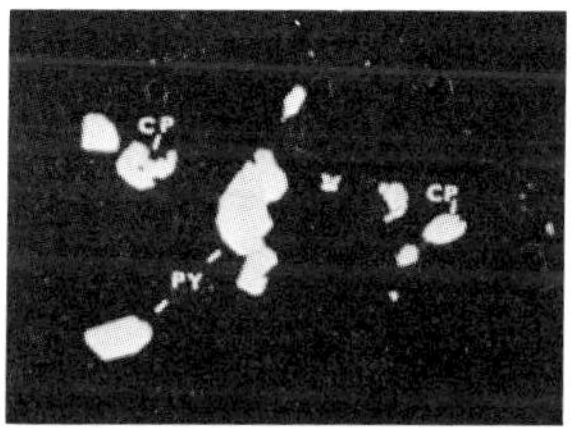

Fig. 3. Photomicrograph showing chalcopyrite (cp) and pyrite (py) dessiminated in rock matrix. Reflected light × 500

The ore minerals are all sulfides. This is important since copper present as oxide or carbonate is not recoverable by flotation. The grain size is a guide to the beneficiation engineer. He can be told that grinding to minus 100 microns should liberate most of the sulfide and a complete knowledge of the size distribution of the sulfides indicates what fraction can be recovered at any given grind.

The disseminated nature of the ore indicates that it is not present in veins but associated with gangue minerals so that some inclusion of the silicates in the concentrate is expected. The gangue mineralogy indicates that biotite and chlorite are most likely to be included in the flotation concentrate. Quartz may also be included if copper levels in the flotation cell exceed ~50 ppm.

In this case, because of the likelihood of obtaining a good concentrate grade (25 to 30 percent) pyrometallurgy will probably be used.

In pyrometallurgy there will only be fringe applications of petrography such as by-product recovery from slag and trouble shooting of plant problems that might arise with furnaces etc. The pyrometallurgical process is principally controlled by the chemistry of the system where appropriate fluxes can be added to produce the desired slag and matte fractions.

Deposit B. This copper deposit is a lower grade deposit. Its petrographic charcteristics are summarized as follows:

Grade	0.4% Cu
Mineralogy	chalcopyrite, bornite, traces of pyrite and molybdenite
Grain size	bimodal for sulfides ~ ⅔ of chalcopyrite with average diameter ~ 200 microns ~ ⅓ of chalcopyrite with average diameter < 40 microns
Texture	vein control of sulfides
Gangue	plagioclase, biotite, chlorite, iron-titanium oxides calcite, anhydrite, orthoclase with chalcopyrite

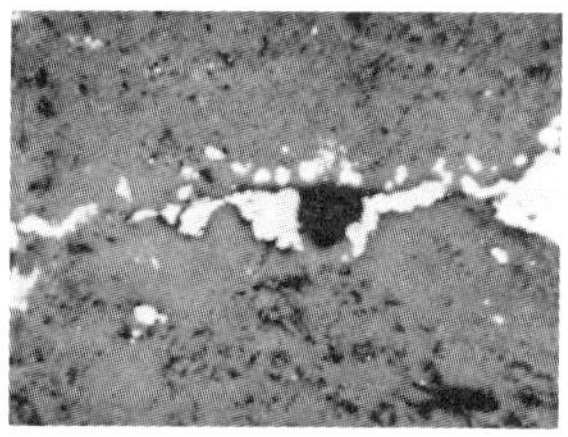

Fig. 4. Photomicrograph of vein chalcopyrite from deposit "B." Reflected light × 500

The texture of the ore from Deposit "B" is illustrated in fig. 4. It shows vein control of the sulfides and the bimodal size distribution of the chalcopyrite.

In considering the information on Deposit "B," it can be predicted that without excessive grinding (that is, to minus 325 mesh) a concentrate suitable for pyrometallurgy would not be obtainable. This is principally because of the bimodal size distribution of the chalcopyrite. If recovery is low and the concentrate grade is also low, then hydrometallurgy will probably be most applicable.

Because hydrometallurgy depends on solubilizing the valuable metals without necessarily destroying the texture of the ore, the size distribution and placement of the sulfide minerals becomes important. The texture as shown in fig. 4 indicated that crushing to minus 100 mesh would expose but not necessarily liberate the sulfides, making them available for leaching.

The petrographic information is critical in choosing an extraction process for a given ore deposit because the mineralogy, composition and texture of the gangue dominate the behavior of the ore in hydrometallurgical processing just as the bulk chemistry dominates in pyrometallurgical processing.

There are various process alternatives within hydrometallurgy for dissolving copper concentrates; these are listed below:

—Sulfate systems
 Concentrated sulfuric acid
 Dilute sulfuric acid and oxygen
 Dilute sulfuric acid and ferric iron
—Ammoniacal systems
—Chloride systems
—Nitric acid systems
—Cyanide system
—Electrochemical systems

The reactivity of the gangue is important because gangue minerals consume active reagents, especially in low grade concentrates. In fact, evaluation of Deposit "B" indicated that an ammoniacal process would be pursued rather than an acid process because of the high levels of calcite and ferromagnesian silicates. In choosing the best processing scheme, a screening procedure is used to test a number of possible systems. One aspect of this is the use of microscopic leaching stages which will be discussed later.

Ilmenite sand deposit

Bulk chemical analysis is limited because it cannot indicate distribution of the elements. In examining an ilmenite sand, a small amount of chromium showed up in the assay. If chromium is present in an ilmenite ore, it can interfere with TiO_2 production even if it is present at levels of ½ percent. The question that must be answered is whether the chromium is present as a highly dispersed trace element in one or more of the major components or whether it occurs concentrated in a separable phase. Large quantities of sample are not needed for this type of study, since the difference in possible modes of occurrence is a fundamental property and not likely to change from place to place. A representative sample of the sand, on the order of several pounds, was satisfactory for this analysis. Metallurgical mounts of the grains were made and studied using various petrographic techniques.

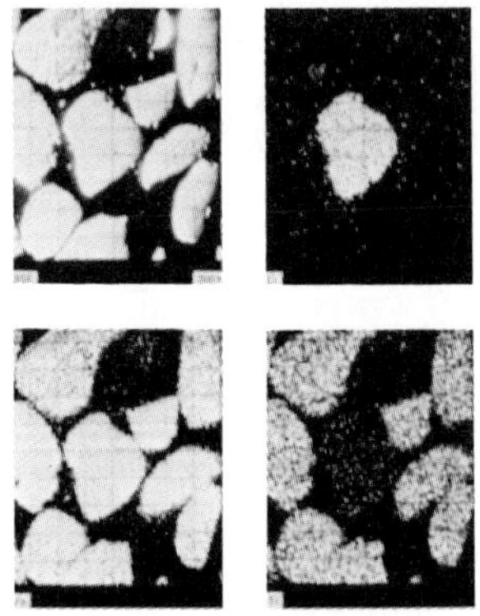

Fig. 5. X-ray scanning pictures showing elemental distribution of Fe, Ti and Cr in ilmenite ore

A petrographic study indicated that chromium is present as a major component of a phase that occurs as only a small percentage of the grains. Fig. 5 shows a qualitative view of the relative chromium and titanium content in a number of grains. This study used the electron probe as a supplement to optical microscopy to determine the distribution of a given element in the ore.

Further study indicated that magnetic separation could be used to separate the chromium bearing phase from the Fe-Ti phases and thus eliminate the processing problem.

Iron ore deposit

An iron ore deposit was developed to provide material for a pelletizing plant. In order to successfully provide this material, it was necessary to know in detail the mineralogical variability of the ore since considerable variations in mineralogy could be present in batches with essentially identical chemical and size analyses.

The plant specifications included chemical and size criteria, but early bench scale pelletizing tests indicated that the mineralogical makeup of the ore had to be known or processing problems would certainly result. Four ore types were recognized:

—*Crustal ore*	hematite in a dense goethite matrix
—*Crystalline goethite*	a pure essentially single phase goethite
—*Specular hematite*	platy crystalline hematite
—*Earthy limonite*	mixture of soft hydrated iron oxides

The various minerals and ore types have properties that cause them to behave differently in various phases of the pelletizing operation. These are differences in grindability, filterability, agglomerating behavior and dehydration at high temperatures. For example, specular hematite is difficult to agglomerate because of its platy shape. The limonites are soft but require a special drying regime because their filterability is reduced due to the high moisture content. Petrographic examination of diamond drill core was used to classify the deposit in terms of its mineralogical zones. From this study, a three dimensional model of the deposit was made to show the distribution of ore types.

A mining plan was evolved using this model which allowed blending of the ore to produce a raw material feed within the practical operating limits of the plant's process control system. A diagrammatic representation of the distribution of ore types (at 2000-ft level) is shown in fig. 6.

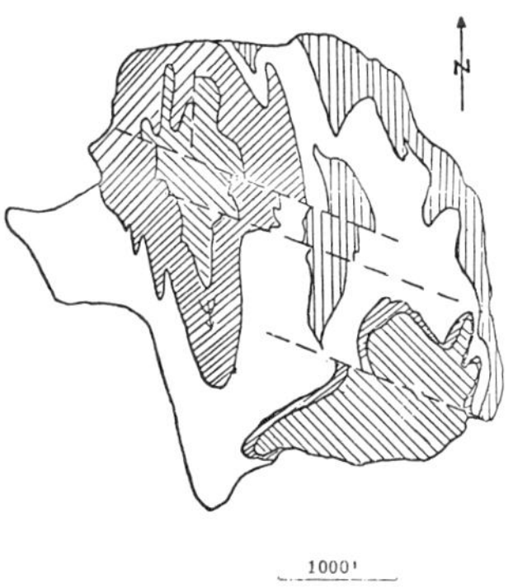

Fig. 6. Iron ore deposit 2000-ft level

Earthy Limonite
Specular Hematite
Crustal Ore
Crystalline Goethite

A general plan of attack resulted from this study. It is one that must be used where variation in chemical analyses alone cannot represent variation in parameters important in ore processing:

1) Contouring of petrographic/physical, mineralogical and chemical properties
2) Definition of physical chemical and petrographic variability within the ore body
3) Selection of mining sites for proper blending at the end of the exploration program and preliminary cost estimates for the development of the deposit can be prepared based on a generalized flow sheet designed for the largest variations expected. The key to successful economic design lies in the blending procedures required to overcome deposit variability in the process operation.

Manganese nodules

Physical beneficiation of manganese nodules (fig. 7) offers a potential means of upgrading the metal values in the ore. Upgrading, of course, enhances the economic attractiveness of any processing scheme. Petrographic studies have shown that there is a potential for upgrading

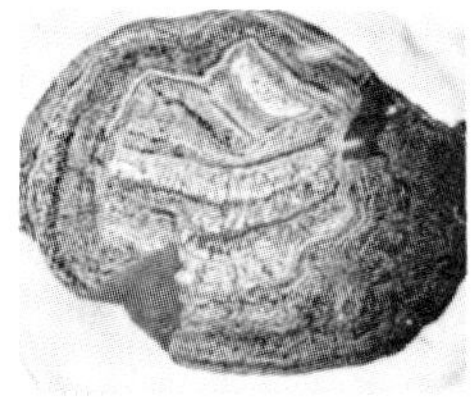

Fig. 7. Typical bonded appearance of a manganese nodule (polished cross section)

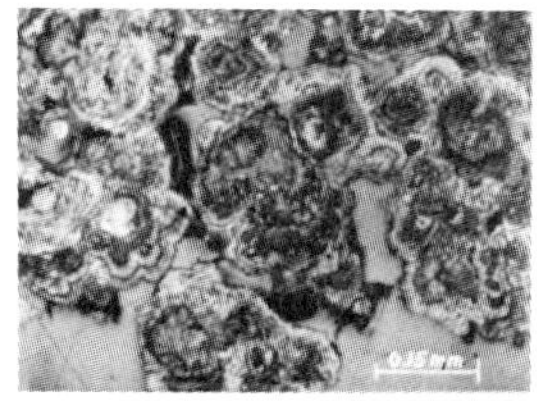

Fig. 8. Photomicrograph of nodule crushed -1/4 in. showing breakage along natural banding

the overall Ni, Cu and Co contents of the nodules, not by fine grinding and concentrating, but by eliminating an undesirable component.

Clays, both illite and montmorillonite mixed with quartz and glassy materials occur in thin layers between the nodule bands and fill natural fractures which cut across the banding. A portion of these clay materials is amendable to removal by physical methods. These include fracturing the ore and washing to remove the loosened clay materials. The nodules pictured in fig. 8 have fractured preferentially along the edges of the natural banding thereby exposing the clays. The finer the crushing the more clay should be exposed, including some of the clay in the fine structure within the bands.

Beneficiation studies

Petrographic study of the interlocking relationships between gangue and sulfide minerals is sometimes necessary in interpreting the results of beneficiation studies.

In beneficiating a nickel-sulfide bearing gabbro, it was found that grinding and froth flotation would produce a concentrate with a high recovery of metal values. However, the concentrate grade was low because it contained large amounts (20 to 70 percent by volume) of gangue.

Petrographic investigation (fig. 9) revealed that the causes of this behavior were related to occurrences of nickel in the ore. Three distinct occurrences were found:

1) Large individual sulfide grains
2) Small nickel sulfides (several microns in size) peppered through some of the silicates
3) Nickel in the silicate phase at a low level.

Fig. 9. Photomicrograph showing fine-grained (10–100 μ) sulfides (black) in plagioclase and large separable sulfide grains

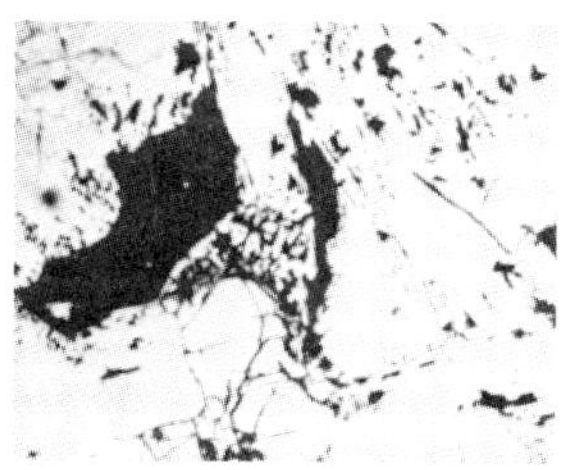

The petrographic evidence therefore shows that there is no hope of achieving a high grade concentrate which is desirable from economic considerations because silicates must be included to obtain high recoveries since some of the sulfide is locked into silicates as fine grains and some is actually present in the silicate structure. Low S to Ni ratios also result from this. These factors led to a choice of a hydrometallurgical processing scheme over a pyrometallurgical process.

Processing

Prediction of process effects on coal. One of the oldest and most successful applications of process petrography has been its use in predicting the properties of coke that will result from studying the petrographic properties of coal.

The carbonization characteristics of the individual entities in the coal determine the nature of the coke produced. Basically, coal consists of reactive and inert materials (fig. 10). The reactives become plastic and solidify to form a bond; inerts remain relatively unaltered and act as aggregate. The strength of the bond formed by the reactives and the ability to assimilate inerts varies with the rank of the coal. A knowledge of the percentages of reactives and inert entities in a coal, together with rank of the reactives, can be used to calculate the strength of the resultant coke.[2]

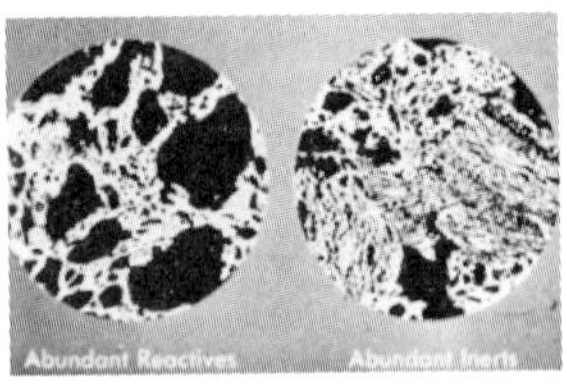

Fig. 10. Effect of abundant reactive and inert entities on coke microstructure. Knowledge of the percentages of these entities is useful in calculating coke strength

Variables such as coal moisture content, pulverization level, oven bulk density, heating rate and oven width, which are independent of the coal itself, have effects on the resultant coke strength. In correlating petrographic data with an acceptable coke-evaluation test, processing variables are minimized. Since the coking characteristics of coals or coal blends can be evaluated petrographically, coals with compensating petrographic qualities can be selected to produce the best coke obtainable from the coals in question.

In all cokes produced, the original coals retain a degree of individuality. Fig. 11 shows the carbon forms produced in a coke from a blend of high-B, high-A and medium-volatile coal. Low rank and oxidized coal produce coke which is structureless and appears the same in all positions of polarized light. As the rank of the coal increases the carbon forms show the development of more ordered or graphite domains. These domains increase in size as the rank of the coal increases. Thus, the carbon forms in coke can be assigned to the rank of the coal from which they were produced.

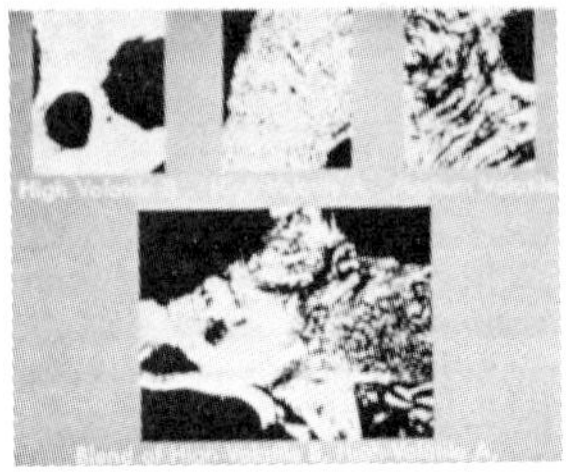

Fig. 11. Carbon forms produced in a blended coke. As coal rank increases, the carbon forms show the development of more graphite domains

Coal hydrogenation and liquefaction. Coal hydrogenation and liquefaction, like other forms of coal utilization, are affected by the nature and properties of the coal selected.

Chemical analyses of coal usually give average chemical properties; that is, the coal is treated as if it were a uniform homogeneous substance. However, in the context of liquefaction and hydrogenation, coal cannot be looked upon as uniform and homogeneous, because some of its components liquefy completely, some only partly and some do not liquefy at all.

Petrography is able to distinguish these various coal components from each other (fig. 12) to select only those coals which promise economic liquid yields and eliminate these coals which cannot be economically liquefied. This approach will help avoid costly experimental bench and pilot plant runs on potential low-yield coals.

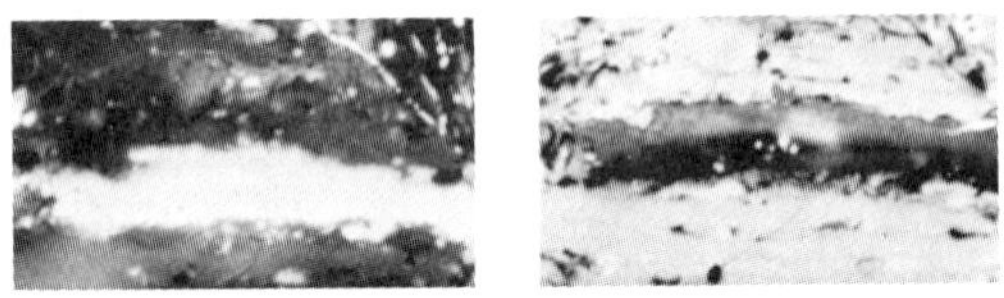

Fig. 12. Photomicrographs of a polished thin section of coal showing same field of view in transmitted (left) and reflected (right) light. ×350. Petrographic study permits identification of those coals most suitable for liquefaction and hydrogenation

The determination of the proportion of non-liquefiable matter in coal is one of the primary purposes for any hydrogenation study. The important variables are temperature, time, pressure, catalyst, nature of vehicle (solvent) and quantity of vehicle. Under ideal conditions, all potentially liquefiable portions of the coal would be converted into oil, water, gases and soluble material. The residue would consist entirely of inert materials such as ash, pyrite, highly carbonified organics and some organic sulfur.

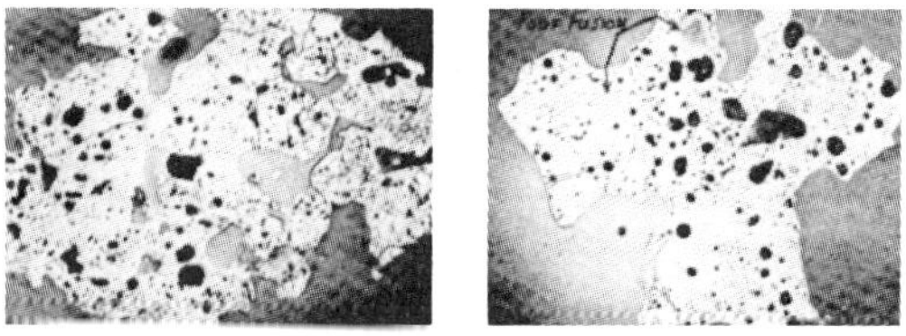

Fig. 13. Photographs of cut and polished surfaces showing cellular structure of two sinters which are chemically similar but have different strengths. × 4. The poorer sinter (right) has distinct areas of weakness resulting from a more uneven distribution of dense and porous areas. Magnetite in the good sinter shows more complete crystal development and greater grain contact to form a stronger interlocking structure

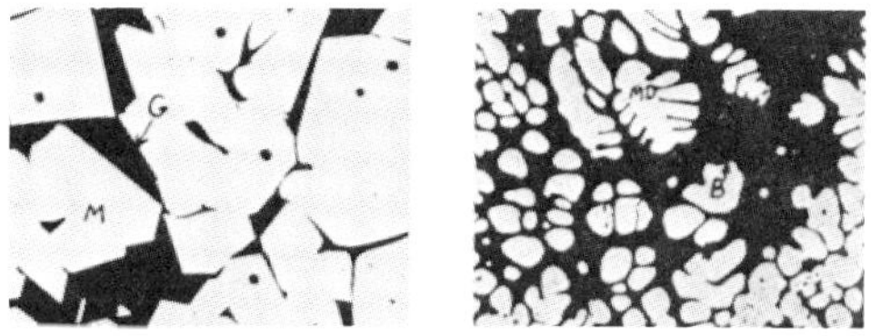

Fig. 14. Photomicrographs of good and poor sinter. M-magnetite, G-clear glass bond phase, MD-magnetite dendrites, and B-granular or finely crystalline bond phase. Variations in sinter strength were attributable solely to differences in extent of fusion. Identification of the problem made it correctable

The U.S. Bureau of Mines performed extensive work in the period approximately 1935-1945 on this subject.[3] The bureau's theoretical liquefaction yields of coal entities is still valid. The original terminology used by the Bureau of Mines is translated in this discussion to conform to present day petrographic terminology. It is important to note here that transmitted light observations of thin coal section are a necessary complement to polished section studies for prediction of liquefaction yields. Observational subleties such as differences in texture and "cloudiness" or translucency are related to liquefaction potential. Such properties cannot be perceived in the study of polished coal surfaces.

The approximate theoretical percent liquefaction yield and residue for the various petrographic entity groups based on the Bureau of Mines study is as follows:

Entity group	*Rank % Ultimate carbon*	*Liquefaction %*	*% Residue*
Vitrinoids	< 89	100	0
Exinoids	< 89	100	0
Resinoids	< 89	100	0
Semi-fusinoids	< 89	75	25
Micrinoids	< 89	40	60
Fusinoids		0	100

Sinter. The steel industry has employed process petrography in solving problems for some time. In the main, their efforts have been to resolve problems that occurred in processing, with only recent interest on prediction of behavior.

In one instance, sinter products from a single steel making facility were chemically similar but differed greatly in strength, a critical parameter. Samples of these sinters were examined microscopically to determine if a relationship existed between structure and sinter strength.

The macroscopic appearance of a typical sinter aggregate from each sample is shown in fig. 13. The poorer sinter shows a more uneven distribution of dense and porous areas, indicating poorer fusion which results in distinct areas of weakness. Microscopic examination of the two distinct areas showed that the crystal development of the magnetite in the good sinter was more complete and that the ends of the grains often contacted one another forming a strong interlocking structure. In the poor sinter, isolated poorly crystallized magnetite grains (fig. 14) were found to be completely surrounded by a granular bond phase resulting in a much weaker aggregate. The variations in the sinter strength as evidenced by the microstructure were attributed solely to differences in the extent of fusion. The better sinter is characterized by extensive fusion and has been sintered at a higher temperature. Identification of the problem as insufficient heating made it possible to correct this situation by proper blending of the coke breeze with ore as well as a slight increase in coke content.

Simulation studies

Hydrometallurgical. At Ledgemont Laboratory, techniques have been devised to observe leaching reactions microscopically as they occur. This is done by simulating various process steps at atmospheric pressure in a leaching stage (fig. 15). The importance of the simulation approach using the leaching stage is that it allows one to understand the mechanisms of leaching and to optimize solution concentrations within a given system.

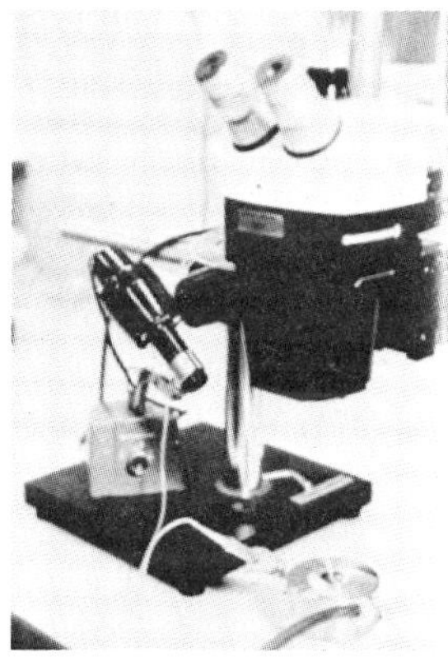

Fig. 15. Leaching stage, atmospheric pressure for temperatures up to 90°C. Technique allows one to understand mechanics of leaching and to optimize solution concentrations within a given system

An example of studying reaction mechanisms is displayed in fig. 16, which is a photomicrograph showing the differential rates of reaction of two intergrown sulfide

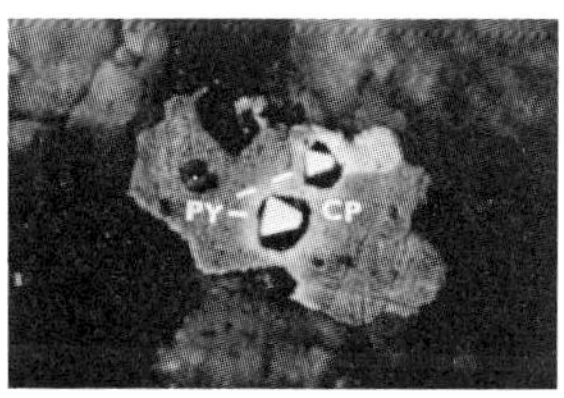

Fig. 16. Photomicrograph showing the differential rates of reaction of two intergrown sulfide minerals, chalcopyrite (cp) and pyrite (py). × 700

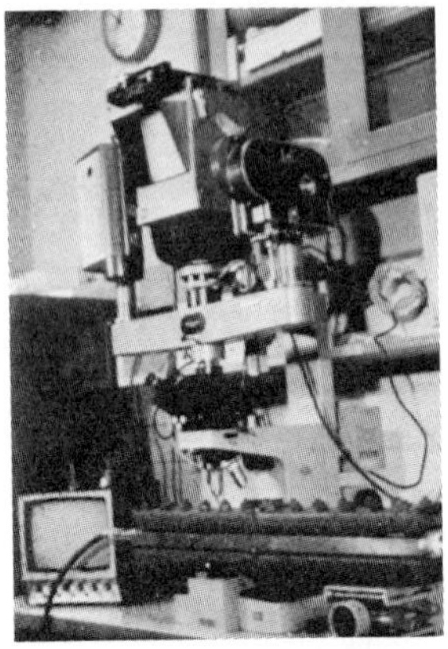

Fig. 17. Special microscopic leaching stage to operate at pressures up to 1000 psi

minerals. The chalcopyrite (outer rim) has reacted more rapidly than the pyrite (inner bright grains). This indicates that even if pyrite were present in large quantities in this ore, it would not consume large amounts of active reagents.

In reactions involving high pressures, a special microscopic leaching stage has been developed (fig. 17). The unit operates at pressures up to 1000 psi and is monitored by remote video tape recording equipment.

Coal gasification. A microscope with pressurized hot stage has been designed and built (fig. 18) at the Ledgemont Laboratory to rapidly heat small coal samples in a controlled atmosphere. Direct observations are made at temperatures up to 2400°F and pressures up to 300 psi. This instrument has been used to simulate steps in a fluidized bed gasification process. This study supplied information valuable in operating a process demonstration unit. It can also be used in simulation studies of pyrometallurgical process steps.

CONTRIBUTION TO PROCESSING PROCEDURE

Because of the intimate knowledge of the process acquired during the development stage, the process petrography group is then in a unique position to assist the chemical process engineers in the startup of the new plant.

In the operations area, the ability to simulate steps in the process is even more important as a diagnostic or predictive tool. It allows changes in process variables to be made on the basis of data acquired through experiments, not be committing the operating plant to a change in con-

Fig. 18. Hot stage microscope for pressures up to 500 psi

ditions which may adversely affect the final product.

In conclusion, process petrography is an important tool in evaluating ore deposits from the exploration stage to the startup of the plant for extracting metal values. It provides necessary information on physical and chemical properties used in selecting processing options and in solving problems during processing.

In hydrometallurgy, petrographic information is critical because the mineralogy, composition and texture of the gangue as well as the ore minerals control the extraction of the metal values.

Future ore bodies will, most likely, be more difficult to process than those now in production because most of the large deposits open to conventional processing have already been found.

In the changing world of today, process petrography is essential in successfully commercializing ore deposits.

References

[1] Roman, J., and B. R. Benner. "The Dissolution of Copper Concentrates," Minerals Sci. Engng, Vol. 5, No. 1, 1973.

[2] Schapiro, N. and R. J. Gray. "The Use of Coal Petrography in Coke-Making," Journal of the Institute of Fuel, No. 6, 1964.

[3] C. H. Fisher et al. "Hydrogenation and Liquefaction of Coal, Part 2. "Effect of Petrographic Composition and Rank of Coal," U.S. Bureau of Mines, Tech. Paper 642, 1942.

AUTHOR CITATION INDEX

SUBJECT INDEX

About the Editor

WILLARD C. LACY, Professor Emeritus and former Foundation professor of geology at James Cook University, received the B.Sc. from DePauw University, the M.Sc. from the University of Illinois, and the Ph.D. from Harvard University. He served as geologist for the Titanium Alloy Manufacturing Company and as petrologist to the chief geologist for the Cerro Corporation in Peru. Before taking the position in Australia, he was visiting professor in economic geology at Harvard, then professor of mining and geological engineering at the University of Arizona.

Professor Lacy has served as consultant for many major companies in areas of economic geology and geotechnics, is author of many papers, and a frequent speaker at professional meetings.